Brian H. Kaye

A Random Walk Through Fractal Dimensions

Distribution:

VCH, P.O. Box 10 11 61, D-69451 Weinheim (Federal Republik of Germany)

Switzerland: VCH P.O. Box, CH-4020 Basel (Switzerland)

United Kingdom and Ireland: VCH (UK) Ltd., 8 Wellington Court, Cambridge CB1 1HZ (England)

USA and Canada: VCH, 220 East 23rd Street, New York, NY 10010-4606 (USA)

Japan: VCH, Eikow Building, 10-9 Hongo 1-chome, Bunkyo-ku, Tokyo 113 (Japan)

ISBN 3-527-29078-8 (VCH, Weinheim) ISBN 1-56081-818-2 (VCH, New York)

Brian H. Kaye

A Random Walk Through Fractal Dimensions

Second Edition

Weinheim · New York · Basel · Cambridge · Tokyo

Professor Brian H. Kaye
Laurentian University
Ramsey Lake Road
Sudbury, Ontario P3E, 268
Canada

First edition 1989
Second edition 1994

Published jointly by
VCH Verlagsgesellschaft, Weinheim (Federal Republic of Germany)
VCH Publishers, New York, NY (USA)

Editorial Director: Dr. Peter Gregory and Dr. Ute Anton
Production Manager: Dipl.-Wirt.-Ing. (FH) Bernd Riedel
Production Director: Maximilian Montkowski
Cover illustration: Newton's Julia (Taken from a series of postcards available commericially from Art Matrix, PO 880-P, Ithaca, NY, 14851, USA. Used by permission of Art Matrix.)

Library of Congress Card No. applied for.

A cataloque record for this book is available from the British Library.

Deutsche Bibliothek Cataloguing-in-Publication Data:

Kaye, Brian H.:
A random walk through fractal dimensions / Brian H. Kaye. –
2. ed. – Weinheim ; Basel (Switzerland) ; Cambridge ; New
York, NY : VCH, 1994

ISBN 3-527-29078-8 (Weinheim ...)
ISBN 1-56081-818-2 (New York)

Printed on acid-free and chlorine-free paper.

Composition: Hagedornsatz GmbH, D-68519 Viernheim
Printing: Colordruck Kurt Weber GmbH, D-69181 Leimen
Bookbinding: Wilh. Osswald + Co., D-67433 Neustadt

Printed in the Federal Republic of Germany

"This book is dedicated to my wife – Phyllis Dew Kaye, who has enriched my walk through the Dimensions of Life."

Biography

Dr. Brian Kaye was born in Hull, Yorkshire, England, in 1932. He obtained his B.Sc., M.Sc., and Ph.D. degrees from London University after studying at the University College of Hull, where he was a George Fredrick Grant Memorial Scholar. After working as a scientific officer at the British Atomic Weapons Research Establishment (Aldermaston) he taught physics at Nottingham Technical College from 1959 to 1963. He then moved to Chicago, where he was a Senior Physicist in the Chemistry Division of the IIT Research Institute (the Research Institute of the Illinois Institute of Technology). There he studied problems as different as why dirt sticks to the fibers of carpet to the design of better propellants for space rockets.

Since 1968 he has been Professor of Physics at Laurentian University in Sudbury, Ontario. He specializes in powder technology, which deals with the manufacture and properties of cosmetics, explosives, powdered metal pigments, drug powders, food powders, and abrasives. He has written a standard text on characterizing powders and authored over 100 scientific papers.

In 1977 his interest in the complex structure of soot involved him in the new subject of fractal geometry, an interest that led to the books "A Random Walk Through Fractal Dimensions" and "Chaos & Complexity. Discovering the Surprising Patterns of Science and Technology". The philosophical side of science has always interested him and has been complemented by his activities as a methodist local preacher in the Sudbury region of Ontario, Canada. He is just as likely to be found holding a service in a protestant church as he is to be lecturing on fractal geometry and chaos theory at the University.

Preface to the Second Edition

The response of readers to the first edition of this book on randomwalk modelling of fractal systems has been most gratifying. It has been a pleasure to receive letters from many people all over the world expressing interest in the subject matter and suggesting improvements for a second edition. Because of pressing demands on my time it was decided that in order to meet deadlines for a second edition the only feasible strategy was to add an extensive bibliography of the rapidly developing applications of fractal geometry.

Indeed, since the first edition of the book it is now becoming apparent that it is useful to differentiate between pure fractal geometry resplendent with Mandelbrot sets and Julian sets, the exotic coloured patterns of which make Joseph's coat of many colours look drab, and the prosaic down to earth measurement of the fractal dimensions of broken rocks and sooty fineparticles. Elsewhere I have explored the basic theory underlying Mandelbrot's exotic coloured set (*Chaos & Complexity,* VCH, 1993) and I continue to collect applications of fractal geometry to materials science and allied subjects, which hopefully will be described in detail at some future date. Accordingly I would like to invite people to write to me if they have new applications of fractal geometry that have escaped my attention. In structuring the bibliography I decided to order the list according to author name, since it would have been too big a task to identify appropriate key words for each paper.

I would like to thank Dr. Ute Anton and Dr. Peter Gregory for their enthusiasm and support in the development of the second edition of this book, and I hope that the topic continues to fascinate many people. As always my thanks go to Garry Clark for his enthusiastic support and work on the bibliography.

Sudbury
December 1993

B. H. Kaye

Preface to the First Edition

I am not sure whether one should write the preface of a book before or after one has completed the book. I have written this preface after I have written the book. Now that it is finished I am rather nervous of some of the simplifications that I have had to make in my presentation of complex ideas. I am reminded of the Italian proverb that "the translator is a traitor". When trying to simplify complex ideas one has to make simple statements which if one were to be making them in a professional context would be surrounded by cautionary statements and other comments on the limitations of the statements. Furthermore fractal geometry is growing so rapidly and ideas are developing so quickly that it is easy to make mistakes – if the reader detects mistakes please remember that "A man who never made a mistake never made anything". After using my first book (on the Characterization of Fineparticles) in a class, a student handed me two pages of corrections to be used in a future edition; hopefully there will not be too extensive a list of corrections to be made in a future edition of this book.

In retrospect, this book appears to be about randomwalk theory as much as about fractal geometry – this re-enforces the dictum I give my students – "scratch a fractal and you will usually find a randomwalk generative model underlying the fractal form".

As I look through some of the recent books on fractal geometry with their brilliant pictures of complex fractal sets my book appears pedestrian; this book is not glamorous, it is intended for a "first reader" in the nuts and bolts of applied fractal geometry. Hopefully readers will bear this intent in mind as they pursue a randomwalk through fractal dimensions.

Sudbury
January 1989

B. H. Kaye

Acknowledgments

First of all I would like to thank Dr. Helmut Grünewald of VCH Publishers who enthusiastically listened to my plans for a possible book an Fractal Dimensions during a visit to Heidelberg. In particular I was pleased that he encouraged me to write in the first person since I believe that this facilitated the presentation of the difficult ideas presented in this textbook. I would also like to thank Dr. Mandelbrot for inviting me to a conference on fractal geometry held in the conference centre of the University of Grenoble in Les Houches France in the early spring of 1985. This conference exposed me to the wide world of applied fractals. One of the pleasures of working in fractal geometry has been the development of a personal acquaintance with Dr. Mandelbrot and his wife Aliette who is a charming supporter of her husband's theories and endeavours.

Over the 10 years in which I have been active in fractal geometry, many students have participated in exploratory experiments to look at new applications of fractal geometry. Several of them are acknowledged in the text of the book. It is not possible to list all of the students who have helped me do develop my ideas on fractal geometry, however, I would like to especially thank student and Research Associate Mr. Garry Clark. Garry has always been very willing to carry out new experiments and has provided much of the data presented in the graphs of this book. He also drew most of the diagrams and co-ordinated the photographs required for this book. Without his sustained effort over a period of 4 years, it is doubtful that the book would have been written. Research Associates Remi Trottier and John Leblanc have been very active in helping me to define my ideas in various discussions and in carrying out experiments to test fractal theories. Barney MacFarlane and Robert Harrison also were of great assistance in generating data for the graphs. In the closing part of the activity Mr. Stephen Horodziejczyk was invaluable in preparing the final text on the word processor. Many typists have worked on converting dictated tapes to final scripts. My daughter Sharon was a major contributor to the editorial and typing process that has produced the book. My other daughter Alison has also contributed to the typing effort as have Mrs. June Talbot, Mrs. Linda Romas, Mrs. Donna Marshall, Mrs. Diane Hrytsak, Miss Vernice Thomas and Miss Leila Lindroos. My colleague Professor Ian Robb has been a constant supporter of my efforts to master fractal geometry and has been a great help in clarifying some of the more difficult aspects of the theory of fractals. Dr. Rizwan Haq of the Physics Department has also maintained a constant interest in the project. I would also like to thank the people who have sent me the literature over the years and who have helped me in my discussions of fractal geometry. In particular I would like to thank Dr. John Davidson of Avon Lake, Ohio, Mr. Alan Flook of Bedford, England, Dr. Englman of Israel and Professor Klinzing of the University of Pittsburgh, who stimulated my activities in fractal dimension through their comments and ideas.

The writing of this book was interrupted by a serious illness and I would like to thank Drs. E. MacCallum and A. Adegbite and the staff of the Neurosurgery ward of the Sudbury General Hospital for their care and attention which made it possible for me to return to my work on this book.

B. H. Kaye, January, 1989

Contents

Word Finder

Coloured Plates

Plate 1 (Figure 3.54). Computer modelling of the growth of a city leads to the simulation of a city outline not unlike that of the sponge fineparticles of Figure 3.1(a) (from M. Batty, "Fractals-Geometry Between Dimensions," **New Sci.**, April 4 (1985) 31-35; reproduced by permission of M. Batty).

Plate 2 (Figure 5.27). The use of colour to label early, medium and late time arrivals of pixels on the growing dendrite illustrates the growth dynamics of the dendritic structures under different electrodeposition conditions. Red pixels denote the early arrival pixels, yellow the pixels that are deposited in the middle stage of growth simulation and green denotes the late-arriving pixels (reproduced by permission of Richard F. Voss/IBM Research).

Plate 3 (Figure 5.48). The transition from non percolating systems to percloating systems at 59.28 lattice occupancy is illustrated by computer simulation studies carried out by R.F. Voss [49, 54].
(a) Cluster build up at lattice occupancy of 59%.
(b) At a lattice occupancy of 59.30% a complete pathway exists from top to bottom of the lattice.
(c) An enlarged portion of (b) illustrates the self similarity of the occupied lattice at higher magnification, an essential feature of fractal systems.
(d) At a lattice occupancy of 59.60% the continuous cluster has started to "mop up" subsidiary clusters existing in the lattice in (a) and (b).
(Reproduced by permission Richard F. Voss/IBM Research).

Plate 4 (Figure 5.51). The meaning of the term lacunarity, used to describe a fractally constructed, partially occupied two-dimensional system, can be appreciated from the above photograph of a 3000 x 3000 units occupied lattice at an occupancy rate of 59.9%. The holes look like an aerial photograph of a region rich in lakes. Lacunarity means, "filled with lakes." (Reproduced by permission of Richard F. Voss/IBM Research.)

Plate 5 (Figure 5.53). The backbone of a fractally constructed cluster is an important property of the cluster. The structure of the backbone of a ramified cluster has been studied extensively by computer simulated by Voss et al. [54].
(a) Lattice cluster at 0.580 occupancy.
(b) Backbones of clusters in (a).
(c) Lattice cluster at 0.593 occupancy.
(d) Backbones of the clusters in (c).
(e) Lattice cluster at 0.610.
(f) Backbones of the clusters in (e).
(Reproduced by permission of Richard F. Voss/IBM Research.)

Plate 6 (Figure 5.60). Colour coded structures of the agglomerates grown by the MDA and CDA by Voss illustrate the stages of the growth of the cluster. Red denotes early stages of growth, yellow the intermediate stage of growth and green the last stages of the simulated growth (reproduced by permission of Richard F. Voss/IBM Research).

Plate 7 (Figure 9.11). Dr. Mandelbrot has suggested that Squig type fractal models may be useful in modelling dynamic cracking of systems [29, 30, 31] (from B.B. Mandelbrot, "The Fractal Geometry of Nature," W.H. Freeman & Co., San Francisco, 1983. Reproduced by permission of B.B. Mandelbrot).

Plate 8 (Figure 10.2.1). The visual beauty of fractal art is one of the reasons for the excitement amongst the general public over the discovery of fractal geometry. (The fractal vista shown in this figure was created using the theories of fractal geometry by Alan Flook. Reproduced by permission of Dr. Alan G. Flook, Unilever Research Laboratory.)

Plate 9 (Figure 10.2.3). This mathematical microbe appearing in the trade literature of the company manufacturing high-resolution microscopes (Tracor Northern) is actually a fractal system used to demonstrate the resolution of the electron microscope (photograph courtesy of Tracor Northern, Middleton, Wisconsin).

Plate 10 (Figure 10.2.5). Fractal designs are beginning to appear on greeting cards.
(a), (b) Front and back of a New Year's Greeting Card. (Pictures from the greeting card used by permission of ACDS Graphics Systems Inc., 100 Edmonton St., 232, Hull (Quebec), Canada, J84 6N2.)
(c) Fractal postcard manufactured by Art Matrix Corporation. (Courtesy of Art Matrix Corp. and Cornell National Supercomputer Facility, P.O. Box 880, Ithaca, NY., 14851-0880.)

Plate 11 (Figure 10.2.6). The Voss mountains, which have appeared in IBM commercial advertising, are a graphic creation out of this world generated by Dr. Richard Voss. (Reproduced by permission of Richard F. Voss/IBM Research.)

Plate 12 (Figure 10.4.3). Many of the pictures generated by satellites surveying the surface of the earth manifest fractal structures which may have some significance in determining the origin of the various features visible in the field of view [10]. (Reprinted from GEOS, Vol. 15, No. 3, Energy, Mines and Resources Canada.)

Plate 13 (Figure 10.4.6). A Fractal crystal in gabbro (a type of rock) located in Munro Township, Ontario, was discovered by Dr. Fowler of Ottawa University (in the language of the geologist, the "fern" is Harristic textured pyroxene) [16].

Plate 14 (Figure 10.7.1). A Fractal front produced by a simulated system of particles diffusing on a 256 by 256 pixel screen [2]. (From B. Sapoval, M. Rosso, J.F. Gouyet and J.R. Colonna, "Dynamics of the Creation of Fractal Object by Diffusion and 1/f Noise," *Solid State Ionics*, 18 and 19, pp. 21-30 (1986), North-Holland Physics Publishing, a division of Elsevier Science Publishers, Physical Sciences & Engineering Division, Amsterdam, The Netherlands. Reproduced by the permission of Elsevier Science Publishers and B. Sapoval.)

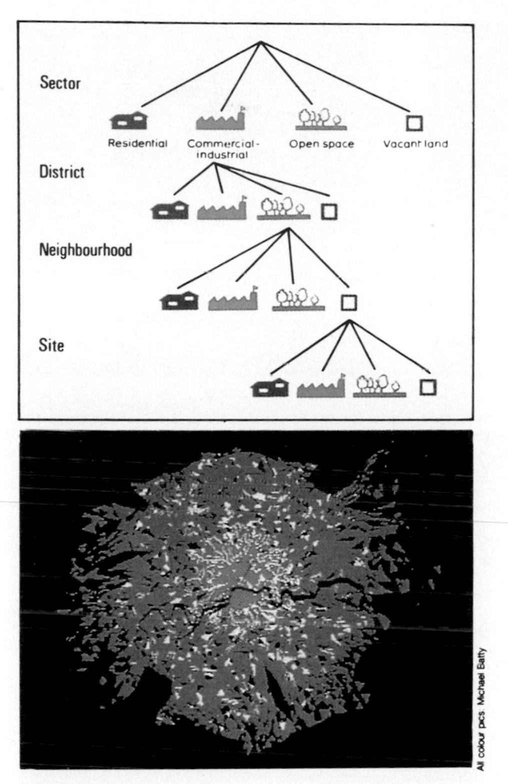

All colour pics. Michael Batty

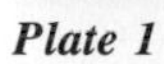

Plate 1

Plate 2

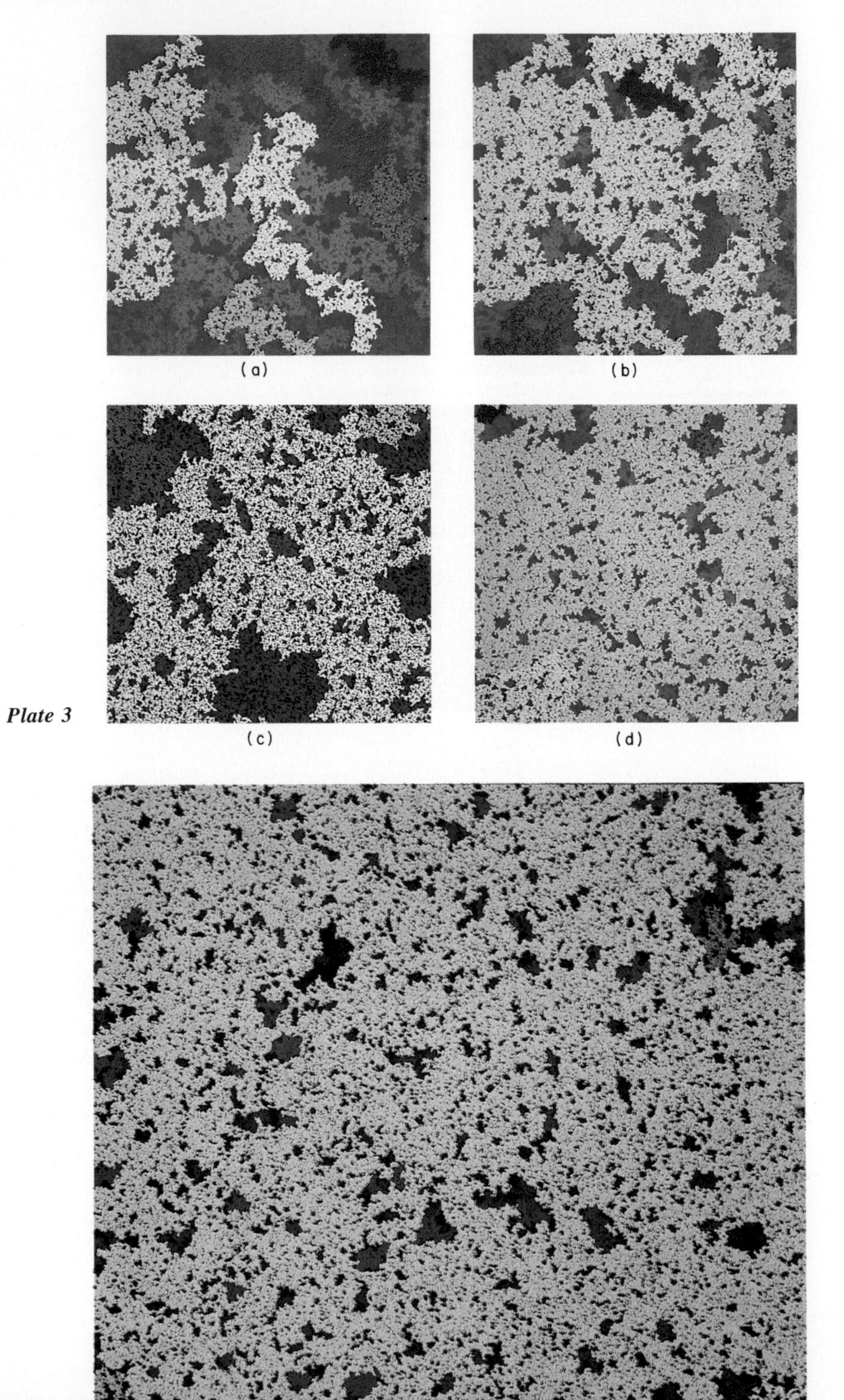

Plate 3

Plate 4

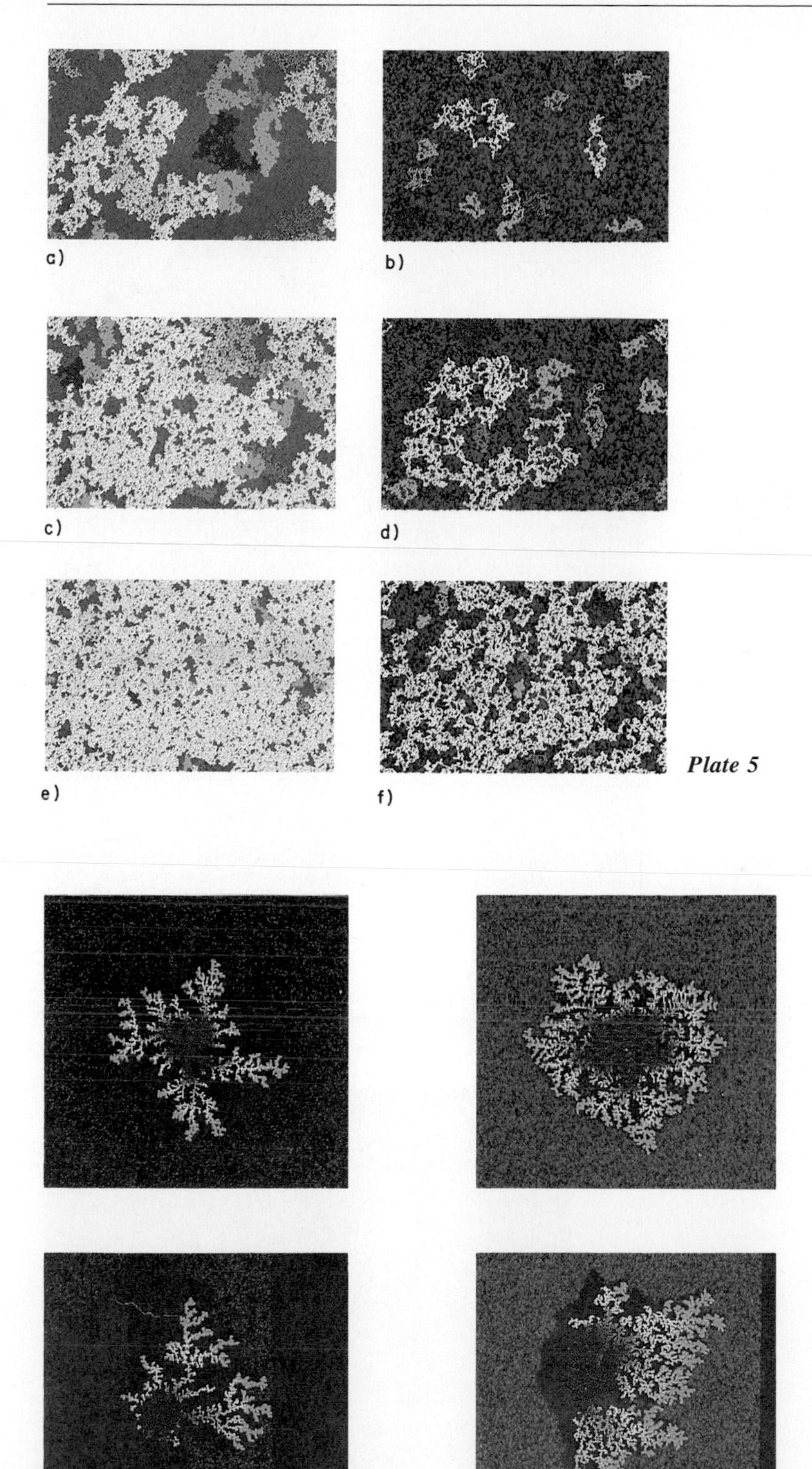

Plate 5

Plate 6

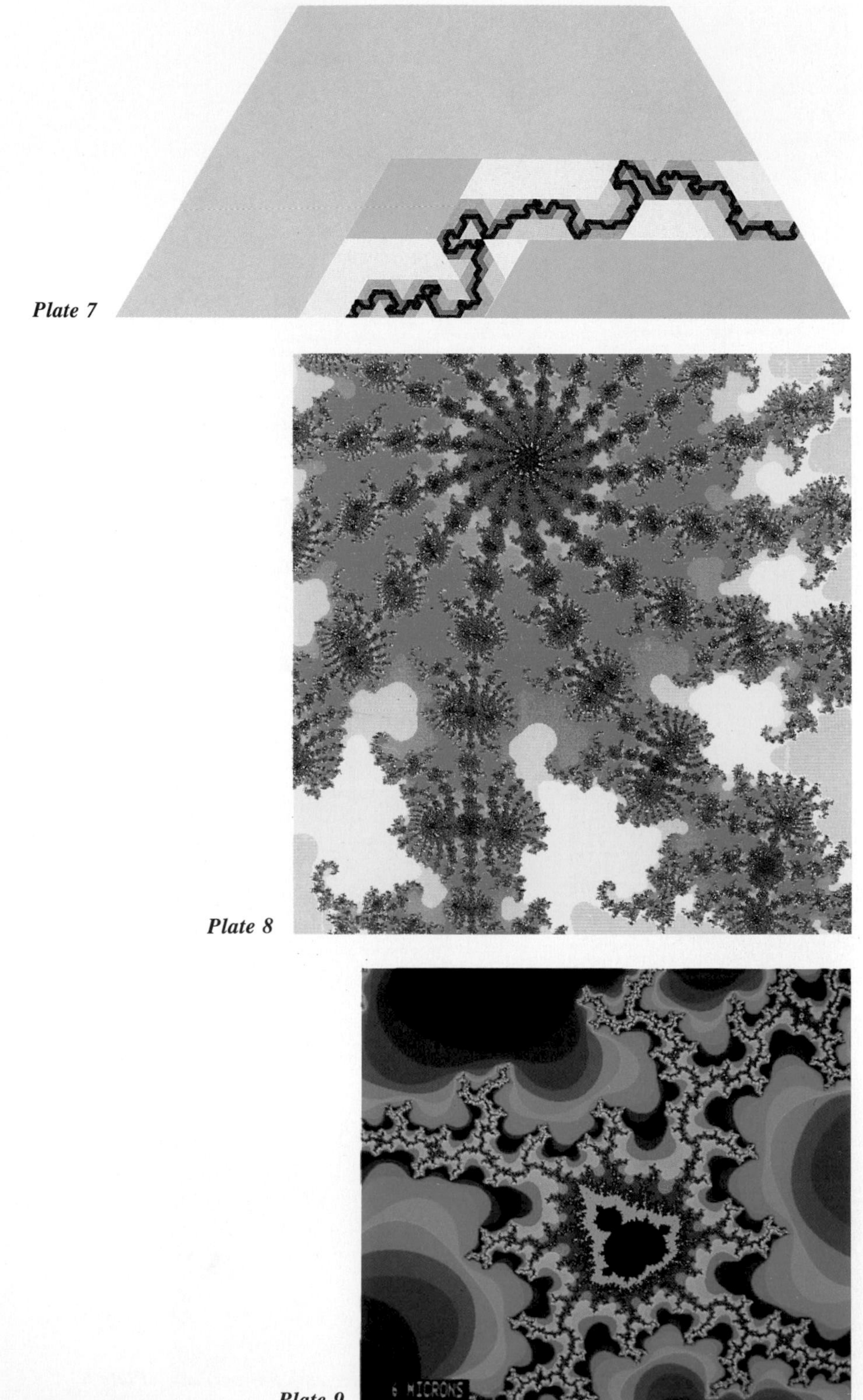

Plate 7

Plate 8

Plate 9

Plate 10

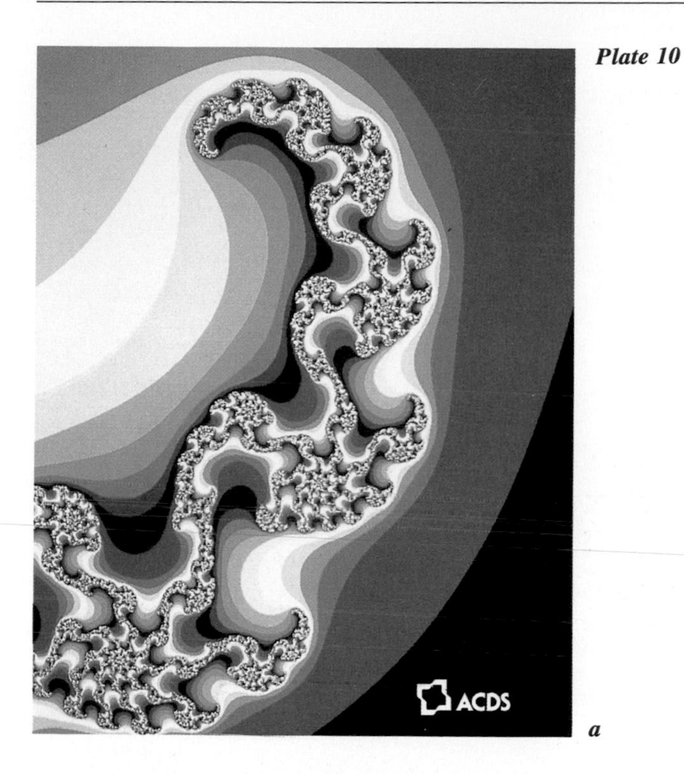

a

b

c

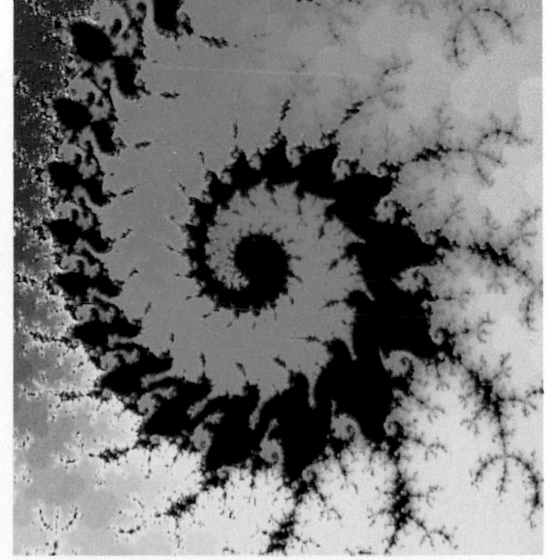

Plate 11

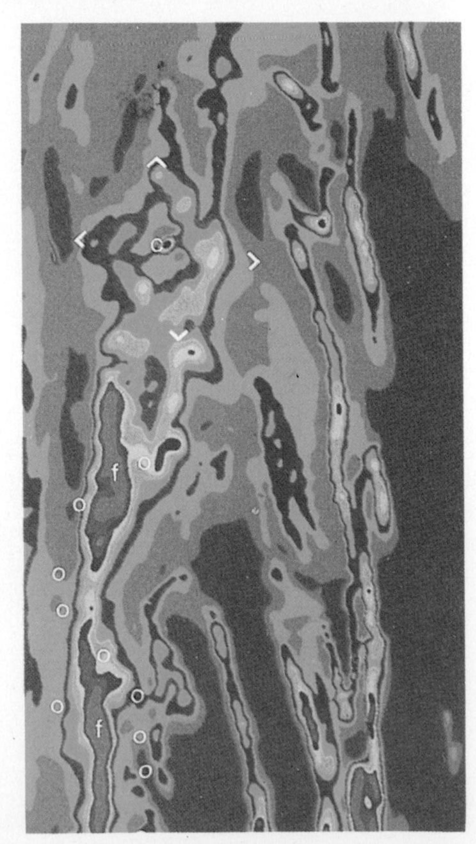

Plate 12

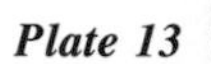

Plate 13

Plate 14

1 A Starting Point for the Randomwalk

Fractal dimensions were introduced to the English-speaking scientific community in 1977 by Benoit Mandelbrot, their inventor, in a book entitled "Fractals: Form, Chance and Dimensions" [1]. I became aware of this book through a brochure sent to me by the Library of Science Book Club, which operates from New Jersey in the United States. In their description of the book, the reviewers for the Library of Science mentioned that Mandelbrot discussed in the book the problem of the length of the coastline of Great Britain. I did not know what a fractal was, but I did remember seeing an article by Mandelbrot on this problem when browsing through the journal *Science* in the 1960s. In his article in *Science* Mandelbrot had drawn attention to some earlier work by Louis Fry Richardson, who had pointed out that a simple question such as, "how long is the coastline of Great Britain?," has no answer apart from an operational description of how one estimates the length of the coastline [2]. For example, if one draws the coastline of Great Britain as shown in Figure 1.1, one can attempt to measure the length of the rugged coastline by striding around the coastline with steps λ to create a polygon whose perimeter is an estimate of the coastline, as shown in Figure 1.2. In Figure 1.2, three different estimates of the coastline of Great Britain, based on polygons of side length λ_1, λ_2 and λ_3, are shown. To obtain the estimates of the coast perimeter in dimension-

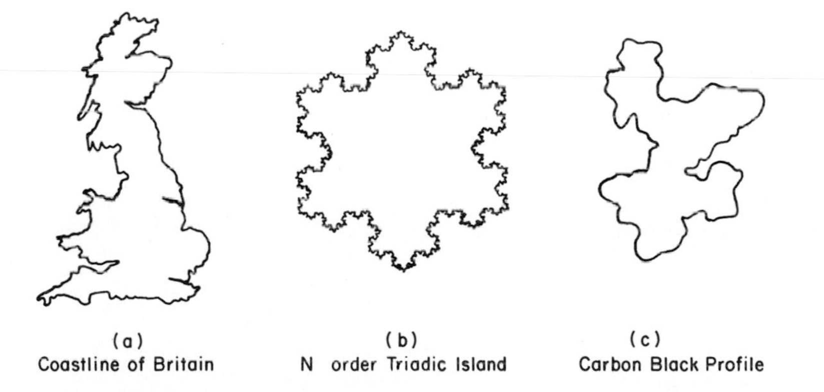

Figure 1.1. Richardson pointed out that the problem of measuring the length of a coastline leads to the paradox that all coastlines are infinite at infinitely small resolution. Mandelbrot linked this paradoxical conclusion to the structure of curves with infinite perimeter such as the *N*th-order Koch's triadic island, the structure of which he described by means of fractal dimensions. A visual comparison of the structure of a carbonblack agglomerate with the coastline of Great Britain and Koch's triadic island suggests that Mandelbrot's fractal dimensions may be useful in describing the structure of the carbonblack profile.
(a) Coastline of Great Britain.
(b) *N*th-order of Koch's triadic island.
(c) Carbonblack profile at high magnification.

less form, the perimeter estimates are normalized by dividing the perimeter of the polygon by the maximum projected length of the island, as illustrated in Figure 1.2.

In the bottom portion of Figure 1.2, the three estimates of the coastline are plotted on a graph having log-log scales. It can be seen that one can draw a straight line through the three points. If one extended the dataline, one would reach the conclusion that the

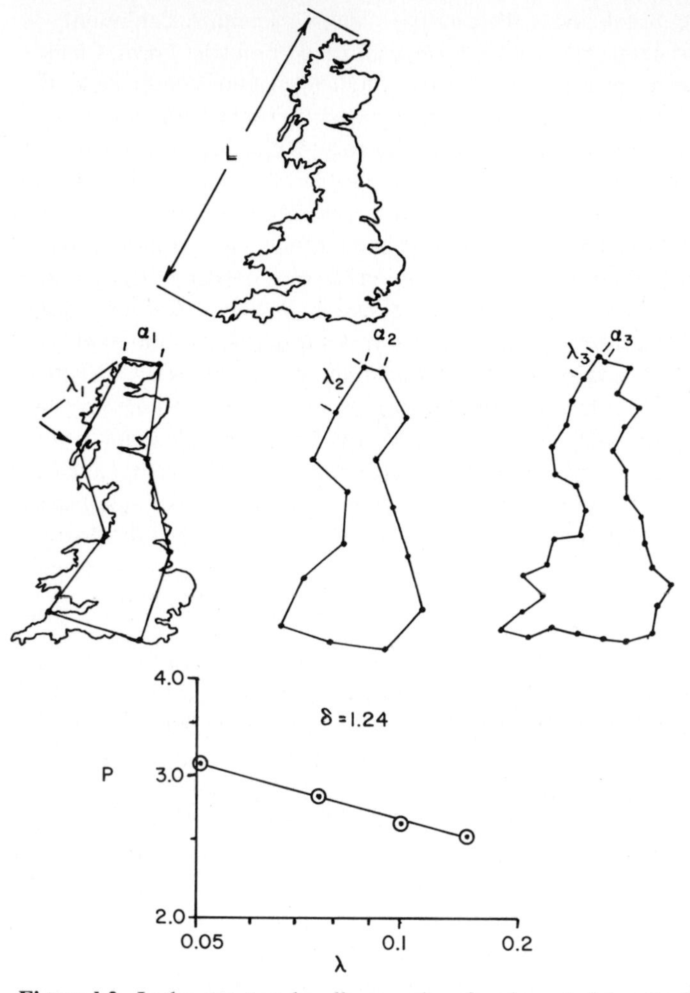

Figure 1.2. In the structured walk procedure for characterizing the fractal dimension of a rugged boundary, a series of polygons of side λ are constructed on the perimeter using a pair of compasses. A plot of P against λ on a log-log plot yields a dataline of slope m where

$$\delta = 1 + |m|$$

if the boundary can be described by a fractal dimension.

λ = side of polygon normalized with respect to maximum projected length of the profile.

P = polygon perimeter of side λ normalized with respect to maximum projected length of the profile.

δ = fractal dimension of the boundary.

L = maximum projected length of the island.

coastline of Great Britain was infinite if one could use smaller and smaller steps in the estimation. Richardson showed that this kind of paradox was inherent in the measurement of any type of coastline and pointed out that any estimate of any coastline should always be linked with a statement of how that estimate was deduced [3].

A vague memory of reading this earlier paper by Mandelbrot on indeterminate coastlines aroused my interest in the book on fractals. Hoping that perhaps the book by Mandelbrot would provide interesting reading, I made a mark in the appropriate box of the request card sent by the book club. I thought no more of the topic of fractals until the book arrived. As I opened the brown-paper parcel containing the book, I was totally unaware of what a difference it was going to make to my professional activities. On a first browse through the book on fractals I saw for the first time Koch's triadic island of the *N*th order. This fascinating mathematical curve is shown in Figure 1.1(b). It has the interesting property that it has an infinite perimeter enclosing a finite area. As I looked at Koch's triadic island, the shape of the carbonblack profile shown in Figure 1.1(c) seemed to float out of my memory and place itself on the page beside Koch's Triadic Island. For many years I had been interested in techniques for characterizing the shape and size of carbonblack fineparticles which are used extensively in industry. Throughout the 1960s and 1970s the resolution and power of electron microscopes and scanning electron microscopes for viewing such fineparticles had been increasing steadily.

Each time a new more powerful electron microscope was used to examine carbonblack fineparticles, scientists were excited by the increased information available through magnification, but also frustrated by the fact that there seemed to be no limit to the increase in detail made visible with the new instruments. In particular, estimates of the perimeter of the profile made at a series of increasing resolutions indicated the paradox that all carbonblack fineparticles would have infinite perimeters, if one could examine them at an infinite scale of resolution! In Figure 1.3, a series of increasingly magnified views of a sponge iron fineparticles are shown. This particular sequence illustrates the enormous amount of detail made visible each time the magnification is increased, and how there seems to be no end to the intricacies of the structure of the fineparticle.

In Mandelbrot's discussion of the *N*th order Koch's triadic island, the statement that it had an infinite perimeter, and that it looked the same at a series of magnifications because the structure was self-similar, led me to recall my experience of trying to reach a finite estimate of the perimeter of a carbonblack at a series of magnifications. The visual comparison between the last carbonblack profile examined and Koch's triadic island created the intuitive idea that any technique used to characterize the structure of the Koch triadic island profile could perhaps be used to describe the ruggedness of the carbonblack profile. Once this idea had suggested itself, it triggered an intensive reading of the material in Mandelbrot's book and a fevered scramble to obtain a first set of data leading to a scientific publication in which fractal dimensions were used to describe fineparticle boundaries [4]. Later in this randomwalk through fractal dimensions, the experiments leading up to my first scientific publication describing fineparticle boundaries by means of fractal dimensions will be reviewed. However, before we can explain the various problems encountered en route to the first successful use of

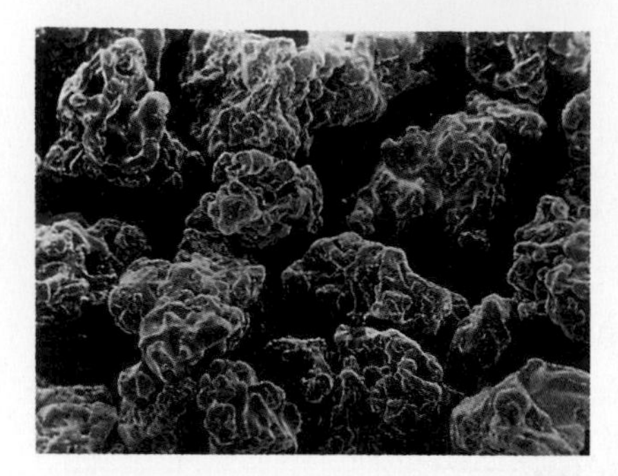

a) Magnified 220 times

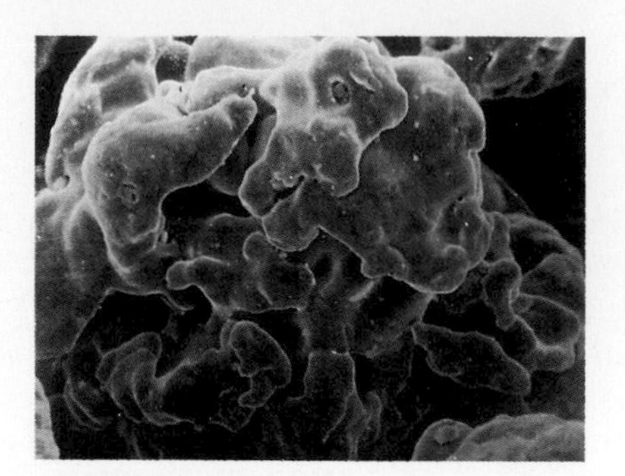

b) Magnified 1100 times

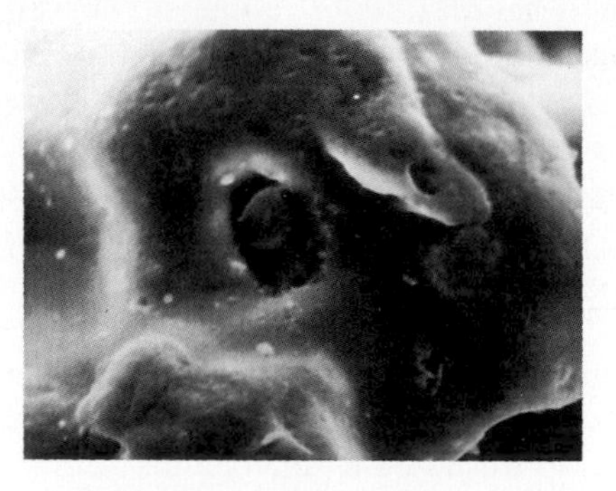

c) Magnified 5500 times

Figure l.3. When the complexity of structure increases with magnification, it may be useful to use fractal dimensions to describe the structure of the fineparticle.

fractals as descriptors of fineparticle boundaries, we need to lay a good deal of groundwork concerning the concept of fractal dimensions and randomwalk theory.

In my attempts to understand Mandelbrot's book, I was handicapped because I had never studied the mathematics of sets, and a great deal of Mandelbrot's discussion hinged on the ability to describe geometric figures using intercepting sets of infinite points. As a student, I had had a brief encounter with Cantorian sets in George Gamow's book "One, Two, Three, Infinity" [5] and had browsed with pleasure through the book "Stories About Sets" by Vilenkin, [6] but both those books had been social encounters. Now I found myself surrounded by an infinite number of Cantorian sets, and by page 100 of Mandelbrot's book I was having mentally to walk up and down the devil's staircase which had an infinite number of steps, but which resulted in a finite climb (see Figure 1.4).

The pictures in Mandelbrot's book, which were so similar to the fineparticle systems I was attempting to characterize, enticed me into the maze of Cantorian sets and through Menger sponges to important conclusions. Essentially, the key idea introduced into applied mathematics by Mandelbrot is that, rugged and indeterminate systems can often

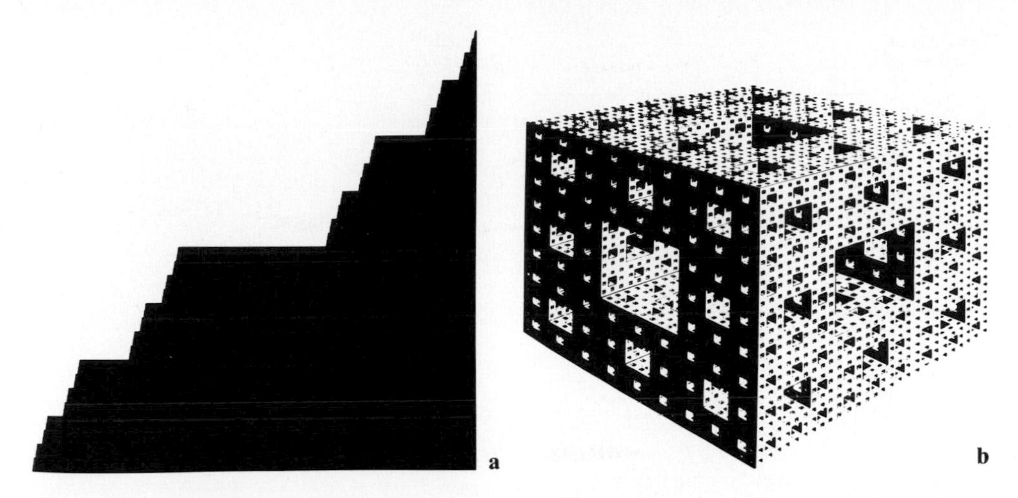

Figure 1.4. Attempting to master fractal geometry without a background in geometric set theory is like trying to stand on every step of the "Devil's staircase" which has an infinite number of steps or is like trying to explore a Menger sponge which has an infinite number of holes of infinitely variable size. The structure of the Devil's staircase leads one to anticipate intuitively that fractals of such boundaries will be useful in describing the chemical physical reactivity of crystals. The Menger sponge suggests that fractal dimensions are useful in describing physical behaviours as diverse as catalysts and controlled-release drug systems.
(a) Devil's staircase.
(b) Menger sponge.
(from B.B. Mandelbrot, "Fractals: Form, Chance and Dimension," W.H. Freeman and Company, San Francisco, 1977, reproduced by permission of B.B. Mandelbrot.)

be described by extending the classical concept of dimensional analysis to include a fractional number that describes the ruggedness of the system in the space spanned by the whole number dimensions encompassing its fractional magnitude. This concept can be illustrated by considering the set of lines shown in Figure 1.5. An important branch of mathematics is topology, which is the study of those properties of geometrical objects which remain unchanged on the continuous transformation of the objects. In ordinary everyday language, this means that topologists would regard the lines in Figure 1.5 as being identical, in that if each of the lines were to be drawn on a rubber sheet one could distort the sheet by stretching it or squeezing it until all of the lines fitted on top of each other. In fact, a popular definition of topology is that it is "rubber sheet geometry" [7]. From a topological point of view, the dimension of all of the lines in Figure 1.5 is 1. These lines have been drawn in two-dimensional space (a flat surface). Mandelbrot suggests that such curves can have their ruggedness described by allocating a fractional number between one and two which will describe their space-filling ability. The difference between the fractional dimension of the lines and their topological dimension describes their ruggedness.

The fractal dimensions of the rugged lines shown in Figure 1.5 were measured by the structured walk technique which will be described later [8, 9]. From the data summa-

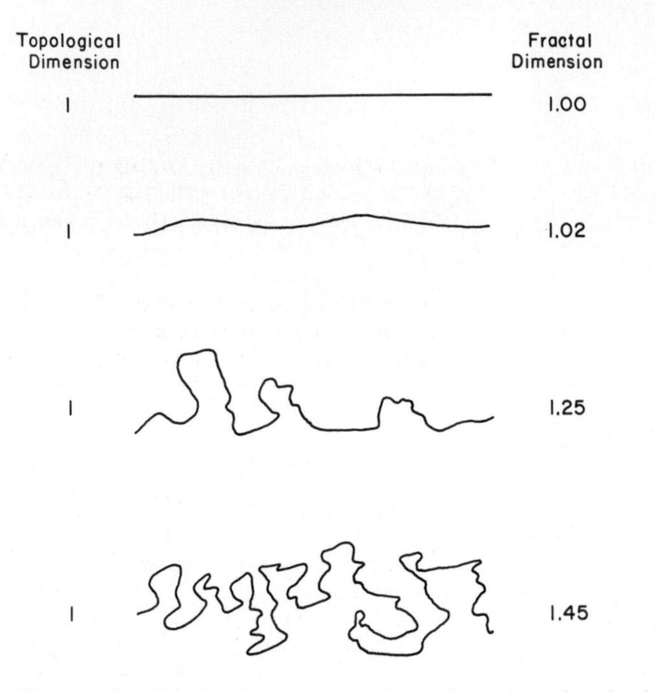

Figure 1.5. The classical concept of the dimension of a physical quantity can be extended by adding fractional quantities related to the ruggedness of a system to the topological dimension.

rized in Figure 1.5, one can see that there is an intuitive link between the fractional dimension allocated to the curve and its apparent structure in two-dimensional space. Using the concept of fractal dimensions, a rugged surface can be given a number between two and three which indicates how the structure fills the space it occupies. Thus, a sponge of 2.4 fractal dimension would occupy space more efficiently and have more surface area than a 2.3 fractal dimension sponge.

Since venturing into fractal dimensions, I have lectured in many places on the fractal dimensions of fineparticle systems. In the question period following the lectures, it became obvious that many people intuitively reject a fractional dimension because their thinking is trapped in the classical framework of one, two and three dimensions.

Mandelbrot anticipated this difficulty in his original discussion of rugged systems, and stressed the fact that our classical view of dimensions is itself an operational perspective. Thus, in his book he discusses the dimensionality of a ball of wool from a variety of perspectives. He points out that, to a fly at a great distance from the ball of wool, the ball of wool appears to be a point with no dimensions. As the fly moves towards the ball of wool, the situation is reached when the ball can be considered to be a two-dimensional target. When the fly is close to the ball, it appears to be a large three-dimensional object. If we could imagine a fly which was small enough to enter the ball, then from the inside perspective, the whole concept of a "ball of wool" disappears, because inside the ball, the fly finds itself flying through a network of ropes which appear to have very complex structures.

As I read the discussion of the operative nature of classical dimension theory as expounded by Mandelbrot, I recalled classroom memories from high-school struggles to cope with one-, two- and three-dimensional descriptions of everyday objects. As a student, I could never accept that a piece of string was one dimensional. To my mind, no matter how thin the string, it was always a three-dimensional object. Many hours of struggle with the concepts of dimensional theory could have been avoided if my teachers had told me that calling a string one dimensional, and a piece of paper two dimensional, was only a matter of convenience, and depended on the perspective of the person using the piece of string or the piece of paper. Thus, a mathematician regards a piece of paper as an infinitely thin structure of two dimensions, whereas a chemist using the paper as a filter needs to regard it as a three-dimensional network of fibres. The definition of the dimensionality of an object depends on the operations performed (either mentally or physically) with the object. The whole number dimensional description of spatial objects is a matter of convenience, not a fundamental attribute of the Universe. In the same way, the fractional dimension of an object which describes its ruggedness is a useful extension of a set of operational definitions of dimensional structure.

The fractional dimension of an object is not unlike the artificial description of the average family in Canada, which for population description purposes can be regarded as having 2.2 children. When confronted with the statement that the average family size in Canada is 2.2, no one attempts to visualize what 0.2 of a child looks like. They accept the fact that it is a useful mathematical description of the average family size.

In the same way, a fractional dimension is a useful description of structure. We learn to give it physical significance by studying different systems which can be usefully given a fractal dimension. Thus, from the set of lines shown in Figure 1.5, one can learn to associate a fractal dimension with a visible ruggedness, and that experience becomes the basis of intuitive interpretation of fractal dimensions quoted for a boundary drawn in two-dimensional space.

It becomes very important when studying fractal dimensions in detail to realize that what appears to be a rugged boundary from one perspective can become a smooth boundary from another perspective, and vice versa. Therefore, when quoting a fractal dimension, it is always necessary to remember that, in the real world, such a fractal dimension is always associated with a given resolution of inspection. It becomes tiresome always to quote the resolution limits for any given fractal dimension and in general discussion the resolution limits are often dropped from the conversation. This verbal shortcut should not lull anyone into thinking that, in the real world, a fractal dimension used to describe a rugged system is independent of the scale of scrutiny.

For example, if the lines in Figure 1.3 were made of smooth nylon thread, then to a fly approaching the thread the structure would appear to be rugged from its perspective in outer space. If the fly could land on the thread and examine the portion of the thread under its feet, it would find that it had a smooth texture and that the surface could be regarded as a traditional two-dimensional surface. Following Mandelbrot's terminology, any boundary which can be described by means of the traditional geometry of smooth surfaces and smooth continuous curves will be described as a Euclidean boundary, because such systems can be described by the geometric systems of Euclid [10].

Since my initial scientific publication on the ruggedness of fineparticle boundaries,

I have been able to tackle many fineparticle problems using fractal dimensions and I have been invited to lecture on the subject in many places [11, 12, 13]. As a result of these various publications and talks, friends urged me to write a book on fractal dimensions. I was initially reluctant to do this, since I did not feel I could write a systematic scientific book describing the theory of fractal dimensions. That book has already been written by Mandelbrot. However, I was tempted to share some of the historical developments of the uses of whole numbers and fractal dimensions with fractal beginners, since these experiences appear to have a certain usefulness in the teaching of applied science. Accordingly, I decided that perhaps I could write a book which was itself a randomwalk through fractal dimensions, hopping from one interesting application to another, hoping that the trail of exploration would introduce others to the fascination of fractal dimensions.

Mandelbrot tells us in his book that he was led to the whole concept of fractal dimensions by his involvement in a problem associated with the signal integrity of messages being sent along a telephone transmission line. The problem he was studying can be understood by considering the information summarized in Figure 1.6. If one looks at the line shown in Figure 1.6(a), the points along the line can represent noise signals received at the end of the transmission line over a period of time.

When studying a pattern of noise signals, such as those illustrated in Figure 1.6(a), Mandelbrot tells us that the apparent clustering of the points on the time line, which he assumed were random events, reminded him of the apparent clustering of crossings on a line of a one-dimensional randomwalk described by Feller in his well known book on probability theory [14].

In Figure 1.6, the basic elements of a simple one-dimensional randomwalk are illustrated. Traditionally, randomwalk theory is introduced to students of physics by considering the progress of a drunk staggering blindly away from a lamp-post. We shall discuss the two-dimensional staggering of a drunk in Chapter 5, but for now we shall discuss the simpler problem of the staggering progress of a drunk trying to leave a lamp-post in one-dimensional space. For a one-dimensional walk, we must consider that the drunk is trying to get away from a lamp-post in a groove. The drunk is presumed to take a series of steps either away from or towards the lamp-post. The essential nature of a simple randomwalk is that the next step to be taken by a drunk does not depend upon the position he has already reached. Thus, even if he is twenty steps away from the lamp-post, in a simple randomwalk model he is just as likely to step away from the lamp-post as he is to step towards it. For the simplest system, we also assume that the drunk always takes strides of the same magnitude. In Figure 1.6(b), the series of steps taken by the drunk in one-dimensional space as simulated using a computer is illustrated. In this graph, the distance from the lamp-post, D, is plotted against the total number of steps taken. When the zig-zag line showing the progress of the drunk crosses the time axis, we know that he has momentarily returned to the lamp-post only to step away in a random manner in his next drunken venture to explore the space around the lamp-post. It will be seen that, even in the short run of the recorded one-dimensional walk shown in Figure 1.6(b), the drunk appears to linger around the lamp-post for a short time before moving further out, where he staggers back and forth for some period before returning to the vicinity of the lamp-post. This apparent clustering of events at the lamp-post is

a) Noise events on a transmission line.

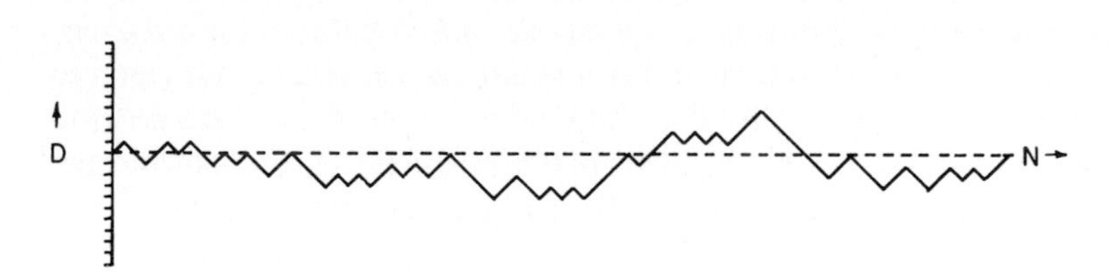

b) A Randomwalk crossing a reference line.

Figure 1.6. Mandelbrot suggested that the sparse random packing of points on a straight line could be described using Cantorian set theory resulting in the allocation of a fractional dimension between 0 and 1 to the pattern of sparse points.
N = Total number of steps taken.
D = Distance from the lamp-post.
(a) Noise events in a transmission line.
(b) Randomwalk crossings of a reference line.

the pattern which suggested to Mandelbrot that the bursts of noise signals in the transmission line could be modelled by randomwalk theory. It also occurred to him that the random packing of the noise event points along the time line, as illustrated in Figure 1.6(a), could be tackled by using some classical mathematical studies of the packing of points on a line when the line in space is not continuously occupied by an infinite number of points.

The packing density of sparse points on a line can be described by the mathematics created by George Cantor (1845-1908), and referred to as Cantorian set theory [15, 16]. Mandelbrot has shown that this is the equivalent of describing the density of sparse, randomly distributed points on a line by means of a fractal dimension between 0 and 1.

In a later chapter of this exploration of fractal dimensions in physical science, we shall explore the physical significance of fractal dimensions between 0 and 1. However, for the time being, it is sufficient to recognize that it was the scattered population of points, generated by a one-dimensional randomwalk crossing a time axis which suggested to Mandelbrot that geometric Cantorian set theory could be used to derive fractional dimensions which would describe the structure of rugged systems in space.

Although a one-dimensional randomwalk sequence plotted in time was the original stimulus to the development of fractal geometry, Mandelbrot chose to introduce the concepts of fractal geometry in his book by considering Brownian motion in two-dimensional space. Brownian motion is the apparently unorganized zig-zagging of tiny fineparticles suspended in a liquid, as viewed through a microscope. This motion was first observed in 1827 by Robert Brown, an English botanist. We now know that the random motion of the fineparticle undergoing Brownian motion is caused by a multiplicity of molecular bombardment of the fineparticle. At any one instance, the forces on the fineparticle are an unbalanced force which pushes the fineparticle in a random direction. Because the molecular bombardment is a random phenomenon with the net force rapidly changing in magnitude and direction, the progress of the bombarded fineparticle is a randomwalk.

Brownian motion was studied in detail by Perrin, and the successive positions of a colloidal fineparticle of radius 0.53 μm, as seen under the microscope and noted every 30 s, is shown in Figure 1.7. A micron (1 μm) is a length measurement unit equal to one millionth of a metre.(10^{-6} m). Given an infinite time period, the zig-zagging colloidal fineparticle would visit every point in two-dimensional space, so that the fractal dimension of the track of the colloidal fineparticle is extremely close to 2, whereas its topological dimension is 1 [17].

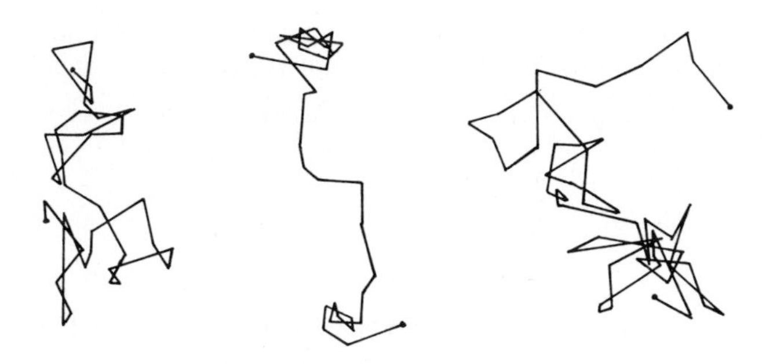

Figure 1.7. Perrin studied the Brownian motion of colloidal fineparticles (two typical tracks of a 1.06 diameter colloidal fineparticles; the position of the fineparticle was noted every 30 s with respect to a grid of side length 3.2 μm) [18].

Throughout Mandelbrot's book, the intimate link between systems which can be described by fractal dimensions and the modelling of such systems using randomwalk theory in 1- to N-dimensional space soon establish the fact that any attempt to understand the fractal dimension description of natural systems, must be based not only on a knowledge of set theory and topology, but also on randomwalk theory. Again, those who would like to understand and use fractal dimensions when studying applied science, but who have a traditional education in mathematics, find it difficult to thread their way through the randomwalks of Mandelbrot's book because randomwalk theory rarely penetrated pre-1970 undergraduate textbooks. Therefore, it became clear in early discussions concerning the possibility of writing an introductory book on fractal dimension methods in applied science that the book would have to include a basic discussion and introduction to randomwalk theory.

Having decided to write a book which was a random exploration of fractal dimensions, the next decision which had to be taken was where to begin the exploration.

From a theoretical point of view, it would be elegant to begin the exploration of fractal dimensions with a study of the packing of sparse points in 0- to 1-dimensional space, and then work up to rugged systems in N-dimensional space. However, a study of the dimensionality of points on a line is difficult to interpret intuitively. Therefore,

I decided to make the first step in the randomwalk a discussion of the use of fractal dimensions to describe fineparticle boundaries, which is where I began my walk through unfamiliar dimensions.

At the end of each chapter, I shall give a plausible reason for the content of the following chapter, but the reader is warned that the choice is entirely random and based upon my whim rather than predicated by any logic.

Since it is my hope that students, in addition to hopeful practitioners of the art of the fractal description of physical systems, can be enticed into meandering through fractal systems, I have found it necessary to explain mathematical concepts and operations in somewhat greater detail than would have been required if I had been writing specifically for practising scientists. Experienced scientists are invited to skip over these patches of explanations and to pursue a shorter randomwalk than that followed by the liberal arts students. The reader is also warned that, like all randomwalks, there are patches of text which appear to have little to do with the topic in hand, in this case fractals. The reader is asked to be patient with such patches, and I promise that, if he reads on, he will eventually discover that what appeared to be irrelevant digression in fact laid important groundwork for exploring the higher realms of fractal dimensions.

References

[1] B.B. Mandelbrot, "Fractals: Form, Chance and Dimension," Freeman, San Francisco, 1977.

[2] B.B. Mandelbrot, "How Long is the Coast of Britain, Statistical Self Similarity and Fractional Dimension," *Science*, 155 (1967) 636-638.

[3] The work carried out by Richardson on the indeterminacy of coastlines was found in his posthumous papers. The most accessible discussion of Richardson's work is in the publications of Mandelbrot, [1, 2].

[4] B.H. Kaye, "Specification of the Ruggedness and/or Texture of a Fineparticle Profile by its Fractal Dimension," *Powder Technol.*, 21 (1978) 1-16.

[5] G. Gamow, "One Two Three Infinity," Bantam Books, New York , 1965.

[6] N.Y. Vilenkin, "Stories About Sets," Academic Press, New York, 1969.

[7] I. Stewart, "Concepts of Modern Mathematics," Pelican Books, Middlesex, England, 1975. A readable introduction to geometric set theory and topology for those unfamiliar with these subjects.

[8] B.H. Kaye, "Direct Characterization of Fineparticles," Wiley, 1981.

[9] B.H. Kaye, "Application of Recent Advances in Fineparticle Science and Technology to Mineral Processing," paper presented at the Symposium "Challenges in Mineral Sciences," 23rd Annual Meeting of the Metallurgical Society, C.I.M., Quebec City, August 19-22, 1984; *Part. Charact.*, 2 (1985) 91-94.

[10] Euclid was a Greek mathematician born in 325 B.C., who wrote the first surviving textbook on geometry. The traditional geometry of triangles and continuous curves is described as Euclidean geometry. There are several non-Euclidean geometries which are based on different assumptions to those made by Euclid when setting up his system. In this book, the term Euclidean is used to describe traditional geometric figures and boundaries, as distinct from the geometry of rugged curves which have no differential functions or which are indeterminate, which in this book is described as fractal geometry.

[11] B.H. Kaye, "The Description of Two Dimensional Rugged Boundaries in Fineparticle Science by Means of Fractal Dimensions," *Powder Technol.*, 46 (1986) 245-254.
[12] B.H. Kaye, "Fractal Description of Fineparticle Systems," Particle Characterization in Technology, Vol. 1, in K. Beddow (Ed.), CRC Press, Boca Raton, FL, 1984, Ch. 5.
[13] B.H. Kaye, "Fractal Description of Fineparticle Systems in *N*-Dimensional Space", paper presented at the 3rd European Symposium on Particle Characterization, Nurenberg, May 9-11, 1984.
[14] W. Feller, "An Introduction to Probability Theory and its Applications," Vol. 1, Wiley, New York, 1950; see reference 1, p. 86.
[15] J.W. Dauben, "George Cantor and the Origins of Trans Finite Set Theory," *Sci. Am.*, June (1983) 122-131.
[16] J.W. Dauben, "George Cantor, His Mathematics and Philosophy of the Infinite," Harvard University Press, Harvard, 1979.
[17] In his book Mandelbrot uses the concept that the ultimate fractal dimension of an infinitely long Brownian motion is 2. The writer feels that, although this is strictly true mathematically, to help keep the physics of the situation in mind it is useful to write the fact that the fractal dimension tends to 2 with infinite time in the form

$$\lim_{t \to \infty} \delta \to 2.$$

This helps to distinguish between a true two dimensional system and one which when "unwound" has a topological dimension of 1, but the coverage of which approaches 2 after infinite time.
[18] B.B. Mandelbrot, "The Fractal Geometry of Nature," Freeman, San Francisco, 1983, p. 13 (revised edition of reference 1).

2 Fractal Description of Fineparticle Boundaries

2.1 The Fractal Dimensions of a Famous Carbonblack Profile

The carbonblack profile shown in Figure 2.1 has become famous in the world of science because many people seeking to use fractal descriptions of fineparticle systems have checked out their own methods of measuring fractals using this profile [1, 2]. This carbonblack profile was originally photographed by Avrom Medalia, who introduced me to the interesting problems of describing the size and structure of carbonblack fineparticles [3]. In Figure 2.1, the original electron micrograph of the profile is shown. It can be seen that it consists of many tiny spheres glued together to form a larger object. An interesting feature of carbonblack agglomerates is that the visible spherical units within any one agglomerate are all the same size. The size of the spherical subunit varies from one type of carbonblack to another, but is the same within any given agglomerate [3].

The reader is warned that there is confusion in the scientific literature with regard to the meaning of agglomerate and aggregate. Thus, Dr. Medalia calls the system shown in Figure 2.1 an "aggregate of quasi-spheres." Throughout our randomwalk through fractal dimensions, we shall often encounter systems built up from visible subunits, so that we must clearly define what is meant by "agglomerate" and "aggregate" in this book. To establish a

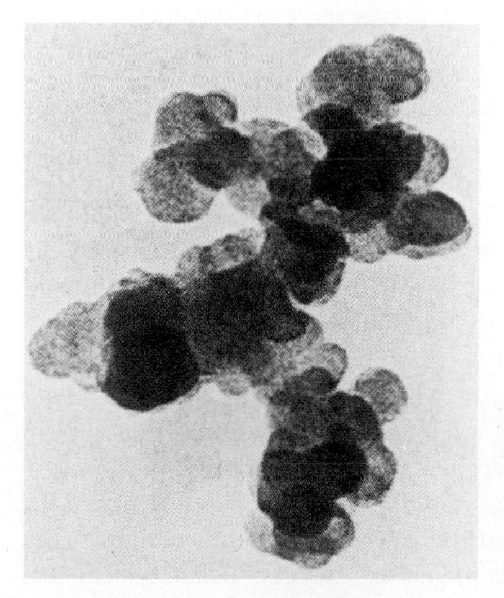

Figure 2.1. A high-magnification electron micrograph of a carbonblack aggregate described by Medalia has been used by many workers to explore the possibility that rugged fineparticle profiles can be described using the logic of fractal dimensions [1, 2, 3]. Reproduced by permission of A.I. Medalia.

reasonable terminology, I have looked at the Latin roots of both words. When an object is made of smaller subunits which are obviously glued together so strongly that they will not come apart when the object is handled, the object is described in this book as an **agglomerate**, because in Latin agglomerate means "made into a globe or a ball." If a group of subunits are only loosely held together so that when they are stirred into a liquid, or used in any other way that they fall apart, then they are described as **aggregates**, because aggregate in Latin means "to bring together like a flock of sheep." Thus, an aggregate is only a loose grouping of units. A flock of sheep will disperse all over the fields at the first opportunity. Whenever the reader encounters agglomerate and aggregate when reading a scientific paper, the meaning of the words in that particular paper should be checked carefully.

Carbonblack is made by burning natural gas with a restricted amount of oxygen, so that carbon is formed in the flame. This carbon is collected on a cold metallic surface. The United States produces more than 2 billion pounds of carbonblack per year. About three quarters of this carbonblack is used by the rubber industry since rubber containing carbonblack is stronger than rubber alone. Carbonblack is also used as a pigment in the manufacture of ink, polishes, paints and of course carbon paper. People who work with carbonblack are very careful about the terms that they use to describe the material. To the ordinary layman, it would seem to be that carbonblack is just a fancy name for soot. However, soot is a general name for any black deposit produced by incomplete combustion of oil and coal. Dirty soot can often contain many carcinogenic chemicals from the incomplete combustion of carbonaceous fuel being burned at any particular device. Carbonblack is produced from natural gas, which does not contain many of the high molecular weight carbon compounds, which are partially degraded in an incomplete combustion process to give carcinogenic chemicals. In the early 1980s, some concern was raised over the safety of carbonblack as used in industry. However, tests seem to indicate that carbonblack produced by the controlled combustion of natural gas does not have dangerous chemicals absorbed into its surface as does the soot produced from, for example, the burning of diesel oil (see the discussion of the structure of diesel soot at the end of this chapter).

In Figure 2.2 several carbonblack profiles photographed by Medalia are shown. Medalia chose to describe the shape of these fineparticles by calculating the size of an ellipse, which would have the same mechanical properties as a thin piece of metal having the same shape as the profiles [3, 4]. Having calculated the size of this ellipse, he then defined two shape factors, which are known as the **anisometry** and **bulkiness** of the fineparticle. The equivalent ellipses calculated by Medalia for the five profiles shown in Figure 2.2 are shown beneath their corresponding profiles. The anisometry of the profile as defined by Medalia is the ratio of the longer to the shorter axis of the ellipse. Medalia defined the bulkiness of the profile as the area of the ellipse divided by the area of the profile. It can be seen from the information summarized in Figure 2.2 that profile (d) is the bulkiest and that profile (e) has the greatest anisometry. Occupational hygienists, who must concern themselves with the dangers posed by respirable dust to workers in mines and factories, have decided from experience that a profile which is three times longer than it is wide is a useful dividing line between "fibres" and "chunky dust." In the discussion of the shape of fineparticles, the length divided by the breadth is called the **aspect ratio** [5]. The aspect ratios of the profiles of Figure 2.2 are almost the same in magnitude as the anisometry ratios, and we see that profile (e) is technically a fibre.

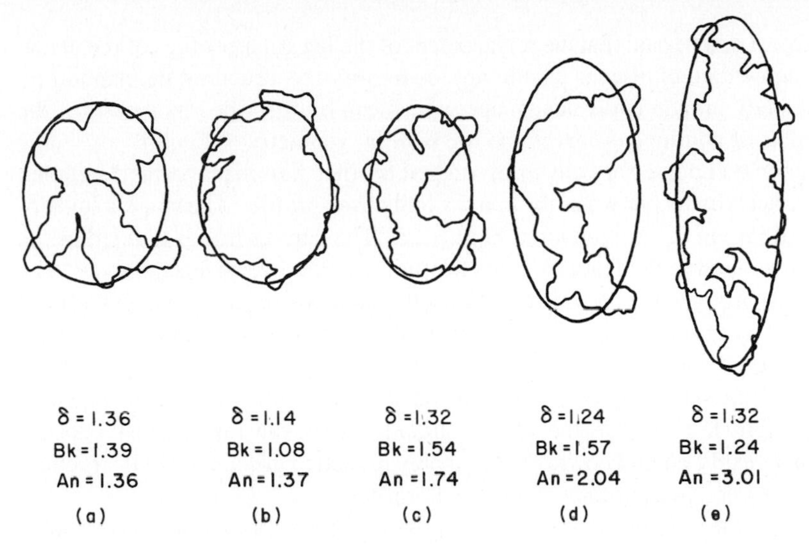

Figure 2.2. Medalia described the structure of carbonblack agglomerates by comparing them to an equivalent ellipse which would have the same mechanical properties as the carbonblack agglomerate. The equivalent ellipses are shown on each of the carbonblack profiles. Using the equivalent ellipse, Medalia defined two shape factors. **Anisometry** is defined as the length of the major axis of the ellipse divided by the length of the minor axis. This quantity is close to the **aspect ratio** of the profile, defined as the maximum projected length of the profile divided by its width. The **bulkiness** of the profile is defined as the area of the equivalent ellipse divided by the area of the profile [3]. Fractal dimensions can also be used to describe these profiles. The magnitude of the fractal dimensions measured over the range $\lambda = 0.08 - 0.32$, units, normalized with respect to the maximum projected length of the profile, is shown for each of the carbon black profiles.
δ = Fractal dimension, An = anisometry, Bk = bulkiness.

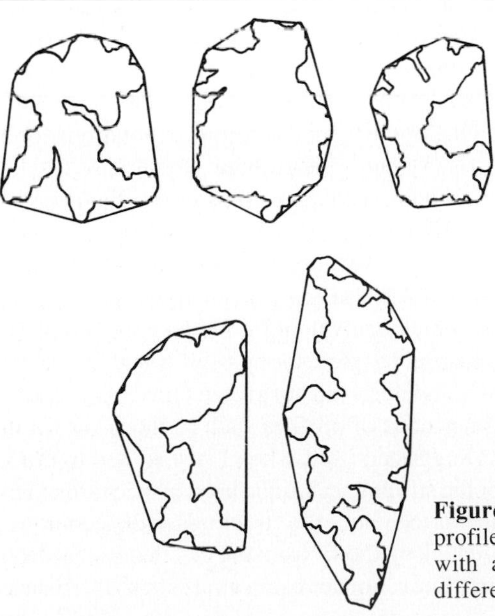

Figure 2.3. The convex hull drawn around a rugged profile is sometimes used to replace a rugged profile with a Euclidian curve, which can be studied by differential calculus and classical geometry.

Philosophically, it can be said that the replacement of the irregular profiles of the actual carbonblack by equivalent ellipses is an attempt to simplify the structural description by transforming the body into an equivalent shape, which can be described by means of the differential calculus of continuous curves and the familiar geometry of Euclid.

Another common technique for converting rugged profiles into manageable Euclidian curves is to construct what is known as the convex hull of the profile. The convex hulls of the five profiles of Figure 2.2 are shown in Figure 2.3. The convex hull is defined as the shortest curve drawn around the features of the profile, which is everywhere convex. For some experimental purposes, the convex hull is a useful reduction in complexity which still retains significant features of the profile. For example, if the profiles were to be falling through a viscous liquid, the outline of the convex hull would probably determine the flow characteristic of the liquid moving around the profile, which in turn would determine how fast the profile would fall through the viscous liquid. Also, the initial gross packing characteristics of a freshly poured powder are probably related to the convex hull structure, rather than to the detailed morphology of a rugged profile.

In spite of the usefulness of such Euclidian curves as equivalent ellipses and convex hulls, in my various studies of carbonblack profiles I was always wary of attempts to simplify the structures in order to give them outlines which made them manageable by available mathematics. I felt that the smoothing out of ruggedness by transforming the profile outline to Euclidian equivalent profiles was a symptom of the "**spherical chicken syndrome**" which sometimes affects scientists. This "disease" diverts them from reality into extensive studies of oversimplified models of real systems. The name of this disease reputedly comes from the historic approach of a physicist asked to study the heat generated by a hut full of chickens. It is said that the aim of this study was to see how much heat was generated and lost by the chickens, and how efficient the cooling-heating system for the chicken house throughout the winter would have to be to keep the chickens happy. Six months after the study began, the physicist was asked if he had solved his problem. "Not yet," he said, "but I am carrying out modelling experiments on a computer to find how much heat is lost from the surface of a spherical chicken." The real feathers, legs and wings of a chicken were too challenging mathematically for the computer, so the physicist replaced the real object with a model that he could handle on a computer. One of the most dangerous temptations for a graduate student studying a particular problem is to change the problem structure so that he can solve it with existing technology, rather than to insist on developing new technology to deal with the reality he is supposed to be studying.

The possibility that the fractal dimension concepts developed by Mandelbrot could be used to describe the real rugged structures of carbonblacks seems to offer an escape from the constraints of spherical chicken modelling, in which equivalent Euclidian curves replace reality. A major problem to be tackled before one could explore the possibilities of the fractal description of the carbonblack profile was that Mandelbrot did not give detailed instructions in his book on how to measure the fractal dimensions of profiles such as those shown in Figures 2.1, and 2.2. The only method which suggested itself when I first started to think about "fractals and fineparticles" was to adapt the striding technique used by Richardson to estimate the extent of a coastline. We shall call this technique the structured walk technique. Other scientists have described this as "the yardstick method," from the fact that Richardson and Mandelbrot discussed the art of making coastline estimates using a yardstick. In a metric age, the yardstick is no longer a commonplace object, and structured walk seems a better name for the experimental technique. In my earlier publications, I called the **structured walk technique** the randomwalk technique. This is not really a good name for the technique and is no longer used by my co-workers and myself. There are several variations in the implementation of the structured walk technique, which have been evolved to cope with fineparticles that either exhibit symmetry or which have deep fissures or extensive bumps. The simplest structured walk technique is known as an inswing structured walk. Other structured walk procedures for exploring rugged boundaries will be described later in this chapter.

Before tackling real carbonblack profiles, I decided to make a first attempt at deducing a fractal dimension by exploring the simpler, randomly drawn rugged profile shown in Figure 2.4, which illustrates the experimental procedure used to explore this rugged profile. Working with an enlarged photograph of the profile, a pair of compasses was used to construct a polygon of side length λ by walking the compasses around the profile. Thus, at the start of the procedure the compass point is placed at point A. One now swings the compasses from outside of the profile until it makes contact with the profile. This contact point becomes the reference point for drawing a straight line of length λ. The compass point is then moved to put the pivot on point B, and the next stride is taken along the coastline to make contact at point C. One now "walks" around the island until one is almost back at the starting point. In any experimental investigation using the structured walk exploration of a rugged profile, the problem of how to complete the walk usually arises. A simple and unbiased method of completing the polygon is to join the point representing the last complete stride to the starting point. The length of this short side needed to complete the polygon is then taken to be a fraction α of the stride length λ. The polygon drawn around the profile is now taken to be the estimate of the perimeter, P_E, at the resolution λ.

In the development of our ideas of fractal geometry, it is necessary to have a clear idea of what is meant by a **dimensionless number**. Physical quantities are said to have dimensions. Thus, a distance between two points is said to have the dimension of length and an area is said to have the dimensions of length times length. In science it has been agreed that we use the shorthand $[L]$ to mean the dimension of length. Using this shorthand, scientists write the statement that "area has the dimensions of length times length" in the form

$$[\text{Area}] = [\text{Length}] \text{ x } [\text{Length}] = [L]^2$$

When we compare two distances with a statement such as "the distance from Toronto to Sudbury is twice as long as the distance from Mactier to Sudbury", we are no longer concerned with actual distances. The statement "twice as far" is said to be dimensionless. In the same way, we can make statements about the relative sizes of two areas by comparing them using a dimensionless statement such as "one area is three times larger than the other." It can be seen that we can make a dimensionless statement about two directly comparable objects by dividing the magnitude of one by the other. When plotting graphs from the data generated in a structured walk exploration of various profiles of different size, it is useful to convert the perimeter estimate and the stride lengths to a dimensionless form by dividing both by some reference length. For reasons discussed later in this chapter, one useful reference length for converting the perimeter estimate and stride lengths used to explore the

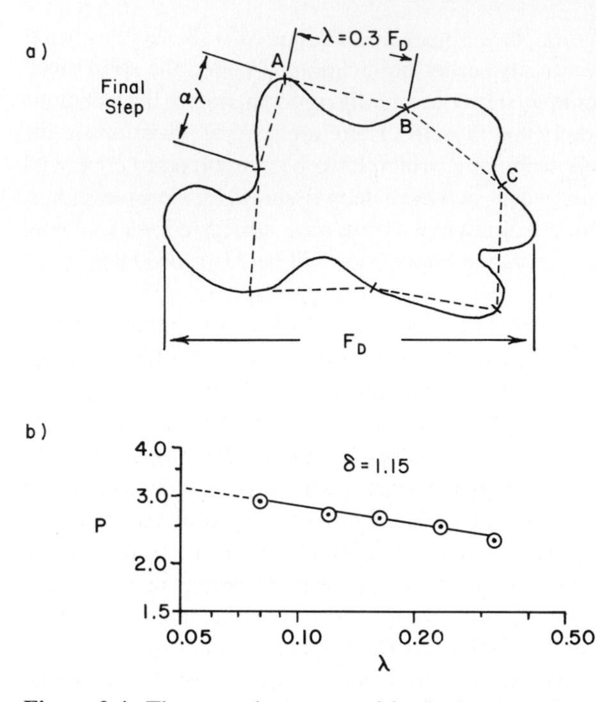

Figure 2.4. The rugged structure of freely drawn randomly rugged profile can be explored using the inswing structured walk technique to construct a polygon on the profile. The perimeter of the polygon becomes the estimate of the perimeter of the profile.
(a) construction procedure for estimating the perimeter of the profile with the equivalent polygon.
(b) a Richardson plot for various perimeter estimates of the profile generated for a decreasing set of λ.
P = perimeter estimate.
λ = stride length.
P and l are normalized with respect to the maximum Feret's diameter.
F_D = maximum Feret's diameter of the profile.
δ = fractal dimension.

rugged profile to a dimensionless form is the maximum projected length of the profile. This quantity is shown in Figure 2.4.

The process of converting quantities such as "length of a profile" into dimensionless form by dividing by a reference length is described by mathematicians as **normalization of the variable**. One of the problems in mathematics encountered by new science students is the fact that ordinary words have special meanings in mathematics. For example, to a new biology student, the "normalization of variables" can seem to be anything but a "normal" practice. In Latin, the word "norma" meant a rule, and it was also the name given to the builder's set-square used to make sure beams and walls were square with respect to each other. In mathematics, **normalization** refers to any rule which is useful to transform variables into a dimensionless form.

Once we have transformed perimeters and stride lengths into dimensionless form, we can compare the structure of tiny carbonblacks with that of the shape of Great Britain, and discuss general features of the profile shape without being worried about kilometres and microns. However, as will be discussed later in this chapter, one can sometimes overlook interesting aspects of the structure of a profile if one uses an inappropriate reference length with which to normalize the estimates of the perimeter of the curve.

For historic reasons, the (maximum projected length of a profile) in any given direction, is known in fineparticle science as the maximum **Feret's diameter** [5]. In my exploration of the profile in Figure 2.1, the first investigative stride length chosen had a magnitude of 0.32 units normalized with respect to the maximum Feret's diameter. The procedure of constructing a polygon using the structured walk was repeated for a series of values of λ down to $\lambda = 0.08$ normalized units. The data for this set of experiments are given in Figure 2.4(b). It can be seen that the data points can be joined together by a line slope of 0.15. Using the logic of Mandelbrot, one is therefore able to deduce that the fractal dimension of the profile over a range of resolutions of $\lambda = 0.08 - 0.32$ normalized units has a fractal dimension of 1.15. Success! The very first attempt to evaluate a fractal dimension of a rugged profile resulted in a fractal descriptor. Only rarely in experimental investigations of real systems does one obtain a first set of data that indicate that the hypothesis being tested, in this case that a rugged profile can be described by a fractal dimension, is reasonable.

The first lecture that I gave on fractal dimensions was to the Particle Size Analysis conference in September 1977 in Salford, UK [6]. As I presented the data shown in Figure 2.4 to the participants of the conference, one could sense the excitement at the fact that a whole new mathematics of rugged boundaries was now available to the fineparticle specialists.

Following the quick success of the adaptation of the structured walk technique to the measurement of the fractal dimension of the rugged profile, the fractal dimension of the carbonblack profile in Figure 2.1 was measured, and it was found to be 1.32 for $\lambda = 0.08 - 0.32$ normalized units (see Figure 2.10). The fractal dimensions of the five profiles shown in Figure 2.2 were measured using the same range of values of $\lambda = 0.08 - 0.32$ and the values are as shown. It can be seen that the magnitude of the fractal dimension seems to match the rugged appearance of the profile. Note, however, that the fractal dimension does not tell us anything about the overall gross shape of the profile. Profiles (c) and (e) have the same ruggedness but obviously have a different overall gross shape, as indicated by their aspect ratios of 1.7 and 3, respectively.

A summary of data from a structured walk exploration of the perimeter of a rugged profile, plotted on log-log scales of the type shown in Figure 2.4(b), is known as a **Richardson plot,** in honour of Lewis Fry Richardson.

2.2 The Dangerous Art of Extrapolation for Predicting Physical Phenomena

When one looks at a dataline such as that shown in Figure 2.4(b), one immediately asks the question of what would one discover if one were to extend the exploration of the profile with larger and smaller λ than those used in the initial exploration. One technique used in science to predict what one might find in an extended investigation of a variable is known as **extrapolation**. This technique assumes that a relationship established by a given set of experiments will persist into regions of behaviour not investigated by actual measurements. Thus one extrapolates, that is, extends the dataline in Figure 2.4(b) by the dotted lines, assuming that the same dataline would be found to exist if new experiments were to be carried out. It has already been pointed out that endless extrapolation of a dataline such as that of Figure 2.4(b), for ever decreasing values of λ, indicates the conclusion that the perimeter is infinite at infinitely small resolutions. Extrapolating the dataline into the opposite region of the graph of large λ is a dangerous procedure which can lead to some false conclusions with regard to the structure of profiles.

To understand the difficulties of experiments on rugged profiles in the coarse resolution region of the Richardson plot, that is for λ greater than 0.3 maximum Feret's diameter, we shall consider the data that one would generate by exploring the structure of a set of ellipses using the inswing structured walk technique. Before making measurements on a set of ellipses, however, let us consider the data which we would generate by using a structured walk technique to explore the perimeter of a circle. Since π, the dimensionless constant encountered throughout geometry and mathematical physics, is defined as the perimeter of a circle divided by its diameter, we will expect to find that we have measured π by using the structured walk technique to explore the circle's structure. Studying a circle in this way may seem to be a trivial experiment, but we can learn a great deal about exploring rugged profiles by looking at the history of the famous dimensionless number π.

No one is sure when the engineers of antiquity began to realize that if one measured the perimeter of a circle and divided it by its diameter, one always, within the accuracy of measurement, arrived at the same number. π has fascinated scientists and engineers ever since it was discovered that its exact value can never be calculated from theory. Even today, large computers are used to search out ever more decimal places in the estimated value of π, in the hope that one day we might discover an exact ending to a seemingly endless number of decimals. π is described by mathematicians as an **irrational number**. This does not mean that it is a crazy number, even though trying to calculate its exact value may have driven some individuals crazy! To understand the mathematical use of the term irrational, we need to know that in Latin the word "ratio" meant "the ability to think and to reason about things; the ability to calculate and figure out mathematics." Originally, a ration was a portion of food calculated by precise mathematics. In everyday speech, a rational person is someone who controls their behaviour by thought, whereas an **irrational** person is someone who acts without reason, someone who acts capriciously, (**capricious** comes from the Latin word for a goat; a capricious thinker leaps about in an illogical series of jumps, like a goat).

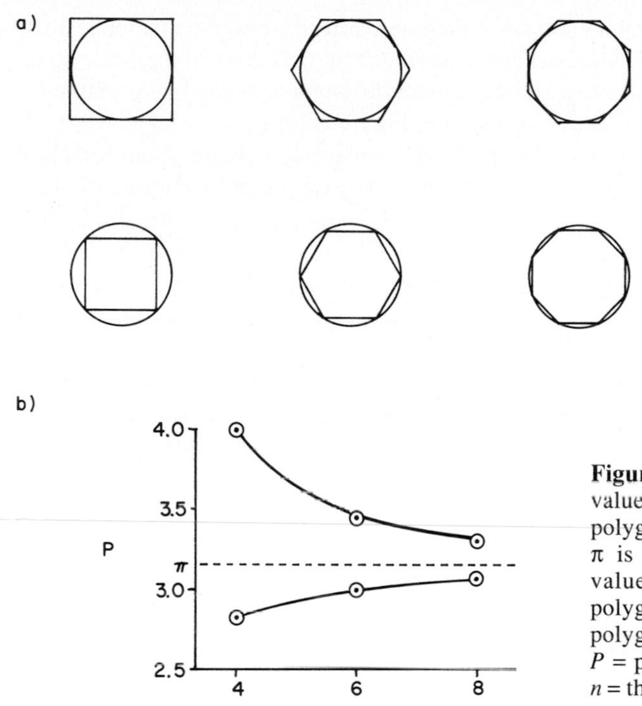

Figure 2.5. Archimedes estimated the value of π by constructing a series of polygons inside and outside the circle. π is estimated from the converging values of the perimeters of the polygons as the number of sides in the polygon increased.
P = perimeter of the polygon.
n = the number of sides in the polygon.

In mathematics, we return to the original meaning of irrational, which can mean "a quantity which cannot be calculated." We shall show in a later chapter that, although we cannot calculate the exact magnitude of π, we can plot its value on a line graph.

In mathematics, the English word **ratio,** derived from the Latin word for rational thinking and exact calculation, came to mean a particular type of calculation, in which one number was divided by another. Thus, we have already defined the aspect ratio of a fineparticle as being calculated from the ratio of the length to the width of the profile. The Greek word for the power of reasoning, the ability to calculate, is "logos." The Greek word for a number was "arithmos." Greek scientists described what Latin mathematicians called "ratio" as "logo arithmos"! This word changed slightly when it came into the English language and it is now written "logarithm." A **logarithm** is any number expressed as a ratio of another. Logarithmic scales used in the graphs of a Richardson plot transform any numbers plotted on the graph into ratios with respect to the quantity which becomes 1 on the axis of the graph. The word logarithm was given to "ratio numbers" by a Scottish scientist, John Napier (1550-1617). Napier invented a method of multiplying and dividing numbers using ratio numbers. He calculated ratio numbers, which became the first logarithmic tables. These log tables were used by generations of students before electronic calculators were invented [7].

One of the earliest attempts to calculate the theoretical value of π is due to the Greek scientist Archimedes, who lived between 287 and 212 B.C. [8, 9]. In his technique, he drew a series of polygons on the outside and the inside of the circle, as shown in Figure 2.5(a). The perimeter of these regular polygons can be calculated by multiplying the side length by the number of sides, and the perimeter estimate is then normalized using the diameter of the circle. The polygon perimeter becomes a better estimate of the perimeter of the circle as the number of sides increases. This is shown by the data for the graph of the perimeter estimates against the number of sides in the polygon shown in Figure 2.5(b).

It can be seen that the two data curves for the outside and inside polygons both converge on the known value of π. The construction of polygons with an exact number of sides becomes a complicated process for polygons with side numbers higher than 8. A few moments of thought should convince the reader that the structured walk technique for exploring a rugged boundary is actually a modification of Archimedes' method for measuring the perimeter of the circle, which avoids the problem of having to construct polygons with an exact number of sides. Thus, if we walk around the circle with a pair of compasses set to a stride length λ normalized with respect to the diameter of the circle, in the same way that we walked around the coastline of Great Britain, then again, as shown by the system shown in Figure 2.6, we end up with the need for a short fractional length to complete our polygon. Again, as in our exploration of the rugged profile in Figure 2.4, our perimeter estimate is given by the equation

$$P = n\lambda + \alpha\lambda$$

If we now plot the perimeter estimates against the stride length on a Richardson plot, we generate the graph shown in Figure 2.6(b). We see that, if we draw a curve through the data points generated by the structured walk exploration of the circle, the curve rapidly becomes parallel to the x axis and asymptotic to a finite perimeter value of 3.14 normalized units, the known value of π for all circles. Note that the dictionary defines an **asymptotic** line as "a line that approaches nearer and nearer to a given curve without ever meeting it."

We can use the information in Figure 2.6 to see how we could tell a robot how to recognize the difference between a Euclidian curve and a fractal boundary. This is important, since any realistic approach to the evaluation of the fractal structure of fineparticle profiles in industry will ultimately have to make use of intelligent robots that can see and recognize the shapes of fineparticles. Using the data in Figure 2.6, we can define a Euclidian curve as one for which perimeter estimates against decreasing values of λ generate a dataline parallel to the x axis, indicating a finite perimeter at high-resolution inspection. This definition of a Euclidian curve may seem to be unnecessarily complicated, but it must be remembered that what is obvious to a human brain often involves a great deal of pattern recognition training, which we do not remember. It is reported that a man blind from birth received an eye transplant in middle age. He never learned to see with his new eye-brain system. For example, he never learned to "tell at a glance" the difference between a square and a triangle. To tell the difference between them, he had to put a finger on one corner and count how many other corners were left. Robots are idiots, and to teach them to tell the difference between a triangle and a square is very difficult, and we would have to develop some experimental strategy for "counting corners" for this type of task. The fact that a structured walk exploration of a boundary results in a final perimeter estimate at high resolution is actually a simple and elegant way of teaching a robot to recognize the difference between a rugged and a Euclidian boundary.

When we use the structured walk technique to explore Euclidian figures more complicated than a circle, we obtain some interesting results. Thus in Figure 2.7 the

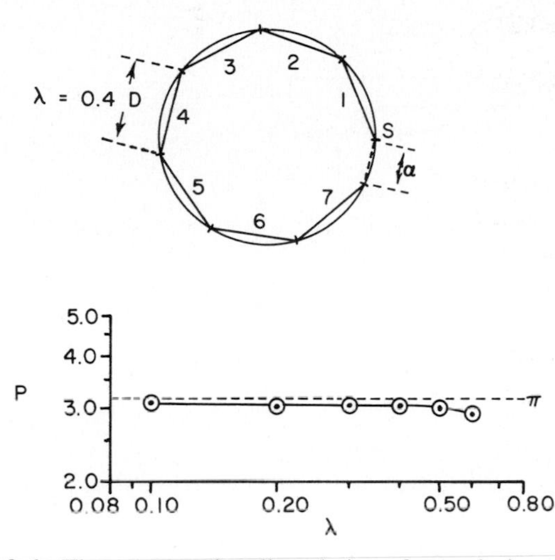

Figure 2.6. The structured walk technique for exploring the magnitude of the perimeter of the profile is a modified version of Archimedes' technique for measuring π.
(a) typical structured walk around a circle leading to a polygon estimate of the perimeter of $P = n\lambda + a$.
(b) Richardson plot of the estimation of p, the perimeter of the circle P expressed as a multiple of its diameter.
P = the perimeter estimate.
λ = the stride length.
P and λ normalized with respect to the diameter of the circle.

type of data one generates by using the structured walk technique to explore a set of ellipses is shown. An interesting feature of these data, which were generated during a laboratory project carried out by a student at Laurentian University, is that the elongated ellipses apparently show a fractal structure for low-resolution exploration, when we know they are Euclidian. Deducing a fractal from the coarse resolution data for elongated profiles results in a fictitious fractal. The fictitious fractal is caused by the fact that for elongated shapes, large-step exploration of the perimeter leads to low values of the perimeter estimate, because of the poor fit of the polygon within the elongated shape. This is shown by the data summarized in Figure 2.7. Dr. D. Avnir of the Hebrew University of Jerusalem in Israel was the first to warn experimentalists that coarse resolution examination of rugged profiles which had high aspect ratios could lead to false fractal deductions from the dataline on a Richardson plot. It is interesting to note that the student who drew the data shown in Figure 2.7 actually drew a line through the data and reported a fictitious fractal of 1.23 for the ellipse. This shows the danger of carrying out an experiment in which one is looking too intently on finding a desirable result. One becomes so intent on finding the expected result that one accepts the data without checking to see if the conclusion corresponds to physical reality. Dr. Avnir, in his original lecture on the dangers of fictitious fractals called such data lines, manifest at coarse resolution exploration of the ellipse in Figure 2.7, having an aspect ratio of 3,

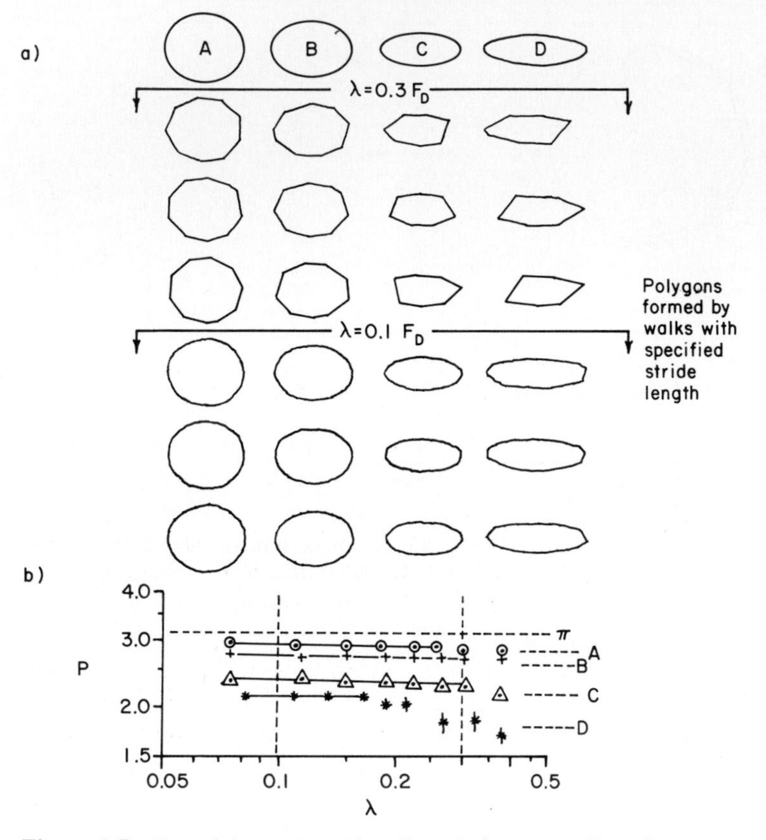

Figure 2.7. Use of the structured walk technique to explore the structure of elongated ellipses can sometimes lead to the discovery of ficticious fractals at coarse resolutions because of the poor fit inside the elongated shape of the polygons constructed with large stride lengths.

"fractal rabbits." This term was derived from the fact that the false fractal appears out of nowhere like a white rabbit out of a magician's hat [10, 11].

After many years of experimentation, I have concluded that it is usually better to restrict ones coarsest resolution steps used to explore a rugged profile to a magnitude of less than 0.3 maximum Feret's diameter, and that if one does attempt to use larger values of λ, one must always check that the conclusions from the graph correspond to physical reality. I have also found that if one is dealing with an aspect ratio greater than 3, one should treat any low-value fractal derived from a Richardson data plot with extreme caution [11].

An interesting feature of the set of data summarized in Figure 2.7 is that, for smooth geometric figures, the values of the normalized perimeters to which the Euclidean portion of the dataline is moving ranges between 3.14 for a circle and 2 for an infinitely thin, infinitely long ellipse. Thus, one can define a shape factor for ellipses which is similar to the number π for a circle, which describes the elongation of any given ellipse. Hence an ellipse which has an aspect ratio of 3 could also be said to be describable by the dimensionless constant π', which has a magnitude of 2.4 normalized units. It should also be noted that, for all of the geometric figures in Figure 2.7, the value of π' has been reached to within an accuracy of 2% by the time the stride magnitude has decreased to $\lambda = 0.1$ normalized units. For the purposes of fineparticle science, where one is often looking at hundreds of profiles within a given population of fineparticles, a classification of shape with 2% accuracy is usually acceptable. Experiments are under way at Laurentian University to teach robots to recognize the shape of an elongated profile by calculating π' from a set of perimeter explorations of Euclidian and rugged profiles.

2.3 Discovering Texture Fractals

In the foregoing discussion, we have explored the dangers of extrapolating from both ends of the dataline in Figure 2.4. Extrapolation in one direction leads us to infinity, whereas extrapolation in the opposite direction may cause us to be seduced into false conclusions by fictitious fractal rabbits. We have also been able to give some practical guidelines as to where not to venture in the Richardson plot, with respect to exploration of profiles with coarse stride lengths. Now we must attempt to discover what happens with real profiles if we venture to make experimental measurements in the region of very small λ stride length explorations. In our early experiments at Laurentian University on using fractal geometry to describe rugged profiles, we constructed the profile shown in Figure 2.8 [5, 12]. This is an agglomerate similar to the carbonblacks in Figures 2.1 and 2.2, but with perfectly circular subunits. From looking at the carbonblack profile, we expected that, when we used values of λ much smaller than the diameter of the constituent spheres, we would find a Euclidean boundary for the agglomerate, because at high resolution it would be a boundary made out of circles rather than a rugged boundary as seen at low resolution. One must remember that a human being with overall vision can always see the rugged agglomerate and the circular subunits at the same time. A robot programmed to use the structured walk exploration technique, can only "see" what it "feels" at any one time with its "fingers" set at λ. For large λ the robot would "see" a rugged profile and at small λ it would "see" a bunch of circles. Only if we give it a sophisticated memory could the robot "see" everything at once in the same way that a human sees the agglomerate and the circles at the same time.

The data generated by a structured walk exploration of the synthetic agglomerate of circles is summarized in Figure 2.8(b). It can be seen that the experimental points can be linked with two distinct lines as shown, but that neither of the two lines indicates a Euclidian boundary.

Initially, we were surprised to find the second dataline with a slope of 1.09 when we were looking for a Euclidian boundary of slope 1. However, we quickly realized that the exploration of the profile at stride lengths of 0.03 – 0.12 normalized units was exploring the way in which the circles were packed together in the synthetic agglomerate, and that we would have to go to even higher resolution steps before we would discover the Euclidian boundary. We now know that, in the language of fineparticle

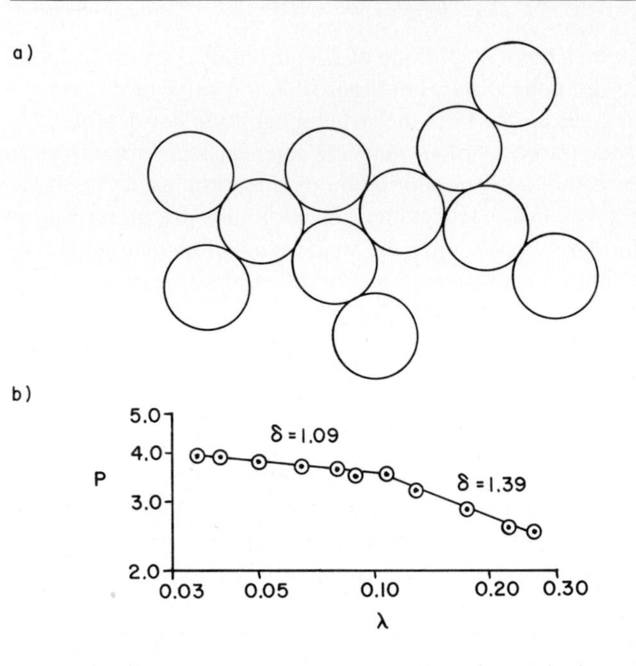

Figure 2.8. The structured walk exploration of a synthetic profile created from the set of circles yielded two fractal slopes rather than the fractal plus Euclidian slopes expected from initial theoretical consideration.
(a) Synthetic agglomerate profile.
(b) Richardson plot for the exploration of the synthetic profile.
P = perimeter estimate.
λ = stride length.
P and λ normalized with respect to the maximum Feret's diameter of the profile.

science, the dataline of slope 1.09 corresponded to an exploration of the texture of the agglomerate as distinct from its structure.

The difference between the terms **texture** and **structure**, as used in fineparticle science, is illustrated in Figure 2.9. The profile used in this figure is one of the carbon-black agglomerates shown in Figure 2.2. When the profile of Figure 2.2 was drawn for that figure, the fact that there were two holes in the agglomerate was suppressed, since the holes were not of interest to us at that point in our exploration of fractals. Topologists describe the structure of profiles by reference to the number of holes within their structure. To a topologist, a cup and a donut have the same shape, and they are described as being of genus 1. In everyday speech, this means that both objects have one hole in their structure. The profile in Figure 2.9 is said to be topologically of genus 2. The overall **topography** or structure of the profile is sometimes described as the morphology of the profile. The word **morphology**, used in science to describe any scientific study of shape and structure, is derived from the Greek word "Morpheus," the name for the God of dreams, who created fantastic shapes in the dreams he gave to mortals. This Greek word has also given us the word morphine for the drug which is supposed to create wonderful visions for its users. The fractal dimension which is

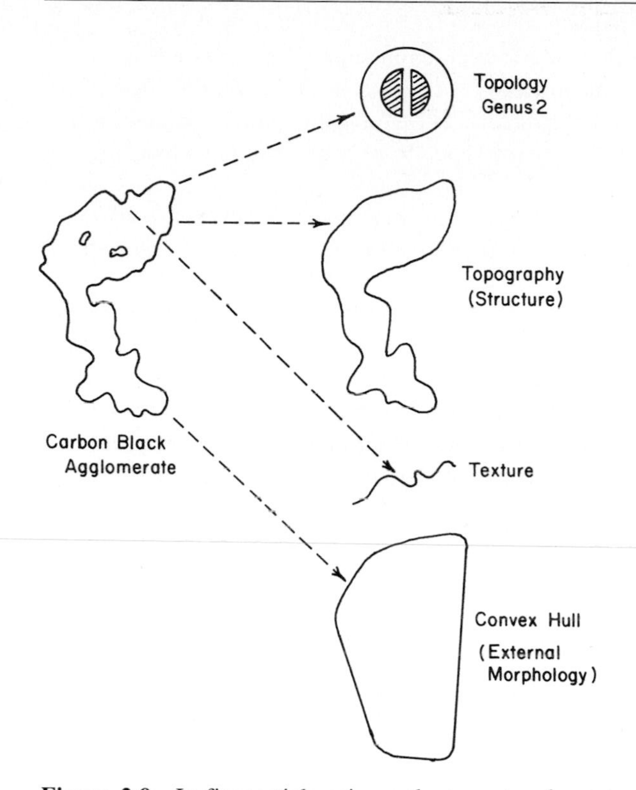

Figure 2.9. In fineparticle science the term topology, texture and topography have specialist meanings.

characteristic of the topography, that is, the overall structure of the agglomerate, is defined as the **structural fractal** and denoted by the symbol δ_S. The agglomerate in Figure 2.8 has a structural fractal of $\delta_S = 1.39$ manifest for λ exploration of 0.12 – 0.28 units normalized with respect to the maximum Feret's diameter. In Figure 2.9, the texture of the profile is illustrated by the line drawing abstracted from the boundary. The fractal dimension, which is useful for describing the texture of the agglomerate, is defined as the **textural fractal.**

The fact that some fineparticles can exhibit fractal structure over all of the ranges of inspection available to the experimentalist poses a problem for the fineparticle specialist when he attempts to specify the size of a fineparticle. Thus, if we look at the carbon-black in Figure 2.9 and attempt to specify the size of the profile, there is no obvious feature of the structure of the profile which corresponds to size. Attempts to specify the size have included calculation of the circle having the same area or the circle having the same perimeter, and the size of the irregular profile is stated to be the diameter of the equivalent circle. It is now obvious that for fineparticles with fractal boundaries, attempts to specify the size by looking at the perimeter of the profile are doomed to failure, since every time the fineparticle is examined at a higher magnification its perimeter and therefore its equivalent size increase. If we consider the structure of

Koch's triadic island, we can see that in a fundamental sense, since the Koch island profile has a finite area even though its boundary is infinite, its size should be described by some dimension based on area not perimeter. Therefore, it is philosophically more correct to specify the size of any profile by reference to its area, which is always finite even if its fractal boundary is infinite.

Similarly, if we consider fineparticles in three-dimensional space, the same difficulty exists, in that attempts to specify the size of the fineparticle by any measurement which includes the surface of a fineparticle showing fractal structure will be essentially indeterminate. In three-dimensional space, the volume of a system exhibiting fractal structure is finite, even if the surface is infinite. It is interesting that over the many years of study, scientists attempting to set up standards for the sizing of fineparticles have come to the conclusion that the fundamental reference value of any fineparticle should be the volume of the fineparticle rather than any exterior dimension or property. Although they do not express their conclusions in the language of fractal geometry, what they have discovered by experience is that it is the volume of a fineparticle which is a determinate quantity, even when boundaries and structures are inherently indeterminate because of fractal features.

In Figure 2.10, an extended investigation of the structure of the carbonblack profile in Figure 2.1 using the structured walk technique is summarized. It can be seen that there are two datalines defining a textural fractal of 1.10 and a structural fractal of 1.32.

When exploring the fractal structure of a rugged boundary, the resolution with which one explores that boundary can be a critical factor in the calculation of the fractal dimension of the boundary. All the initial work on fineparticle boundaries undertaken at Laurentian University was carried out manually, using a pair of compasses to carry out a structured walk technique. It can be appreciated that this work is tedious, and it is obvious that one would attempt to restrict the number of data points that one generated in a Richardson plot. Before we discovered the reality of textural and structural fractals for carbonblacks, some of the earlier experiments were carried out with a small number of values of λ over the stride size range 0.02 – 0.32 normalized units. If one only had a few data points over this region, one can inadvertently smooth out the difference between the texture and the structural fractal. Thus, in our early studies of the carbonblack in Figure 2.1, we actually averaged out the two different fractals with our small number of data to obtain an overall fractal of 1.18 [13]. Flook, who developed the use of an automated image analysis technique known as erosion-dilation logic, which will be discussed in Section 2.4, was able to carry out a much more intense and resolved study of the profile. He pointed out that our reported fractal dimension of 1.18 was an average of the textural and structural fractals of approximately the magnitudes shown by a more detailed study of the profile [1]. Tentative conclusions based on preliminary explorations of a system always carry the hazard of failure to distinguish fine detail because of scatter in the data. As we shall show later in this chapter, the powerful computer systems now being used to measure the fractal boundaries of fineparticle profiles reduce the hazard of "detail washout" of the type that we unfortunately encountered in our early study of the carbonblack profile in Figure 2.1.

In Figure 1.5, a series of lines of different ruggedness were presented to help build an intuitive appreciation of what a line of a given fractal dimension looked like when

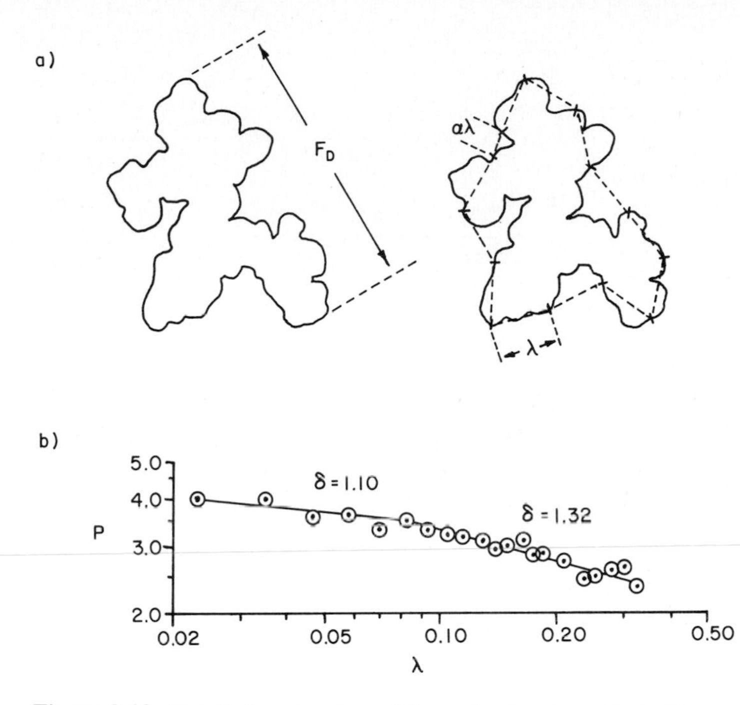

Figure 2.10. Detailed exploration of the carbonblack profile in Figure 2.1 can be used to deduce the textural fractal and the structural fractal describing the ruggedness of the boundary at different levels of inspection.
P = perimeter estimate.
λ = stride length.
P and λ normalized with respect to the maximum Feret's diameter of the profile.

it was encountered in fractal geometry. In the same way, students at Laurentian University constructed and characterized a series of synthetic agglomerates having different structural fractals. These agglomerates together with their structural fractals are shown in Figure 2.11.

Figure 2.12 shows data for the structured walk exploration of the synthetic agglomerate made from glass spheres, built to simulate a diesel exhaust soot fineparticle. This synthetic agglomerate was constructed with all the spheres lying in the same plane. Many carbonblack and soot agglomerates exhibit a flaky structure, such as that exhibited by snowflakes, so that this model diesel soot fineparticle is not as artificial as it may seem to the casual reader. It can be seen that the data for the synthetic soot fineparticle exhibits three distinct datalines in the Richardson plot shown in Figure 2.12. These datalines can be used to calculate a structural fractal of 1.22 and a textural fractal of 1.15, and it can be seen that a Euclidian boundary is manifest for λ smaller than 0.04 normalized units [14].

In our original discussion of the process of normalizing variables so that perimeter and stride lengths could be expressed in dimensionless form, we hinted that, although the maximum Feret's diameter was a useful normalization reference length, similar to

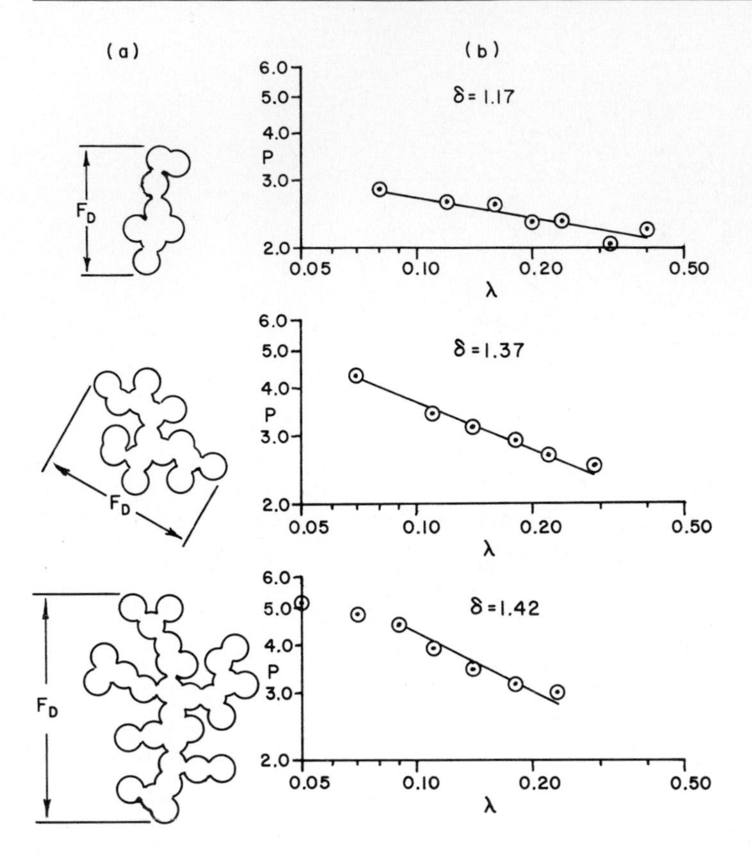

Figure 2.11. The physical significance of the magnitude of structural and textural fractals used to describe agglomerates can be appreciated by studying a series of synthetic agglomerates of circles.
(a) Synthetic agglomerates of varied structural fractal dimension.
(b) Synthetic texture fractals of variously packed circles.

the use of the diameter to create a dimensionless value of π for circles, other normalization procedures were sometimes useful. In the case of an agglomerate constituted from obviously equi-sized subunits, it is sometimes useful to use the dimension of a subunit as a normalization length. Thus,. in Figure 2.12, the upper scale marked λ' shows the stride lengths normalized with respect to the sphere diameter. It is interesting that the changeover to the textural fractal occurs as we start to use stride lengths just smaller than the diameter of the spheres. This aspect of the data in Figure 2.12 will be discussed in Chapter 3, when we explore the usefulness of fractal description of fineparticle boundaries.

In discussing the fractal properties of objects, it is useful to distinguish between ideal and natural fractals. In order to illustrate the difference between the two types of fractal boundaries, we shall now discuss the structure of Koch's triadic island in some detail. This geometric shape was introduced briefly in Chapter 1 in Figure 1.1. In Figure 2.13, the way in which the figure is constructed from an equilateral triangle is illustrated.

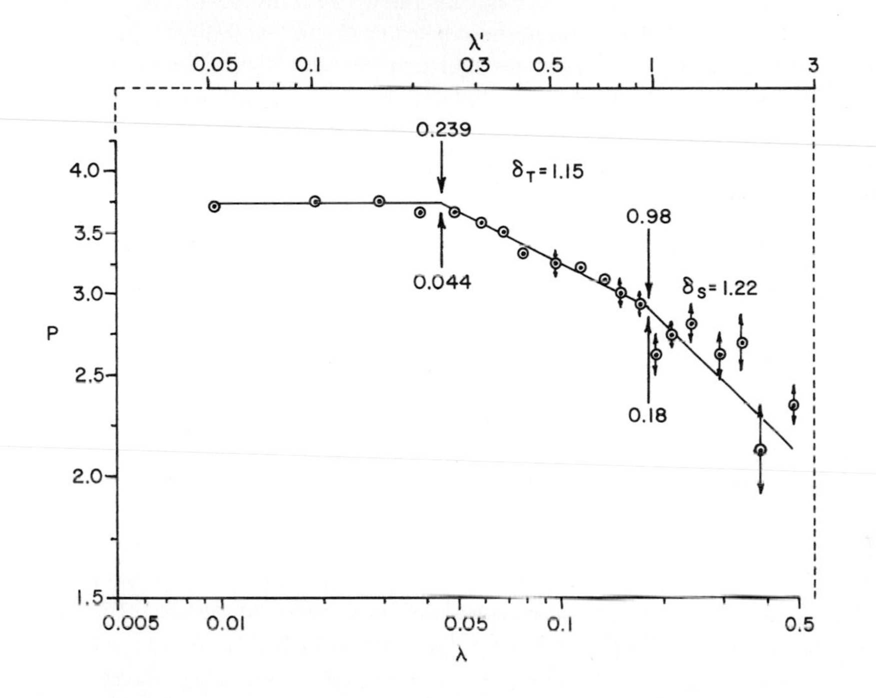

Figure 2.12. A simulated diesel exhaust manifests three different types of boundary at three different levels of inspection.
δ_S is the structural fractal = 1.22 for λ = 0.18 – 0.4; δ_T = 1.15 is the textural fractal for λ = 0.044 – 0.18. A Euclidian boundary exists for 1 smaller than 0.044.
P = perimeter estimate.
λ = stride length.
P and λ normalized with respect to maximum Feret's diameter.
λ = stride length, normalized with respect to the diameter of the spherical subunit of the agglomerate. Each data point represents the average of five different polygons for a given stride magnitude, each polygon being initiated at a different point on the perimeter. The arrows on the data point show the range of the five different estimates at any one given stride length.

From this diagram, it should be noted that we have given new mathematical names to an equilateral triangle and the Star of David. To mathematicians, these familiar shapes are known as the first- and second-order Koch's triadic island, respectively. The geometric construction for transforming the equilateral triangle into the Star of David involves dividing a side of the triangle into three parts. Then the middle part is removed and the gap closed with a "tent" having sides equal to one third of the side of the original triangle. This geometric construction increases the length of each side of the triangle by a factor of 4/3. The process is repeated on each side of the Star of David to give us the third-order Koch's triadic island.

Mathematicians call a recipe for carrying out a calculation or a geometric construction an **algorithm**. Thus, a mathematician would say that the construction algorithm for drawing the *N*th-order Koch's triadic island is illustrated in Figure 2.13. When the general reader browses through the literature on computer science and meets the term algorithm for the first time, he often assumes that it must be a distant relative of a logarithm. The two words are completely unrelated. The term algorithm comes from English attempts to pronounce the name of a famous Arabic mathematician, (780-850 A.D). Al-Kawarizmi.who wrote a book entitled "Ilm Al-jabr Wa'l Muqabalah." Literally translated, this title means "The Science of Transposition and Cancellation." In this book, Al-Kawarizmi had invented the branch of mathematics which we now know as algebra. This term was invented from a corruption of the Arabic term in the middle of the title of Al-Kawarizmi's book, "Aljabr." Students who do not like algebra are tempted to think that modern textbooks on algebra are still written in Arabic! In honour of Al-Kawarizmi's great contributions to mathematics, mathematicians started to describe any recipe for carrying out a calculation or a geometric construction as an algorithm.

If we continue the construction algorithm for increasing the perimeter of Koch's triadic island until we run out of fine enough pens and patience, we eventually achieve what is known as the *N*th-order Koch's triadic island. This mathematical figure is described by Vileinkin as "prickly all over." It has no differential function and has an infinite perimeter enclosing a finite area. (In his book, Mandelbrot discusses the problems of making drawings of infinite perimeters and discusses his technique for printing out diagrams such as Koch's triadic island of the *N*th order. The Koch's triadic island perimeter is by definition an **ideal fractal curve.** If one were to take a portion of the perimeter of Koch's triadic island and look at it under a microscope, the magnified portion would look exactly the same as the original large part of the boundary. This means that if we were to look at a picture of a portion of an ideal fractal boundary, we would not know the magnification of the photograph from the ruggedness of the boundary, since it always looks the same at every magnification. This "look-alike" feature of an ideal fractal boundary at various magnifications is described by the mathematician as **self-similarity.** Thus an important property of an ideal fractal curve is that it is self-similar at any magnification.

The physical significance of the mathematical statement that an ideal fractal is self-similar can be appreciated by looking at the data summarized in Figure 2.14. The various order Koch triadic islands in Figure 2.13 were explored using the inswing structured walk technique. It can be seen that, as the order of the Koch island tends to

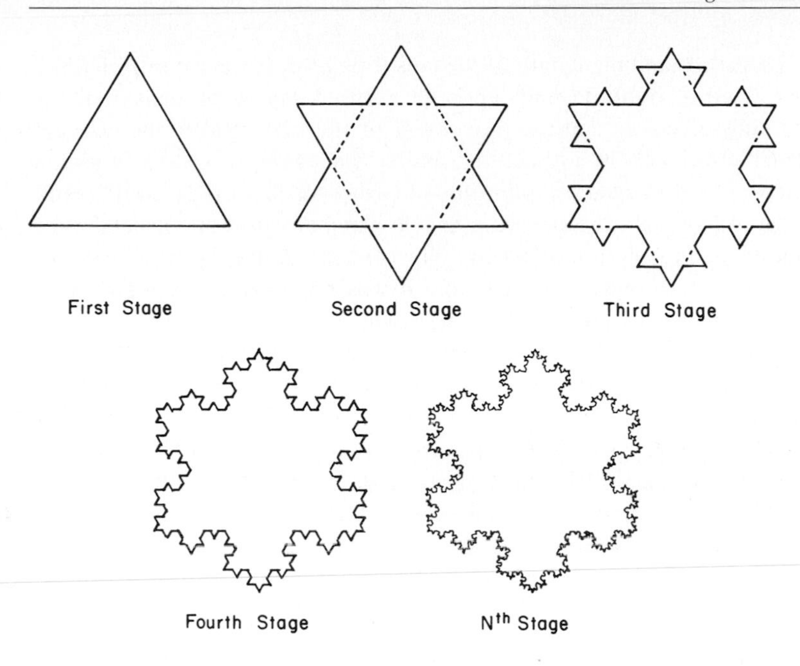

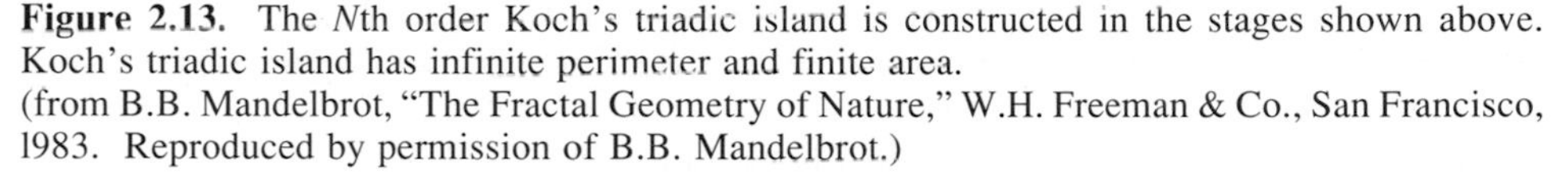

Figure 2.13. The *N*th order Koch's triadic island is constructed in the stages shown above. Koch's triadic island has infinite perimeter and finite area.
(from B.B. Mandelbrot, "The Fractal Geometry of Nature," W.H. Freeman & Co., San Francisco, 1983. Reproduced by permission of B.B. Mandelbrot.)

N, the fractal region of the Richardson plot increases and eventually becomes one long straight line tending to infinity, with a slope related to the fractal dimension $\delta = 1.26$, the value of the fractal dimension of the Nth-order Koch triadic island defined by Mandelbrot. (Note that 4/3 is the increase in perimeter achieved each time the side of a triangle is opened up by removing 1/3 of the line, which is replaced by the construction of a "tent," and that log 4/3 = 1.26). Hence the physical significance of the fact that an ideal fractal curve is self-similar is that the Richardson exploration of such a profile leads to single dataline slope at any level of experimental investigation. We can now state that an important difference between a natural and an ideal fractal is that the **natural fractal** can only be used to describe a real boundary over a specified range of resolution, and that any natural boundary may exhibit different fractal boundaries and/or Euclidian boundaries at various levels of resolution [15]. We have made this basic statement with respect to rugged line boundaries, but the same statement holds true for rugged surfaces and other fractal systems. An ideal fractal surface looks the same at all magnifications.

Mandelbrot, in his original discussion of his development of fractal dimensions, pointed out that in the real world any fractal description of a natural boundary would have inspection limits. Thus, in his discussion of the ruggedness of natural coastlines, he pointed out that the whole concept of a coastline disappears when we start to use measuring units of less than several hundred metres. In my home town, the rise and fall of the tide is over 7.5 metres and the length of the coastline of the estuary on which my home city (Hull, England) is built, depends upon the time of day when measurements are made. The surging of waves and the movement of the tides makes the concept "coastline" undefinable at levels of inspection smaller than units of several hundred metres. In the same way, any intensive exploration of carbonblack profiles would result in the whole concept of boundary disappearing as one came down to the inspection of an electron micrograph at a resolution of several ångstroms. An **ångstrom** is a unit of measure, used by scientists for many years, before it was replaced by the metric unit **nanometre.** The ångstrom was named after the Swedish physicist A.J. Ångstrom (1814-1874). Although the unit is no longer used in modern science, many of the older science books were published when the unit was in use, so that one needs to be familiar with its magnitude. It is useful to remember that:

one micron equals 10 000 ångstroms (1 μm = 10^4 Å)
1 micron equals 1000 nanometre (1 μm = 10^3 nm)
1 million microns equals 1 metre; (10^6 μm = 1 m)
10 nanometres equals 1 ångstrom (10 nm = 1 Å)

At a resolution of tens of ångstroms, the layers of carbon atoms can be seen on the fringe of the carbonblack sphere and the boundary of the sphere becomes fuzzy. At this high resolution, just what constitutes the boundary of the carbon sphere is no longer clear (see, however, the measurement of fuzzy boundaries described in Section 2.4).

In psychology, the **Rorschach** test is used to explore the personality of someone undergoing psychoanalysis. The test was invented in 1920 by a Swiss psychiatrist, Herman Rorschach [16]. In the Rorschach test, a person is presented with a randomly constructed colourful ink blot and tells the psychologist what comes to mind when he looks at the blot. (It is sometimes said that the Rorschach interpretation made by the psychologist on the basis of the observer's reaction to the blot tells us more about the psychologist than it does about the person who took the test!) The many pictures in Mandelbrot's book act like Rorschach blots to entice the scientist into fractal dimensions. The computer-drawn pictures of ideal fractal systems challenge the observer to recognize regions of his own science which might be studied profitably by using fractal dimensions. When I first saw the series of profiles illustrating the construction algorithm for Koch's triadic island, as shown in Figure 2.13, not only did the *N*th-order island suggest that I apply fractal dimensions to carbonblack, but the various intermediate profiles of first, second, third and fourth order suggested to me that fractal dimensions could probably be used to study crystals formed in a solution by chemical reaction and in the study of crushed rocks.

The structured walk explorations of the various Koch's triadic islands in Figure 2.13 summarized in Figure 2.14 illustrate how the range of a natural fractal is important information for the scientist. Thus, it is obvious that the length of the dataline increases as the order of the Koch triadic island increases. The third-order Koch triadic island starts to look like a crystal agglomerate and the range of the fractal slope in Figure 2.13 changes to Euclidian at approximately the size of the subsidiary triangles defining the limits of the third-order crystal. If the islands in Figure 2.13 are thought of as representing crystals, then as the crystallite size in a precipitated agglomerate of crystals decreases, the range of the natural fractal describing the precipitate would increase in

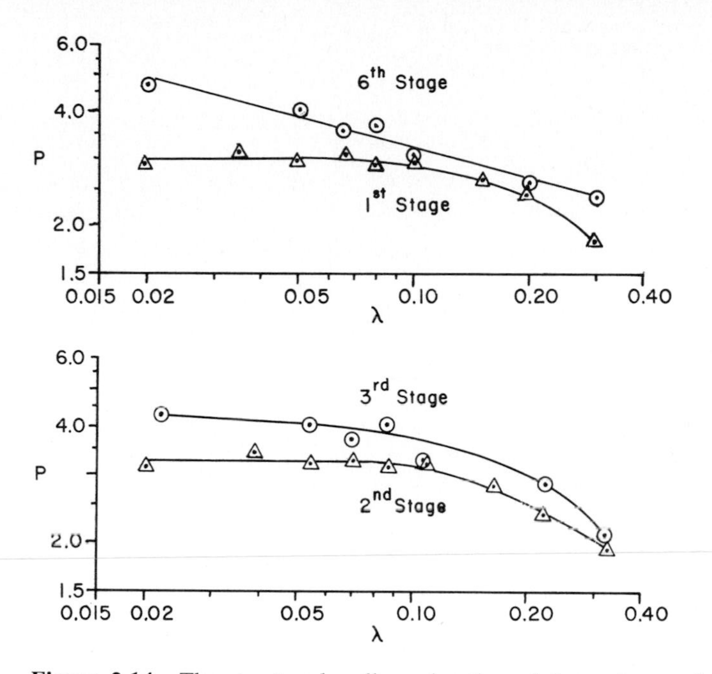

Figure 2.14. The structured walk exploration of the various orders of Koch's triadic island illustrates the difference between a natural and an ideal fractal curve.

length. Thus, if one were to study crystals forming in a precipitate, the fractal dimension used to describe the rugged structure of the agglomerate crystals would describe the shape of the agglomerate and the range of the fractal dimension would generate information on the crystallite size within the agglomerate (see discussion of barytes crystals in Chapter 3).

Mandelbrot realized that many natural boundaries would not satisfy the requirement that the curve defining the boundary appear to be self-similar at any magnification before it could be considered to be a fractal. He therefore proposed a less restrictive definition of fractal structure, which could be used when describing natural boundaries. He said that many natural boundaries may be **statistically self-similar,** as distinct from exactly self-similar. We can explore the physical significance of this statement for rugged fineparticle boundaries by looking at a series of profiles, first described by Flook [1], shown in Figure 2.15. This type of rugged profile is encountered in fineparticle science when looking at crushed rocks produced in the mining industry. Describing such rugged fineparticles is an obvious application of the fractal logic of Mandelbrot. Statistical self-similarity means that although the magnified structure of any local part of the boundary is not exactly self-similar, any one section of the boundary looks like any other "snippet" taken from different parts of the profile. Thus, snippets of the rugged profiles have been abstracted from each closed boundary in Figure 2.15 and placed by the side of the appropriate profile. It can be seen that, on the average (that is, statistically), the snippets look the same as each other. In geographical terms, self-

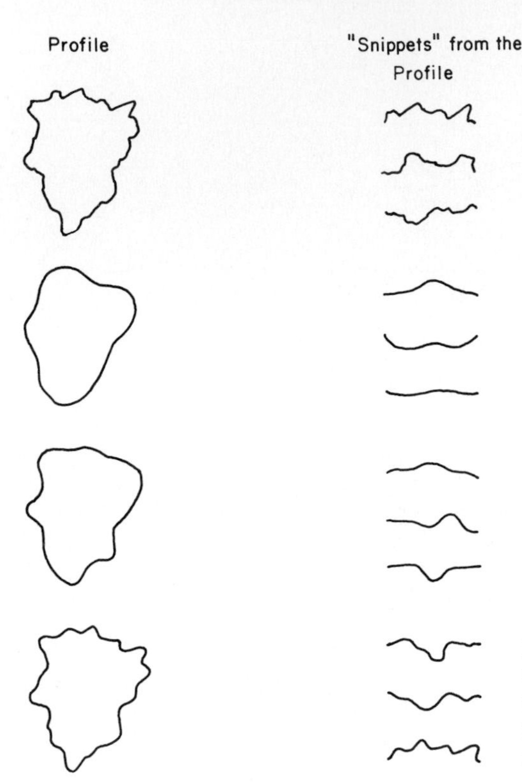

Figure 2.15. A series of profiles, first described by Flook, can be used to illustrate the meaning of statistical self-similarity used in the description of the structure of natural fractals [17].

similarity means that when we are on a coastline, the coastline may not exhibit exact mathematical similarity at all magnifications, but the structure of any one part of the coast would, on average, be the same as any other part of the island boundary. Thus, on a statistically self-similar coastline, we would know that we were on an island, but inspection of the structure of the coastline would not tell us where we were on the island. The concept of statistical self-similarity will be discussed several times throughout this book as we encounter different fractal systems.

In Figure 2.16, the Richardson plots for the structured walk exploration of the Flook profiles in Figure 2.15 are presented. It can be seen that the structural fractals of the boundaries are essentially of the same order, but what the eye sees as increasingly wiggly boundaries actually represents an increase in the range of resolution over which the boundary manifests the fractal structure, which at most is the same for all the boundaries. The information summarized in Figure 2.15 cautions us that although in general, all things being equal, the visual ruggedness of a line corresponds to the physical meaning of the fractal dimension the quantitative evaluation of boundaries is a complex problem which involves not only the magnitude of the fractal dimension used to describe the system but also the range over which that fractal dimension is an operative description parameter. The fractal dimension of a line is a measure of the line's ability to fill space.

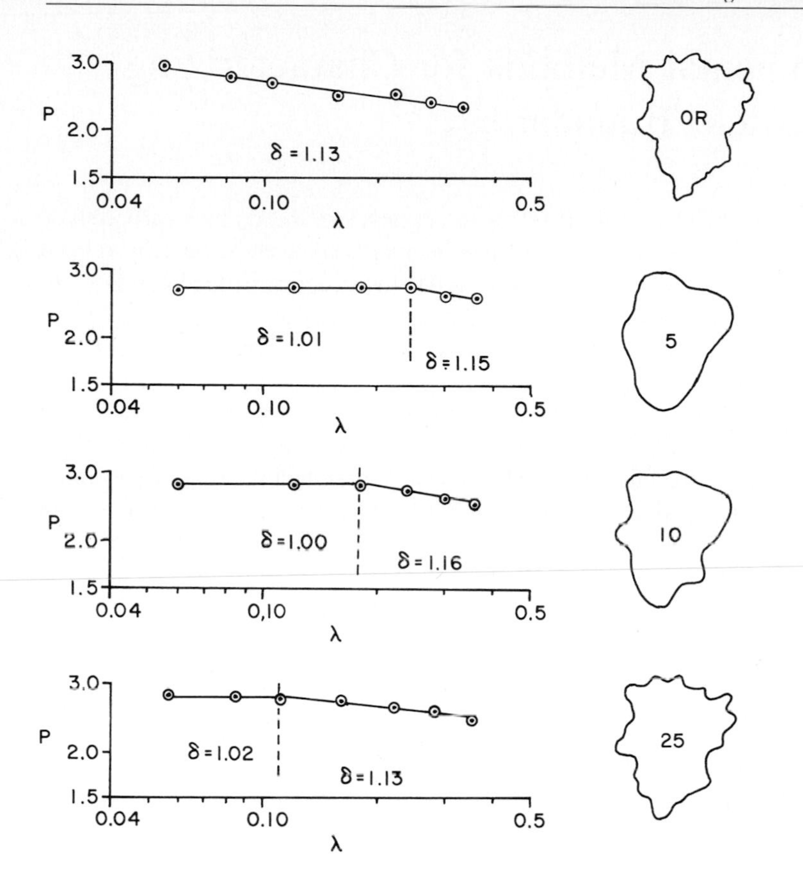

Figure 2.16. Exploration of the Flook profiles by the structured walk technique illustrates the physical significance of the range over which a fractal dimension can be used to describe the structure of the system [15].

Our discovery of textural fractals has helped us to penetrate deeply into the logic of fractal dimensions and, in particular, the difference between textural and structural fractals has helped us to differentiate operationally between natural and ideal fractal boundaries. We have also explored some of the pitfalls awaiting the beginning experimentalist. In the next section, we shall explore in detail some experimental procedures for characterizing the fractal dimension of rugged fineparticle boundaries.

2.4 Experimental Methods for Characterizing Fineparticle Boundaries

In our discussion of the history of attempts to characterize the famous carbonblack profile in Figure 2.1, we considered how one may inadvertently average out textural and structural fractals because of a lack of detailed data on the structure of the boundary. In the same way, it is sometimes possible to suppress significant information in the structure of a boundary if one averages out data for the whole boundary. This is illustrated by looking at the measurements that we would make on the coastline of Great Britain if we divided the island into two parts with a line drawn parallel to the maximum Feret's diameter of the outline. The data for the separate exploration of the two halves of the coastline are summarized in Figure 2.17. It can be seen that the two halves of the coastline are describable by two fractal dimensions, the west coast of Great Britain being much more rugged than the east coast. This is significant information. The magnitude of the two fractals can be linked to the different erosive forces operating on the two sections of the coastline. It may also be linked to the fact that, in general,. the west side of Great Britain is more mountainous than the east coast. The fractal dimension measured for the whole of the coastline of Great Britain is $\delta = 1.25$. This value is seen to be intermediate between the high fractal of the west coast and the low fractal of the east coast. It is very difficult to teach automated machinery to differentiate between regional fractals on an overall boundary. For this reason, one should exercise caution when handling over a measurement task in fineparticle science to a robot. Often, however, a given population of fineparticles has similar overall characteristics. Therefore, one can often inspect the profiles by eye before carrying out detailed measurements and, from this visual inspection, give some general guidance to a robotic system before it starts to evaluate thousands of fineparticle profiles.

Early in our experimental investigations of fineparticle profiles, we came to realize that the presence of a deep fissure in the profile or the presence of symmetry in the profile could cause problems in the interpretation of data summarized on a Richardson plot. The difficulties caused by the presence of a deep fissure can be illustrated by considering the profile shown in Figure 2.18. The fineparticle profile shown is that of a piece of debris found in lubricating oil. Fineparticle specialists can look at the accumulating debris in lubricating oil and predict engine wear and performance from the shape and size of the fineparticles in the oil. For historic reasons, the experimental study of procedures for separating wear debris from lubricating oil and the subsequent characterization of the shape and size of the fineparticles is known as **Ferrography** [18, 19]. The technique uses a high gradient magnetic field to retrieve magnetic fineparticles from the oil. This retrieval technique usually captures most wear fineparticles, since even fragments of plastic and clumps of carbonaceous dirt (soot) will have sufficient magnetic content from associated metal fragments to be captured by the magnetic field. The technique was originally developed for looking at the wear of complex jet engines, but is now applied to many other systems. It has been used to inspect the fluid removed from human joints and to study the wear of metal and/or ceramic parts implanted in

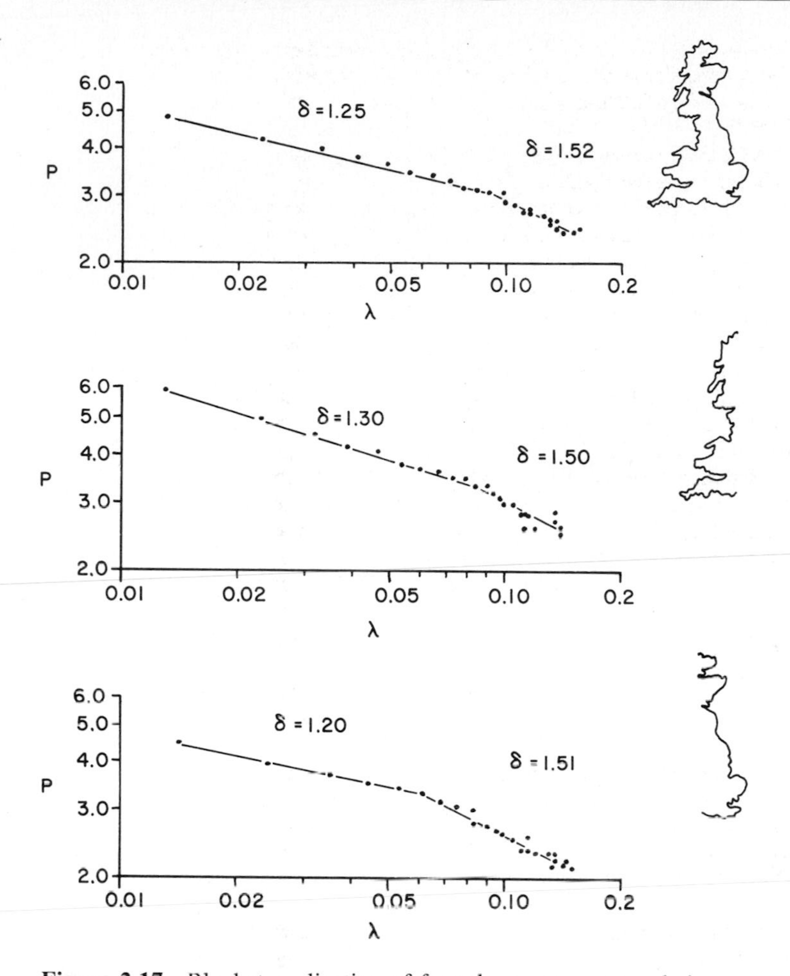

Figure 2.17. Blanket application of fractal measurement technique to a rugged boundary can sometimes suppress important information on the structures of portions of the boundary. This is illustrated for the case of the coastline of Great Britain which has a more rugged structure on the western side of the island than on the east.
P = perimeter estimates.
λ = stride length.
P and λ are normalized with respect to the maximum projected length of the island.
δ = fractal dimension.

damaged human joints [20]. For example, an early study of the shape of metal fragments found in the fluid from a human joint that contained metal replacement parts led to a realization that the failure to remove the fragments made when the surgeon drilled the metal part to install it in the human joint was causing excessive wear. New clean-up procedures for removing the fragments made by the surgeon's drill when installing the metal joint overcame the problem [20].

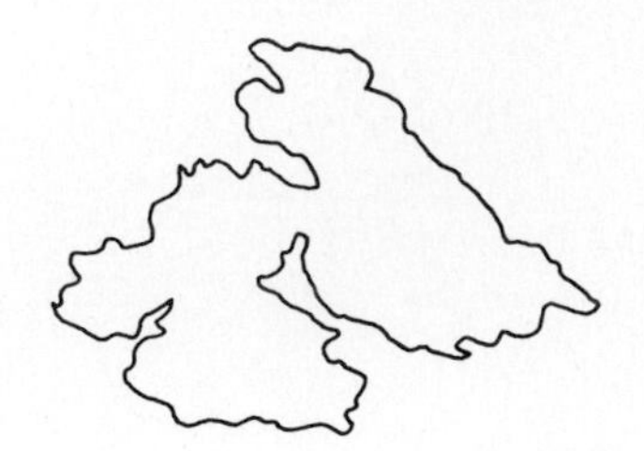

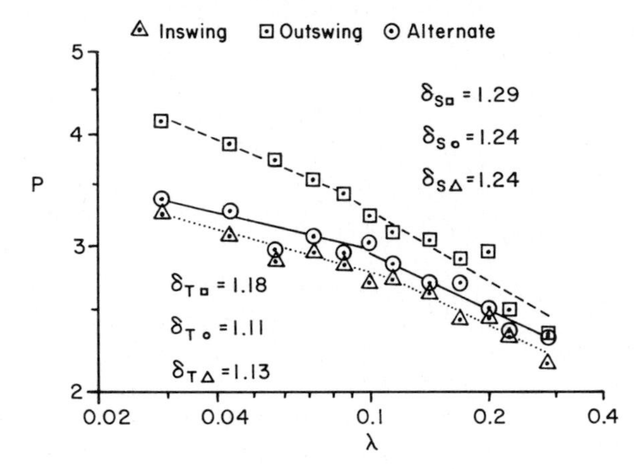

Figure 2.18. The presence of deep fissures in a profile can distort a linear relationship anticipated in a Richardson plot of perimeter estimates against stride length used to explore the profile. The exact exploration technique used to explore the profile can affect the answer deduced from a structured walk exploration of the profile.
P = perimeter estimate
λ = stride length
P and λ are normalized with respect to the maximum projected length of the island.
δ = fractal dimension.

It can be can seen that the wear fineparticle shown in Figure 2.18 has two deep fissures that almost cut the profile into two parts. When exploring this type of profile using the inswing structured walk technique, one has to come down to a stride length comparable to that of the access throat to the fissure before one starts to explore it in any detail. This results in a jump in the measured perimeter at a value of λ which is close to the size of the throat. This jump in the value of the estimated perimeter will often be interpreted as scatter on the data line of the Richardson plot when evaluating a rugged structure. When averaged out with the other data, the data from the fissure will result in a value of fractal different from that which would be deduced from measurements made by a slightly different approach to the exploration of the profile. In Figure 2.18, the data resulting from perimeter exploration of the wear debris fragment made by three different variations of the structured walk technique are summarized. The three techniques used are:

(1) the inswing structured walk technique, in which one always pivots the compasses from the outside of the profile inwards towards the first encounter with the boundary;
(2) the outward swing structured walk technique, in which one always starts with the exploring compass point within the profile swinging towards an encounter with the boundary from within the profile;
(3) the alternate swing structured walk technique, in which swings towards the next encounter with the boundary are made inwards and outwards on alternate steps.

As can be seen from the data in Figure 2.18 the alternate swing method leads to an intermediate estimate of the fractal compared with that evaluated by the outswing and inswing techniques. We should emphasize again that an automated evaluation of the profile often involves a robotic system which cannot "see" the fissures in the profile boundary. The use of all three variants of the structured walk technique for exploring the profile, together with a comparison of the resulting datalines for the various techniques, can be used as an algorithm for training a robot to discover deep fissures within a profile.

Some of the earlier attempts to measure the fractal dimensions of geometric figures ran into difficulties when research workers attempted to check their experimental methods by measuring the fractal dimension of the boundary of Koch's triadic island. The experimental data appeared to show wide scatter in the experimentally determined data points. Intensive study of this scatter revealed a periodicity in the data, as shown in Figure 2.19, reported by Flook [21]. It is now understood that this periodicity is generated by the fact that Koch's triadic island has a symmetrical structure corresponding to the six-pointed symmetry of the original Star of David. In Figure 2.20, the data for

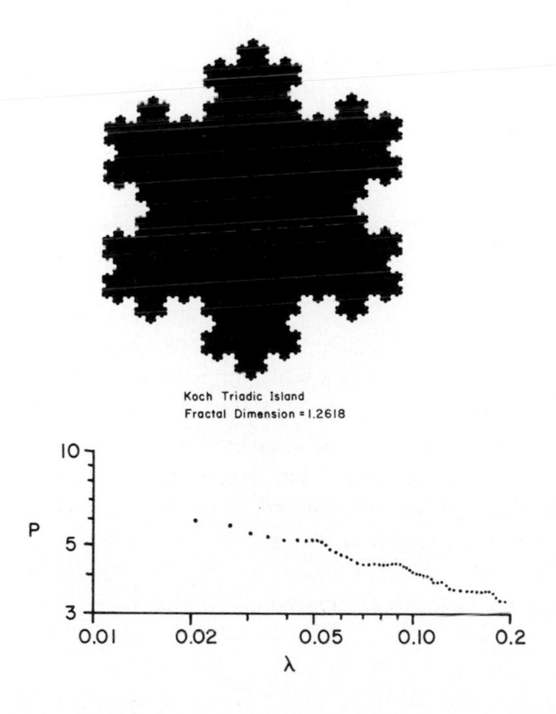

Figure 2.19. Presence of symmetry in an otherwise rugged boundary can result in a cyclic disturbance to the data on the Richardson plot exploration of the profile. This can be regarded either as a source of error or alternatively as a technique for detecting symmetry in a profile [21].
P = perimeter estimate.
λ = stride length.
P and λ are normalized with respect to the maximum projected length of the island.
δ = fractal dimension.
(from B.B. Mandelbrot, "The Fractal Geometry of Nature," W.H. Freeman & Co., San Francisco, 1983. Reproduced by permission of B.B. Mandelbrot.)

the inswing and alternate swing structured walk procedures for an exploration of Koch's triadic island are summarized. It can be seen that again the alternate swing strategy suppresses the problem caused by the symmetry of the profile. Again, the type of information generated by the studies summarized in Figure 2.20 can be used to program a robot to detect any symmetry present in an overall rugged profile.

The structured walk technique using the manual swing of a compass was useful in the early days of the study of rugged fineparticle boundaries, but it is a slow and tedious technique. Everyone interested in fractal dimensions should explore a few boundaries for themselves using a pair of compasses, since using the technique helps to build up an appreciation of the physical reality of the ruggedness of the boundary. However, anyone who measures fractals by manual walking of compasses around a profile will soon yearn for automated methods for evaluating fractal dimensions of rugged boundaries. We shall now discuss some automated procedures for evaluating the fractal dimensions. These techniques will have applications to other rugged two-dimensional boundaries, but we shall illustrate the logic of the automated methods by carrying out experiments on the familiar carbonblack profile in Figure 2.1.

In all our previous discussions about estimating the perimeter of an irregular profile, we have talked about constructing polygons with sides of equal length. There is really no fundamental advantage in using equi-sided polygons to explore a rugged profile,

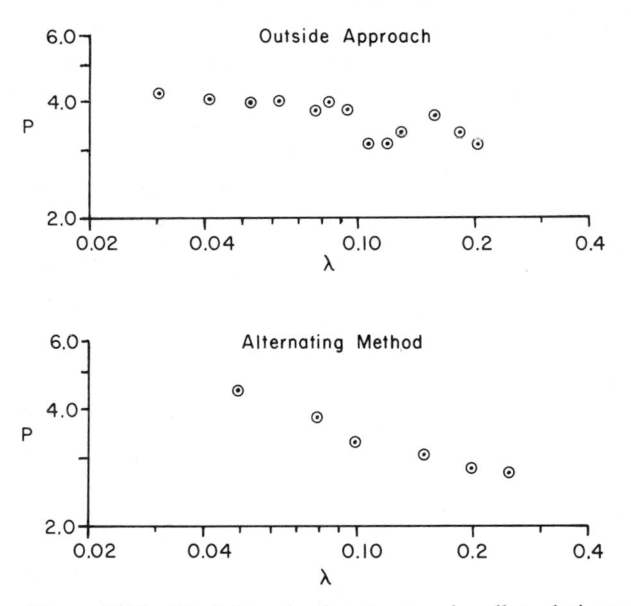

Figure 2.20. Variations in the structured walk technique can be used either to exploit or to suppress the effects of symmetry in a profile, as shown by data obtained by exploring the *N*th-order Koch's triadic island.
P = perimeter estimate.
λ = stride length.
P and λ are normalized with respect to the maximum projected length of the island.
δ = fractal dimension.

provided that one varies the length of the polygon sides constructed on a profile in an unbiased manner. An unbiased construction algorithm for constructing an irregular polygon as a boundary estimate is to lay a set of parallel lines over the profile and use the intercepts of the parallel lines with the profile as a set of points to be linked up by the polygon. This construction procedure is illustrated in Figure 2.21. If the spacing of the lines is λ, we can calculate the length of the side of the polygon spanning two intersect points from the well known Pythagoras theorem that

$$d_n^2 = (x_{n+1} - x_n)^2 + (y_{n+1} - y_n)^2$$

(Pythagoras was a Greek mathematician who lived between 582 and 497 B.C.). Calculating the length of the side of the polygon side from this equation seems to be unnecessarily complicated when we could measure the length directly using a ruler. However, the importance of the technique for calculating side length using the Pythagoras theorem lies in the fact that many commercial image analyzers used to measure the dimensions of a fineparticle profile, use a TV camera to turn an image into a set of lines just like those shown in Figure 2.21. These automated image systems store the locations of the ends of the line representing the image in its memory, so that one can program the computer to calculate the perimeter of the Pythagorean polygon. The fractal dimension of the profile can then be calculated by measuring the perimeter of the polygon at a series of line spacings λ. On the Richardson plot the perimeter estimates are plotted against the line separation λ, as shown in the lower portion of Figure 2.21. In actual practice, the computer system would only take one set of data at the smallest value of λ and then would calculate the perimeter estimates at the other values of λ by using every other line and then every third line, etc., to simulate the effect of coarser and coarser resolution examination. It can be seen by comparing the data in Figures 2.21 and 2.10 that the Pythagorean polygon technique yields the same answer for the fractal dimension of the boundary. (Note that only the structural fractal of the carbonblack profile has been measured by the data summarized in Figure 2.21). The algorithm for calculating the fractal dimension of a profile from a sequence of Pythagorean polygons constructed on a line set generated by commercial image analysers was first suggested to me by Dr. A. Reid of CSIRO, Melbourne, Australia, when he visited Laurentian University in the summer of 1980. This method of measuring fractals could be implemented on several of the currently available automated image analysers [22, 23, 24, 25].

One of the first companies to offer a direct procedure for the automated calculation of fractals from image analysis data was Kontron in Germany. A fractal calculation algorithm, based upon a knowledge of perimeter co-ordinates of a profile for use with the Kontron equipment, was first described by Schwarz and Exner [2, 26]. In their technique, an enlarged image of the profile is placed on a graphics pad of the type available with many minicomputers. By tracing around the perimeter with the appropriate stylus, the coordinates of many points on the perimeter are transferred to the memory. In Figure 2.22, a slight variation of the technique is illustrated. In this technique, we have measured off equidistant points all along the profile and measured the co-ordinates of each point. A little consideration will convince the reader that this is slightly different from using a fine-mesh co-ordinate system on a graphics pad to fix the co-ordinates, that is, the location of a point on the perimeter with respect to a fixed grid.

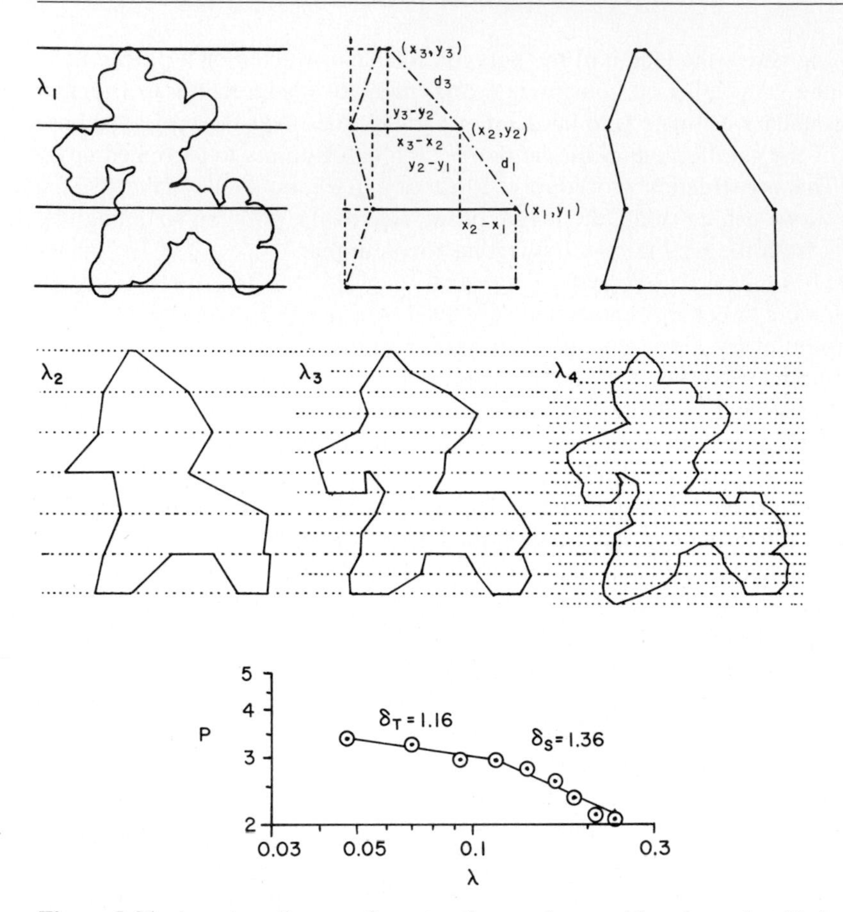

Figure 2.21. A systematic way of constructing a polygon with an irregular side length on a profile to form an unbiased estimator of the profile boundary is to link up the intersection points on the profile generated when a set of parallel lines is laid over the profile. The lengths of the sides of these polygons can be calculated using the Pythagoras theorem, linking the sizes of the three sides of a right-angled triangle.
P = perimeter estimate based on the Pythagorean polygon
λ = line separation
P and λ are normalized with respect to $F_{D,}$ the maximum Feret's diameter of the profile.
δ = fractal dimension.

However, the difference is slight and the calculation procedure is exactly the same as for the procedure illustrated in Figure 2.22. To construct a polygon on the profile, an even number of paces along the profile are counted and a line is drawn between the beginning and the end of the paced-out distance. Thus, if we decide to take ten paces along the profile we would start to construct a polygon of the type shown in Figure 2.22. Again, the paced-out distance and the perimeter estimates are normalized with respect to the Ferret's diameter of the profile. To generate data for evaluating the fractal structure of the boundary, a series of polygons, using a series of paced-out distances

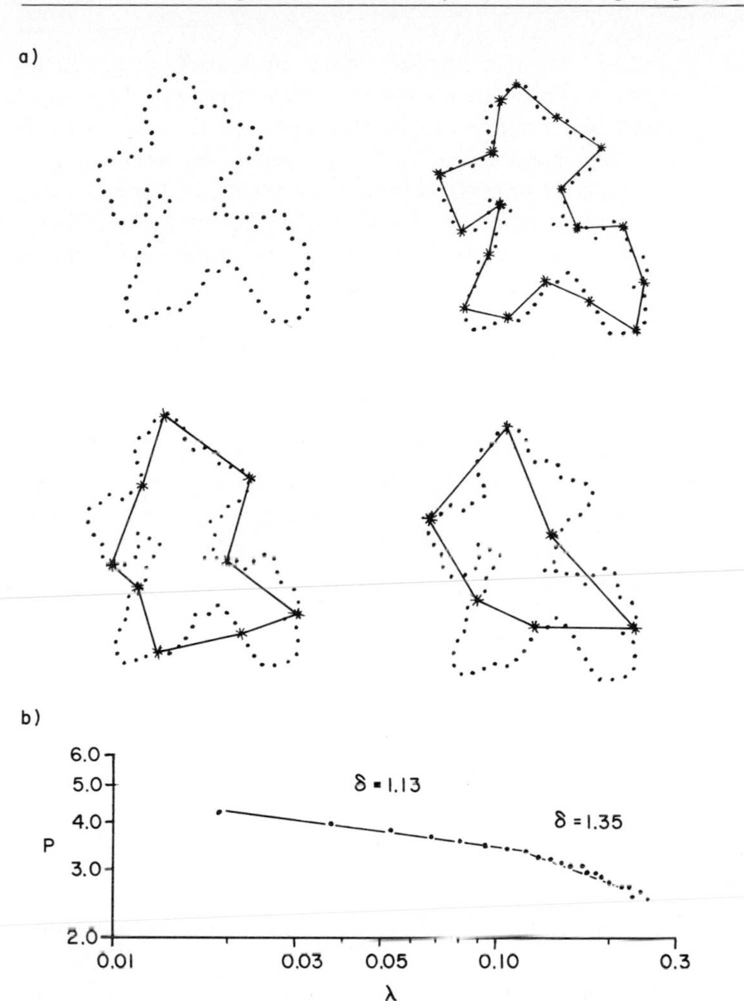

Figure 2.22. An alternative algorithm for constructing an irregularly sided polygon on a rugged profile is to mark out the perimeter with a set of equispaced points. One then constructs a side of the polygon by marking off a number of paces along the profile.
P = perimeter estimate based on equispaced constructed plygon where λ equals the magnitude of the equispaced distance along the perimeter.
P and λ are normalized with respect to F_D, the maximum Feret's diameter of the profile.
δ = fractal dimension.

along the profile, are constructed. The Richardson plot for the equipaced polygon perimeter estimates against the paced-out resolution factor λ is shown in Figure 2.22. It can be seen that the slope of the line yields the same fractal dimension for the structural fractal of the carbonblack profile as the other techniques. Once the perimeter coordinates have been transferred to the memory of the computer linked to the graphics pad, all further calculations proceed automatically with subsequent printout of the fractal dimension. At Laurentian University, students have successfully programmed Commodore and Apple computers to measure the fractal dimension of rugged profiles using the procedures illustrated and summarized in Figure 2.22, although the small computers take a long time to calculate a fractal by the paced-out polygon technique. One could transfer rapidly a whole set of profiles to the memory of the computer and leave it to "chug away" on its own time schedule. It would then print out the fractal data when its computing tasks have been completed (see the discussion of the measurements of cosmic dust profiles in Chapter 3).

All of the image analysis procedures outlined in the forgoing paragraphs ultimately require a machine or a human to decide on the exact location of the boundary of a profile. Thus, the student using a graphic pad looks at the image of his profile and, by the act of drawing around the profile, fixes the boundary of the profile. In the logic of the television camera-computer analysis system, somewhere in the electronic logic of the machine the computer has to decide exactly when the electron beam of the television camera crosses the boundary of the profile. When we use a pair of compasses, the human operator must decide when the exploring point of the compass meets the profile. Unfortunately, many fractal boundaries are fuzzy. When scrutinizing a fuzzy boundary, one does not know exactly where the boundary is located or even, in some cases just exactly what the boundary is supposed to look like. When we look at a cloud from the ground, it appears to have a boundary. That boundary, however, is being decided by our eye-brain system. Anyone who has flown up into a cloud in an aircraft knows that the beginning and ending of the boundary of a cloud is a fuzzy, indistinct phenomenon. We shall now describe a technique for measuring the fractal boundary of an object which has fuzzy limits. The technique for evaluating the fractal structure of a fuzzy boundary is based on some original mathematical logic for estimating the length of indeterminate curves, such as that of Koch's triadic island, originally put forward by Minkowski. The history of Minkowski logic was discussed by Mandelbrot in his discussion of fractal curves in his book. Using Minkowski logic, one imagines that one has drawn a circle around each point on the fuzzy boundary and that these circles merge to form a ribbon overlaying the boundary. Because the ribbon, when straightened out, looks like a sausage, this technique for measuring the length of fuzzy curves is known as Minkowski's sausage logic. One can measure the area of this merged set of circles without knowing how many circles are in the ribbon. One then divides the length of the ribbon by the width of the underlying circles to obtain an estimate of the length of the profile. In this technique, the width of the circle is the resolution parameter at which one is estimating the length of the indeterminate boundary. Mandelbrot (see Chapter 1 references [1, 18]) pointed out that this is like measuring the coastline of Great Britain by having someone lay touching rubber tires around the boundary of the coastline. One does not need to know the location of the coastline underneath the tire; one only needs to know that somewhere each tire touches the coastline. If one were to carry out this procedure all around the coastline, by measuring the area covered by the tires to obtain the area of the ribbon and then dividing the area of the ribbon by the width of the tire, the length of the coastline can be estimated. The procedure is illustrated in Figure 2.23. In the lower portion of this diagram, the tire ribbon has been straightened out so that one can see both the origin of Minkowski's sausage and also that one does not need to know

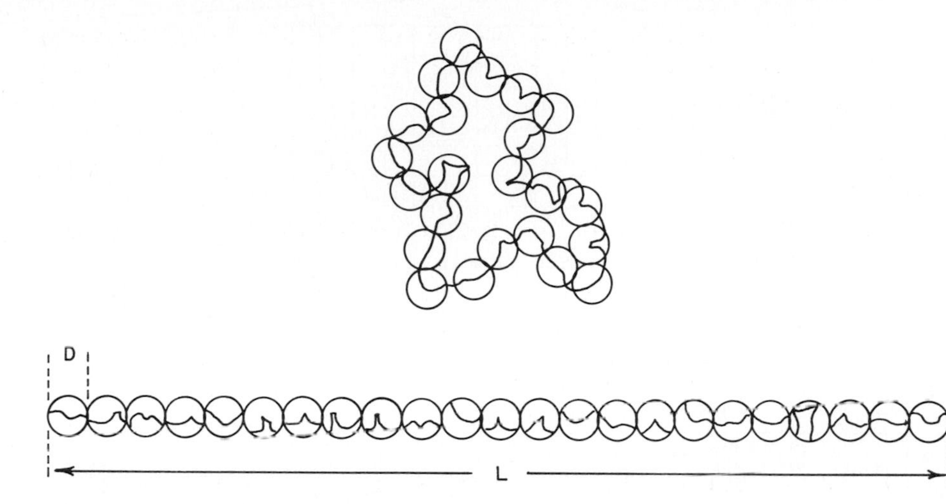

Figure 2.23. Minkowski's sausage logic can be used to estimate the length of an indeterminate boundary at a given level of resolution.

whether the boundaries within the local circles are sharp or fuzzy. (Note, however, that Mandelbrot gives a different origin to the term Minkowski sausage in his description of the mathematical procedure, see page 33 of his book).

Two different techniques for using Minkowski sausage-type logic for evaluating fractals have been described in the scientific literature. The two methods are closely related but were developed separately. These two procedures are known as **mosaic amalgamation** and **erosion-dilation logic.** We shall describe the mosaic amalgamation logic first, since it is easier to illustrate graphically than the erosion dilation logic procedure. From one point of view, we can say that the mosaic amalgamation logic uses square tires to achieve Minkowski's sausage system. The profile to be evaluated is overlaid by a rectangular grid, as shown in Figure 2.24. The square elements of the grid that sit on the boundary can be regarded as square tires thrown on to the perimeter of the profile. One does not have to decide where the perimeter is, only whether it is in or out of any square. To estimate the length of the profile boundary, one now counts the number of squares that are on the boundary and the area of the ribbon replacing the actual profile is calculated as

$$\text{Area} = n\lambda^2$$

where λ is the length of the grid spacing. To estimate the length of the perimeter, one divides this area by the width of the square tire, which results in an estimate of the perimeter being λ. By placing a square grid over the profile, one actually transforms the image into a mosaic of square tiles. Fineparticle scientists are often interested in the area of a profile and its perimeter when examined at a given resolution represented by λ. An unbiased estimate of the area is the number of tiles entirely within the profile plus half of the tiles on the boundary. Estimating the area in this way avoids all the agonizing over where a fuzzy perimeter begins and ends. It also avoids having to guess at the area of the profile by comparing it with standard areas, a technique widely used in finepar-

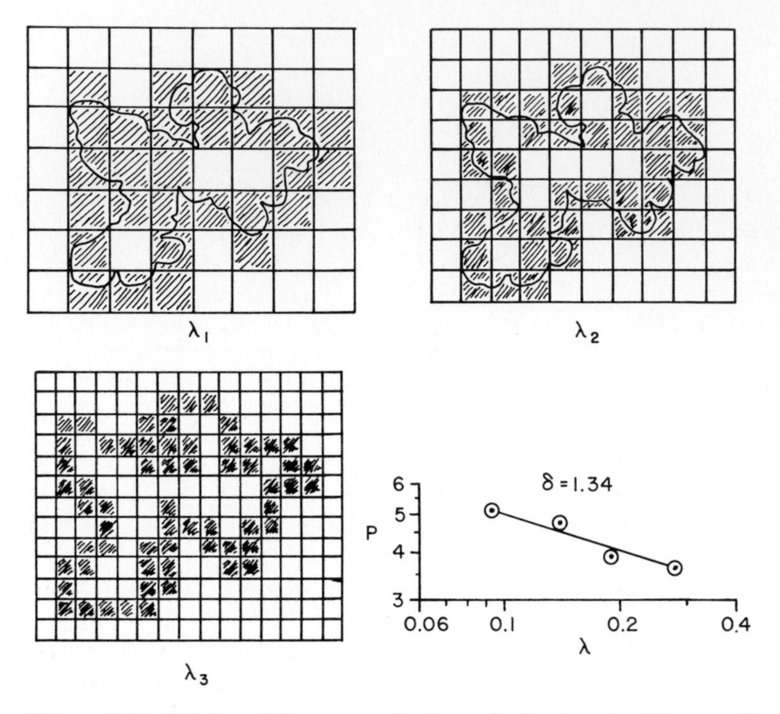

Figure 2.24. Minkowski's sausage logic can be implemented by transforming an image into a mosaic and regarding elements (the individual tiles) of the mosaic as being square tires laid around the boundary. The mosaic transformation can be used to set up a procedure for evaluating the fractal structure of the boundary by a technique known as mosaic amalgamation.
P = perimeter estimate.
λ = side length of mosaic tile.
P and λ normalized with respect to F_D, the maximum Feret's diameter.
δ = fractal dimension.

ticle science (see the discussion of area measurement in reference 5). We shall use this logic for measuring the area within a rugged profile in Chapter 3 when we discuss the fractal dimension of clouds.

To evaluate the fractal dimension of the boundary, one converts the profile into a series of mosaics of ever decreasing tile size. One then constructs a Richardson plot of perimeter estimates against tile size width.

In Figure 2.24, the fractal dimension of the carbonblack profile already studied many times is shown along with appearance of several mosaics used in the evaluation of the fractal dimension. A new generation of television cameras known as charged coupled devices (CCD) convert the image directly into a mosaic and store the mosaic image in a memory by storing the location of each tile within the mosaic along with a description of whether it is inside, outside or on the boundary of the profile. Implementation of mosaic transformation logic to calculate the fractal dimension of a fineparticle boundary will be a simple algorithm with CCD television cameras.

Many graphic systems for displaying scientific information display images as mosaics. In computer language, each small tile in the mosaic is known as a **pixel.** This word is a contraction of the term "picture element." In practice, implementation of the mosaic transformation procedures for calculating fractal dimensions would be undertaken in a sequence opposite to that outlined in the foregoing paragraphs. In an automated system, the image to be evaluated would be transformed into a mosaic at the smallest possible pixel size available. The effect of looking at the profile with larger and larger pixels would then be created in the computer memory by amalgamating information from sets of adjacent tiles. It is for this reason that the technique has been given the term mosaic amalgamation. The mosaic amalgamation technique was originally devised as a graphical technique for manual evaluation of photographs, but will probably be implemented in image analysers for use in fineparticle science which are based on CCD television camera systems.

A persistent problem in the early days of the development of image analysers for evaluating fineparticle systems was the fact that problems occurred in the implementation of linescan logic if a scan line crossed two lobes of a convoluted profile. This problem is illustrated in Figure 2.25(a). It was difficult to decide if two short lines generated by a scan line belonged to the same profile or represented encounters with two separate profiles. Another problem was that many fineparticle profiles have holes in them which cause problems for linescan logic-based image analysis systems. The logic problem caused by holes is illustrated in Figure 2.25(a). Another major problem for linescan logic systems used to study fineparticle systems occurred when counting the number of fineparticles in the field of view if some of the fineparticles touched each other. Thus, if two spherical fineparticles overlapped each other, as shown in Figure 2.25(b), how could the robot tell that there were two fineparticles and not one?

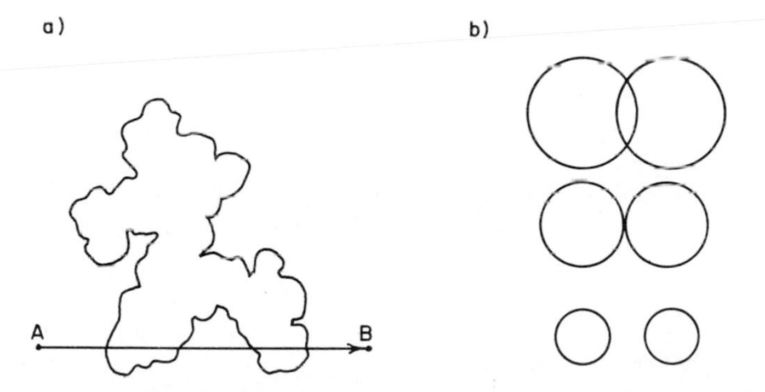

Figure 2.25. Linescan logic systems used in the image analysers of the 1970s had trouble interpreting inspection tracks that moved over a deep fissure in the profile. Also, they sometimes reported a single fineparticle when the human eye could clearly see that two fineparticles were touching and represented two separate entities.
(a) Track AB generates two short intercepts of the profile. In the absence of any other information, the logic of the image analyser would report two small fineparticles rather than one large-lobed fineparticle.
(b) Two droplets which just overlapped in a field of view would be reported by an automatic image analyser as being one large fineparticle.

To overcome the problems of interpreting tracks made across convoluted profiles, and when inspecting fineparticles which might be touching rather than fused together, the developers of commercial image analysers for fineparticle technology developed what is known as erosion-dilation logic. The word erosion comes from two Latin words, "ex" meaning from and "roder" meaning to gnaw. When an object is eroded, it shrinks as if it is being eaten away by a pack of rodents. To see it, an agglomerate was made up of easily separatable parts and the computer processing part of an image analyser was provided with logic that could erode an image mathematically. To do this, the image of a profile is stored as an array of points in a memory of a computer. Then, a routine is devised which strips one pixel at a time, all the way around the boundary. After the erosion of one pixel all around the boundary is completed, the robot inspects the remaining array of points to see if erosion of the profile resulted in two or one group of image points. In Figure 2.26, the breakdown of the carbonblack profile in Figure 2.1 when subjected to pixel erosion in the Dapple image analyser is shown [28].

It can be seen that after nine erosions the profile has started to break up into three main clusters. It almost breaks down into four subunits after eleven erosions but even down to fourteen erosions the three subunits of the original carbonblack are clearly visible. We interpret this to mean that the original carbonblack was probably formed by the collision of three or four growing clusters of unit spheres (see the discussion of diesel smoke structure in Chapter 3).

When using erosion logic in an image analyser, one has to decide how many erosions are legitimate before deciding that the agglomerate is broken down into real subunits. For example, if we were only looking for fineparticles that just touched, we would probably set some criterion which, if it did not break down after three erosions, was a chance clustering of fineparticles in the image. However, this type of decision has to be based upon experience. There are no absolute criteria that permit the scientist to decide what is a fused agglomerate or a clustering which is a chance juxtaposition of profiles.

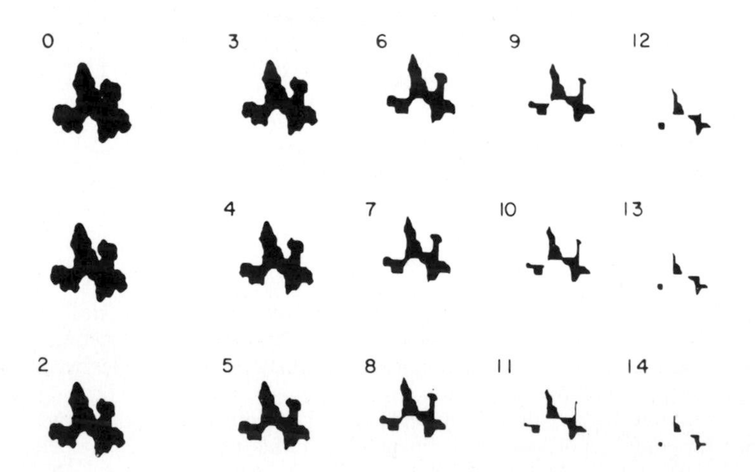

Figure 2.26. Repeated pixel erosion of the image of the profile in Figure 2.1 suggests that the cluster was originally formed by the collision of three or four subunits.

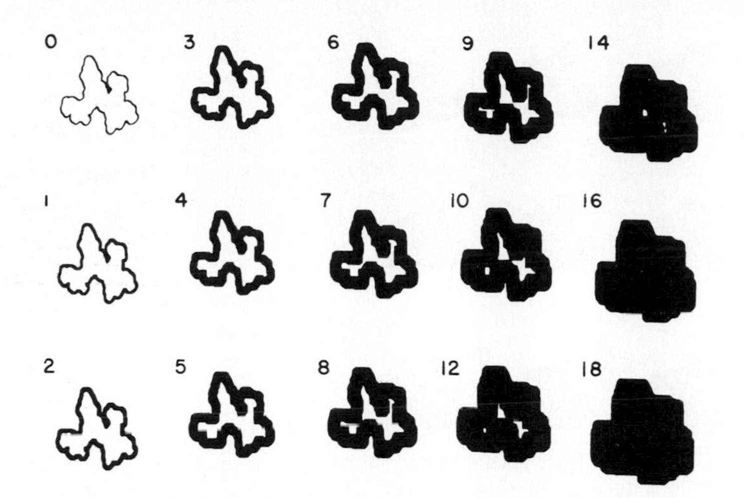

Figure 2.27. Dilation logic can be used to fill in internal holes and or deep fissures in the image profile to avoid logic problems generated when a line scan inspection routine crosses a hole or a fissure in the profile.

The counterpart of erosion logic, dilation, was provided for linescan image analysers to close up holes and close up deep fissures in the profile, before counting the number of fineparticles on a microscope slide. The word dilate means to make wider. Thus, the pupil of the eyes of a scientist would dilate if he really saw a pack of electronic rodents nibbling away at the pixels of an image profile. The process of dilation is essentially that of mosaic amalgamation discussed earlier in this chapter. In sophisticated image analysers, the image is transformed into hundreds of thousands of pixel points. The amalgamation of these pixels to produce a coarser mosaic all takes place in the memory of the computer, with the resulting image being printed out in mosaic form. In Figure 2.27, the dilation of the image of our carbonblack profile boundary as generated in the Dapple image analyser, is shown. It can be seen that by eighteen dilations all of the fissures of the original shape have been closed down. Before sizing this profile, one has to erode it back down to the size of the original image. In Figure 2.28, the erosion of the profile back to its original size is shown. At the bottom of Figure 2.28, the dilated-eroded profile is shown alongside the original profile for comparative purposes. It can be seen that the new profile is simple in structure.

In Figure 2.29, a series of dilated images of a simulated carbonblack profile, as reported by Flook [29], are shown. Alan Flook attended the Particle Size Analysis 1977 Conference on Fineparticle Science referred to earlier and was one of those who saw immediately the potential for fractal description of fineparticle boundaries. Over a cup of coffee, Flook sketched on a piece of paper how easy it would be to measure the fractal dimension of a profile using the dilation logic available on the Quantimet image analyser [22]. He pointed out that the dilated boundary was actually a Minkowski-sausage/ribbon, and that progressive dilation of the boundary corresponded to using bigger and better types to locate the irregular boundary. (Note that the dilation

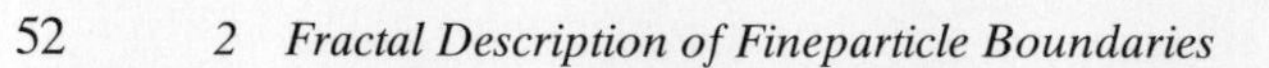

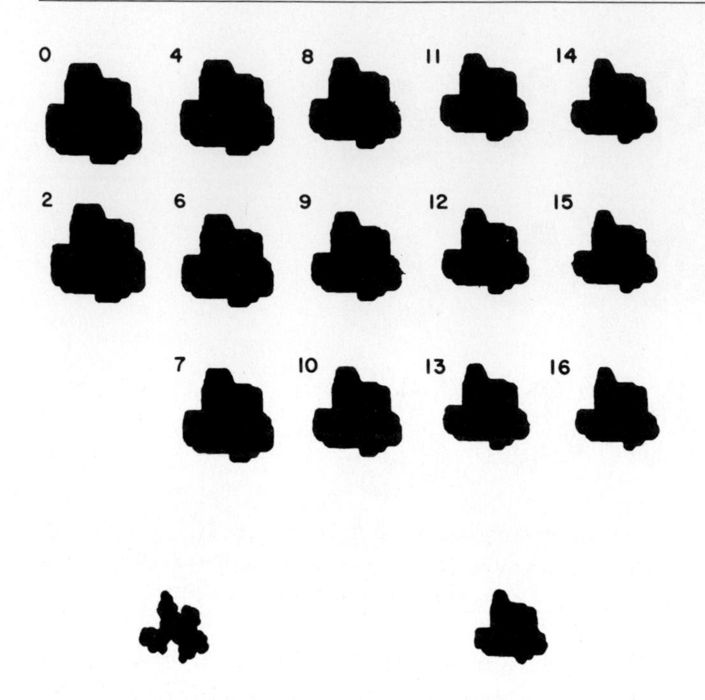

Figure 2.28. Erosion of the dilated image in Figure 2.27 results in a simple profile which is easier for robotic inspection by linescan logic cameras.

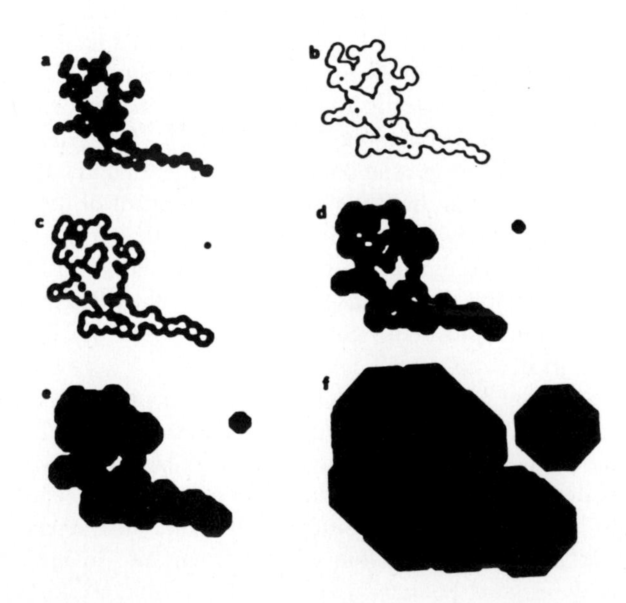

Figure 2.29. Flook was the first to use dilation logic to measure the fractal dimension of an irregular profile.

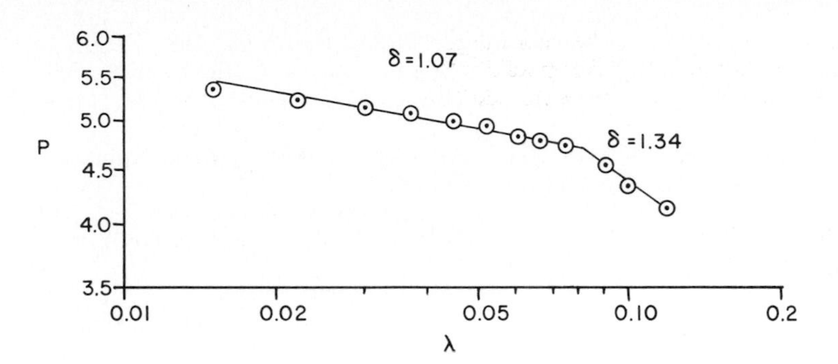

Figure 2.30. Dilation logic processing of the carbonblack profile in Figure 2.1 results in the manifestation of the same fractal dimensions of the profile as evaluated by other image analysis procedures.

structuring element in Figure 2.29 is an octagon; other structuring elements, e.g. mosaic tiles, can be used in the transformation of the image). The logic element required on a Quantimet 720 image analyser to enable the research worker to use dilation logic is known as the 2D Amender module.

Flook realized that the measurement of the perimeter of the profile at each stage of dilation, plotted against the size of the structuring element, would yield the fractal dimension of the profile. Within a very short time, Flook was able to publish his method for measuring fractal dimensions with dilation logic [1]. He has since published several scientific papers on techniques for measuring fractal dimensions [29, 30].

In Figure 2.30, the data for the measurement of the fractal dimension of the carbonblack profile in Figure 2.1 by dilation logic using the Dapple image analyser are presented. Again it can be seen that the value of the structural fractal dimension measured by dilation logic is the same as that obtained by other experimental procedures. Readers interested in the detailed theory of the logic underlying dilation-erosion operations in automated image analysis should read the development of theory presented by Serra [31].

References

[1] A. G. Flook, "The Use of Dilation Logic on the Quantimet to Achieve Fractal Dimension Characterization of Texture and Structured Profiles," *Powder Technol.*, 21 (1978) 295-298.

[2] H. Schwarz and H. E. Exner, "The Implementation of the Concept of Fractal Dimensions on a Semi-Automatic Image Analyzer," *Powder Technol.*, 27 (1980) 207-213.

[3] A. I. Medalia and G. J. Hornik, "Pattern Recognition Problems in the Study of Carbonblack," *Pattern Recognition*, 4 (1975) 155.

[4] A. I. Medalia, "Dynamic Shape Factors of Particles," *Powder Technol.*, 4 (1970-71) 117-138.

[5] B. H. Kaye, "Direct Characterization of Fineparticles," Wiley, New York, 1981.
[6] B. H. Kaye, "Characterization of the Surface Area of a Fineparticle Profile by its Fractal Dimension," M.J. Groves (Ed.), in "Proceedings of Particle Size Analysis," Heyden, London, 1977, 250-259.
[7] See entry on Napier in I. Asimov's "Biographical Encyclopedia of Science and Technology," Revised Edition, Doubleday, Garden City, NY, 1972.
[8] For an interesting review of Archimedes' technique for estimating the value of π, see D.J. Fink, "Computers and the Human Mind," No. 43 in "Science Study Series," Anchor Books, Doubleday, Garden City, NY, 1966.
[9] P. Beckmann, "A History of p," Golem Press, Boulder, CO, 1970.
[10] Dr. Avnir's observations on the reality of fractal rabbits formed part of his presentation at the conference "Fractal Aspects of Materials: Metals and Catalyst Surfaces, Powder and Aggregates," organized as part of the annual meeting of the Materials Research Society, November 26-27, 1984. The book of extended abstracts of this Symposium edited by B.B. Mandelbrot and D. Passoja is available from the Materials Research Society, 9800 McKnight Road, Suite 327, Pittsburgh, PA, 15237, U.S.A.
[11] The reality of fractal rabbits has been explored in a set of lecture notes prepared by the author and used when presenting workshops on fractals. These lecture notes are entitled "**Fractals Rabbits are Real - Sometimes!**" These lectures were given for the first time at a workshop on fractal dimensions held during the conference on Powder Technology at Rosemont, Illinois, May, 1984. The lecture notes are available in booklet form directly from the author at Laurentian University. Many of the experimental problems of evaluating fractal dimensions from structured walk exploration techniques are discussed in an informal manner in these lecture notes.
[12] B.H. Kaye, "Fractal Dimensions and Fineparticle Science," in "Proceedings of the Powder Technology Conference," Rosemont, Illinois, published by Cahners Exposition Group, 1978.
[13] B.H. Kaye, "Specification of the Ruggedness and/or Texture of a Fineparticle Profile by its Fractal Dimension," *Powder Technol.*, 21 (1978) 1-16.
[14] B.H. Kaye, "The Description of Two Dimensional Rugged Boundaries in Fineparticle Science by Means of Fractal Dimensions," *Powder Technol.*, 46 (1986) 245-254.
[15] The multi-fractal structure of natural boundaries was discussed in detail by B.H. Kaye, "Multi-Fractal Description of a Rugged Fineparticle Profile," *Part. Charact.*, 1 (1984) 14-21.
[16] For a discussion of the use of the Rorschach in psychology, see J.V. McConnell, "Understanding Human Behaviour," Second Edition, Holt, Rhinehart and Winston, New York, 1977.
[17] A. Flook, "Fourier Analysis of Particle Shape," in Proceedings of the Fourth Particle Size Analysis Conference, Loughborough University of Technology, December 21-24, 1981, N.G. Stanley Wood and T. Allen, (Eds.), "Particle Size Analysis 1981-1982," Wiley-Heyden, London.
[18] D. Scott, W.W. Seifert, and V.C. Westcott, "The Particles of Wear," *Sci. Am.*, 230 (5) (1974) 88-97.
[19] D. Scott, "Debris Examination – a Prognostic Approach to Failure Prevention," WEAR, 34 (1975) 15-22.
[20] C. Evans, "How Human Joints Wear - Learning From Machine Methods," *New Sci.*, November 9 (1978), 444-445.
[21] A.G. Flook, "Fractal Dimensions - Their Evaluation and Their Significance in Stereological Measurements," *Acta Stereol.*, Vol. 1, No. 1, Jan. (1982) 79-87.(Proceedings of the Third European Symposium on Stereology, part 2).
[22] The Quantimet image analysis system, used by Flook, is manufactured by Cambridge Instruments Ltd. of Melbourne, Royston, Hertfordshire SG8 6EG, UK.
[23] The computerized image analysis system known as the Omnicon System was manufactured and marketed by Bausch and Lomb of Rochester, New York. In 1985 the division of the company marketing this system was acquired by another company.

[24] A whole range of image analysis systems are manufactured by Ernst Leitz, B633 Wetzlar, Postfach 2020, FRG.

[25] An image analysis system known as Magiscan is manufactured by Joyce Loebl Corp., Marquis Way, Team Valley, Gateshead NE11 0QW, England. Several image analysis systems have been developed by Carl Zeiss, Oberkochen, Wurtt, FRG.

[26] The Kontron system used by Schwarz and Exner is manufactured by Kontron Messgerate, Breslauer Strasse 2, D-8057 Eching, FRG.

[27] The logic of the mosaic amalgamation procedure was outlined briefly in reference 13 and developed more fully in an unpublished set of lecture notes which circulated extensively amongst those who took up the challenge of developing systems for evaluating fractals. In the publications of these workers these lecture notes are referred to by the title B. H. Kaye, "Sequential Mosaic Amalgamation as a Strategy for Evaluating Fractal Dimensions of Fineparticle Profile." Laboratory Report No. 21, Institute for Fineparticle Research, Laurentian University, Sudbury, Ontario, 1978. The material presented in this chapter summarizes all of the essential information presented in that early set of lecture notes.

[28] The Dapple System is available from Dapple Systems, P.O. Box 2160, Sunnyvale, CA, 94087, U.S.A.

[29] A.G. Flook, K. Leschonski and W. Hufnagl., (Eds.) "The Characterization of Textured and Structured Particle Profiles by Automated Measurement of Their Fractal Dimensions," in Proceedings of Partech Symposium, published by Drukerei Heinrich Schuster, Nurnberg, 1979, 591-599.

[30] A.G. Flook, "A Comparison of Quantitative Methods of Shape Characterization," *Acta Stereol.,* Vol. 3, part 2 (1984) 159-164.

[31] J. Serra, "Stereology and Structuring Elements," *J. Microsc.*, 95, Part 1 (1972) 93-103.

3 What Use are Fractals?

3.1 Elegance and Utility of Fractal Dimensions

In the early 1980s, when I attended scientific conferences, acquaintances and friends would sometimes ask me, "What are you working on these days?" When I told them "fractals," they used to ask (some politely and others not so politely), "What are fractals?" Out would come the sketch pad and after a quick discussion about the structure of rugged rocks and convoluted carbonblacks, I would convince them that fractal dimensions could be used to describe rugged boundaries. At this point of the discussion, the next inevitable question was, "What use are fractals?" I was sometimes tempted to answer that fractals did not have to be useful: the use of fractal dimensions is an elegant technique for describing rugged systems, and elegance is a virtue worth discovering for its own beauty. However, answering the question "What use are fractals?," by referring to their elegance would not normally promote interest in the subject. In a world where applied science must be funded by grants from utilitarian oriented agencies, it is best to answer this question by quoting direct applications and hoping that individuals enticed into exploring fractal geometry by utilitarian motivation will learn to enjoy, as well as use, the theorems of fractal geometry. Therefore, we shall occasionally pause, as in this chapter, to examine the utility of ideas encountered as we proceed along the various stages of our randomwalk.

3.2 Fractal Description of Powder Metal Grains and Special Metal Crystals

An important branch of **metallurgy** (defined as the study of the properties of metals, their possible uses and how to separate them from the ores in which they may be found) is powder metallurgy. In **powder metallurgy**, machine parts and other pieces of fabricated metal are made by pressing powder together into a desired shape. The powder grains are then fused together by a process known as **sintering.** This word, coined by German scientists, is related to the English word cinder. It is used to describe the fusing of metal powder by the application of heat, without melting the individual powder grains. When I graduated from university in 1955, the British government required all science graduates either to serve for 2 years in the army or to volunteer for work of national importance. At the time that I was facing this choice, I was approached by M.J. Donaldson, Superintendent of Analytical Chemistry at the British Atomic Weapons Research Establishment, to join his staff and make a comprehensive study of the techniques for characterizing fineparticles. He made this offer because my research for my M.Sc. thesis involved a study of the droplet sizes in an emulsion of water and oil, and I had expressed an interest in continuing to study droplet and dust sizing techniques. On joining the staff of Mr. Donaldson, I was given two major tasks. One was to study the size distribution of beryllium powder, which was being used to make metal parts for the atomic bomb and to fabricate neutron moderator rods for experimental nuclear reactors. The other task was to characterize the size and shape of the fineparticles of uranium dioxide and plutonium dioxide used to fabricate fuel rods for nuclear reactors.

It is very difficult to cast parts from molten beryllium because, as it crystallizes, it forms large crystals which chip out of the surface of a part if that part has to be machined or polished. Therefore, the moderator rods were being made from compressed powder which was being sintered to produce high-density beryllium rods. It was known that the density of the rods achievable by the sintering process was dependent upon the shape and size distribution of the beryllium powder grains. I was given the task of characterizing the shape and size of beryllium powder grains produced by different processes. This was a very difficult task since beryllium is extremely toxic; inhaled fineparticles of beryllium cause a lung disease called **berylliosis**, and powder grains present in a surface wound on the body prevent healing and cause severe ulceration. All the beryllium powders had to be studied and handled in controlled, closed environments. My study of beryllium powders led me into a general study of powder metallurgy and the health dangers created by inhaled dust.

Metal powders are made in various ways, and the way in which they are made determines the structure of the grains and their surface energy. The ease with which a powder will sinter depends upon the surface energy of the powder. In Figure 3.1, the profiles of three different types of metal powders produced by three different processes are shown. The highly irregular profiles in Figure 3.1(a) are produced by electrolysis, and this production process will be discussed later in this section. This type of electrolytically produced powder has a high surface energy and sinters rapidly. Unfortunately, because of their very rugged outlines, the compacting of this type of powder into a mold is difficult, and the metallurgist must often make a choice between the density of the compact he can achieve on sintering and the rate at which the sintering consolidation proceeds.

The type of powder shown in Figure 3.1(b) can be made by crushing a compound of copper to produce small grains which are then chemically purified to remove the other chemicals in the compound. The basic outer profile is produced by the crushing and grinding process, but the chemical purification of the copper results in a porous material with high surface energy. The basically compact grains can often be consolidated to a relatively low porosity by direct application of pressure and then the high surface energy results in efficient sintering of the compact. The iron oxide grains photographed at various magnifications and presented in Figure 1.3 are made by this type of process. The pictures in Figure 1.3 demonstrate the complexity of the internal structure of powder metal grains produced by this type of process. In a later chapter, we shall

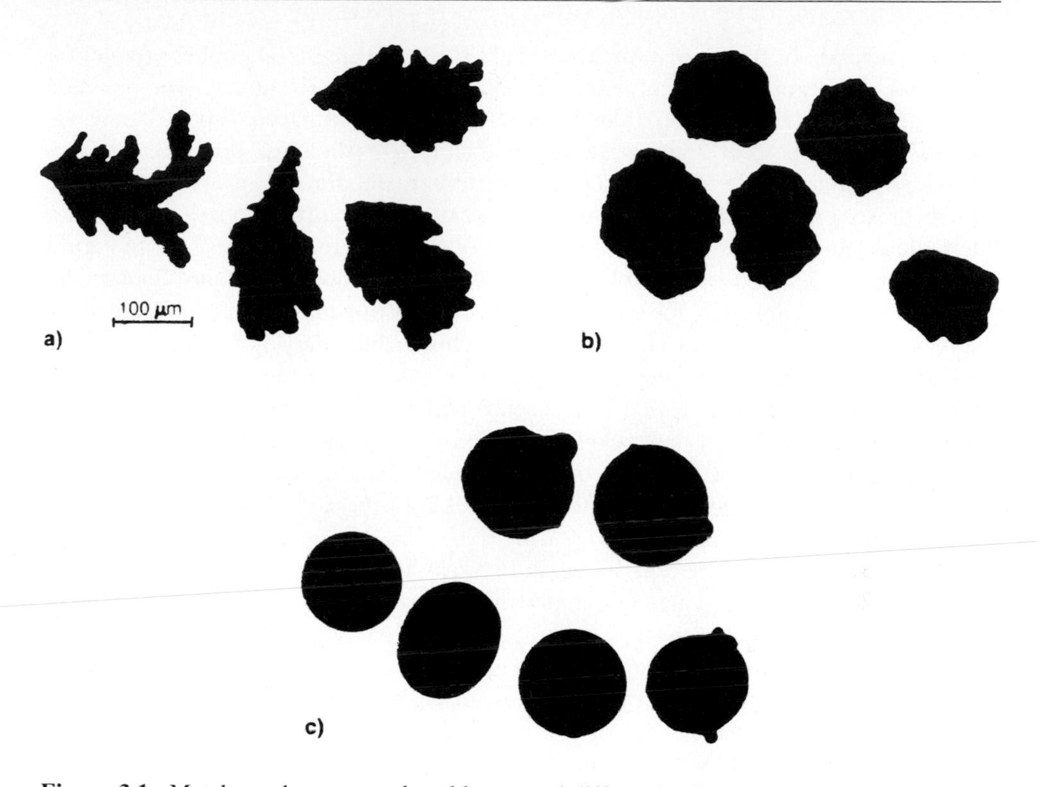

Figure 3.1. Metal powders are produced by several different processes. The shape and size of the metal grains in a powder can be related to the production process.
(a) Highly irregular metal grains are produced by an electrolytic deposition process.
(b) Chemical reduction of a crushed compound of the metal produces rounded grains with high internal structure.
(c) The breakup of a jet of molten metal produces a type of metal powder known as metal shot. The process for producing metal shot is known as atomization or nebulization of the metal [1].

discuss the measurement of the internal structure of such metal grains by mercury intrusion porosimetry, and we shall discover that the internal structure of the grain can be described by means of a sponge fractal having a magnitude between 2 and 3 classical dimensions.

The third type of metal powder grain shown in Figure 3.1(c) is produced by the disintegration of a jet of liquid metal with subsequent cooling of the droplets as they fall down a tall cooling tower. This production method is described by the metallurgist as a **shotting process.** If the shot fineparticles cool slowly in the tall tower, the resulting metal grains are almost spherical. The shotting process is used to produce lead shot used in shotgun ammunition.

In recent years there has been a demand for highly irregular metal grains as a "scintillating paint" pigment. Thus, the modern automobile paints contain flake aluminium pigments which give the scintillating effect in the finished paint. One way to produce this type of pigment is to chill rapidly the contorted droplets produced by the turbulent breakup of the jet of molten metal [2, 3]. In Figure 3.2(a), a photograph of a typical set of irregular-shaped aluminium shot metal grains are shown. The powders shown in Figure 3.2(a) were provided by Mr. J. Thompson of Alcoa Centre, Pennysylvania, USA. The powder grains were approximately 100 μm long. It has been shown that the profiles in Figure 3.2(a) can be described by fractal dimensions when examined at coarse resolution, but that, on high-resolution examination the boundaries of the shot fineparticles are Euclidian because of the effect of surface tension forces [4]. In Figure 3.2(b), a typical Richardson plot for a structured walk exploration of a shot fineparticle is shown. It is probable that, the magnitude of the inspection unit at which the boundary changes over from a rugged structure to a Euclidian boundary is a function of the cooling rate experienced by the fineparticle as it falls down the cooling tower [4]. Aluminium powder is used as fuel in spacecraft rocket engines and the rate at which it burns and how it packs in the rocket containers is very important in predicting the performance of the engine. The fractal dimension of the aluminium fineparticles will probably be useful in predicting the flow and packing behaviour of the powder and its chemical reactivity.

Aluminium powder of the type shown in Figure 3.2(a) can be flattened in a ball mill to produce flake aluminium pigment. In the dry state, fine aluminium powders are very explosive, and the fractal of the surface structure of the irregular aluminium is probably useful in predicting how easily the fine metal grains become airborne and whether or not they will burn rapidly in a cloud of metal dust to produce an explosion. The speed at which a cloud of burning metal dust will explode is determined by a complex interaction of the generation of heat by the burning fineparticles and the absorption of radiant heat by non-burning grains to raise them to their ignition temperature so that they in turn will burn. The burning rate and the absorption of radiant heat will probably be a function of the fractal dimensions of the powder metal grains.

One of the first physical systems to which Mandelbrot applied fractal logic was the study of the turbulent flow of a liquid. If the aluminium shot fineparticles produced by the breakup of the jet were rapidly chilled, perhaps the structure of the droplets as they froze would capture the complex patterns of turbulence and give information on how the jet broke up under the pressure and flow conditions when the molten metal was ejected from the nozzle. Perhaps the aluminium shot particles shown in Figure 3.2(a) are frozen messengers carrying in their rugged profiles information about the turbulence in the jet when they were formed. It would be very interesting to carry out a study of the different shapes produced by turbulent breakup of a molten jet of metal subjected to different cooling conditions

In Mandelbrot's book (see Chapter 1 reference [18]), there is a picture which illustrates the construction algorithm for changing a square into an indented island with a fractal dimension of 1.73. Early stages in the construction of this shape are shown in Figure 3.3. This complicated Koch's island is of interest to fineparticle scientists since, as will be shown in a later chapter, some natural smoke fineparticles can have fractals of the order of 1.72 and bear a striking resemblance to this particular Koch's island. However, when I first encountered profile (i) in Figure 3.3(a) when browsing through Mandelbrot's book, it reminded me of the type of structure that develops when a surface is being corroded by chemicals. As I looked at the picture, I did not see a geometric

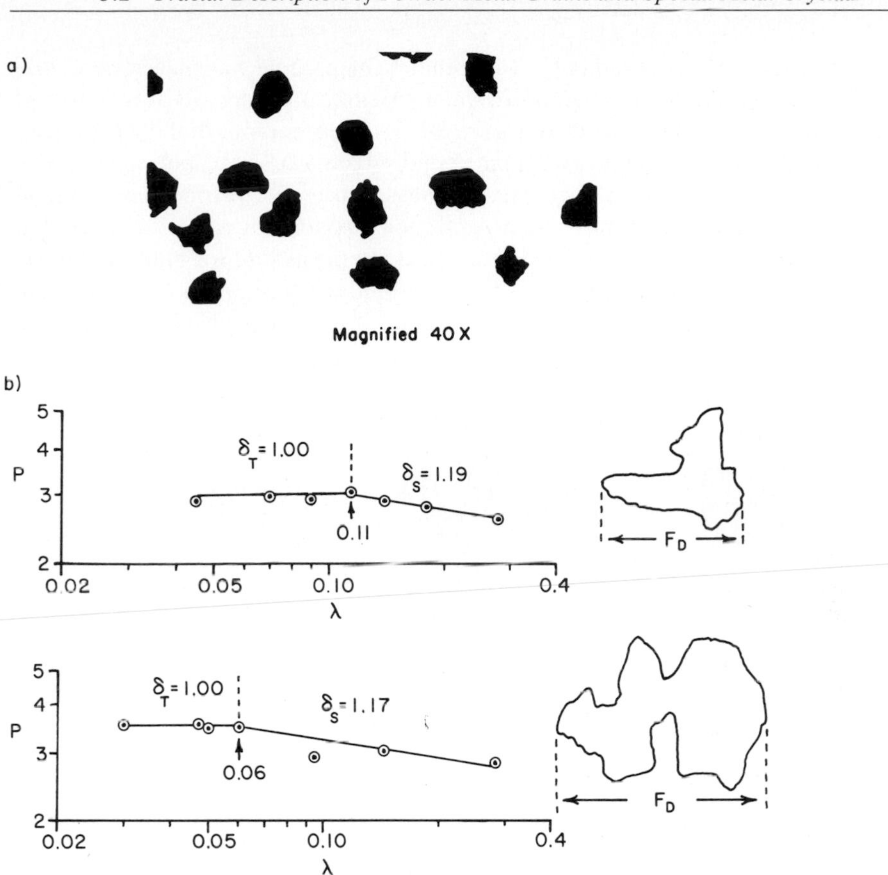

Figure 3.2. The structure of irregularly structured aluminium shot is such that it can be described by a fractal dimension at coarse resolution, even though at high resolution it manifests a Euclidian boundary.
(a) A typical set of irregular aluminium shot fineparticles of average size 100 μm.
(b) Richardson plot for the exploration of two aluminium shot profiles by the inward structured walk exploration technique.
P = perimeter estimate.
λ = exploration stride length.
P and λ normalized with respect to maximum Feret's diameter of the profile.
δ_S = structural fractal dimension and δ_T = textural fractal dimension.

construction transforming a square; I saw acid eating into the profile, with the buildup of a chemical crust as a canopy above the eroded surface. To me, chemical crusts on eroded surfaces look like fractals, but I have yet to carry out any experimental investigation of the fractal structure of a corrosion crust. I suspect the fractal of a corrosion crust will be related to the rate and mode of corrosion attack.

Although we have yet to study corrosion crusts, my students and I have carried out an experiment which demonstrates that fractals can be used to describe in a quantitative manner the progress of erosion when a fineparticle is placed in an acid solution. In one of the experiments that we carried out, an aluminium fineparticle was placed on a slide underneath a microscope and photographed over a series of time intervals as it dissolved under the attack of hydrochloric acid. In Figure 3.3(b) the appearance of the profile after 234 s is shown. It can be seen that over a range of $\lambda = 0.08 - 0.3$ units normalized with respect to the maximum Feret's diameter, the structure of the profile has a fractal dimension of 1.05 but that at inspection resolutions smaller than 0.08 a Euclidian boundary is manifest. After the acid had been attacking the profile for more than 1000 s, the Euclidian portion was no longer visible at resolutions as small as 0.05 units and the fractal dimension of the boundary had increased to 1.10. The Richardson plot of the exploration of this eroded profile is shown in Figure 3.3(c).

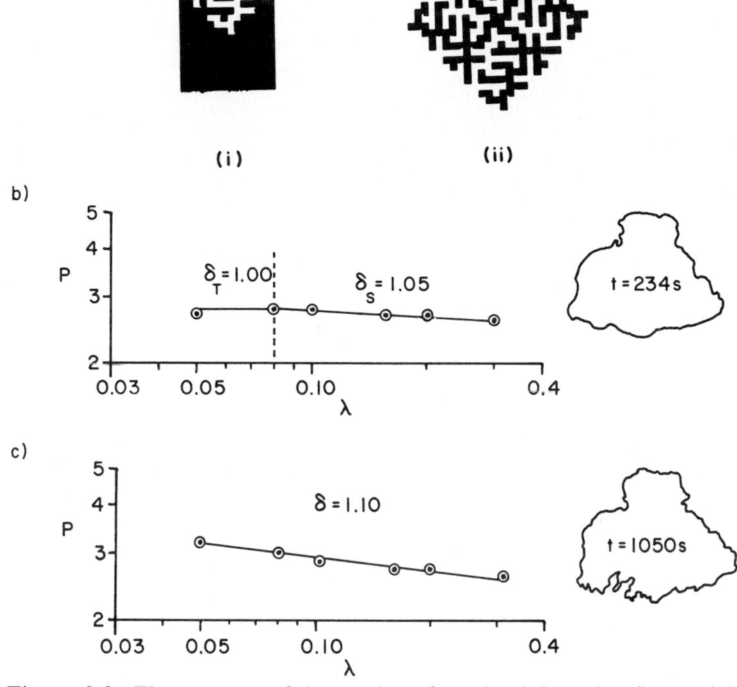

Figure 3.3. The progress of the erosion of an aluminium shot fineparticle by acid attack can be quantified using fractal dimensions.
(a) The appearance of various stages in the construction of a Koch's triadic island with a fractal dimension of 1.73 suggests that fractal dimension may be useful in a study of corrosion and erosion phenomena (reproduced with permission of B.B. Mandelbrot, "The Fractal Geometry of Nature," W.H. Freeman and Company, San Francisco, 1983).
(b) As the acid attack proceeds, the fractal dimension of the boundary changes and the Euclidian portion of the boundary is no longer obvious after prolonged acid attack.
(c) Richardson plot of the exploration of the eroded profile.
P = perimeter estimates by the inward swing structured walk technique
λ = exploration stride length
P and λ are normalized with respect to the maximum Feret's diameter.
δ = fractal dimension.

One of the surprising aspects of our study of eroding fineparticles was that we had expected the fractal dimension of the rugged boundary to increase steadily from time zero. In fact, the magnitude of the structural fractal of the boundary decreased for the first 400 s of acid attack and then started to increase. We interpreted this as being due to the fact that when the fineparticle was placed in the acid it had a few sharp prominences which dissolved quickly to give a smoother profile before the acid really began to bite into the profile. The whole sequence of changes in profile shape with time under acid attack for one shot fineparticle is shown in Figure 3.4(a) and the measured structural fractal dimensions of the various profiles are summarized in Figure 3.4(b) [4].

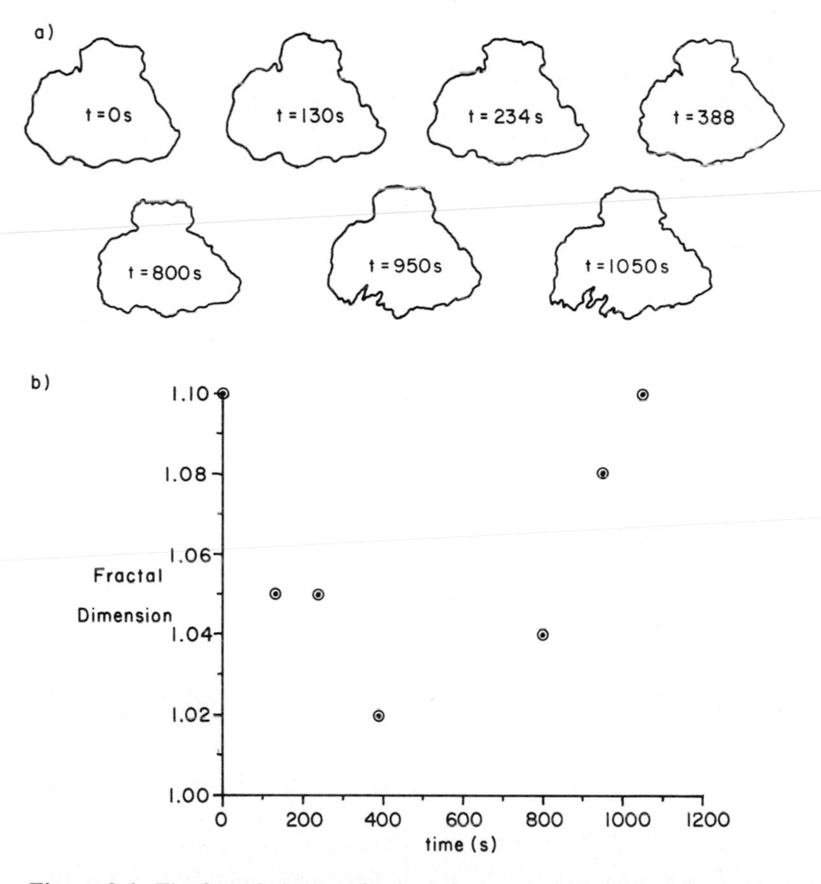

Figure 3.4. The fractal characterization of an eroded aluminium fineparticle would indicate that initially the acid attacks the few promontories on the boundary of the profile since they are more easily dissolved. This causes the fractal dimension of the profile to drop in the initial stages of erosion but, once the erosion phenomenon becomes firmly established, the fractal dimension of the boundary increases.
(a) The profile changes visibly as acid attacks the boundary over a period of time.
(b) The changes in the fractal dimensions of the profile undergoing acid attack indicate the initial smoothing of the boundary by the dissolution of a few rugged features, with a subsequent increase in the overall ruggedness of the profile, as the acid bites into the profile.

Similar work, using different analytical techniques, has been carried out by Beddow and co-workers [5].

Professor Keith Beddow and co-workers have studied extensively the shape and size of powder metal grains [6]. One of their earliest studies was concerned with the comparison of the shape of metal oxide grains, similar to those in Figure 3.1(b), and metal powders produced by a shotting process. Two sets of profiles studied extensively by Professor Beddow are presented in Figure 3.5. Powder metallurgists describe the shotting process for producing metal powders by the alternative name of atomization. This name is unfortunate, since the process does not reduce the powder to atoms. I prefer to call the **atomization** process **nebulization** from the fact that the Greek word "nebula" means "cloud," since the breakup of the molten metal jet produces a cloud of metal fineparticles. However, the name atomization is firmly established in metallurgy so we shall use it in the discussion. Atomized powder tends to be irregular because of the distortion of the shape of the molten metal drop by the forces of gravity as it falls through the air of the cooling tower, and doublets and triplets are also formed because of the collision of the semi-molten droplets before they reach the base of the tower. The overall shape of the atomized metal powders can be described using the **chunkiness** of the profile, a dimensionless ratio defined by the relationship

$$\text{chunkiness} = \frac{\text{width}}{\text{length}} = \frac{[W]}{[L]}$$

This is obviously the reciprocal of the aspect ratio. The chunkiness has the advantage with respect to the aspect ratio that, when plotting data graphically, it ranges from 1 to 0 whereas the aspect ratio can vary from 1 to infinity. Use of the aspect ratio can require extensive range graph paper for the presentation of data on a powder containing a wide range of shapes.

Another useful measure of fineparticle structure is a dimensionless number known as the **compactness.** The definition of the compactness of a profile is illustrated in Figure 3.5(c). To describe the range of shapes present within the two metal powders, one can draw a graph of the compactness of the metal grain versus the chunkiness of the profiles. The description of the structure of the profiles in Figure 3.5 by these two ratios is illustrated in Figure 3.6(a). When plotted in this way, the data for the two types of powder (sponge iron and copper shot) intermingle in the plane of the diagram. However, the ruggedness of sponge iron fineparticles can be described by a structural fractal dimension and we can plot a three-dimensional graph using the fractal dimension, chunkiness and compactness of the type sketched in Figure 3.6(b). A three-dimensional graph of the type shown in Figure 3.6(b) is referred to as a three-dimensional dataspace. In such a dataspace, the data points for the rough profiles are clearly separated from the atomized fineparticles, as illustrated in Figure 3.6. The graph in Figure 3.6 is only a sketch of a three dimensional dataspace since the real graph for the profiles in Figure 3.5 is too complex to display in this introductory discussion of fractal dimensions.

It has already been pointed out that a sponge-type fineparticle will have a sponge fractal descriptive of an internal structure. If we carried out an intensive study of this type of metal grain, we would have to store information on structural fractals, sponge fractals, chunkiness, compactness and the size of the individual grains in the computer memory. The computer scientist refers to such a storage system as a six-dimensional

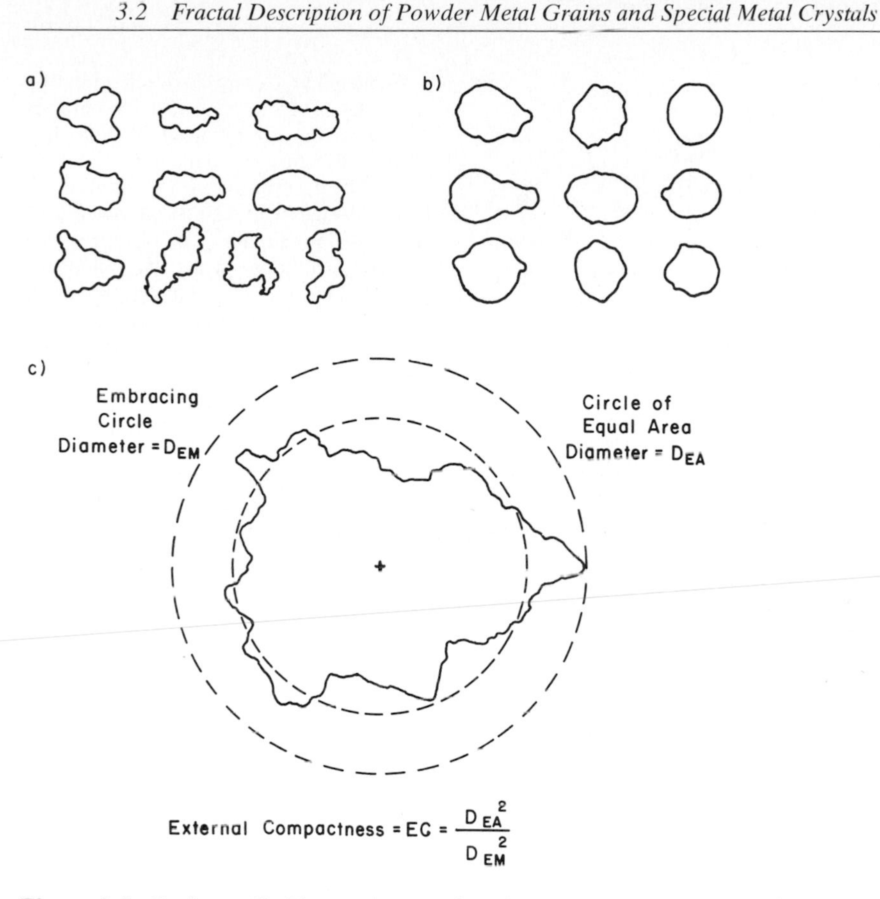

Figure 3.5. Professor Beddow and co-workers have made an extensive study of the difference in shape between sponge iron powder grains produced by chemical reduction techniques and copper shot fineparticles formed by the disintegration of a jet of molten copper [6].
(a) sponge iron metal grain profiles.
(b) copper shot fineparticle profiles.
(c) The compactness of a profile is a useful dimensionless ratio for describing the shape of a fineparticle.

dataspace. One cannot draw a graph on paper representing all six sets of data in six-dimensional space – it is difficult enough to draw three dimensions on a two-dimensional piece of paper! Any three-dimensional drawing of dataspaces will have to be a reduced description taken from the six-dimensional dataspace in the computer memory. Unfortunately, many popular discussions of the dimensionality of dataspaces leave out the word **"dataspace,"** and one is told that one is working in six dimensions. Invitations to work in the sixth dimension frighten most people away from any subject and they are left wondering what a sixth dimension looks like. Most people could appreciate what was meant by "six layers of storage" in a computer, and could learn to accept the description of six data layers as constituting six-dimensional dataspace, whereas they tend to feel inadequate (dumb!) when they meet scientists who can work in six dimensions, since average people drop out intellectually in any discussions beyond the three

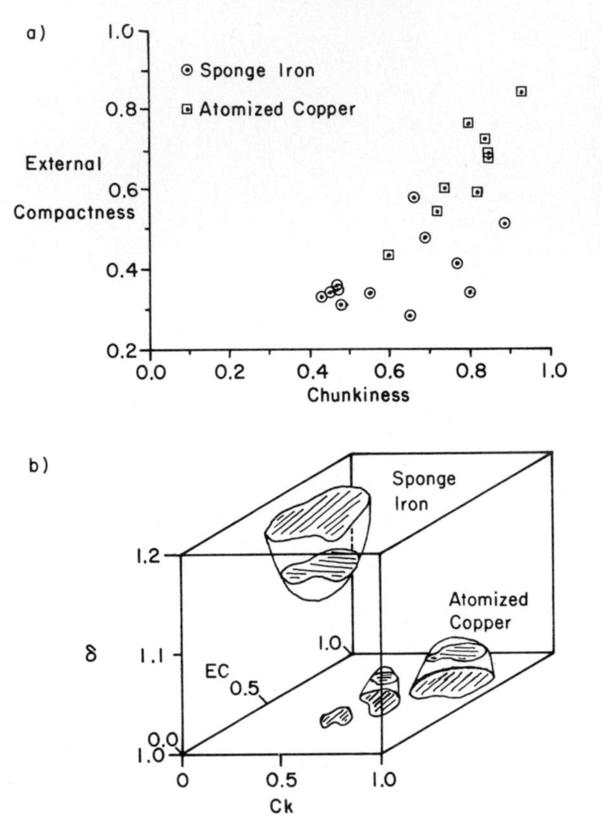

Figure 3.6. Complete description of a certain fineparticle can sometimes require the presentation of data in a three-dimensional dataspace.
(a) The description of the chunkiness-compactness of the Beddow metal profile in Figure 3.5 obscures the difference between the two powders.
(b) In three-dimensional dataspace, the roughness of the sponge iron fineparticles separates them in the dataspace from the much lower roughness atomized metal powders. *Ec* is the external compactness of the profiles (see Figure 3.5) and *Ck* is the chunkiness of the profiles described as their width divided by their length.
δ = the fractal dimension of the profiles of the metal oxide fineparticles.

dimensions of physical space. Some technical discussions of the uses of fractals refer to working in four-, five- and six-dimensional space; often what is meant is dataspace rather than physical space.

The fineparticles with the highest fractal dimension shown in Figure 3.1 are described as electrolytic metal powder. The process of electrolysis is illustrated in Figure 3.7. An electrical voltage is applied across the electrodes to cause an electric current to flow between the electrodes. The passage of the electric current causes copper to be deposited at one electrode with the release of chemicals at the other electrode. The word **electrolysis** is made up of two root words, "electro" and the Latin word "lysein," which means to unleash to free. Thus, electrolysis frees the copper ions, Cu^{2+} from their captivity with the sulphate ion, $SO^{2-}$and deposits them at the cathode. One of the problems involved in depositing a coating on any object by means of electrolysis is that sometimes the deposit forms in a flaky manner and can be rubbed from the surface after electrolysis, unless treated in a special way. In the production of electrolytic metal powders, one deliberately selects the deposition conditions so that one can ensure that the coating on the electrode is very flaky and can be rubbed off to make the powder of the type shown in Figure 3.1(a).

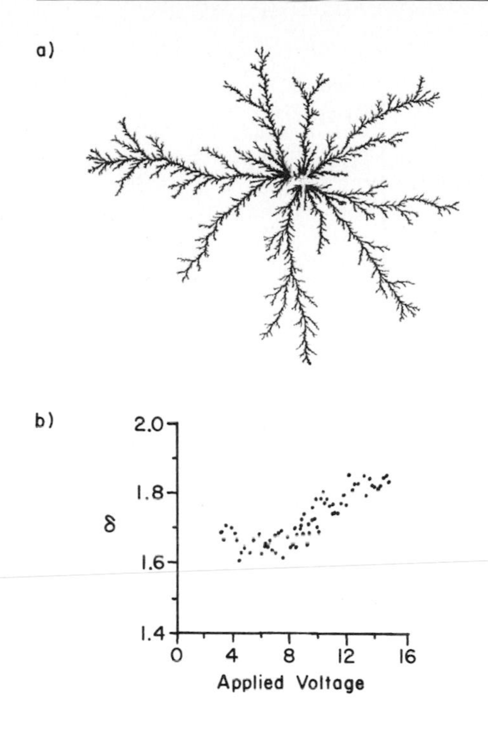

Figure 3.7. Fractals could be of importance in the design of efficient electrical cells for generating electricity from chemical reaction and in the design of electric storage batteries.
(a) Basic system of an electric cell for generating electricity or for the separation of material by electrolysis.
(b) The classical electric storage battery used to deliver electric power in an automobile, releases energy from the chemical reactions which occur when a current flows between two electrodes, one a porous lead plate and the other a porous lead oxide plate.
(c) Electrolytically deposited flakes of metal can fall off the plates during the charging of the battery, forming a sludge at the bottom of the cell which shorts out the system.
(d) Growth of dendritic crystals between the electrodes can cause electrical failure of the battery.

Matsushita et al. [7] studied the growth of two-dimensional metal crystals formed during electrolysis of a metal solution. They carried out their experiments in a shallow dish 20 cm in diameter, in which was placed a cylindrical anode 2.5 cm high, 3 mm thick and 17 cm across. The dish was filled to a depth of 4 mm with an aqueous solution of zinc sulphate. This solution is described as the electrolyte used in the electrolysis cell (any solution which conducts electricity is known as an **electrolyte**). A thin layer of an organic compound, n-butylacetate, was then poured on to the zinc sulphate solution. A carbon rod which acted as a cathode, was set near the centre of the dish with its tip placed just at the interface between the aqueous solution and the organic compound. An electric current was passed between the carbon cathode and the zinc metal ring anode through the electrolyte. This current deposited zinc on the tip of the carbon rod. In this special electrolysis cell the zinc crystal grew outwards from the cathode towards the anode as a thin, flaky, essentially two-dimensional crystal. This type of crystal is referred to by those who study electrolysis as a **metal leaf.** A typical metal leaf crystal grown by Matsushita et al. is shown in Figure 3.8(a). It can be seen that this crystal bears a striking resemblance to the fractal geometric section (ii) shown in Figure 3.3(a). The measured fractal dimension of the metal crystal shown in Figure 3.8(a) is 1.66. Matsushita et al. showed that the fractal dimension of the crystal grown in their electric cell could be changed by altering the voltage applied to the cathode. In Figure 3.8(b), the change in the fractal dimensions of metal crystals grown using different applied voltages are shown [7].

Electric storage batteries of the type used in automobiles are special electrolytic cells which can be used to deliver electric current and then can have their operational

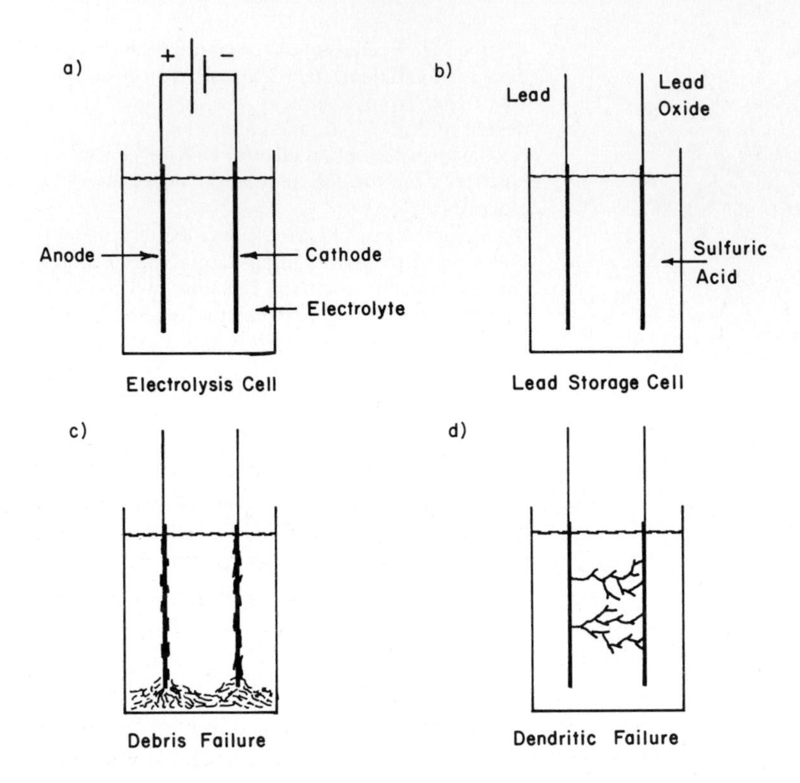

Figure 3.8. Two-dimensional metal crystals can be grown in specially designed electrolysis cells. Two-dimensional crystals grown by electrolytic deposition show a dendritic structure with high fractal dimensions.
(a) Typical metal crystal grown by Matsushista et. al. [7]. Reproduced by permission of Dr. Matsushista.
(b) The fractal dimension of a crystal grown by electrolysis depends on the applied voltage used to deposit the crystals.
δ = fractal dimension of the crystals.
V = applied voltage used to deposit the crytals [7].

conditions reversed to replace and store this electrical energy. There are many different types of electrical storage batteries now in use or being developed for use in automobiles and the discussion given here will be limited to a description of the well known lead-acid battery. The basic system of the lead-acid battery system is illustrated in Figure 3.7(b). For simplicity, we have shown only two electrodes, whereas a real battery would have many electrodes [8].

In the Faurer process for making the electrodes of electric storage batteries, a mixture of red lead powder (a form of lead oxide) and sulphuric acid is pressed to make a strong grid which can then be dried for use in the battery. Two similar plates are immersed in weak sulphuric acid then an electric current is passed through the electric cell. Owing to the passage of the electric current, one electrode is attacked by hydrogen released from the electrolyte to form a lead sponge. At the other plate, the lead is oxidized by the sulphuric acid to form a sponge of lead oxide. Sponge electrodes are used in automobile batteries to increase the surface activity of the electrodes, which in turn increases the length of time that one can take useful levels of electric current out of the storage battery. When the storage battery is used to deliver electric current, the lead oxide plate is attacked by hydrogen, formed by the decomposition of the sulphuric acid solution between the electrodes. This ultimately leads to the formation of lead sulphate on the electrode. At the negative plate of the storage battery, the sulphate ions from the sulphuric acid form lead sulphate directly. When one starts to charge the electric battery, the chemical reactions run in reverse and the layer of spongy lead is redeposited on the negative plate. There are two major failure problems with this type of electric storage battery. One problem is caused by the regenerated spongy lead failing to stick to the electrode. The flakes of metal fall from the electrode to the bottom of the cell formed by the two electrodes, where they can form a sludge which will short out the electric plates, as illustrated in Figure 3.7(c). The other problem is that the lead deposited on the cathode as the storage battery is recharged can form a dendritic-type crystal similar to that grown by Matsushita et al. and shown in Figure 3.7(a). A highly dendritic lead crystal can obviously sprout across the electrodes [Figure 3.7(d)] to short out the cell electrically, making it useless. It may be that a fruitful area of research would be to extend the work of Matsushita et al. by examining the growth of lead dendrites on storage battery plates to see if one can alter the growth of the crystals, perhaps by adding a chemical to the electrolyte to make them of lower fractal structure so that they do not reach across the electrode so quickly. Again, one may be able to examine the fractal of deposited flakes to see if one can modify the electrolyte, again by the addition of chemicals, to lower the fractal of the flakes to make them more dense and more adherent to the electrode plate.

In the computer industry, there is a great deal of interest in forming thin films of metal on silicon chips to provide conductive paths between the solid-state devices making up the computer chip. However, it is not easy to start the growth of crystals on a smooth, clean surface when one bombards that surface with metal vapour. What usually happens is that there are certain points on the surface which have physical properties that enable the crystal to start growing at that point. These points are called nucleation centres (nucleus means "tiny nut" in Latin; nuclear physics deals with the "nuts" inside the atoms of elements; **nucleation centres** are "little nuts" on the surface on which crystals start to form). When a liquid is being heated to its boiling point, bubbles will start to form on scratches on the surface of the container or on the edges of a few small, sharp stones ("boiling chips") placed in a liquid to make it easy for the bubbles to form, rather than for the liquid to reach its boiling point suddenly and bubble explosively. The scratches on the container and the edges on the stones are nucleating centres on which bubbles form. In Figure 3.9, a dendritic crystal deposited on a quartz surface by a process known as sputtering is shown. The crystal was grown by Elam et al. and is described as a "sputter-deposited Niobium Germanium thin film" [9]. The similarity between this dendritic crystal and the zinc crystals shown in Figure 3.8 is obvious. However, it will be noticed that the branches towards the outside of the crystal appears to be thicker, as if they had grown "leaves" on their "twigs." The individual crystalites of the dendritic crystals shown in Figure 3.9 were between 2000 and 5000 Å thick and

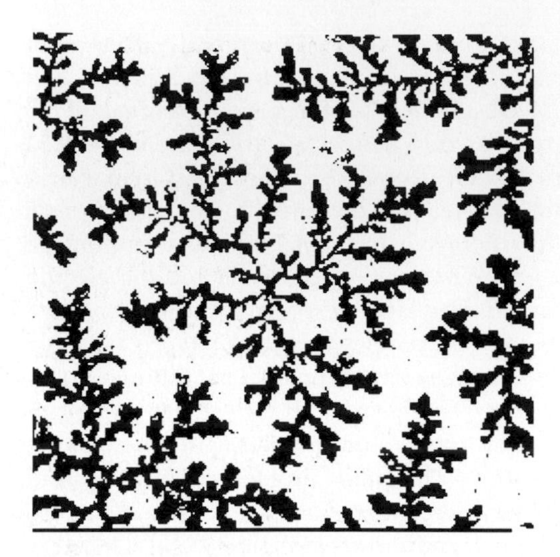

Figure 3.9. Under certain operating conditions, attempts to produce thin films on a quartz crystal by sputtering a niobium target operating in a mixture of argon and germane gases under very low pressure results in the growth of dendritic crystals from nucleating points on the quartz crystal. This type of dendritic crystal has a fractal dimension of 1.7 for the "branches" in the centre of the crystal and 1.88 for the outer "twigs with leaves" [9].

consisted of small grains of average size 0.1 μm, with the larger grains on the ends of the branches having an average size of 1 μm. Elam et al. showed that the branches of the crystal before it thickened out had a fractal dimension of 1.7 but that the outer branches had a measured fractal dimension of 1.88. The reason for this difference will be considered in Chapter 4 when we discuss computer simulation of the growth of dendritic crystals. For the time being it is sufficient to note that, as Elam et al. pointed out, a study of the fractal dimensions of various crystals deposited by the sputtering process will help one understand how crystals form and grow so that one can be more efficient in the commercial growth of thin films of metals in the large-scale manufacture of computer chips [9].

The reader will probably have noticed the strong resemblance between the crystals of Matsushita et al. and Elam et al. and frost patterns growing on a window pane and the structure of snowflakes. The process by which "frost, ferns and leaves" grow on a window pane is similar to that used by Elam et al. in that the crystal of frost starts on a nucleating centre on the glass. It then grows by deposition on the growing crystal of water molecules contained in the air above the window pane. "Jack Frost" was painting fractal scenery long before we learned to simulate it on computers! The growth of snowflakes and their fractal structure will be discussed in Chapter 4.

An important branch of applied science is colloid science. **Colloid science** deals with fineparticles which are so tiny that their behaviour is totally controlled by surface forces. Thus, a suspension of colloidal gold fineparticles will never settle under gravity; the gold colloidal fineparticles are kept in suspension by Brownian motion. In broad terms, we can consider fineparticles smaller than 0.1 μm to be collodial fineparticles. Collodial science and fineparticle science overlap in several systems. The term colloid comes from two Greek words, "Kolla" meaning glue and "idos" meaning form or shape. The first colloids that were studied scientifically were "glue-like" in their physical behaviour, hence the name. Recent studies have shown that many collodial fineparticles are themselves made up of much smaller units, agglomerated into larger units which display fractal structure. Thus, in Figure 3.10, a typical colloidal aggregate described by Weitz and Oliveria is shown [10, 11].

In Figure 3.11, a series of photographs of a special type of rugged nickel fineparticles at various magnifications are shown. This nickel pigment powder is used to make composite materials for the electromagnetic shielding around sensitive instruments. Of all of the metal fineparticles discussed so far in this section, the metal grains in Figure 3.11 probably have the widest range of fractal structures. The nickel pigment powder has to be loaded into the plastic forming the composite material at a sufficient concentration to provide electrical conducting paths within the composite material. It can be shown that long, thin pigment fineparticles establish conducting paths through the composite materials at lower solids concentrations than more spherical pigment fineparticles [12, 13] (this fact is discussed in more detail in Chapter 5 when the design of specially reflecting paint for spacecraft is explored). On the other hand, the strength of the composite material is probably related to the ability of the pigment surface to offer a high surface area to the bonding plastic. Thus, the fractal dimension of the surface of the pigment probably determines the strength of the composite material, and the fractal dimension of the individual pigments, together with a form factor such as the aspect ratio or the chunkiness, could probably be used to predict the amount of pigment powder which must be added to the material to create the required properties in the composite material. The amount of pigment added into the composite material is usually measured in terms of the volume fraction of pigment in the final material. The **volume fraction of pigment** is also known as the **pigment loading.**

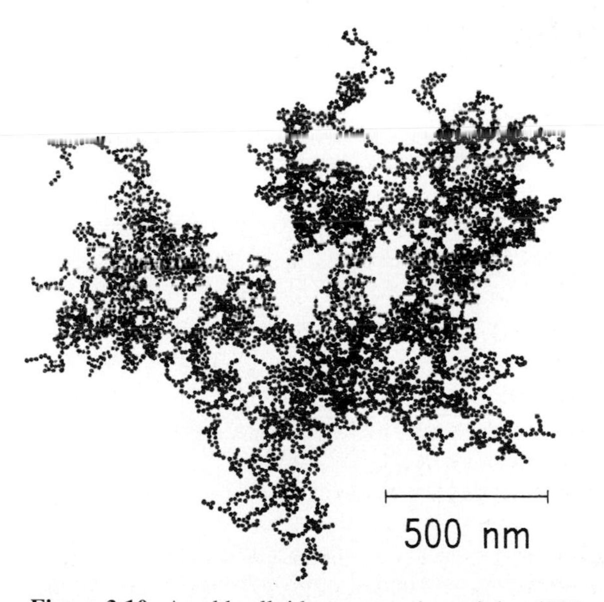

Figure 3.10. A gold colloid aggregate (containing 4739 sub-spheres!) was found to have a fractal dimension of 1.75 by Weitz and Oliveria [10].
(from D.A. Weitz and M. Oliveria, "Fractal Structures Formed by Kinetic Aggregation of Aqueous Gold Colloids," *Phys. Rev. Lett.*, 52 (1984) 1433-1436; reproduced by permission of D.A. Weitz and Physical Review Letters.)

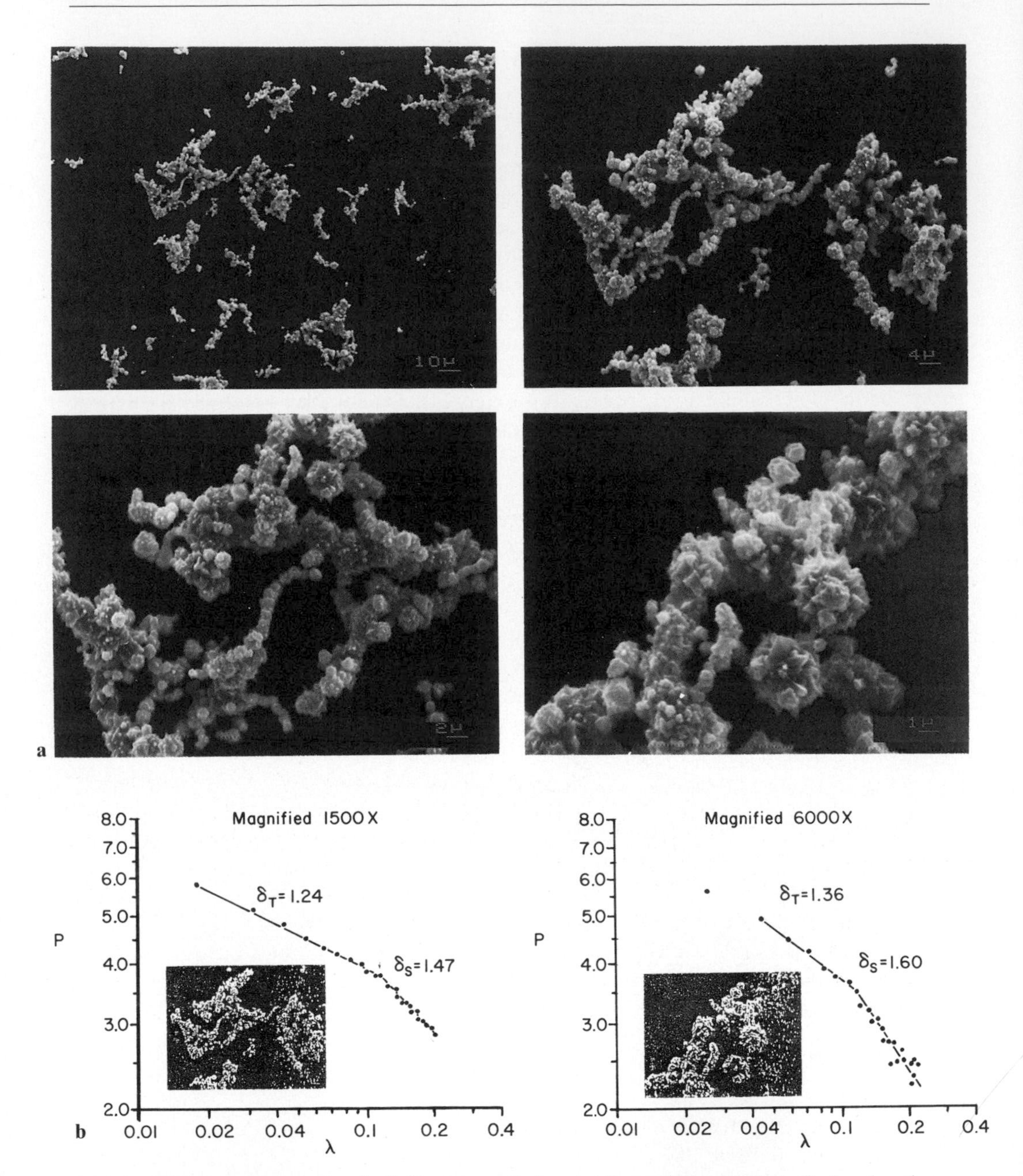

Figure 3.11. The fractal of an individual rugged pigment fineparticle and that of a local region of the boundary of the rugged pigment at high magnification can be measured by the inswing structural walk technique.
(a) Pigment profiles at various magnifications.
(b) The fractal dimension of an individual pigment particle and the portion of the pigment boundary at higher magnification. (Photographs of pigment profiles provided by and used with the permission of INCO Limited, Sudbury).

3.3 Fractals and the Flow of Dry Powders

An important problem faced by the powder metallurgist when making parts from metal powders is the feeding of the powder into the mold in which the part is going to be sintered. The scientific study of the flow properties of different materials is described as **rheology.** This word was derived from the Greek word for a flowing stream, "rheos." Dry powder rheology is a very difficult subject because of all the factors that can affect the flow properties of a powder. Some of these factors are absorbed moisture, electrostatic forces, ruggedness of the profiles of the powder grains, size distribution of the grains and the fact that, as a powder flows into a mold, in many cases air must flow out. This exiting air can blow the finest fineparticles out of the mold. Powder metallurgists have known for many years that the shape of a powder affects its flow properties but, in the absence of efficient techniques for describing the shape of a powder, studies of the affect of shape on powder flow have been rudimentary. However, now that fractal geometry is available for describing the ruggedness of fineparticle profiles, several scientists have begun to study the link between the fractal description of the structure of fineparticles and the rheological properties of a powder. Thus, Dr. Peleg of the University of Massachusetts is looking at the flow properties of spray-dried food powders and attempting to link the fractal dimension of the powder grains with the flow and packing properties of the powder [14, 15].

In the **spray drying** process, used widely in the pharmaceutical, food and chemical industries, a **slurry** (which is defined as a concentrated suspension of fineparticles), or a concentrated solution of a substance, is sprayed at the top of a tower in much the same way as lead shot is made by disintegrating molten lead at the top of a shotting tower. As the droplets of the slurry or solution fall down the tower, their moisture evaporates and a dry granule is collected at the bottom of the tower. The granule is usually porous, although in some processes the drying conditions are adjusted so that a relatively solid grain is achieved by the process. In the case of instant coffee and detergent powders, the manufacturer aims to produce an open structured granule which can dissolve easily in a liquid. It is not clear at the time of writing whether or not the ruggedness fractal dimension of the profile of a spray-dried granule is closely related to the sponge fractal of a porous grain. Studies to resolve this problem and to link the flow properties of a powder with the fractal dimensions of the individual fineparticles in the powder are said to be under way in several research institutions [16].

Dr. Peleg has pointed out that the structural fractal dimension reported in the scientific literature for powder grains, as distinct from smoke fineparticles, appears to have an upper limit of 1.36. Dr. Peleg points out that this may well be due to the fact that spidery-type powder grains, with fractal dimensions of the order of 1.68 and higher, would not be mechanically stable. If stressed, the thin arms of the "spider" fractal structure would fracture to make denser granules of lower fractal dimension [15].

Dr. Peleg and colleagues have quantified the structural fractals of instant coffee grains and are studying the relationship between the flow of coffee powder and the fractal structure of the individual grains. In Figure 3.12(a), the profiles of several coffee grains are shown together with their measured structural fractals. In Figure 3.12(b), the fractal structures of three Tide detergent grains are shown.

It appears likely that fractal description of powder grains will be very important in the pharmaceutical industry since the structural fractal may well be predictive of both the flow behaviour and packing behaviour of the powder in a tablet-making machine. The sponge fractal of the porous grains will probably be descriptive of the **biological availability** of the drug made in the spray-dried form. Under regulations governing the manufacture of drugs, the manufacturer must not only demonstrate that a therapeutic dose is contained in a tablet, but must also show that this tablet can disintegrate and deliver the drug in a usable form to the body. The availability of the drug to the body

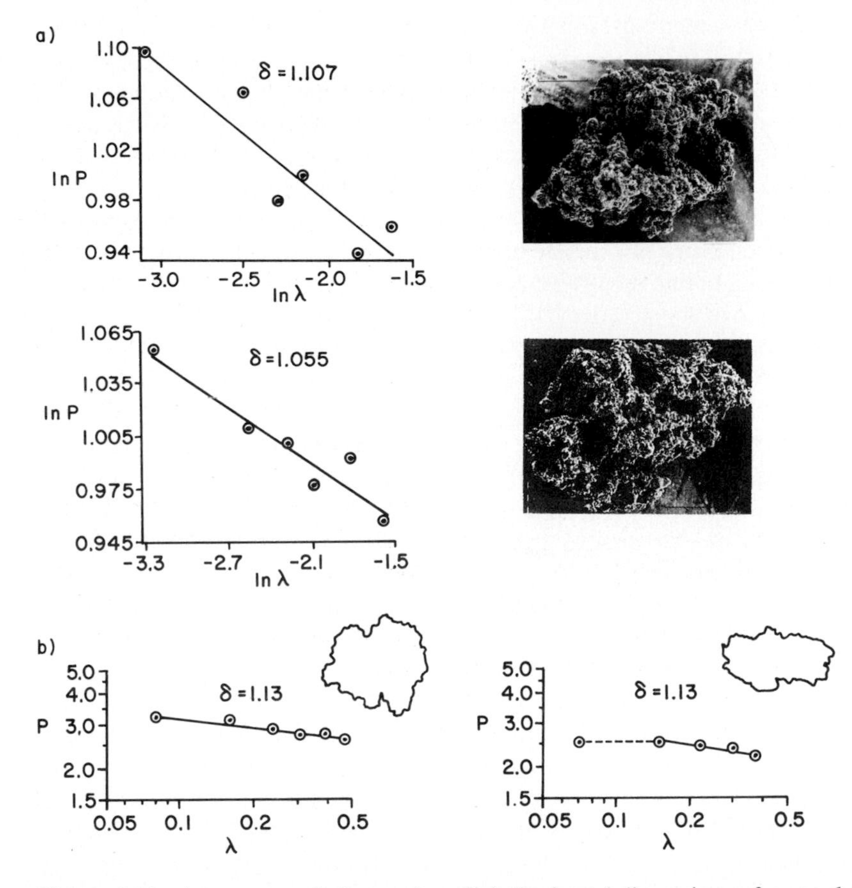

Figure 3.12. Attempts are being made to link the fractal dimensions of spray-dried powders to their rheological properties.
(a) Instant coffee grains studied by Peleg and Normand have structural fractal dimensions ranging from 1.05 to 1.2.
(b) Tide detergent is a spray-dried product, and the individual grains have fractal dimensions similar to those reported for instant coffee grains by Peleg and Normand. (Measurements on the Tide profiles were carried out by J. Leblanc of Laurentian University).

system of the patient when the tablet disintegrates is known as the bioavailability of the drug delivery system (see also discussions of sponge fractals and targeted controlled release drugs in Chapter 7).

3.4 Fractals in the Mining Industry

Another industry where it would appear that fractal geometry has a lot to offer the research scientist is the mining industry. Here, pieces of ore are taken out of the ground and crushed to release the valuable **mineral grains** from the rock holding the mineral grains together to form the piece of ore. The rock holding the mineral grains together is known as the **rock matrix.** In Figure 3.13, the type of structure which exists in a typical piece of ore is illustrated. For our purposes, we shall assume that we wish to obtain the mineral A from amongst the other minerals [17].

The release of the valuable mineral grains from a matrix of other rock species present in a piece of ore is known as **mineral liberation.** The study of the rate at which the mineral is released from the rock matrix by any mining process is known as a study of liberation kinetics (the Greek word "kine" means movement; liberation kinetics is the study of the progress of a liberation process. "Kinema" has given us the English word cinema for the name of the place where people go to watch moving pictures). Usually, the first step in a mining liberation process is the crushing of the ore rocks to break them up into smaller fineparticles. At this stage, the spreading of a crack through the rock is an important factor in the liberation kinetics. Although crack propagation is the basic process leading to the liberation of the valuable mineral grains from the rock matrix, there have not been many detailed studies of crack propagation through a rock. This is probably because of the great difficulty in describing the structure of cracks and how they propagate through a rock. The mining industry consumes enormous amounts of energy in the liberation of mineral grains from crushed ore. The current common practice for liberating valuable mineral grains trapped in a piece of ore rock is to grind a set of ore fineparticles in some type of a grinding mill. The liberation process is controlled by looking at the size and constitution of the fineparticles leaving the grinding mill [17 – 19]. The emerging material is examined to see if the mineral grains have been liberated or if most of them are still composite agglomerates of mineral and worthless rock. The industrial term for the fragments of the rock material which are of no commercial value and which are eventually dumped at the end of the liberation process is **gangue.** This term comes from the German word used to describe the vein of ore which is mined in the mine. It is related to the English word gangway, describing a path from one place to another. The gangue fineparticles are also referred to as **tailings.** Vast quantities of rock tailings have to be disposed of in many mining processes in **tailing ponds,** large open slurry dumps retained by specially built dams [19].

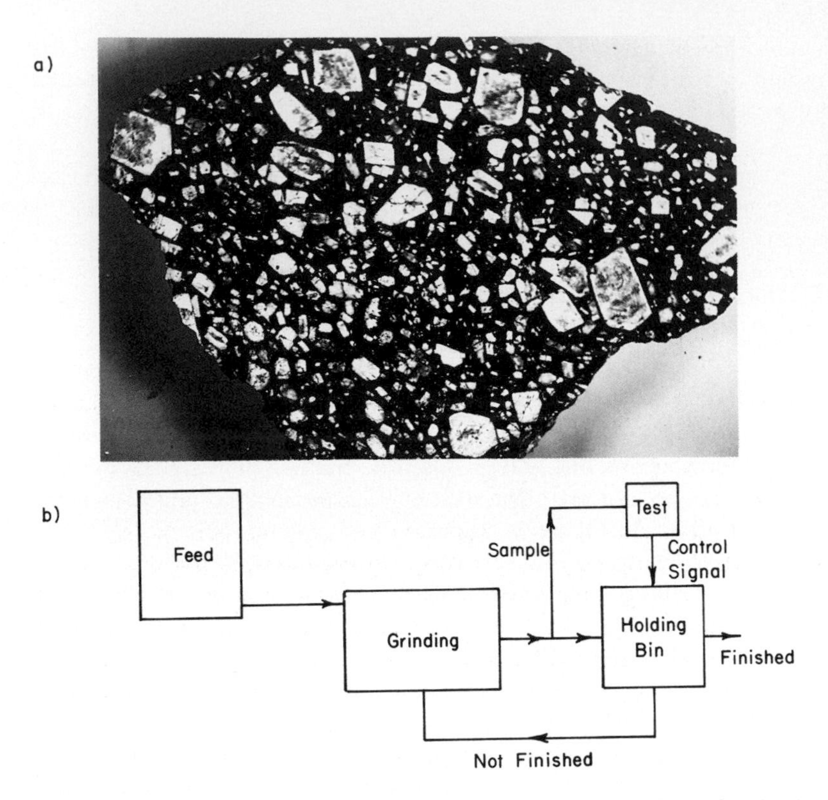

Figure 3.13. In the mineral processing industry, the primary piece of rock taken from the ground is termed an ore fineparticle. This contains valuable mineral grains which have to be released from the waste rock holding the mineral grains together.
(a) A section through a typical ore specimen shows the various types of rock species present in the ore fineparticle.
(b) It is common practice to use a grinding circuit to liberate the valuable mineral from the waste rock.

If the valuable mineral has not been liberated sufficiently from the gangue, the material leaving the grinding mill is fed back to the mill for further grinding. At the same time, a signal is sent to the controller putting material into the mill to slow down the feed rate of the raw crushed rock in order to increase the time spent by the rock in the grinding mill. This type of process is illustrated by a flow chart in Figure 3.13(b). The overall system of rock feed, grinding process, inspection of product and recirculation of product with any subsequent fractionation of the material leaving the grinding mill is known as a **grinding circuit.** Controlling the grinding circuit by studying the constitution of the material leaving the mill, with adjustments to the movement of the product and the feed to the mill, is known as **feedback control** of the grinding circuit. In feedback control, the basic idea is that the information on the product being manufactured is fed back to the controlling unit. One of the great problems with feedback control is that if the mill is overgrinding to produce unnecessarily fine rock fragments, then a great deal of energy is being wasted and inspection of the material emerging from the grinding circuit is too late to save energy in the grinding process. The mining industry would like to replace feedback control of grinding circuits with **feedforward control.** In feedforward control, a rock specimen is crushed and, from a study of the fragments in the specimen, the engineer can predict how long the ore rocks must remain in the grinding circuit and predict the size distribution of the product he will achieve by the grinding process. Two important stages are necessary in the use of liberation kinetics to achieve feedforward control of a grinding circuit. One is a study of the shape and size of the mineral grains within a rock specimen. Techniques for evaluating mineral dispersion in a rock matrix are discussed in Chapter 7. The other is the characterization of the way in which the specimen cracks under stress to produce fragments. Until recently, there have been very few techniques for describing a crack, but cracks look very like the whiskery crystals grown by Matsushita et al. shown in Figure 3.8(a) or, the crystals grown by sputtering by Elam et al. shown in Figure 3.9. Fractal description of such crystals suggests that fractals can be used to describe cracks in rocks and other materials. Work is currently proceeding at Laurentian University into the description of the crack structure in fractured ore specimens by fractal dimensions [20]. Attempts will be made to relate the crack structure in a stressed piece of ore to the structural properties of the ore specimen and the way in which stress and strain are applied to the rock piece (see the discussion of crack modelling in Chapter 9). It is hoped that these studies will eventually lead to improved techniques for feedforward control of grinding circuits.

One approach to the characterization of crack structure in a study of liberation kinetics is to look at freshly shattered specimens of materials. The fractal boundary exhibited by freshly generated fineparticles will be related to the fractal structure of the crack that caused disintegration of the larger piece of material from which the fragments were formed.

One way in which it may be possible to improve the efficiency of ore grinding is to consider the possibility of thermally stressing the surface of a piece of ore to initiate crack formation which then spreads during the grinding process. It is well known that pieces of plastic and rubber which are pliable at room temperature can be chilled in liquid nitrogen until they become brittle. A popular demonstration used by physics teachers is to take a rubber ball and immerse it in liquid nitrogen. The chilled ball is then dropped on to a hard surface, where it shatters. An important aspect of this demonstration is often overlooked, namely that the rapid chilling of the ball not only changes the ball from a pliable solid to a hard solid but that the thermal shock of the chilling process creates multiple cracks in the surface of the ball. It is probably the presence of these cracks rather than the brittleness of the ball which creates the sensational disintegration of the ball as it shatters on the surface.

In Figure 3.14, some fragments of a plastic cup cooled in liquid nitrogen and dropped on to the floor from the height of 1 m are shown. The measured structural fractal dimension of these fragments is indicated. The chilling of a material to temperatures as low as those of liquid air and liquid nitrogen before grinding the material is known in the engineering industry as **cryogrinding.** The Greek word "kryos" meant frost. In English, it has given us the word crystal since the first obvious crystals studied by

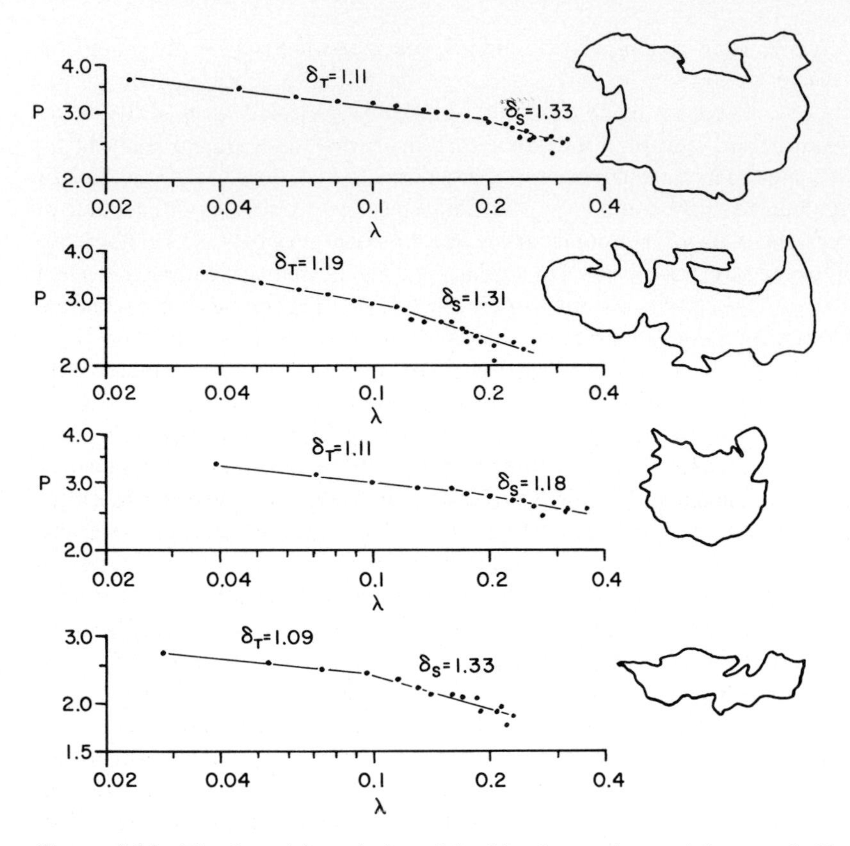

Figure 3.14. The fractal boundaries of freshly shattered material are probably related to the cracks which propagated through the material to generate the fragments. The fragments shown above were generated by cooling a plastic cup to liquid nitrogen temperature and then allowing the cup to shatter by dropping it from a height of 1 m on to a hard surface.

Greeks were crystals of frost. The "cryo" root word in cryogrinding is from the same Greek word. Cryogrinding is used in the grinding of spices so that the grinding process does not raise the temperature of the fineparticles to such a level that the aromatic oils essential to the flavour and aroma of spices are evaporated from the surface of the fineparticles being ground.

One grinding device used to produce fineparticles is a **ball mill.** In this device, the powder whose grains are to be reduced in size is mixed with some larger balls of a hard substance, such as steel or hardened ceramic, and the mixture is tumbled around in a cylinder as illustrated in Figure 3.15. The mixture of powder and balls is known as the **millcharge**. The balls are lifted up by the rotation of the cylinder of the ball mill and at a certain height they cascade down and crush fineparticles which are trapped between the falling balls and balls lower down in the mill. If the freshly generated fragments of plastic shown in Figure 3.14 were being cryoground in the ball mill, then as they traveled around in the ball mill after being formed by the initial act of shattering, it is most likely that the jagged features of the profiles would be removed by a process that engineers call **attrition**.

Attrition means rubbing two things together. The word comes from the Latin verb "terere" meaning to rub. The same root word is found in the English word tribulation, used to describe a hardship or something which distresses us. Tribulation comes from the Latin word for a sledge, "tribulum," used to rub out corn in a process similar to threshing. This word has also given us the scientific term **tribology**, which is the study of the wear of surfaces that rub together. Thus ferrography, described earlier, is a branch of tribology in which one tries to understand wear conditions in such things as the bearings of jet engines from the fragments produced in the lubricating oil.

If we wish to study the relationship between the fractal dimension of a freshly fragmented fineparticle and the crack structure which caused the failure of the material producing the fragment, we must avoid attrition in subsequent handling of the fineparticles before we measure their fractal dimension. For this reason, if we wish to understand the failure mechanisms which crack open a piece of ore rock from the fractal dimension of fragments, it is too late to look at the fragments emerging from a ball milling process since they will have been polished by attrition.

Some of the readers will recognize that the ball mill illustrated in Figure 3.15 is very similar to the polishing device used by individuals who like to make their own jewelry from polished semi-precious stones. In the polishing process, the **lapidarian** (a technical term for the person who polishes stones, from the Latin word "lapis," meaning a stone) tumbles the stone to be polished with sharp sand or other polishing agents. By attrition, the tumbling sand removes all the roughness from the surface of the stone. Using the language of fractal geometry, we can state that the aim of the polishing process is to reduce the fractal dimension of the profiles of the polished stones to 1.0. If a lapidarian looked at the series of profiles shown in Figure 2.15, the profiles would look like a series of snapshots of a rugged profile being polished down to a smooth, brilliant surface. It has been shown that fractal dimensions can be used to quantify the progress of the polishing of a surface. Thus, in Figure 3.16, successive profilometer traces for a copper surface being polished are shown. The fractal dimension of the profilometer traces as measured by the structured walk technique are shown with each of the profilometer traces (a **profilometer** is a device which measures the ruggedness

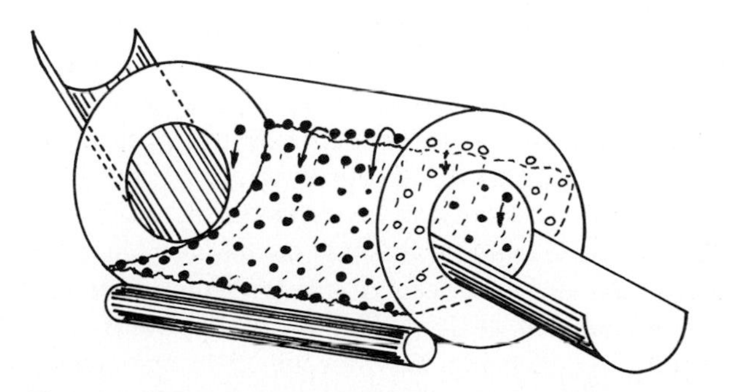

Figure 3.15. A ball mill is used to crush ore fineparticles by letting hard balls cascade down on to moving rock material in a rotating cylinder.

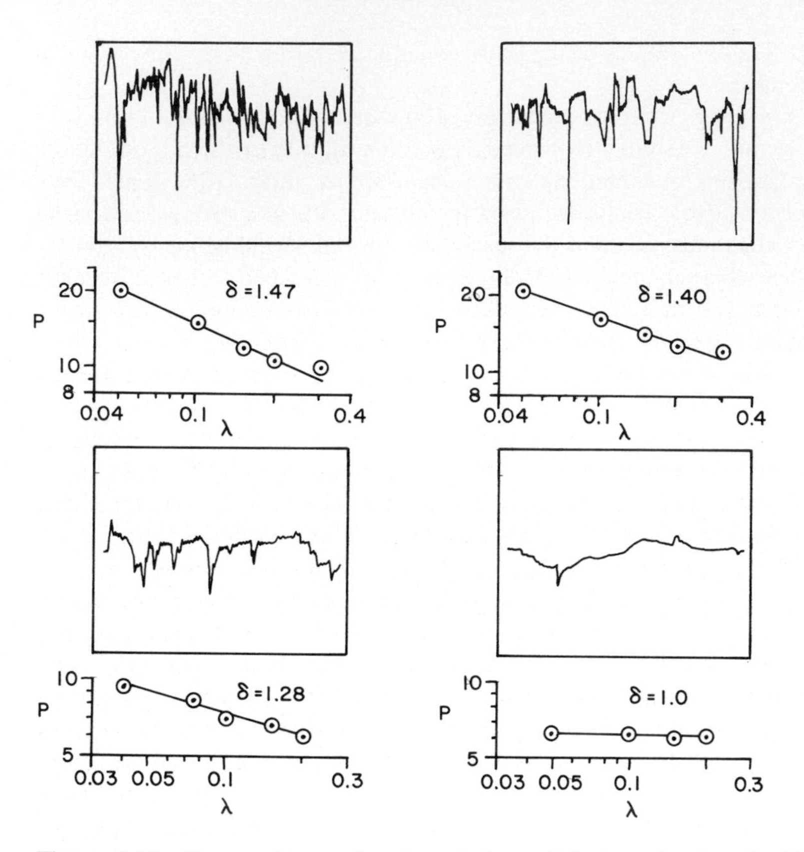

Figure 3.16. The roughness of surfaces being polished can be described by means of fractal dimensions. The four rugged lines shown represent the sequence of profilometer traces taken during the polishing of a copper surface. The fractal dimensions for the profilometer traces were measured using the inswing structured walk technique [22, 23].

of a surface by measuring the movement of a diamond needle up and down as it traverses the surface) [21, 22].

In mining research, one should be able to look at the various forces contributing to the size reduction and fracture occurring in a ball mill by taking a piece of ore and shattering it by applying stress of known magnitude, with subsequent measurement of the fractal dimension of the freshly generated fractal fragments. These freshly generated fragments could then be tumbled in a ball mill with or without sharp sand, to reduce the ruggedness of the fragments by attrition, without further size reduction of the fragments by stress cracking. Over a period of time, it may be possible to reduce the fractal dimension of the tumbling rock fragments achieved by attrition polishing to that of fragments emerging from an industrial ball mill, so that one could study how important attrition inside a ball mill is in reducing the ruggedness of the freshly produced fragments. One could carry out this type of study using thermal stressing of the original ore lumps, to see if such treatment could improve the efficiency of crushing and grinding systems by altering the way in which cracks form in the ore specimen when treated in various ways.

A new branch of mining technology which is growing at a fast rate is known as hydrometallurgical extraction technology. In **hydrometallurgical technology,** valuable minerals are retrieved from ores by a process known as leaching. **Leaching** comes from an old English word, "leccan," meaning to wet or irrigate. In technical English, leaching is the movement of liquid through a porous body. In the leaching process, water containing chemicals or bacteria trickles through the ore body, either in the form of a crushed pile of ore on the surface of a mine or by injecting water containing chemicals or bacteria into cracks in a vein of valuable minerals in the ground. The chemical or bacterial action dissolves the valuable minerals to the water, which is then treated chemically to recover the valuable mineral.

Sometimes the leaching process also happens inadvertently. For example, water trickling down through the rock tailings dumped from a uranium mine has been found to leach residual radium salts from the tailings dumps, and the subsequent movement of dissolved radium salts into the river systems close to the mine created a public hazard. In this case, bacteria in the water attack the iron pyrites present in the rock tailings forming sulphuric acid. The sulphuric acid then dissolved the trace quantities of radium in the uranium ore tailings with the subsequent movement of radium sulphate into the river system. Mining engineers working with the same uranium ores from which the tailing dumps were created are attempting to leach the uranium out of the ground by cracking the ore body with explosives and feeding a suspension of bacteria into the rocks to obtain the release of valuable minerals in the water moving through the cracked ore body. A critical aspect of the deliberate use of hydrometallurgy to recover valuable minerals from underground strata is the way the rocks crack under the impact of explosive charges detonated underground. The mining engineer now has a great interest in learning how to specify the structure of cracks created by explosive stress, so that he can relate the cracks produced to the explosive technology he is using to open up the underground mineral body to bacterial attack. West and Shlesinger have used random-walk techniques to model the fractal structure of a body failing under impact stress produced by a high-velocity projectile. Their theory is probably applicable to the study of how rock fails in the vicinity of a drill impact when drilling rock [24].

The fact that fragments emerging from grinding processes used in the mining industry still exhibit a fractal structure, in spite of attrition from multiple tumbling in the ball mill, is illustrated by the rock tailing fineparticles shown in Figure 3.17(a). The fractal dimensions of the various rock profiles are indicated by the side of each profile. An important property of a slurry of fineparticles is its viscosity. The **viscosity** of a fluid or a suspension can be regarded as a measure of the internal friction of the fluid when one attempts to move the fluid. In other words, it is a scientific description of the "stickiness" of a fluid. Water has low viscosity; molasses and treacle have high viscosities. A complete discussion of the viscosity of suspensions is too difficult to include in this book, but the effect of the fractal dimension of rugged fineparticles on the viscosity of a slurry can be illustrated by some data generated by Syed Kaab Akhter [25]. Akhter measured the viscosity of suspensions of nickel ore tailings, of the type shown in Figure 3.17(a), in a viscous mineral oil using a rotating co-axial cylinder viscometer [26].

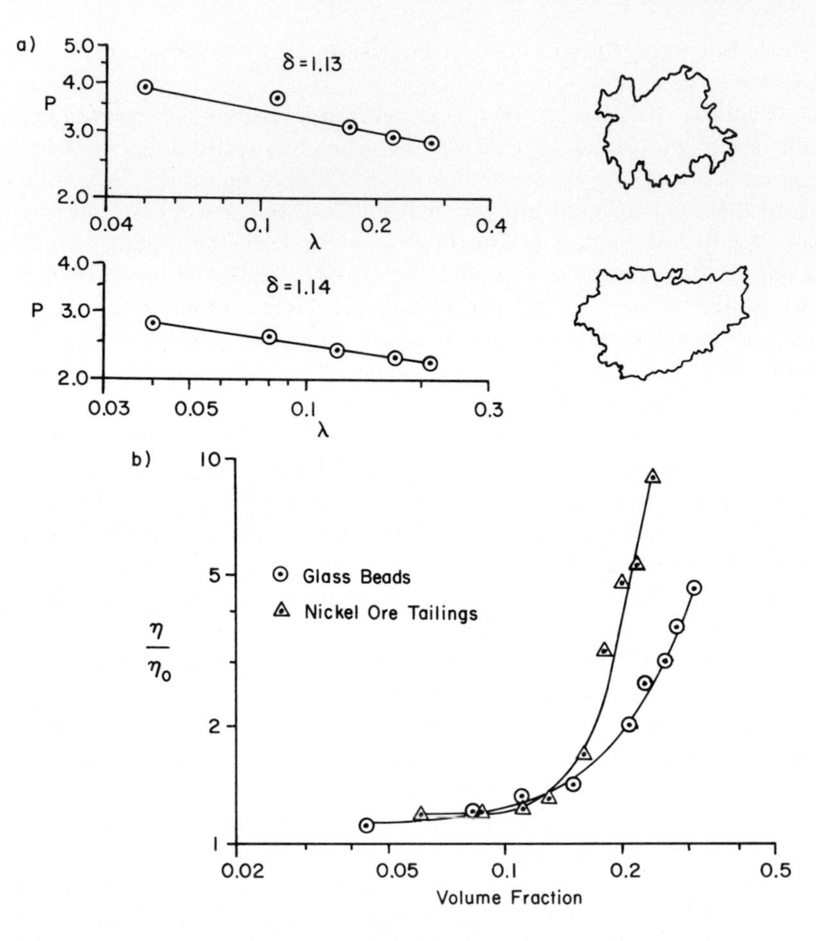

Figure 3.17. The viscosity of an industrial slurry of mine tailings can be related to the ruggedness of the profiles of the suspended fineparticles.
(a) Fractal dimensions can be used to describe the ruggedness of the ore tailing fineparticles.
(b) A comparison of the measured viscosities of slurries of various concentrations of nickel ore tailings and glass beads shows that effects of shape and ruggedness on the viscosity of the suspension becomes important after a solids concentration of 0.15 volume fraction [25].

In his experiments, the viscosity of suspensions containing different amounts of nickel ore tailings were compared with the viscosity of suspensions containing different amounts of spherical glass beads. In Figure 3.17(b), a set of data for the viscosity of various slurries of nickel ore tailings and glass beads are reproduced from Akhter's M.Sc. Thesis. The concentration of rock tailings in the suspension is expressed as a volume fraction of the slurry; thus, a slurry of 0.1 nickel ore tailings is 10% by volume tailings. The measured viscosity of the slurry of rock tailings is divided by the viscosity of the mineral oil without any rock tailings in suspension to obtain the effective viscosity of the slurry in dimensionless terms. It can be seen from the data in Figure 3.17(b) that up to a volume fraction of approximately 0.15, the viscosity of the slurry of mono-sized glass beads is almost the same as that of the rock tailings. However, above that volume fraction the effective viscosity of the rock tailings increases more rapidly than that of comparable slurries of glass beads. Thus, at a volume fraction of rock tailings of 0.22, the viscosity of the rock tailing suspension is nearly five times greater than that of the same volume fraction of glass beads in oil [note that both scales of the viscosity-volume concentration graph in Figure 3.15(b) are logarithmic]. The explanation of the large difference between the viscosity of rock tailing suspensions and that of the glass beads is that a rugged rock fineparticle immobilizes fluid in the immediate vicinity of their rugged "ins and outs" of the profile. At higher solid concentrations, as the fineparticles are forced to move in close proximity to each other, this immobilization of fluid by the rugged contours has a much greater effect than when they are widely separated in low concentration slurries. It is to be expected that the difference between the viscosities of the rock tailings and the spherical glass beads will be related both to the overall shape of the rock tailings, as quantified by a form factor such as the aspect ratio, and to the fractal dimension of the boundaries of the profiles.

Fractal dimensions will probably be very useful in describing the effect of ruggedness of the suspended fineparticles on the viscosity of industrial slurries. In the mining industry, vast quantities of tailings slurries must be pumped around a mineral processing plant and the power required to pump the slurry depends upon the effective viscosity of the slurry. Graphs as shown in Figure 3.17(b) can be used to predict power requirements when moving slurries. A data curve such as that in Figure 3.17 suggests that, other things being equal, it requires less power to pump twice as much suspension at a volume fraction of tailings of 0.15 than it does to pump a single quantity at a volume fraction of solids 0.30. On the other hand, returning the larger quantities of the water to the mines after dumping the tailings may counterbalance the savings made in pumping the less concentrated tailings suspension out of the mine.

As already mentioned, an important element in the operation of any grinding circuit with feedback control is an examination of the size of the fineparticles leaving the grinding mill. In the mineral processing industry, there is a great drive to make the inspection of the fineparticles and the provision of the necessary information for feedback control of the grinding completely automated [27, 28]. The term **on-line size analysis** is used to describe automated sizing procedures forming part of an automated grinding circuit. A widely used on-line size analysis procedure in use in the mineral processing industry, and many other fineparticle producing industries, is a technique known as eriometry. In **eriometry**, the size of the fineparticles is deduced from the diffraction pattern generated by passing a laser light beam through a random array of the fineparticles. The name comes from the fact that Thomas Young (1773-1829), who established the wave theory of light and was the first to study the diffraction properties of light, developed a technique for measuring the fineness of wool being imported from Australia, which depended upon the diffraction pattern generated by monochromatic light passed through a sample of the wool. Young named his original instrument an eriometer because the Greek word for wool is "erios." In modern technical English the term eriometry is used to describe any instrument in which size is deduced from the pattern of diffracted light generated by fineparticles and/or fibres.

In Figure 3.18, the diffraction pattern generated by a regular array and a random array

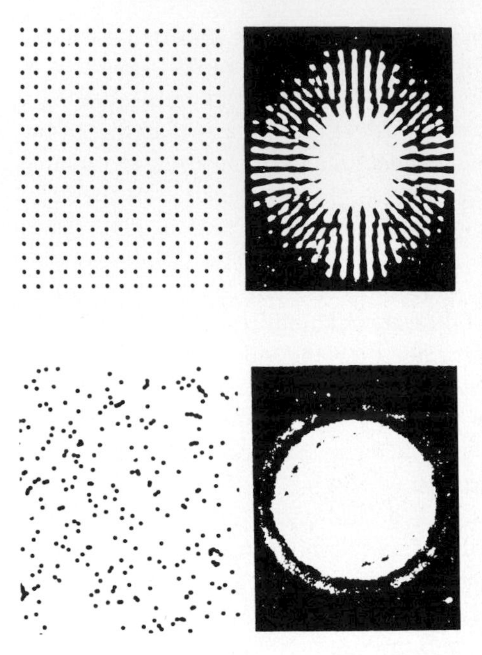

Figure 3.18. The diffraction pattern of a regular array of mono-sized spheres contains an interference pattern generated by the regular structure of the diffracting objects. The diffraction pattern of a random array of mono-sized fineparticles is the same as the diffraction pattern for a single sphere, with the strength of the pattern being n times that of the diffraction pattern generated by the single scattering sphere, where n is the number of scattering spheres in the random array. (Used by permission of the Ealing Scientific Corporation and taken from "The Ealing-Hoover Diffraction Plates – Theory and Application" by Richard B. Hoover and an Ealing Science Teaching Catalogue 73/74.)

of mono-sized spheres is shown. It can be seen that the pattern generated by the random array is the same as the pattern which will be generated by the single sphere, but that the strength of the light pattern is n times that of a single particle, where n is the number of diffracting fineparticles [29]. The size of the fineparticle can be deduced from the position of the light and dark rings of the diffraction pattern. If one generates the diffraction pattern of a random array of spheres of different sizes, a complex pattern generated by the overlap of the various patterns for each size of the sphere would be generated by the passage of laser light through the system. In commercial eriometers, computers are used to break down the energy contributions from each size present in the measured energy levels of a complex diffraction pattern. Often the interpretive theory used in commercial instruments assumes that all of the diffracting fineparticles are spherical (spherical chicken modelling!). This is a dangerous interpretive path to follow, since the shape and texture of the fineparticle scattering the light can make important contributions to the spatial distribution of the energy in the diffraction pattern [30].

The effect that the structure and texture of a fineparticle can have on the spatial distribution of scattered light in a group diffraction pattern is illustrated by the data summarized in Figure 3.19. The light-scattering properties of Koch's triadic island of various orders, as described in Figure 2.13, have been measured experimentally by Leblanc [31]. In Figure 3.20, the diffraction patterns of various geometric shapes are shown for information purposes, taken from Murphy's M.Sc. Thesis [32]. From the diffraction pattern of the geometric shapes, it can be seen that edges send light further out from the centre of the diffraction pattern than the diffraction which would be formed by a simple shape. If a robot looks at a diffraction pattern not knowing, that sharp edges

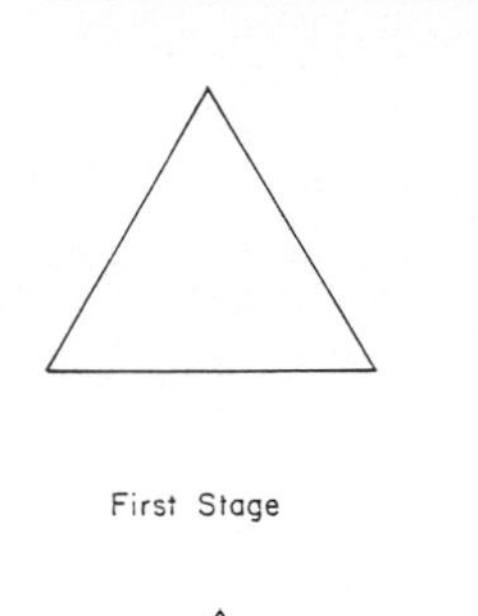

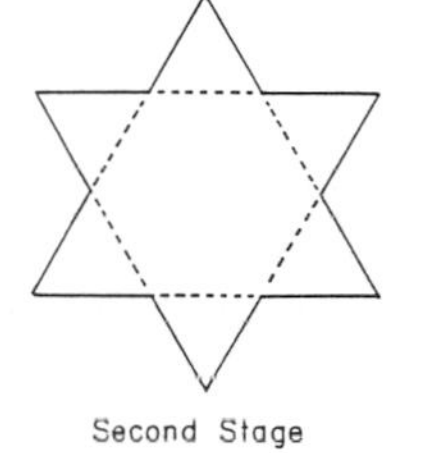

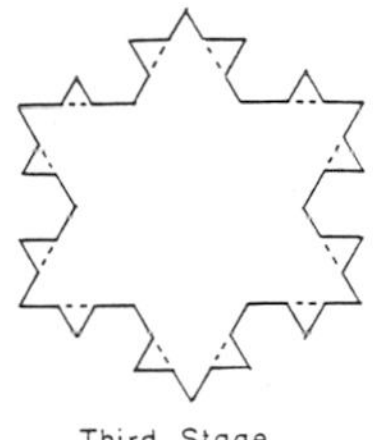

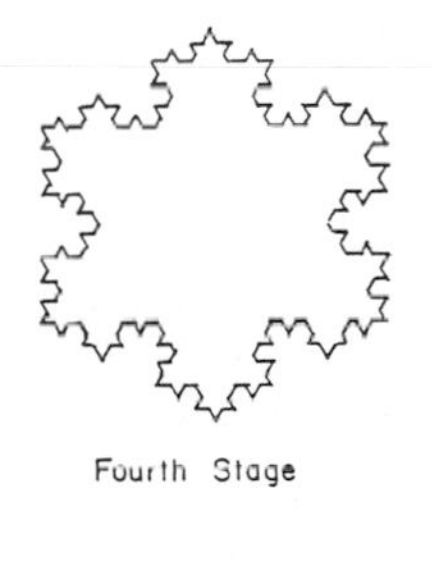

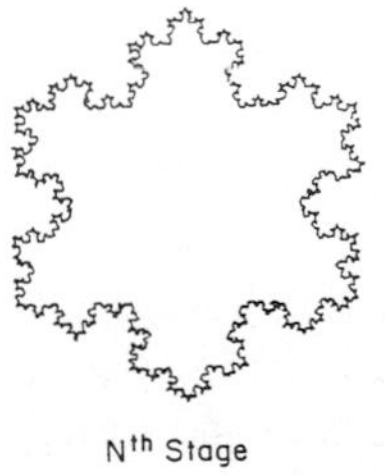

Figure 3.19. The diffraction patterns of a sequence of Koch's triadic islands of increasing order illustrates the effect of texture on the diffraction pattern [31] (reproduced by permission of J. Leblanc).

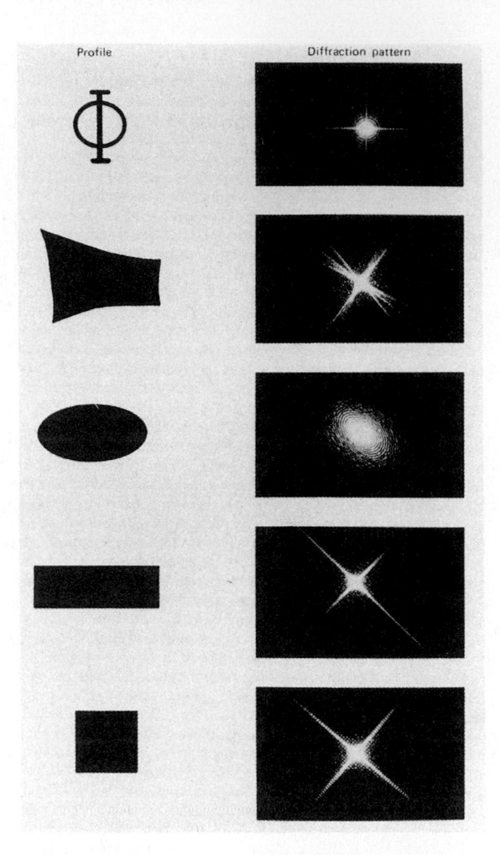

Figure 3.20. The diffraction patterns of sharp-edged fineparticles were obtained in a study of the optical properties of fineparticles carried out by Murphy [32].

are present in the profiles scattering the light, it will interpret energy further out from the centre as being due to smaller fineparticles than those which scattered the light near the centre of the pattern. Commercial eriometers are essentially robots, which are not usually programmed to look for edge diffraction patterns as distinct from the diffraction energy pattern made by spheres. They have been taught to interpret all the diffraction patterns as if the fineparticles in the energy beam were spheres. The effect of edges and fineparticle texture on scattered light pattern is illustrated by the diffraction patterns of the various Koch's triadic islands shown in Figure 3.19. First of all, the diffraction pattern of the simple triangle has edge diffraction energy which is represented by the six lines of light extending from the relatively dense central pattern. All of the photographs shown in Figure 3.19 were photographed under the same conditions and the diffraction profiles were photographic negatives all of the same size used on the same optical bench. Thus, although one cannot measure the energy distribution in the diffraction pattern quantitatively from the photographs in Figure 3.19, one can gain an appreciation of the relative effects of the various structural features of the Koch's triadic island on the diffraction pattern of that island.

As the order of the island increases, the basic shift in geometric structure is that the perimeter becomes more rugged. It can be seen that this leads to a general decrease in the sharpness of the edge-dependent features of the diffraction pattern, but coupled with this is an increase in the energy scattered further out from the centre of the pattern by the texture of the Koch islands. Thus, the edge diffraction effect on the lower order profiles will generate misleading energy patterns, leading the machine to report a false smaller size range of fineparticles present in the pattern. Things become worse as the ruggedness of the boundary increases. This information is important to those who would like to use eriometers to characterize fineparticles in an industrial on-line situation. The diffraction patterns of the Koch's triadic island warns the technologist that, if he is dealing with fineparticles that have fractal boundaries, the eriometer will generate light energy distribution information which may be misinterpreted by the logic of the machine. The way in which the texture of a fineparticle can completely alter the scattering pattern of the fineparticle as compared with its spherical equivalent has been discussed by Bickel et al. [33], and readers interested in pursuing the subject in depth should consult their paper.

One industrial situation where the diffracting power of the texture of individual fineparticles may generate misleading results when studying fineparticles in suspension is in industries where chemicals are precipitated as crystals and one wishes to study the growth of the crystals in suspension. It has already been pointed out that various types of crystals bear a strong resemblance to the Koch islands of different fractal dimension and order. In Figure 3.21, a cluster of inorganic crystals known as barytes (barium sulphate) and its structural boundary fractal are shown. This is an example of how perfectly cubic subunits can agglomerate to produce a fused whole which exhibits fractal properties [21].

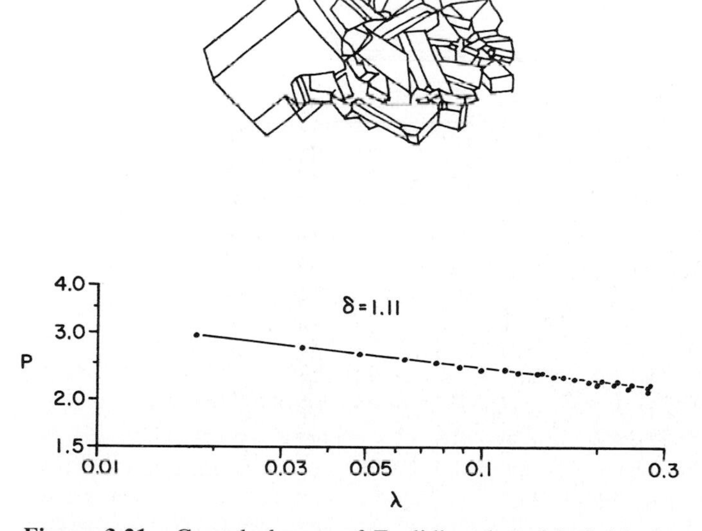

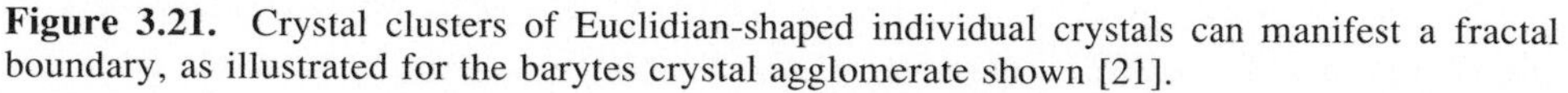
Figure 3.21. Crystal clusters of Euclidian-shaped individual crystals can manifest a fractal boundary, as illustrated for the barytes crystal agglomerate shown [21].

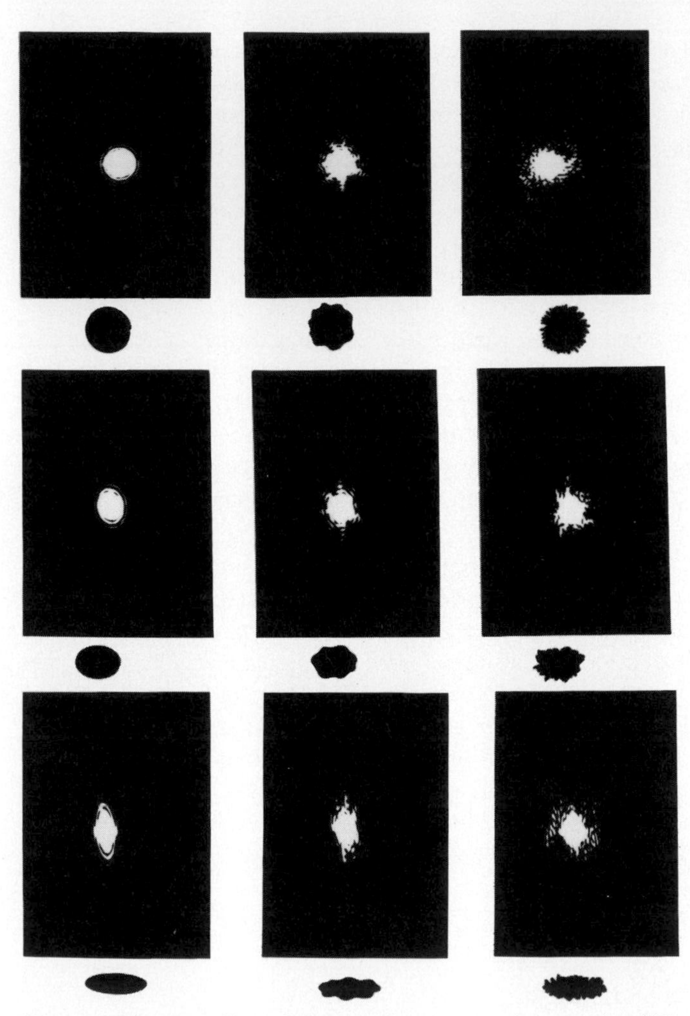

Figure 3.22. The effect of shape and texture on the complex diffraction pattern of a fineparticle is illustrated by a set of data generated in a study at Laurentian University to provide correction factors for the eriometeric evaluation of industrial slurries of irregularly shaped fineparticles [31, 34].

It may be that the study of the fractal dimension of a series of fineparticles which are to be characterized in an industrial situation using an eriometer, may result in the evolution of a simple correction factor for the interpretation of the diffraction patterns based on the fractal dimension of the profiles of the boundaries. In Figure 3.22 the diffraction patterns of another series of fineparticle profiles are shown. These diffraction patterns illustrate the interacting effect of shape and fractal boundaries on the light-scattering properties of fineparticles [31].

The data on the Koch's triadic islands were generated by Leblanc when exploring the possibility that the fractal dimension of the boundary may be measurable from the diffraction pattern generated by the boundary. Unfortunately, as can be seen from the diagrams in Figures 3.19 and 3.21, the interpretation of the diffraction pattern generated by a fineparticle with a fractal boundary is complex because of the interaction of shape, size and texture effects in the diffraction pattern [33].

Another area of research where fractal dimensions will be of interest to the mining engineer is in the clarification of effluent water from the mines. Often the effluent water is stored in large tanks to enable the suspended solids to settle out of the water. To accelerate this process, it is usual to add flocculents to the water. **Flocculents** are chemicals which create large, loosely structured aggregates of the suspended solids (the term aggregate is used here rather than agglomerate since simple acts such as stirring the water can often break up the aggregates). Loosely structured aggregates are described as **flocs** because they look like pieces of wool floating in the water; the Latin word "floccus" meant a lock of wool. The flocs settle out of the water quicker than single fineparticles, and it takes less time to clean the water of suspended solids by sedimentation than in the absence of the flocculating chemicals.

The mechanisms of flocculation are of great interest in predicting the efficiency of any given flocculating agent. The growth kinetics of flocs have been studied in depth by several scientists. In particular, Sutherland and Goodarz-Nia have simulated floc growth on a computer [35 – 38]. In Figure 3.23, two different sets of simulated flocs

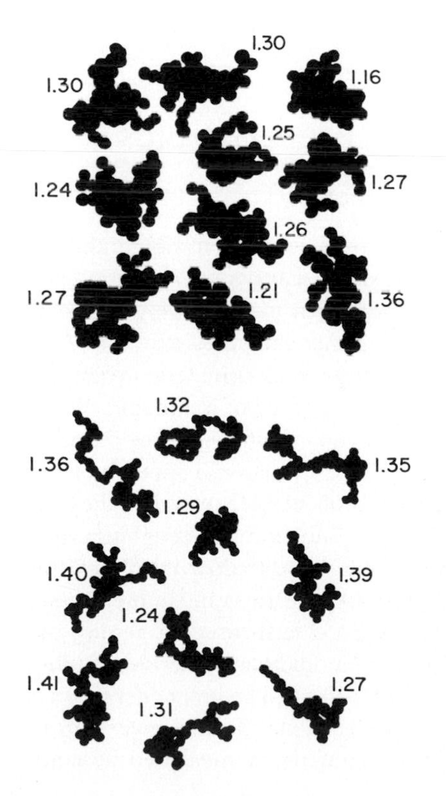

Figure 3.23. Sutherland and Goodarz-Nia simulated the growth of flocs using computers. The structure of their simulated flocs can be described using fractal dimensions [35 – 38, 42] (reprinted with permission from *Chemical Engineering Science*, Vol. 26, D.M. Sutherland and I. Goodarz-Nia, "Floc Simulation – The Effect of Collision Sequence," 1971, Pergamon Press plc).

grown by Sutherland and Goodarz-Nia by computer are shown. The aggregates in Figure 3.23(a) grew by the accumulation of single fineparticles colliding with the growing aggregate and those in Figure 3.23(b) grew by single particle capture and by cluster collision. Those in Figure 3.23(a) were generated by a modified version of the same type of growth mechanisms underlying the flocs in Figure 3.23(b). These flocs were simulated before fractal geometry was invented, and Sutherland and Goodarz-Nia found it difficult to quantify the structure of these flocs. The floc profiles in Figure 3.23 obviously cry out for fractal description and they have been examined "fractally" by several workers [39 – 42]. The fractal dimensions of the various types of floc, as measured by erosion dilation logic, are shown by the profiles in Figure 3.23. It seems reasonable to suggest that the fractal dimensions of real flocs created by chemical flocculants in a settling tank, compared with the measured fractal dimensions of the flocs in Figure 3.23, would enable the mining engineer to interpret the mechanisms by which any given flocculating agent was achieving the flocculation of a given suspension. The growth mechanisms of the various types of flocs shown in Figure 3.23 are discussed in Chapter 4.

It is interesting that in the second chapter of his original book (see Chapter 1 reference [1]), Mandelbrot quotes Jean Perrin's, discussion written in 1906, of the indeterminate nature of flocculated soap flakes precipitated out of soap solution. Mandelbrot points out that in this perceptive discussion of a "soap floc," Perrin was groping towards a fractal description of a natural phenomenon which has become part of fractal geometry.

3.5 Fractal Structure of Cosmic Fineparticles

In our discussion of the possible utility of fractal descriptors in the mining industry, it was pointed out that freshly shattered pieces of rock are probably much more rugged than those that emerge from crushing and grinding equipment in the mining industry, since attrition will round off the fractal structure of the freshly generated material. One place in the solar system where we can find freshly shattered rock fragments which are not subjected to polishing by erosion of any kind is on the surface of the moon. When the Apollo 12 mission brought back some dust from the lunar surface, it was found to contain fragments of various shapes. In particular, one type of material appeared to be very rugged. One of the pioneers of fineparticle science, Professor Heywood, at the time a Professor in Chemical Engineering at the University of Technology, Loughborough, UK., was asked to characterize the lunar dust. In Figure 3.24, four of the fragments photographed by Professor Heywood are shown [43]. At the time he studied them, Professor Heywood described these as being **scoracious**, a term used in geology to describe rugged type rocks with many holes in them. Scoracious is derived from a Greek root word which can be politely translated as "dung-like." The lunar dust profiles in Figure 3.24 can obviously be described by fractal dimensions. In the lower part of Figure 3.24, the fractal dimensions of the four lunar fragments as measured by equi-

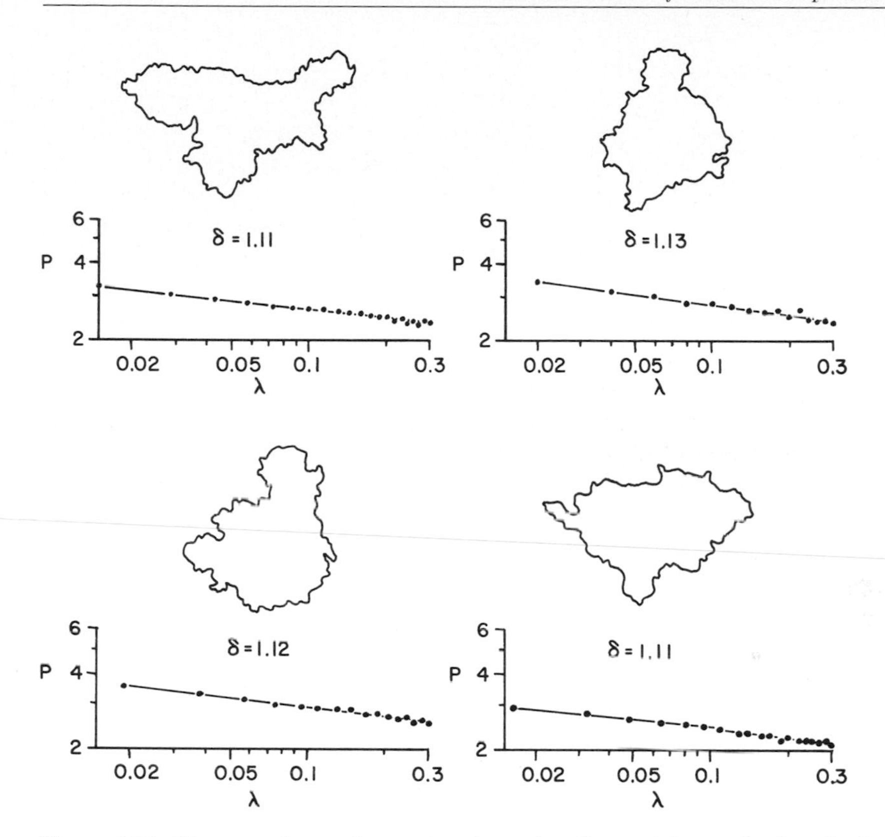

Figure 3.24. The ruggedness of scoracious lunar dust fineparticles can be described using fractal dimensions [44].

paced polygon logic are shown [44]. It would be interesting to see if the fractal dimensions of these fragments are similar to those produced by freshly shattered rock upon the earth.

It was my privilege to know Professor Heywood, who encouraged me in my studies of the properties of fineparticles. Professor Heywood loved science and was never happier than when carrying out new investigations. I think his attitude is well expressed by his comment on his own study of lunar dust. "This is a fascinating project (looking at the dust from the moon). Although only preliminary observations have at present been made, one is tempted to spend many hours just looking through the microscope at these unique particles and having feelings which must be akin to those of Hooke when he first examined terrestrial sand particles through the microscope" [43]. Professor Heywood had a restless enquiring mind. He told me that he was fascinated by the problems of measuring the shape of fineparticles and that one day, when he sat on the beach watching the waves move the gravel around, he could not just sit and look but started to make measurements of the shape of the gravel on the beach. Professor Heywood would have loved fractal geometry. In 1984 his widow, Dr. Francis Hey-

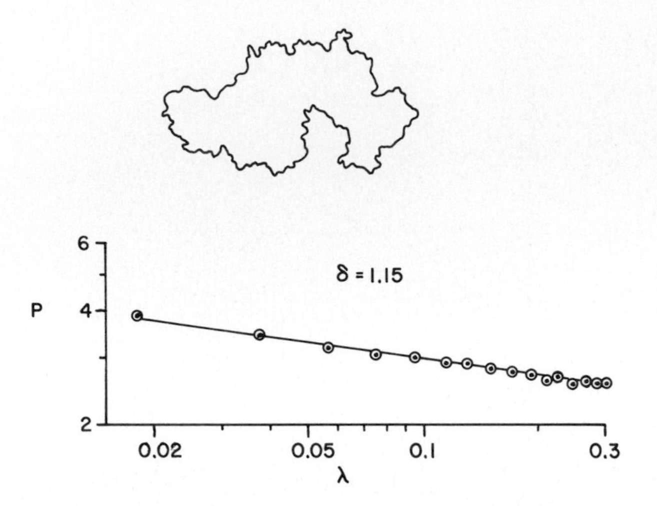

Figure 3.25. Cosmic dust fineparticles collected by Dr. Brownlee using high-flying aircraft exhibit fractal structure [45, 44].

wood, donated all of the publications of Professor Heywood to the fineparticle group at Laurentian University. These extensive publications, dating back to early 1932, contain much material which illustrate his pioneering efforts to develop fineparticle science.

Another type of cosmic fineparticle that shows fractal structure is the cosmic dust material collected by Professor Brownlee. In a series of experiments aimed at studying the ancient cosmic dust of the Universe, Professor Brownlee flew U2 aircraft at high altitudes, with sweeper collectors on the wings of the aircraft to capture cosmic dust settling into the earth's atmosphere from outer space [45, 46]. The cosmic dust appears to be formed by the agglomeration of tiny fineparticles as shown in Figure 3.25. The fractal dimension of Dr. Brownlee's cosmic dust, as measured at Laurentian University using the equipaced polygon technique, is shown by the side of the cosmic fineparticle [44]. It may well be that the fractal structure of such agglomerates will help scientists to interpret the way in which the agglomerates form (see the discussion in Chapter 4 of growth mechanisms yielding different agglomerate structures).

3.6 Fractal Structure of Some Types of Sand Grains

In our earlier discussion of the possible utility of fractal descriptors in the mining industry, it was mentioned that it would be interesting to study the decrease in the structural fractal dimensions of a fresh rock fragment subjected to attrition. This type of experiment has been going on in nature for hundreds of thousands of years in the process by which natural forces generate sand from quartz rocks. The initial process in

Figure 3.26. In some parts of the world sand grains are made up of calcium carbonate. Whalley and Orford evaluated the fractal structure of this type of sand grain. Four different views of the same calcium carbonate sand grain are shown. Whalley and Orford characterized the structure of this calcium carbonate sand grain by means of fractal dimensions. (From: Whalley & Orford (1982) Courtesy of Dr. Brian Whalley.)

the formation of quartz sand is the fracture of quartz rocks. Other types of sand grains are formed from other types of rock, but generally sand grains from softer rocks are disintegrated relatively quickly by the forces of the rivers and the oceans. Most of the beach sand with which we are familiar is composed of ancient quartz grains rounded by the pummeling action of the movements of the ocean. The profiles of rounded beach sand are usually too smooth to be described by means of fractal dimensions, but the use of fractal dimensions to describe some irregular types of sand grains has been pioneered by Whalley and Orford [46, 47].

In Figure 3.26, four different views of a calcium carbonate (limestone) beach grain is shown. Whalley and Orford reported that this particular sand grain had a structural fractal dimension ranging from 1.06 – 1.13 when it was viewed in different positions, and that it had a textural fractal of 1.05 – 1.18. The internal structure of this type of sand grain could probably be characterized by a sponge fractal, but Whalley and Orford did not carry out mercury intrusion studies on this type of sand grain in their original study. In Figure 3.27, some other interesting fineparticles studied by Whalley and Orford are shown. In Figures 3.27 (a) (i) and (ii), two different views of a freshly fractured quartz grain are shown. It is this type of fragment, that becomes rounded by wave action to

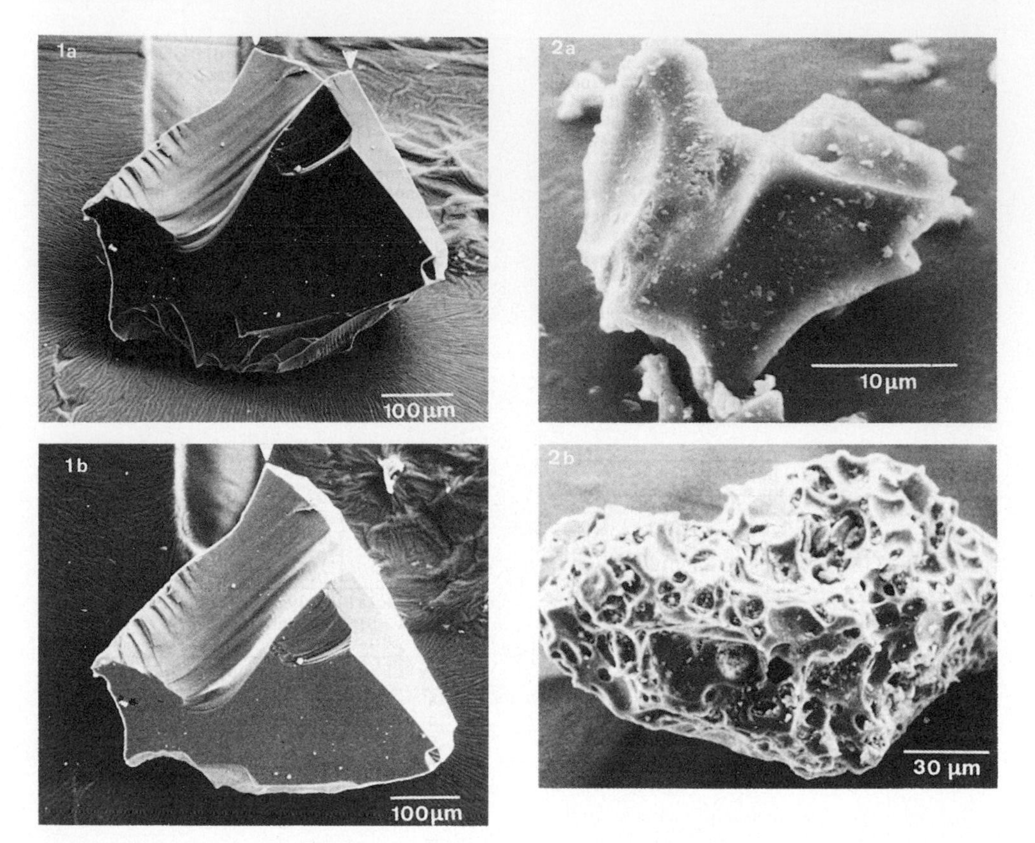

Figure 3.27. Whalley and Orford used fractal dimensions to characterize both freshly shattered quartz grains and pyroclastic fineparticles. (From: Whalley & Orford (1982) Courtesy of Dr. Brian Whalley.)

form the rounded sand grains of a typical silica sand beach (quartz is a crystalline form of silica, chemical formula SiO2). In Figure 3.27(b) and (c), two types of fineparticle emitted out during the eruption of the volcano Mount St. Helens in the U.S.A. (1980) are shown. This type of fineparticle is known as a **pyroclastic** fineparticle. This word comes from two Greek roots, "pyro," meaning fire, and "klaein," meaning to break. Thus the idea is that pyroclastic dust is formed by rocks that are broken by fire to give the fragments thrown out of the volcano. Whalley and Orford reported that structural fractals of the non-porous pyroclastic fineparticle was 1.07 and that it had a textural fractal of 1.02. The porous structured pyroclastic particle was found to have a structural fractal of 1.05 and a textural fractal of 1.04. In their various publications, Whalley and Orford indicated that fractal dimensions will be very useful in the description of rugged sand grains.

3.7 Fractal Structure of Some Respirable Dusts

3.7.1 What is the Technical Meaning of Respirable Dust?

As indicated at the beginning of this chapter, one of the first powders I studied in detail at the beginning of my career in applied science was a uranium dioxide powder used to fabricate fuel rods for nuclear reactors. The fabrication technology used to make the fuel rods was a blend of powder metallurgy and ceramic technology.

Ceramics is defined in one technical dictionary as "Pertaining to products or industries involving the use of clay or other silicates." Another dictionary gives the following information: "**Ceramic**," "pertaining to pottery from KERAMOS -the Greek word for the clay used by potters." Both of these definitions have been overtaken by 20th century technology. Increasingly modern industry is turning to space-age ceramics to replace metals in the fabrication of the machines and aircraft of tomorrow and to increase the performance efficiency of internal combustion engines. A useful definition of a modern ceramic material is that it is a rigid body fabricated by the fusion of metal oxides and/or silicates, carbides and other chemically related compounds. Ceramics have a high resistance to thermal distortion, low thermal expansion coefficients and are usually poor conductors of electricity. Modern ceramics technology is probably one of the fastest growing areas of applied science and ceramics and composite materials will become the dominant materials technology in the 21st century. Current trends indicate that many ceramic raw materials will be very fine powders (the maximum permitted size of powder grains will be less than 2 μm in diameter) so that they will have enough surface energy to fuse in the molds used to form ceramic parts. Although such fine powders constitute optimum raw materials from an "energy of fusion" technology point of view, they will also pose a potential respirable dust hazard to those who handle the powders en route to the molding process.

Further, because of the way in which such fine powders are made, there is every reason to believe that assessing the hazard posed by ceramic powders may be more complex than those posed by the classical industrial dust hazards studied by occupational hygienists for the past 50 years. The reason for this increasing complexity of respirable hazards posed by newer ceramic raw material is the more complex shape of some ceramic fineparticles. Mine dusts, such as freshly shattered quartz dust in a gold mine, coal dust in coal mining and asbestos fibres, have relatively simple geometric shapes; powders used in the ceramic industry often have a fractal structure. This is illustrated by the two types of fineparticles shown in Figure 3.28. It can be seen that the coal fragments have a simple structure, whereas the structure of the thorium dioxide fineparticles can be described by fractal dimensions [31].

For reasons that will be discussed at the end of Chapter 5, the appearance of the thorium dioxide powders indicates that it was formed as either a fume or a precipitate, and it is useful to discuss why many modern ceramic powders will probably be

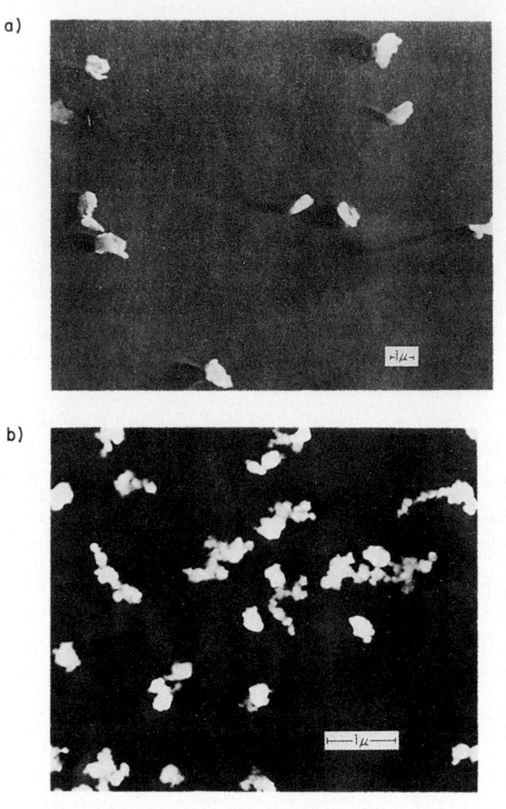

Figure 3.28. Coal dust has individual grains of relatively simple structure. Some of the newer ceramic raw materials manifest a fractal structure (photographs reproduced from reference 56).
(a) Coal dust.
(b) Thorium dioxide powder used in the fabrication of nuclear reactor fuel rods. (T. Mercer, P.E. Morrow and W. Stöber, (Editors), "Assessment of Airborne Particles," 1972. Courtesy of Charles C. Thomas, Publisher, Springfield, Illinois.)

precipitates or fumes. Scientists who have studied the breakage of solids suspect that powder grains smaller than 2 μm in diameter cannot be created by the crushing of larger solid grains, because regions in a solid material smaller than 2 μm are likely to be flawless crystals of high strength. Solid materials break up as cracks leap from flaw to flaw in the body. Much more energy is required to break regular crystals than is required to drive cracks from flaw to flaw. This is illustrated by the fact that it is relatively easy to crack diamonds with a sharp blow, but a flawless diamond is the hardest substance known to man, (cleaning diamonds in an ultrasonic bath can be dangerous because gemstone diamonds in engagement rings have been known to crack apart in such cleaning devices!).

Because it is difficult to produce powders smaller than a few microns by crushing and grinding, ceramicists (and nuclear engineers) make their powders by chemical action. One chemical action route used to prepare fine powders involves precipitation when chemical reaction in a liquid produces fine crystals. The fine primary crystals produced by precipitation usually agglomerate by turbulent collision and/or continued crystallization. Precipitated agglomerates will usually manifest a fractal structure (see Chapter 5).

A **fume** is fine smoke. The word comes from the Latin word for smoke, "fumus." To **perfume** a place originally meant to fill it with smoke. **Fumigation** means to disinfect an area with smoke (when a person uses perfume, are they disinfecting themselves?). A widely used chemical product in industry is described as **fumed silica** and we can learn a great deal about fumes from a brief study of how fumed silica is manufactured [48].

The process for making fumed silica is illustrated schematically in Figure 3.29. Silicon tetrachloride is burnt in a flame of oxygen and hydrogen and the chemical reaction indicated in the diagram takes place in the flame. In the hottest part of the flame primary spherical fineparticles of silica are produced and these cool very rapidly as soon as they leave the hottest region of the flame. The primary spheres are 0.007 – 0.014 μm in diameter. In the cooler part of the flame, these primary spheres collide to form agglomerates and, when cooled, these "fluffy" agglomerates tangle with each other to form relatively large aggregates which are of low cohesive strength and can be dispersed in subsequent processing. The primary agglomerates are very firmly fused and cannot be broken apart easily. In Figure 3.30, the detailed structures of the primary agglomerates are illustrated in a sequence of high-resolution electron micrographs. The fractal structure of these agglomerates is obvious. The relationship between the fractal structure and the collision kinetics inside the flame will be discussed later in this chapter.

The secondary random agglomeration of primary spheres or single crystals produced in the initial stages of a combustion or precipitation process is a common feature of

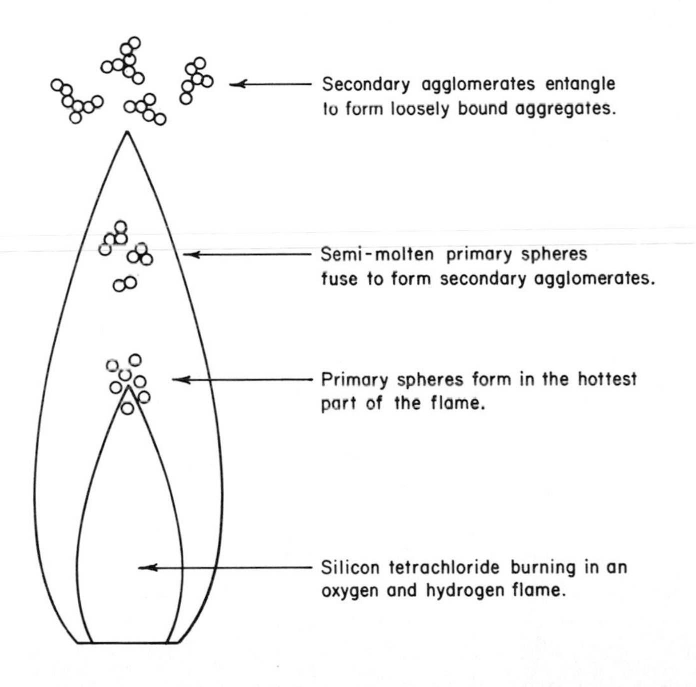

Figure 3.29. In the formation of fumed silica, primary spheres created by the combustion of silicon tetrachloride cool rapidly (in a few thousandths of a second) and form strong agglomerates as they collide with each other in the cooler part of the flame. These primary agglomerates then aggregate to form loose clusters which can be dispersed by relatively mild shear forces [48] (reproduced by permission of Cabot Corporation, CAB-O-SIL Division, Tuscola, IL).

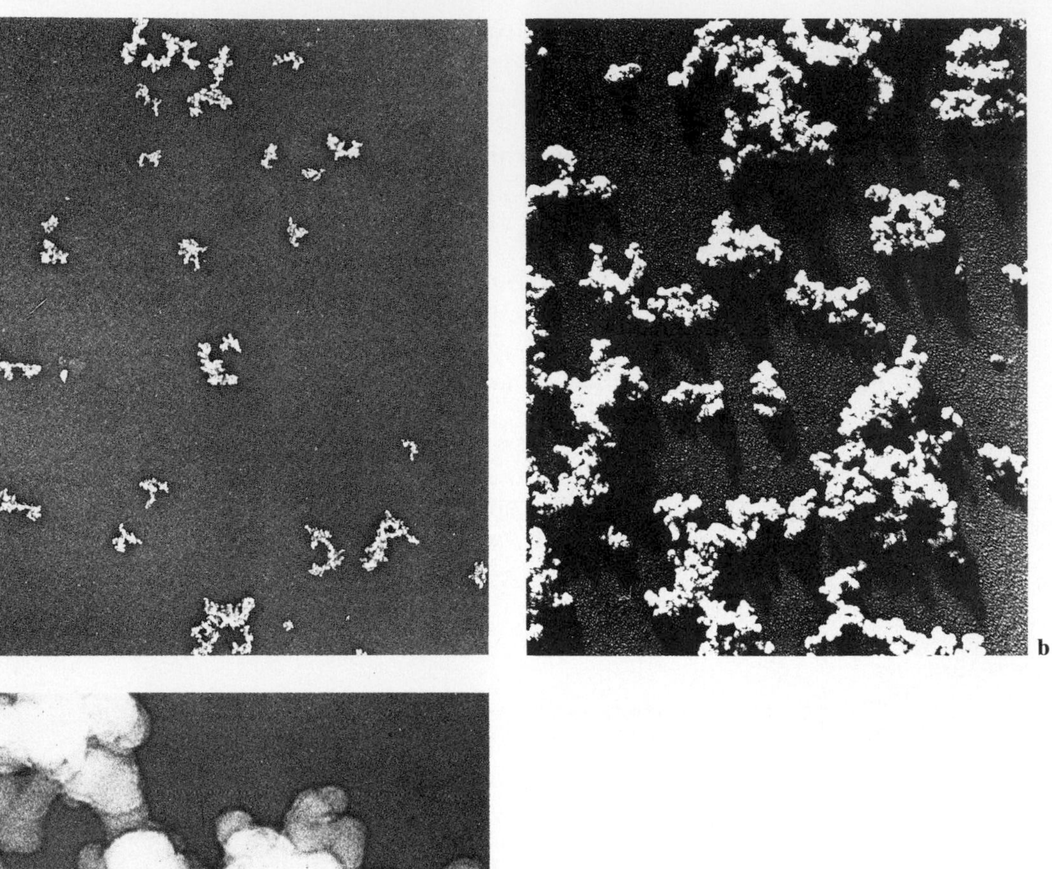

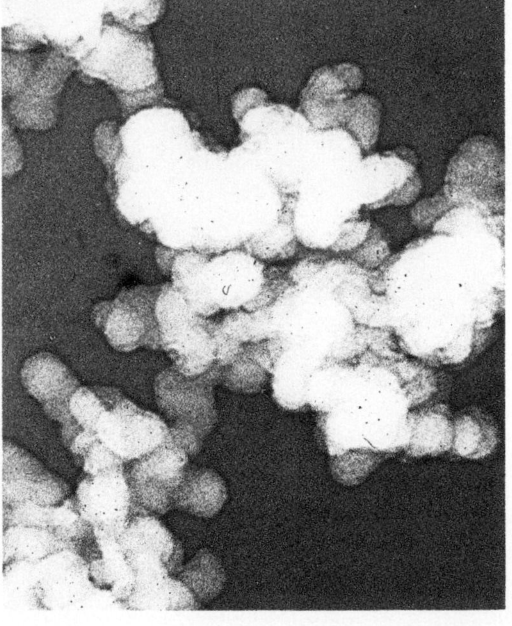

Figure 3.30. Fumed silica fineparticle agglomerates manifest a fractal structure, as shown by this set of electron micrographs [48] (reproduced by permission of Cabot Corporation, CAB-O-SIL Division, Tuscola, IL).

many production processes for generating fineparticles. Therefore, we can anticipate that products formed by such processes will often manifest fractal structure.

In order to understand why it is more difficult to assess the potential hazard to the lungs from inhalation of respirable dust of fractal structure than the dangers posed by simple dusts, such as the coal dust shown in Figure 3.28(a), it is necessary to discuss how dust invades the lungs and the problems that deposited dust can create once inside them. In Figure 3.31, a simplified sketch of the structure of a human lung is shown [52] (the fractal structure of a lung will be described in Signpost 5 in Chapter 10). In Figure 3.32, the basic hydrodynamic problems to be faced in calculating the movement of fineparticles into and out of the lung are illustrated. First of all, in the mouth and nose region of the human body, only fineparticles below approximately 200 μm in diameter can penetrate into the upper regions of the lung system. As the fineparticles move down beyond the mouth region, the passages of the lung continually divide into narrower and narrower passages with subsequent changes in the flow behaviour in the air in the narrowing passages (in Chapter 10, it will be shown that the branched structure of the lung can be described by means of a "finger fractal"). Calculating the sedimentation behaviour of a dust fineparticle in moving air involves looking at the two main forces operating on the fineparticle.

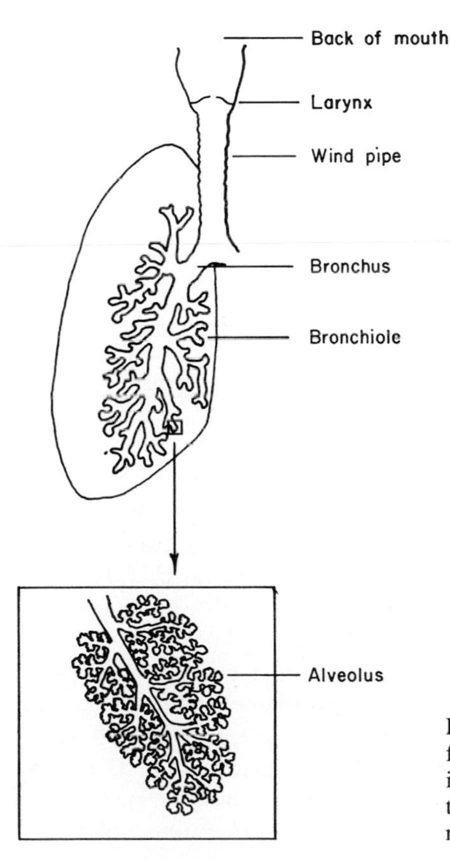

Figure 3.31. Assessing the capacity of dust fineparticles to penetrate the depths of a human lung is a problem in hydrodynamics in which one studies the movement of the dust fineparticles in the air moving in the passages of the lung [52].

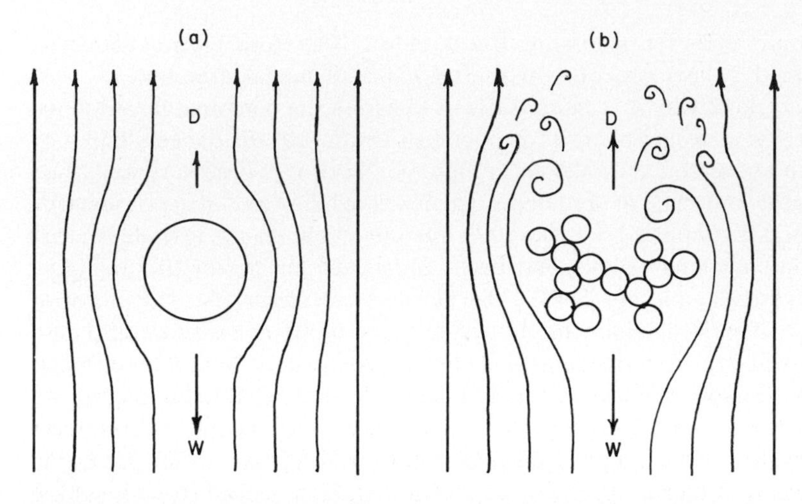

Figure 3.32. Calculating the dynamics of lung penetration for spherical dust fineparticles is not technically difficult, but predicting the dynamics of fractally structured dust in the lungs is not feasible within the current state of technology. The aerodynamic diameter of irregular structured dust fineparticles must be measured experimentally [49 – 51].
(a) Smooth air flow around a sphere.
(b) Air flow around a fractally structured object may be too complex to describe easily.
D = Drag force
W = Weight

First of all, the weight (W) of the fineparticle tends to drive it down into the lung. On the other hand, the drag of air moving past the fineparticle (D) tends to entrain it with the moving fluid. Below a certain size of fineparticle, the drag forces created by the moving air become a major factor in predicting the behaviour of the fineparticle. Mathematicians and physicists have solved the problem of dust dynamics in the lung for the penetration of a hard, smooth, single sphere 3.32(a) into the lung, but predicting the dynamic behaviour of an agglomerate 3.32(b) that looks like a bunch of grapes is a very difficult problem.

Several scientists have built special devices for studying the dynamics of irregularly shaped fineparticles in moving streams of fluid and have presented data relating to the overall size of the fineparticle to its dynamic behaviour. When reporting the dynamic behaviour of irregularly shaped fineparticles, scientists have found it useful to define a quantity known as the **aerodynamic diameter.** This quantity is the size of a smooth sphere of unit density which settles in a viscous fluid at the same velocity as the irregular dust fineparticle. The aerodynamic diameters of an irregular fineparticles can be measured directly with several commercially available instruments [50, 51, 53, 54].

Timbrell [50], who developed one of the commercially available devices for measuring aerodynamic diameter directly, devised an elegant technique for demonstrating the correspondence between the physical structure of an aerosol fineparticle and its aerodynamic diameter. He used the fact that dried droplets of **shellac**, a lacquer used to give a good finish to furniture, had a density of approximately 1.0. Therefore, spheres of shellac have physical dimensions identical with their aerodynamic diameters. Therefore, if one injects a small number of shellac spheres into a cloud of aerosol fineparticles to be fractionated in the aerosol spectrometer, the shellac spheres deposited with the dust fineparticles of a given aerodynamic diameter give a direct impression of the difference between physical magnitude and aerodynamic diameter for any particular dust fineparticle. Thus, in Figure 3.33, three sets of fractionated industrial dusts are shown together with the comparable shellac spheres. Timbrell used such data to demonstrate that for long, straight fibres their diameter was the dominating factor to be used in predicting their aerodynamic behaviour.

When studying dusts of relatively simple structure, it is usually stated that non-toxic dusts having aerodynamic diameters larger than 5 μm do not pose a special threat to the lungs. This is because such larger dust fineparticles are deposited on the linings of the air passageways of the lung which are coated with mucus and fine hair-like cells known as **cilia**. Dust deposited on the cilia is carried up to the outer regions of the lung by the rhythmic movement of the cilia in much the same way as an injured person is passed over the heads of a crowd on their arms. It is suspected that one of the reasons why industrial workers who smoke are much more prone to lung diseases than non-smokers, is that the nicotine in the tobacco smoke paralyses the cilia on the walls of the lung, preventing them from clearing the lung passages of larger dust fineparticles. Thus, the nicotine neutralizes the first line of defence of the lung against inhaled dust. The tiny air sacs at the end of the branching passageways of the lung do not have the cilia mucus protective system and therefore dust that penetrates into the air sacs can no longer be removed by the primary defences of the lung. Experimental work has shown that chunky dust fineparticles smaller than 5 μm in aerodynamic diameter penetrate to the air sacs. The air sacs, or **alveoli** as they are described by the medical specialist, are several hundred microns across so that the danger from the respirable dust does not come from physical blocking of the air sacs by the inhaled dust.

In industrial hygiene, **respirable dust** is defined as dust having aerodynamic diameters less than 5 μm. When the dust is finally deposited in the air sacs, there is much evidence to indicate that the surface of the dust fineparticle is an important property. For example, one of the reasons why a combination of asbestos dust and cigarette smoke appears to be particularly deadly is that the cancer-causing chemicals (carcinogens) present in the cigarette smoke are adsorbed in a special way on the surface of the asbestos fibres. Activated chemically by their oriented adsorption on the asbestos, the carcinogens attack the lung much more vigorously than would be the case if their molecules were suspended in the air in the absence of any dust fineparticle.

It is known that the shape of the fineparticles is involved in the deposition kinetics of the dust moving through the airways of the lung. We shall call the probability of lodging on the walls of the lung, the **lodgability factor.**

For a simple geometric shape such as a fibre or relatively chunky dust fineparticle, the aerodynamic diameter, the lodgability factor and the surface reactivity are all related in a relatively simple manner to the observed dimensions of the dust fineparticle. However, when we come to consider a respirable dust such as those present in a fumed smoke such as diesel exhaust soot, there is no simple way to relate the lodgability and surface reactivity of the dust to the aerodynamic diameter. For such fineparticles, it is

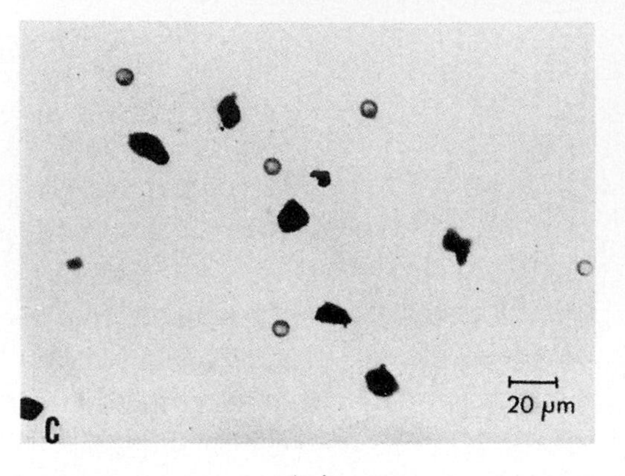

(a)

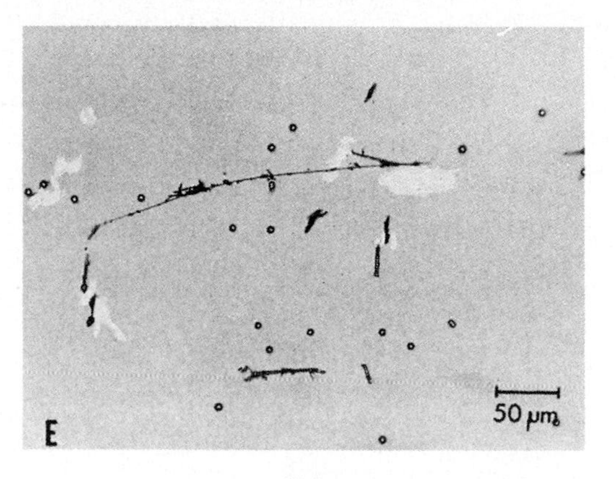

(b)

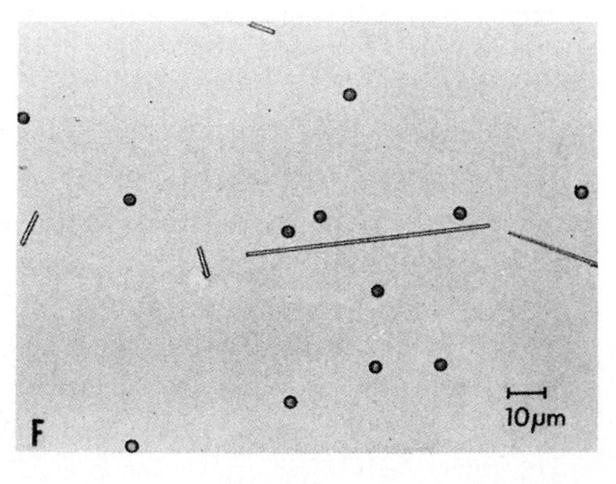

(c)

Figure 3.33. Shellac spheres have physical diameters identical with their aerodynamic diameters and can be used to illustrate the aerodynamic diameter of fractionated dust in a direct manner [54].
(a) Coal dust fineparticles all of the same aerodynamic diameter.
(b) Amosite asbestos fibres having the same aerodynamic diameters.
(c) Glass fibres with the same aerodynamic diameters.
In all sets of dust fineparticles the spherical fineparticles are shellac spheres. (T. Mercer, P.E. Morrow and W. Stober, (Editors), "Assessment of Airborne Particles," 1972. Courtesy of Charles C. Thomas, Publisher, Springfield, Illinois.)

necessary to carry out experimental studies to examine the capture rates of open-structured fumes, such as the soot of diesel exhaust and other special dusts (see Figures 3.39 and 3.40). It seems probable that the lodgability factor, linking deposition rates to the structure of dusts exhibiting fractal structure, can probably be related to the fractal structure of the agglomerate. At Laurentian University, we are currently building models of fractally structured agglomerates and measuring their behaviour in a viscous fluid. We hope to be able to link the fractal dimension of the agglomerated structure to the physical behaviour of the agglomerate in a viscous fluid [49].

When a fractally agglomerated dust fineparticle is deposited on the surface of the lung, it is probable that the fractal structure of the dust fineparticle can be related to the potential for lung damage inherent in the dust. In the next few paragraphs, we shall briefly explore the potential uses of fractal logic for describing specific dust hazards. We shall discover that in some situations the fractal structure indicates increased respirable hazards compared with simple dusts. In other situations, the fact that a dust grain has a fractal structure may indicate a reduced hazard for lung penetration and hazard for that particular substance in comparison with similar chemically constituted grains but which have different physical structure.

3.7.2 Is Fumed Silica a Respirable Hazard?

Perhaps one of the best known industrial diseases which is associated with the inhalation of respirable silica dust is **silicosis**. In the early days of the development of occupational hygiene, many of the original workers were chemists. They tended to focus on the chemical constitution of a respirable dust, looking for chemical reactions of the dust with the lung as the initiator of the lung disease. This is unfortunate, since in many situations it is the physical state of the chemical compound which is responsible for the hazard posed by the respirable dust. In the case of silica, it is most certainly the physical state and dimensions of the respirable dust which initiate disease problems. Most beach sands are silica grains, but playing on the beach does not expose the individual to an increased risk of silicosis.

If one takes a close look at the regulations governing the hazard of silica to industrial workers, one discovers that it is fully recognized that the industrial dust danger comes from inhaling freshly shattered crystalline quartz and two other crystalline forms of silicon dioxide. Other forms of silica fineparticles, although meeting the definition of respirable dust, do not constitute a hazard to the lung. In 1984 we had a project at Laurentian University funded by the Canadian government, entitled "Hotline on Science and Society." The project made it possible for people to call into the University to ask questions about scientific aspects of everyday life that either interested them or were a source of concern. One of the questions that we received during this program concerned the danger of silicosis from the inhalation of fine powder generated when pouring artificial coffee creamer powder and/or sugar substitute into a cup of coffee or tea. The enquirer had noticed that the label of contents of these items often indicates that the powder contains up to 1% by weight of silicon dioxide. Furthermore, the enquirer had noticed that a puff of powder escapes into the air when one pours the ingredients out of such a packet into a drink, and that one obviously inhales the powder because one can taste "sweetness" in the air. I was able to assure the enquirer that in this case there was no danger of silicosis, since the powder in question was what is known as **amorphous silica**, the type of silica powder shown in Figure 3.30.

Fumed silica powders are often added to food products and pharmaceutical powders to improve their flow properties. The high surface area of the powder mops up the moisture which may be present when the product is packaged, and also prevents the caking of the powder under the pressure that it experiences during storage and transportation. The silica powder prevents caking under pressure because, amongst other things, the tiny agglomerates increase the friction between the food powder grains. Fumed silica of this type is described in the food industry and in pharmaceutical technology as a **flow agent.** Because the tiny spheres making up the agglomerates of fumed silica are formed extremely rapidly in the flame process as they move out of the hot region of the flame into the cooler regions, the individual spheres are not crystals. They cool so rapidly that they consolidate to a glassy state similar to that of ordinary window glass. This type of structure, which is the opposite of crystalline structure, is known as **amorphous structure**. In Greek mythology, Morpheus was the god of dreams, who created fantastic shapes in the dreams of mortals. Early in the history of chemistry, the term "morpheus" (which has given us the name of the drug morphine, a drug that creates hallucinations in the person taking the drug), was used to create a term to describe the scientific study of shape. This term is **morphology. Amorphous** in Greek means "without shape." When molecules are in an amorphous body, the intermolecular bonds are not as strong as when they are in a crystal structure and therefore less energy is required to break up the bonds holding the molecules in the solid. This in turn means that substances are more soluble in an amorphous form than in a crystalline form. Because of this fact, the human body can probably dissolve amorphous silica used as a flow agent in food powders. Furthermore, the high surface area of the open-structured agglomerate probably makes it easier for the body to break down the substance. In North America, government regulations permit up to 1% by weight of fumed silica in food and pharmaceutical systems [55].

3.7.3 Dust from Nuclear Reactor Systems

In Figure 3.34, two different types of uranium dioxide dust are shown. The fineparticles shown in Figure 3.34(a) were photographed by Kotrappa. He took uranium dioxide powder as provided by the manufacturer and passed it twice through a high-speed pulverizing mill. He then dried the powder and dispersed it into a test chamber using a device known as a Wright dust feeder. He sampled this dust using an instrument that deposited the fineparticles with the same aerodynamic diameter on the same part of a strip of metal used to collect the fineparticles [56].

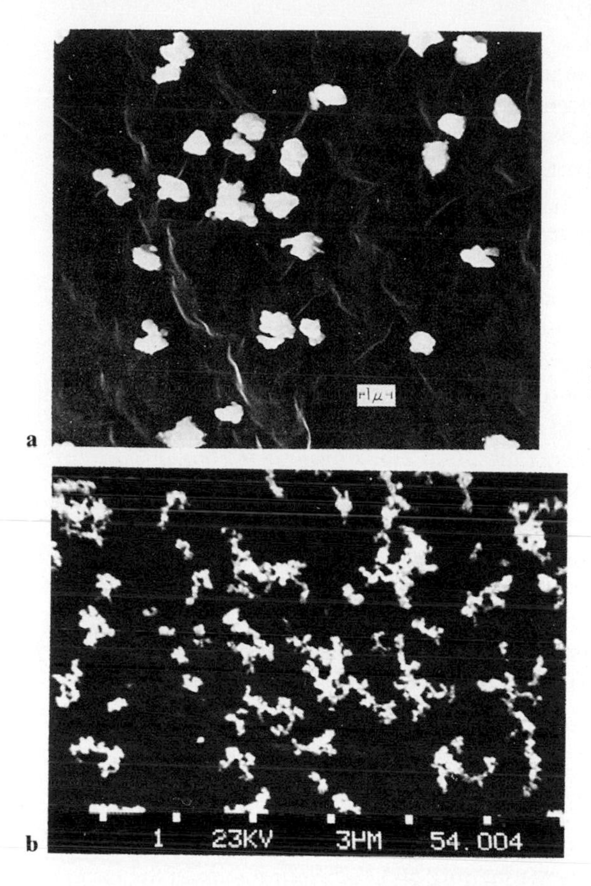

Figure 3.34. Two different samples of uranium dioxide dust.
(a) Uranium dioxide subjected to processing by an attrition mill and described by Kotrappa [56]. (T. Mercer, P.E. Morrow and W. Stöber, (Editors), "Assessment of Airborne Particles," 1972. Courtesy of Charles C. Thomas Publisher, Springfield, Illnois.)
(b) "Meltdown fume" studied by Zeller [57] (from W. Zeller, "Direct Measurement of Aerosol Shape Factors," *Aerosol Science and Technology*. Reproduced by permission of W. Zeller).

The dust fineparticles in Figure 3.34(a) all have the same aerodynamic diameter of 2.5 μm. In the case of uranium dioxide fineparticles which have a high density, it can be seen from the scale bar shown in Figure 3.34(a) that the aerodynamic diameter is much larger than the physical diameter for dense chunky fineparticles. It can also be seen that several of the fineparticles look as if they started life similar to the thorium dioxide powders in Figure 3.28(b), but that the attrition mill shattered the fluffier fineparticles to leave chunky residues. Thus, several of the uranium dioxide profiles in Figure 3.34(a) show rudimentary fractal structure, but do not have fractal dimensions anywhere close to those of the thorium dioxide grains in Figure 3.28(b). It should be noticed that all of the fineparticles in Figure 3.28(b) have the same aerodynamic diameter of 1.03 μm. For chain-like structures such as in Figure 3.28(b), the width of the chain-like aggregate is about the same size as the diameters of the dense chunky profiles. This is similar to the finding of Timbrell [54] that for a fibrous fineparticle the diameter of the fibre is approximately equal to the diameter of a chunky fineparticle having the same aerodynamic diameter. Note, however, that for fineparticles showing pronounced fractal structure, their physical size is much greater than that of chunky fineparticles having the same aerodynamic diameter. This is an important problem when determining how dangerous a uranium dioxide powder is to the person who might inhale a fineparticle which becomes lodged in his lung. A large, fractally structured agglomerate represents much more uranium dioxide than a chunky fineparticle of the same aerodynamic diameter. Therefore, knowing the number of fineparticles of a given aerodynamic diameter inhaled does not lead to a realistic estimate of the amount of radioactive material inhaled if the dust is fractally structured.

One of the public concerns over the potential health hazards of nuclear reactors is that a failure of a nuclear reactor involving a "**meltdown**" may spread radioactive dust over the areas around the nuclear reactor. In a meltdown reactor accident, it is assumed that in some way the cooling system of the reactor fails and the central part of the reactor melts. Such a melting could be accompanied by the fuming of the metals to produce a dangerous toxic dust. Zeller has studied the type of fume that could be given off from a melting nuclear reactor fuel rod. The fineparticles shown in Figure 3.34(b) were produced by Dr. Zeller in a study of the evaporation and subsequent condensation of uranium dioxide fumes produced by melting a uranium dioxide pellet in an electric arc [57].

The fractal structure of the fumed fineparticles is obvious. The fineparticles bear a striking similarity to the fumed silica fineparticles in Figure 3.30(b). We can see from Figure 3.34 that any quantitative study of the respirable dust health hazards and the aerodynamics of the dispersion of a dust cloud created by fuming nuclear reactors will involve fractal dimension characterization of the fumes.

3.7.4 Fuse Fumes and Welding Dust

Some types of electrical fuses work by the melting of a connecting wire in the circuit when the current burden of the circuit is too high for safety. Anyone who has been close to such a fuse when it has "blown" will remember the acrid puff of smoke coming from the wire. Zeller has studied the type of fineparticles generated by the explosive melting of a wire. In Figure 3.35 some platinum oxide fumes generated by the evaporation of a platinum wire are shown. The fractal structure of these fume fineparticles is obvious.

It is probable that any metal-working process in which metal is cut or welded using a hot flame generates fractally structured fumes. Thus, in Figure 3.36 fineparticles collected during a welding operation are shown [60]. There are probably two or three types of fumed fineparticles present in this welding dust. It is probable that a better understanding of the health hazard presented by welding dust requires a fractal characterization of the structure of the dust.

Welders exposed to the inhalation of fumes suffer from an illness known as "**welders fever**." This could be caused by very high surface area, open-structured, amorphous metal oxide dusts, which dissolve quickly in the fluids of the body. The body's fight against such dissolved fumes could give rise to the fever observed in those who work with the fumes. Artists who make stained glass windows could also suffer health problems from inhaled metallic fumes that condense out of the vapour created by their fabrication techniques.

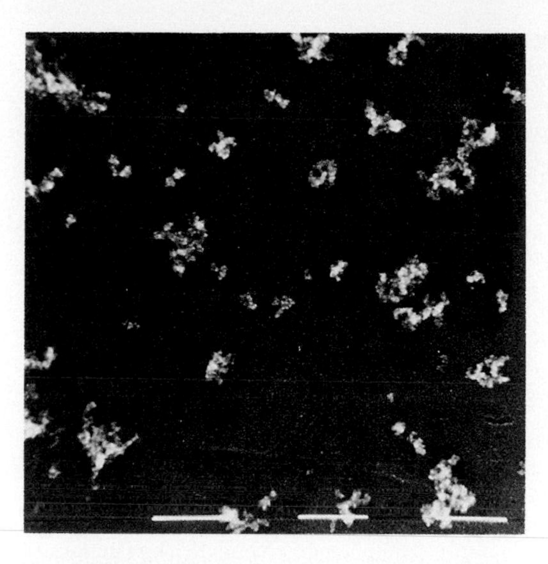

Figure 3.35. Exploding metal wires (electrical fuses) and welding fumes exhibit fractal structure characteristic of smoke-formed agglomerates such as the platinum oxide fumes shown above (magnification x 40 000) [57].

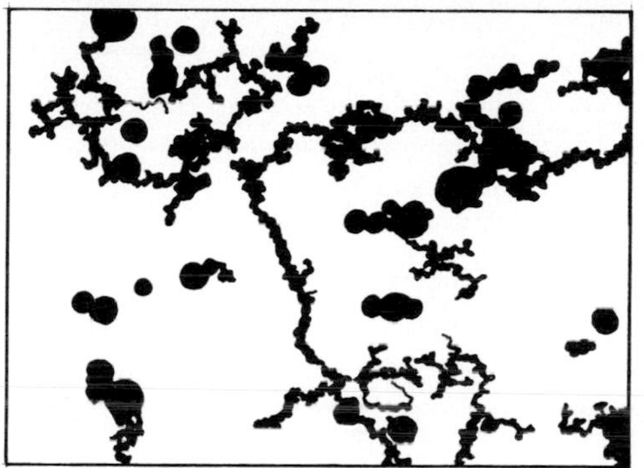

Figure 3.36. Welding dust. The quenched evaporated metal generated during the act of welding produces fractally structured fumes. Combustion products from the welding arc also appears to produce fractally structured soot fineparticles intertwined with the metal fumes in this electron micrograph reported by Stern [60]. (Richard M. Stern (Unpublished) [60].)

3.7.5 Characteristics of Dust Generated by Explosions

There is a great deal of interest in the type of dust generated by explosions from the point of view that fineparticles generated in this way constitutes some of the dust formed during underground mining. Military authorities must do their best to protect soldiers who have to move through clouds of dust and smoke generated by explosions. Also, the type of dust encountered in off-road areas must be studied by military experts so that they know how to protect engines from potentially dangerous dust stirred up by the movement of vehicles and smoke from detonating shells.

In Figure 3.37, some dust created by the detonation of an explosive charge is shown [58]. It can be seen that there are two types of fractal structures present in this dust. First there are obviously some rugged structures produced by fragmentation of objects, and second, there appear to be agglomerates formed by the collision of primary fragments.

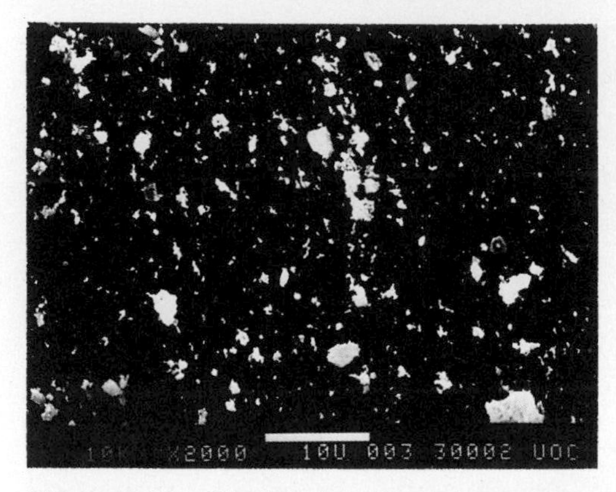

Figure 3.37. Pinnick dust from detonation. Military vehicles must move through smoke generated by the detonation of explosives. This dust can be drawn into the engine of a tank and cause catastrophic failure. The same type of dust is presumably generated when blasting ore bodies in a mine. In this scanning electron micrograph of dust produced by detonating high explosive, components of the dust manifest fractal structure (photograph provided by Pinnick and co-workers and used with permission of Dr. Pinnick).

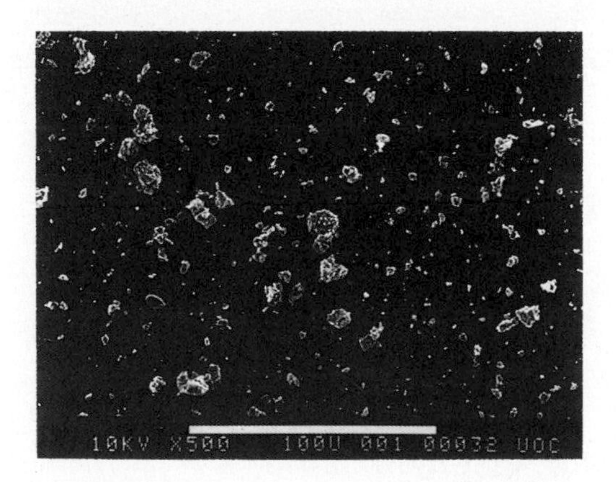

Figure 3.38. Road dust from a location in New Mexico (unpaved road) exhibits a fractal structure (photograph provided by Pinnick and co-workers and used with permission of Dr. Pinnick).

Obviously, understanding the dynamics and potential dangers of such dust will involve an assessment of the fractal structure of the individual fineparticles.

The dust generated by trucks driving across open land rather than paved surfaces is a familiar sight in such situations as diverse as tanks moving across the sand of a desert or heavy tractors moving across dried-out fields. One of the problems of operating helicopters safely is the protection of the engines from the dust thrown up by the downthrust of the blades when the helicopter lands. Obviously, the type of dust generated in any given situation depends upon the fineparticles to be found on the surface over which the vehicle is moving. In Figure 3.38, some dust created by the movement of heavy trucks over unpaved roads in New Mexico is shown. Again, fractal structure is evident [58]. Air drawn into the engines of vehicles and helicopters which must operate on dusty, off road conditions must have the potentially dangerous fineparticles removed before it enters the combustion chamber. The efficient design of appropriate filters and/or cyclones must take into account the fractal structure of the **off road dust** [59].

3.7.6 Diesel Soot and Fumed Pigments

In recent years, there has been increasing concern over the health hazard posed by the soot in diesel engine exhaust fumes [60]. It has been widely recognized that incomplete combustion of the diesel oil gives rise to many carcinogenic chemicals. What has not received equal recognition is the fact that the potential danger from such chemicals is enhanced when they are adsorbed on the soot fineparticles which are subsequently breathed in and become lodged in the lung system. In Figure 3.39, a high-magnification scanning electron micrograph of diesel soot produced by the free burning of diesel oil is shown. It is clear that the soot forms in essentially the same way as that illustrated for the formation of fumed silica in Figure 3.30. Unburned carbon leaving the hottest part of the flame condenses to form spheres (it has been reported that some of the spheres are hollow) [61]. These spheres then agglomerate to form the soot fineparticle. The way in which the soot grows leads to an obvious fractal structure as illustrated in Figure 3.39.

Different burning conditions can alter the structure of the agglomerate, making it either chain-like or grape clustered, but the basic structure remains fractal; only the magnitude of the fractal dimension and the gross shape of the agglomerate changes.

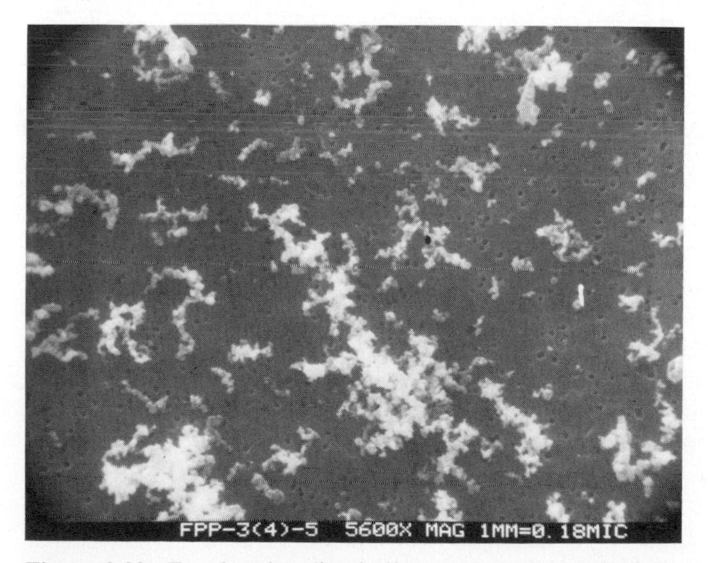

Figure 3.39. Free-burning diesel oil generates a soot which manifests a fractal structure [58].

The highly open fractal structure of the diesel soot means that it has a high surface area for adsorbing the carcinogenic chemicals. Therefore, a simple knowledge of the aerodynamic diameter of the diesel agglomerate would give no indication of the amount of the adsorbed chemicals present on the surface of the agglomerate unless the physical dimensions and fractal dimension of the agglomerate are known. Solving problems concerning the dynamics of the diesel exhaust soot in the human lung will also require a knowledge of the dynamics of fractally structured agglomerates.

I first suggested that the structure of diesel exhaust soot could be described by means of fractal dimensions in 1981. Thus, in Figure 3.40 the Richardson plot for a diesel exhaust fineparticle is shown. The actual diesel exhaust was originally described by Kittelson and the fractal measurements were presented in a review paper published in 1981. In Figure 3.40(b), two other diesel exhaust agglomerates photographed by Kittelson are shown. An interesting feature of this photograph is that the shadows of the agglomerate created by vapour shadowing are clearly delineated. Perhaps the fractal structure of the shadow compared with the fractal structure of the projected agglomerate might give some information on the three dimensional structure of the agglomerate (the possible significance of the fractal structure exhibited by diesel exhaust fumes is discussed in Chapter 5).

Another respirable dust hazard closely related to the problem of diesel exhaust soot is that posed by fineparticles in automobile exhausts. The soot is particularly dangerous when burning leaded gasoline. Tiny lead fineparticles in the exhaust soot gain ready access to the human body through the respiratory tract. Recent studies have shown that household dust is rich in potentially dangerous leaded dust from automobile exhausts if the houses are near to a major arterial highway.

In Figure 3.41, a fineparticle captured from automobile exhaust which contains lead compounds is shown. In the original photograph, the constituent spheres forming the soot agglomerate can be seen clearly. The magnification of the original photograph was of the order of 300 000. The photograph of this exhaust fineparticle was taken by Dr. Cheng of the University of New York [64].

The open structure of soot produced by burning oil is an advantage when it comes to looking for military security behind a smoke screen. The more open-structured, that is the higher the fractal dimension of the soot fineparticles, the more opaque is the soot cloud and the more persistent is the cloud, since dense agglomerates will settle faster than open-structured agglomerates. Therefore, from a military point of view, a smoke screen should consist of individual fineparticles of maximum fractal dimension that can be created by the combustion process.

Another area of technology where maximum fractal structure in fineparticles is an advantage is in the preparation of white pigments. The more open the structure of the individual fineparticles, the higher is the scattering power of an individual pigment fineparticle. Therefore, for a given size of pigment fineparticle, the higher the fractal dimension, the greater is the whiteness power of the pigment. It is not surprising that some of the best pigments, such as titanium dioxide, are made by a fuming process. The fractal structure of freshly prepared titanium dioxide is illustrated in Figure 3.42. The size of the individual components in the agglomerates and the fractal structure of the agglomerates can be changed by heat treatment [65]. The paint industry has focused

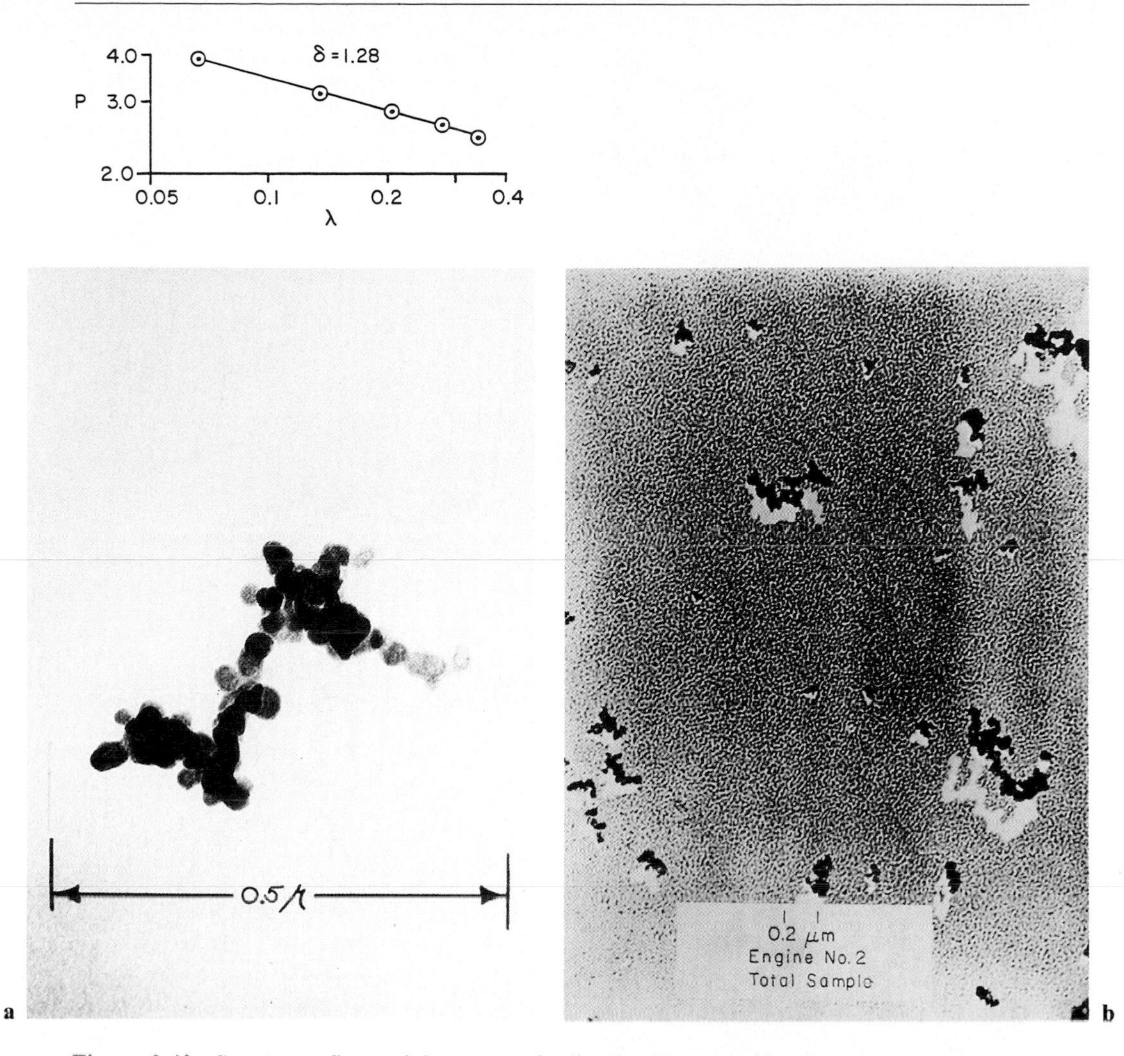

Figure 3.40. Some soot fineparticles present in diesel exhaust manifest fractal structure.
(a) Richardson plot for an exploration of a chain like soot agglomerate originally photographed by Kittelson [62, 63]. (Kittelson, D.B. and D.F. Dolan, "Diesel Exhaust Aerosols," presented at the Symposium on **Aerosol Generation and Exposure Facilities,** Honolulu, Hawaii, 1979, Generation of Aerosols and Facilities for Exposure Experiments, Ed.: K. Willeke, Ann Arbor Science Press (1980) pp. 337-359. Reproduced by permission of D.B. Kittelson.)
(b) Shadow cast scanning electron microscope pictures of diesel exhausts photographed by Kittelson [63]. (Dolan, D.F., D.B. Kittelson and D. Pui, "Diesel Exhaust Particle Size Distribution Measurement Technique," SAE Paper No. 800187 (1980). Reproduced by permission of D.B. Kittelson.)

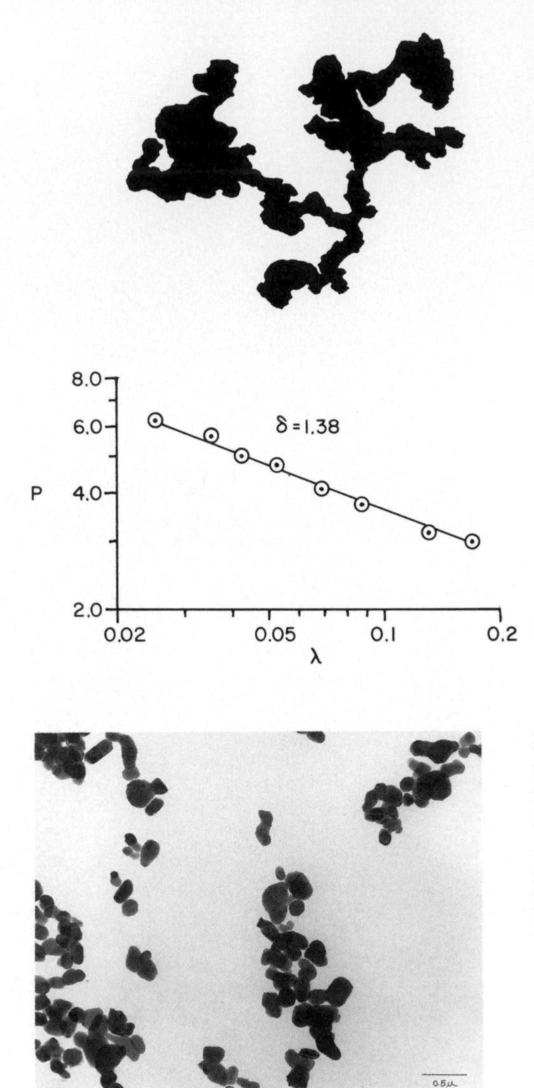

Figure 3.41. Lead containing exhaust fineparticles from an automobile effluent manifests the fractal structure evaluated by Leblanc [31]. Reproduced by permission of J. Leblanc.

Figure 3.42. Optical pigments such as titanium dioxide, used as a white pigment in many different systems ranging from toilet soap to household paint, are manufactured by a fuming process. The resulting pigment manifests fractal structure [65]. (Reproduced by permission of T.I. Brownbridge, Kerr-McGee Corporation, Oklahoma City, OK.)

attention on maximizing pigment opacity by reducing the size of the pigment; perhaps they need to pay more attention to changes in manufacturing technology which will optimize the fractal structure.

A theoretical study of the light scattering properties of fractally structured pigments would be extremely complicated. Therefore, it is likely that the first steps towards understanding the relationship between the fractal dimension of a pigment fineparticle and its light-scattering properties will be an experimental study of the optical properties of pigments fumed in different ways to generate pigments of different fractal structure.

3.7.7 Fractal Specimens of Flyash

When coal is burned to provide energy, some of the ash present in the original piece of coal escapes from the stack of the power station as a very fine material which is known as **flyash**. The structure of flyash varies considerably from one power station to another. The structure of the flyash depends on the burning temperature, the type of coal used in the power station and the pretreatment of the coal before it is sent to the burning chamber. If the burning process is not efficient, some of the material leaving the power station stack represents unburned fuel. Some of the impurities in the original coal are actually small pieces of sand trapped in the decaying organic material which turns into coal when buried in the earth for aeons of time. At the temperatures inside the furnace of the power station, some of this ancient sand is melted to form tiny glass spheres.

Many flyash fineparticles have a fractal structure formed from agglomerated glass spheres. The utility and/or health hazard of any particulate flyash may depend upon that fractal structure.

a

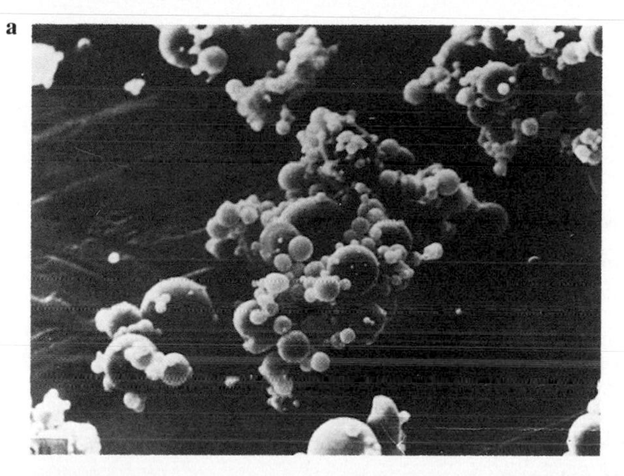

b

Figure 3.43. Some flyash fineparticles manifest fractal structure.
(a) Flyash from an electrical power station burning pulverized coal in a conventional furnace. The flyash agglomerate is made of agglomerate glassy spheres [66]. ("The Measurement of Particle Size of Pulverized Fuel Ashes" by J.G. Cabrera and C.J. Hopkins, Edited by N. Stanley-Wood from The Particle Size Analysis Conference, Loughborough University of Technology, 1983. Reprinted by permission of John Wiley & Sons, Ltd.)
(b) Fisher et al. showed that some flyash spheres are hollow spheres which in turn are filled with spheres forming a cascading structure of miniature spheres with a theoretically infinite surface area within any one sphere. These spheres have been termed plethospheres [67].

The electrostatic properties of the flyash determine the efficiency with which they can be captured by electrostatic precipitators. (The basic concept of an electrostatic precipitator is discussed in Chapter 4). The fractal structure of the flyash will be definitely related to the electrical properties of the fineparticle of flyash. It may be that a study of the fractal structure of some types of flyash may be important in developing a better understanding of the functioning of electrostatic precipitators. Also, as will be discussed at greater length in Chapter 4, the fractal structure of an agglomerate formed in the fuming or combustion process may give important hints as to the generative forces at work in the forming of the agglomerate.

Various workers have published photographs of different types of flyash which have obvious fractal structure. In Figure 3.43(a), a fractally structured agglomerate of glassy spheres present in flyash studied by Cabrera and Hopkins [66] is shown. A particularly fascinating piece of flyash, photographed by Fisher et al. [67] is shown in Figure 3.43(b). They were able to show that many of the spheres in flyash were hollow spheres which were filled with hollow spheres, ad infinitum. Mandelbrot has commented on this particular type of system, which approximates to the behaviour of an ideal fractal structure in three dimensional space. A special name, **plethosphere**, has been given to this type of flyash fineparticle, an example of which is shown in Figure 3.43(b). The name comes from the Greek word meaning "full," thus indicating that plethospheres are full of spheres which are full of spheres, and so on ad infinitum [67, 69].

Obviously, predicting the dynamics of the different types of flyash shown in Figure 3.43 in the lung is a difficult problem. Fractal evaluation of the flyash structures will obviously help in the elucidation of their physical properties.

Many experiments have been made to develop a fuel system for using pulverized coal which can be used in existing oil burning power stations. In one system, finely

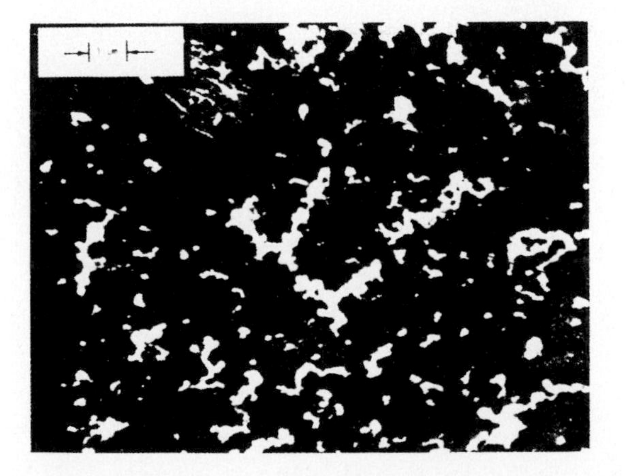

Figure 3.44. Flyash generated by the combustion of coal oil slurry generates flyash manifesting fractal structure. Photograph taken by Kirschner et al. [70] and reproduced with permission of Dr. Kirschner. (Reprinted by permission of Elsevier Science Publishing Company from "Toxicologic and Physiochemical Characterization of High Temperature Emissions." By F.R. Kirchner, P.F. Dunn and C.B. Reed, Aerosol Science and Technology, Vol. 2, 1983, pp. 389-400. Copyright 1983 by Elsevier Science.)

pulverized coal is mixed with oil to form a slurry which can be sprayed into oil-burning combustion chambers. When the coal is finely pulverized, many of the impurities which cause air pollution problems in the combustion of the untreated coal can be removed before mixing it with the oil.

The cleaned, pulverized coal is smaller than coal used in the existing pulverized fuel power stations. The coal oil slurry can be burned in existing oil burning power stations with only relatively small modifications of the jets feeding the fuel into the burning chamber. There is some concern that this type of fuel may produce a different type of flyash to that produced by conventional pulverized fuel stations. In Figure 3.44, some

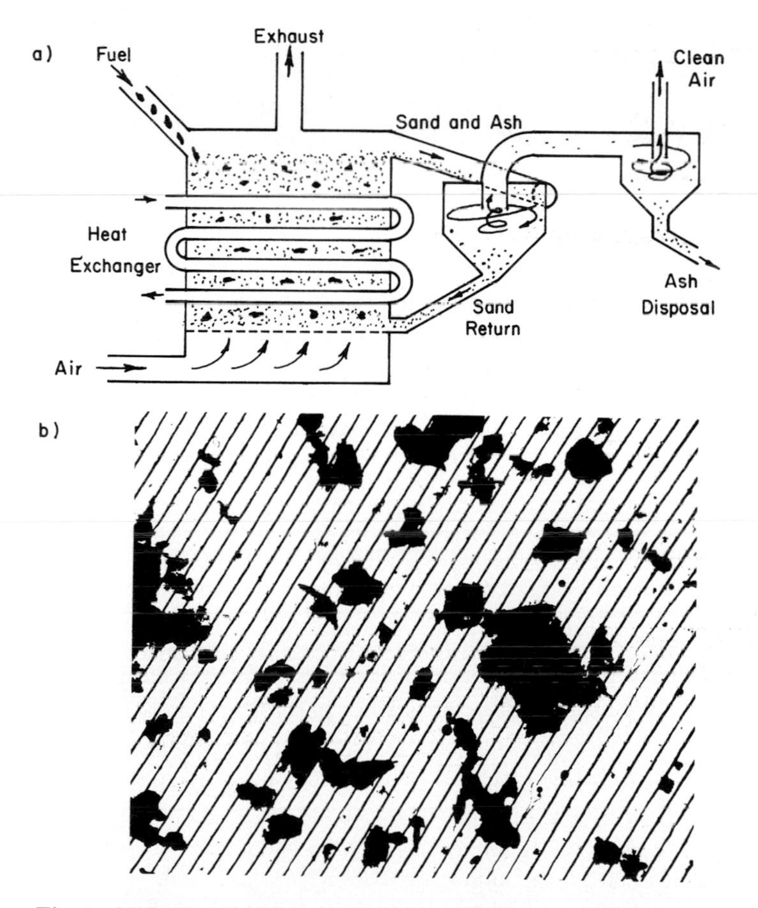

Figure 3.45. Fluidized bed combustion of pulverized coal results in a different type of flyash to that produced by high-temperature combustion in conventional pulverized fuel systems.
(a) Basic system of a fluidized bed combustion system for burning pulverized coal.
(b) Flyash generated by the combustion of coal in a fluidized bed system. (Reprinted by permission of the Elsevier Sciencc Publishing Company, from "Size Distribution of Fineparticle Emissions from a Steam Plant with a Fluidized Bed Coal Combuster" by Y.S. Cheng, R.L. Carpenter, E.B. Barr and C.H. Hobbs., Aerosol Science and Technology, Vol. 4, 1985, pp. 175-189. Copyright 1985 by Elsevier Science Publishing Inc.)

flyash collected from the combustion of a coal slurry, photographed by Kirchner et al. [69, 70], is shown. The fractal structure of this type of flyash is obvious. The scale bar illustrates that this dust is well within the respirable hazard range.

Fluidized bed combustion is a new technique for burning pulverized coal. In this combustion system, the coal to be burned is added to a bed of sand, or other suitable powder, which is made to behave like a fluid by passing air up through a porous metal plate system forming the base of the combustion chamber. A fluidized bed combustion system is shown in Figure 3.45(a). The sand is preheated with a propane burner so that when the pulverized fuel is dumped into the fluidized bed, the hot sand ignites the coal and the constant movement of the sand ensures an efficient burning rate. The tubes containing the water to be turned into steam by the heat generated by the burning of the coal are placed directly in the fluidized bed. Because heat transfer from the sand-burning coal system is more efficient than in a traditional burning system, one can operate the burning system below the glassification temperature of the ash in the coal.

Furthermore, if one adds limestone powder to the fluidized bed, it can react directly with the sulphur in the coal to produce a sulphate which can later be removed from the powder bed to be used as fertilizer. Thus, instead of pollution problems from emitted sulphur compounds, one obtains a useful by-product from the combustion process. However, because the combustion temperature in a fluidized bed system is kept below the melting point of the flyash, scientists are not too sure that all the potentially noxious compounds given off by the burning coal are destroyed by the heat of the furnace. Furthermore, the flyash will have a different structure from that made by melting the ash during the combustion process. The type of flyash generated by the combustion of pulverized coal in a fluidized bed system has been studied by Cheng et al. [71]. In Figure 3.45(b), the type of flyash generated during the combustion process studied by Cheng et al. is shown. It can be seen that the flyash differs in structure from that formed in a combustion process operating above the melting point of the ash, but that the ash still demonstrates a fractal structure. It may be that the difference in the fractal structure of flyash from different combustion systems may give a quantitative indication of the structural differences of the different flyash systems.

3.8 Polymer Grains and Rubber Crumbs

In Figure 3.46, the profile of a poly vinyl chloride (PVC) powder grain is shown. This grain has been produced by a polymerization process. It can be seen that it manifests the fractal structure over a range of inspections summarized by the dataline on the Richardson plot shown. PVC powders are manufactured in large tonnages, and variations in the polymerization process are probably manifest as changes in fractal structure of the resultant agglomerated PVC fineparticle.

In Figure 3.47, the outline of a rubber crumb produced by a process designed to facilitate the recycling of rubber is shown [72]. The fractal dimensions of such rubber crumbs may be useful in describing the dry powder flow of the rubber crumbs. They may also be related to the compressive strain which could be sustained by an assembly of crumbs when under pressure.

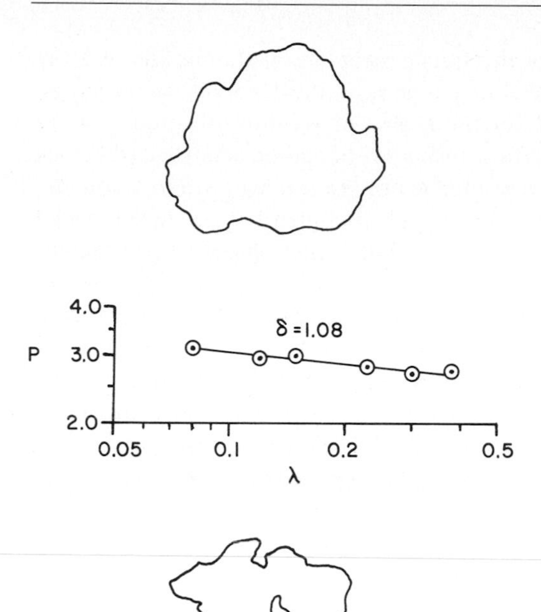

Figure 3.46. Some grains of PVC powder manufactured by a polymerization process manifest fractal structures (photograph of PVC grain provided by J. Davidson of B.F. Goodrich and reproduced with permission of Dr. Davidson).

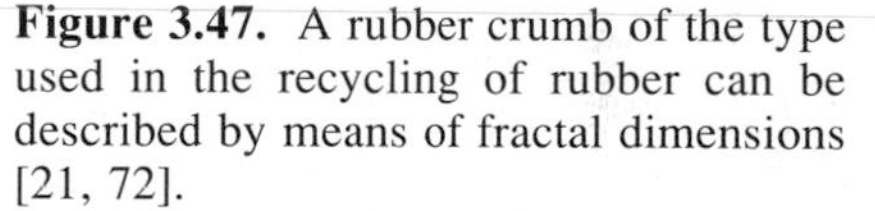

Figure 3.47. A rubber crumb of the type used in the recycling of rubber can be described by means of fractal dimensions [21, 72].
P = estimated perimeter, λ = stride length, P and λ normalized with respect to maximum Feret's diameter. $\delta = 1.09$ for range of investigation shown.

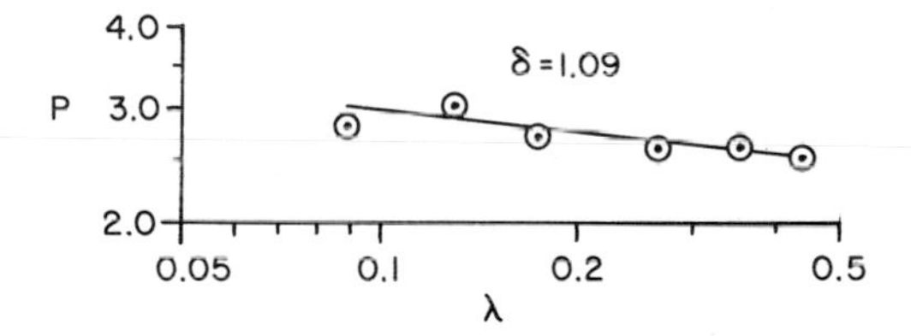

3.9 Fineparticle Look-Alikes

The title of this section shows the bias I have when I look at profiles. My first reaction is to see a fineparticle even if the system is not a fineparticle. However, the intention of the title is also to anticipate how experience gained with fineparticle profiles has found applications in many other areas of science. It also gives us an excuse for peeping ahead to look at the various areas of applied science which we will visit as we pursue our randomwalk through fractal dimensions.

We have already commented on the fact that the outline of an island looks like a fineparticle profile. Lakes also resemble fineparticles when viewed on a map or from an aircraft. Laurentian University is built on the shores of Lake Ramsey. In Figure 3.48, the Richardson plot for an exploration of the fractal structure of the outline of Lake Ramsey is shown. It can be seen that the ruggedness of the lake shoreline can be described by the fractal dimension 1.12 for the scales of scrutiny illustrated in the Richardson plot. It is possible that this fractal dimension can be related to the rock formations in which the lake has been formed and the erosive forces at work to structure the lake with the observed rugged boundary. In Chapter 10 Signpost 4 we shall consider the fractal dimensions of clusters of lakes, clouds, archipelagoes and other geofractals.

As fractals started to become known in the scientific community, and when it was learned that we were measuring fractals in the Physics Department, several colleagues in the biological sciences discussed the possibility of measuring the fractals of biological systems. For example, one of the problems affecting the health of animals that eat grass is that their teeth become worn. Thus, the fractal structure of a photograph of the teeth will decline as the teeth are eroded by the grass. The active agent in the grass that erodes the teeth is tiny crystals of silica which are known as **phytoliths**. This word literally means plant stones. A study of the link between the wear of teeth and the phytolith content of the diet would seem to be a good project but, because of the pressure of other work, experimental studies were not carried out (see the discussion of the fractal dimensions of old and new mountain ranges in Chapter 10 Signpost 4). Another colleague was concerned with the measurement of many leaf profiles to determine the effect of growth conditions on the profile and also the possibility of looking at leaf damage in quantitative terms from the distorted profile of damaged leaves. To investigate the possibility of using fractal dimensions to characterize the structure of some leaves, damaged and undamaged, exploratory experiments were carried out on oak and maple tree leaves.

Structured walk explorations of the fractal structure of two maple leaves were carried out. The data for this experiment are summarized in Figure 3.49. It can be seen that to a first approximation, both leaves manifest fractal structure for the range of scrutiny investigated for the two profiles. In Figure 3.50, data from a similar experiment

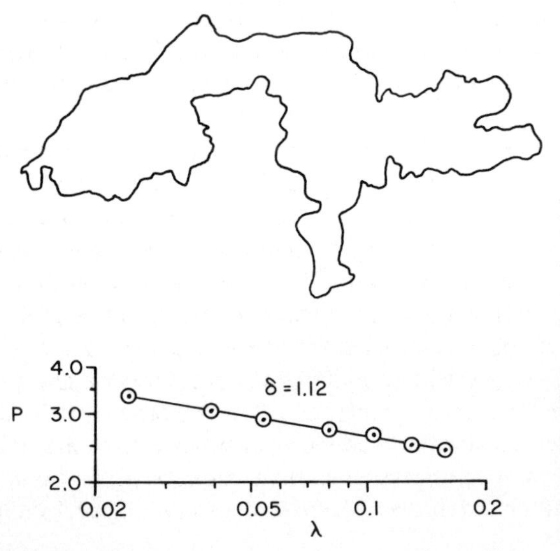

Figure 3.48. Lake profiles can manifest fractal structure as illustrated by the outline of Lake Ramsey from Laurentian University in Sudbury, Canada. P = the perimeter estimate, λ = the stride size used in the structured walk exploration of the profile, P and λ are normalized with respect to the maximum projected length of the lake and δ the fractal dimension of the lake shoreline = 1.12 for the resolution limits illustrated in the Richardson plot shown.

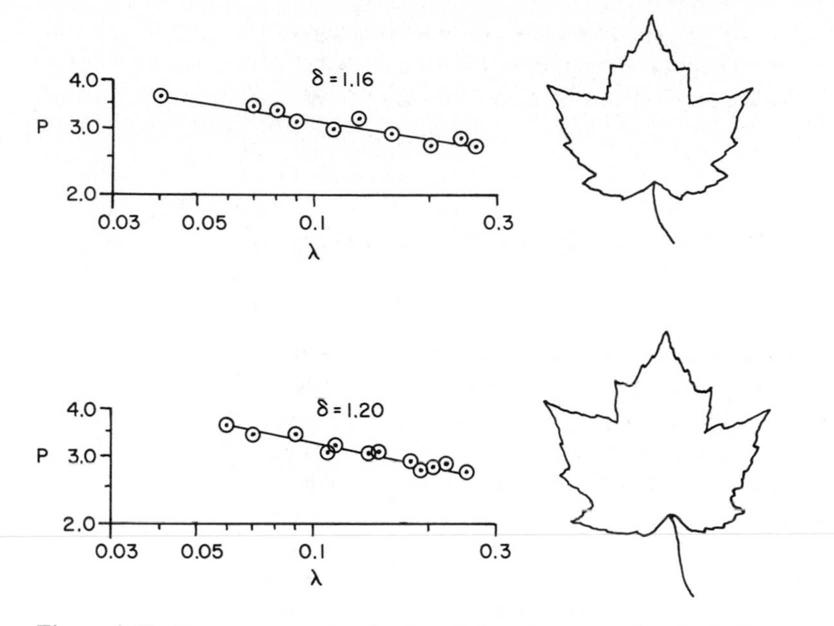

Figure 3.49. From an operational point of view, it appears that the leaf structure of a maple leaf can be described by a fractal dimension over a wide range of scrutiny.

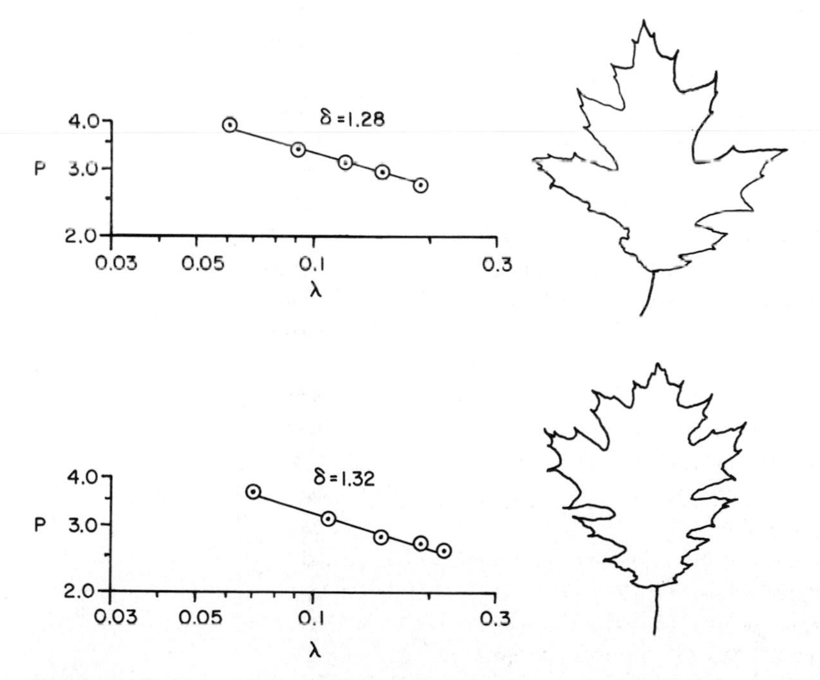

Figure 3.50. Two oak leaves from the same oak tree appear to have essentially the same fractal structure.

performed on two oak leaves from the same oak tree are summarized. Again, it can be seen that for operational purposes, it is possible to describe the leaves as having fractal structure. Next, the fractal structure of a leaf from a different species of oak tree was investigated and the data are shown in Figure 3.51. It can be seen that a fractal

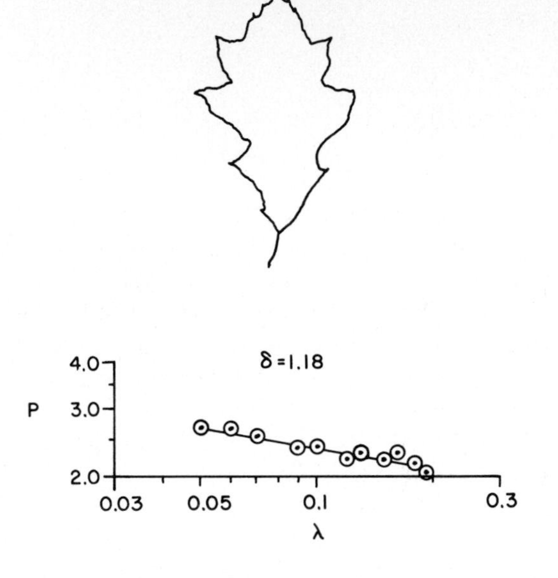

Figure 3.51. The leaf from a different type of oak tree to that shown in Figure 3.50 appears to have a different fractal structure.

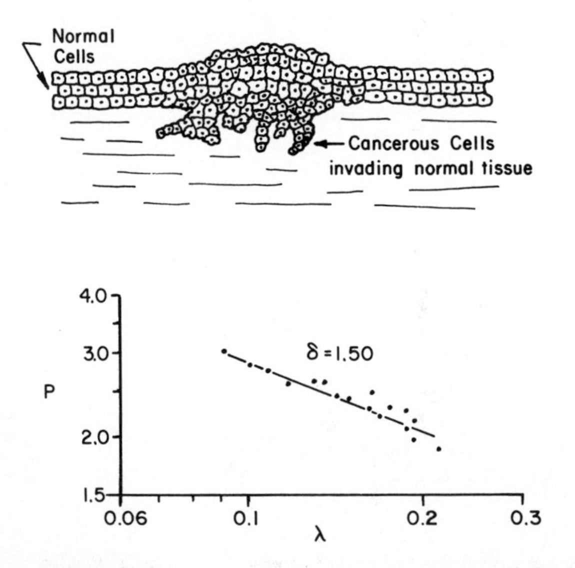

Figure 3.52. A cancerous growth can be characterized by a fractal dimension.

dimension can be used to describe the structure of the leaf. Experiments are under way to investigate the possibility of using fractal dimensions to measure the distortion in the structure of a leaf caused by a biological and/or chemical attack.

Dr. A.E. Martens, then Vice-President of Bausch and Lomb, Analytical Instruments Division, was one of the first to recognize the importance of fractal geometry for describing irregular shaped objects in automated image analysis. In a conversation that I had with Dr. Martens in late 1977, he suggested that fractal dimensions may prove useful in characterizing the structure of cancerous growths in human and other tissue [73].

I do not know of any actual experiments aimed at characterizing cancerous tissue, but the potential of the method for describing such growths can be appreciated from the system shown in Figure 3.52. The term **cancer** is an ancient name coming from the Latin word for a crab. The tentacles of the cancerous growth that bites into healthy flesh look like the claws of a crab. The system shown in Figure 3.52 is from a textbook describing the appearance of a cancer starting to spread out from a surface layer of cells into normal tissue. The fractal dimension of the cancerous tissue shown is 1.50. An exploration of the fractal dimension of cancerous growths as viewed through the microscope with quantitative evaluation of any modification of the fractal dimension due to treatment of the growth would appear to be a fruitful area of research.

In Figure 3.53, is shown the outline of a Purkinje brain cell is shown. In the original diagram of this cell, presented by Llinas, nerve junctions (synapses) at the lobe type protuberances were shown [74]. Although such a cell is not an ideal fractal, from an operational point of view it is obvious from the Richardson plot shown in Figure 3.53 that the cell can be described by a fractal dimension, at least for the range covered. The applications of fractal dimensions to the description of biological systems has been pioneered by Rigaut and co-workers [75 – 79].

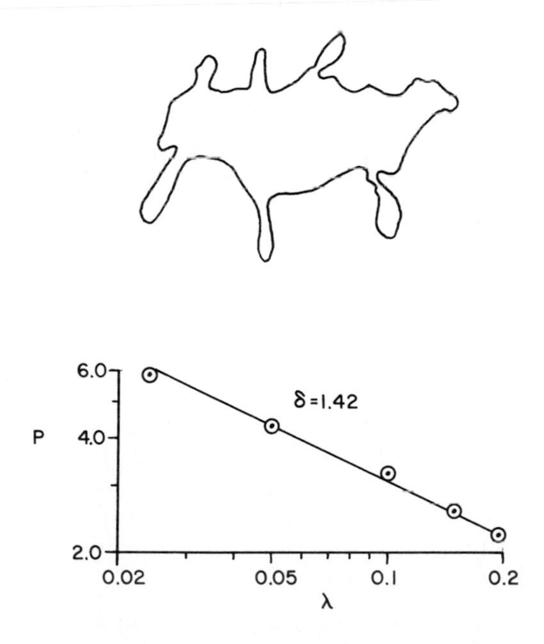

Figure 3.53. Although not ideal fractals, some biological cells exhibit a ruggedness which from an operational point of view can be described by a fractal dimension. Thus the outline of a Purkinje cell can be described by a fractal dimension of 1.42 over a range of resolutions from 0.20 to 0.02 Feret's diameter exploration of the profile.
P is the perimeter estimate and λ the stride for the structured walk exploration of the profile, P and λ are normalized with respect to the maximum Feret's diameter.

In an interesting general article on fractal geometry, Dr. Batty, Professor of Town Planning in the University of Wales Institute of Science and Technology, Cardiff, Wales, points out that the outline of urban areas on a map look like fractal objects [80]. The profile he used to illustrate this fact in his article looks like any one of the sponge iron fineparticles which we studied at the start of this chapter. Dr. Batty indicated that fractal geometry can be expected to find many applications in a study of the morphology of cities and greater urban areas. In Figure 3.54 (see Plate 1 at the beginning of the book), a city profile used by Dr. Batty to discuss the hierarchical structure of a model for predicting a city growth is shown. Professor Batty pointed out that the computer-simulated city shown in Figure 3.54 is not unlike a satellite photograph of London, England.

The profile shown in Figure 3.55(a) could be the oldest fractal in the solar system [81] (note that in the second edition of his book [see Chapter 1 reference 18] Mandelbrot shows an intriguing picture plate C1 of God designing fractals into the solar system). As can be seen from Figure 3.55(b), the outline of this hole, found in a meteorite, has a fractal dimension of 1.48 for the range of values of lambda investigated. It is probable that the fractal dimension of this hole can be related to the formation dynamics operative in forming the meteorite.

If the reader finds it a little difficult to become excited about holes in meteorites, the general topic of holes in powder metallurgical compacts manufactured using the powder metals shown in Figure 3.1 is a general source of interest to metallurgists. Thus, Figure 3.56 share some holes found in a powder metal compact produced by the sintering process. The fractal dimensions of holes are of interest in predicting the residual channels through the powder metal compact. A study of holes will help to see how to predict the structure of sintered metal materials.

It is said that if one undergoes a randomwalk in a forest, where one cannot see where one is going, eventually one returns to the spot at which one started. We started with the powder metal grains in Figure 3.1 and we have now come randomly back to our

a

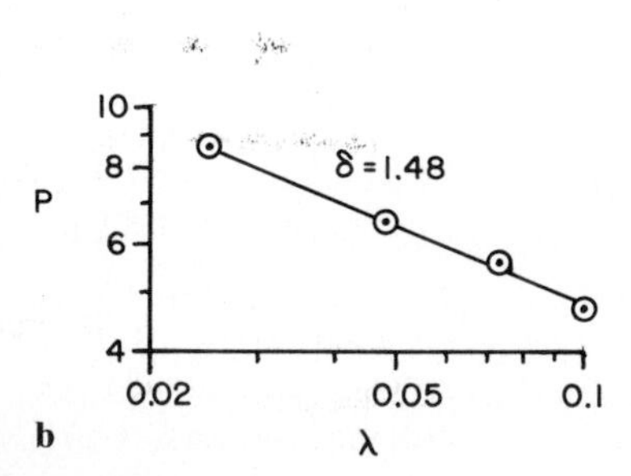

b

Figure 3.55. Holes can often manifest a fractal structure. (a) profile of a hole found in a section through a meteorite. (b) Richardson plot for the meteorite hole.

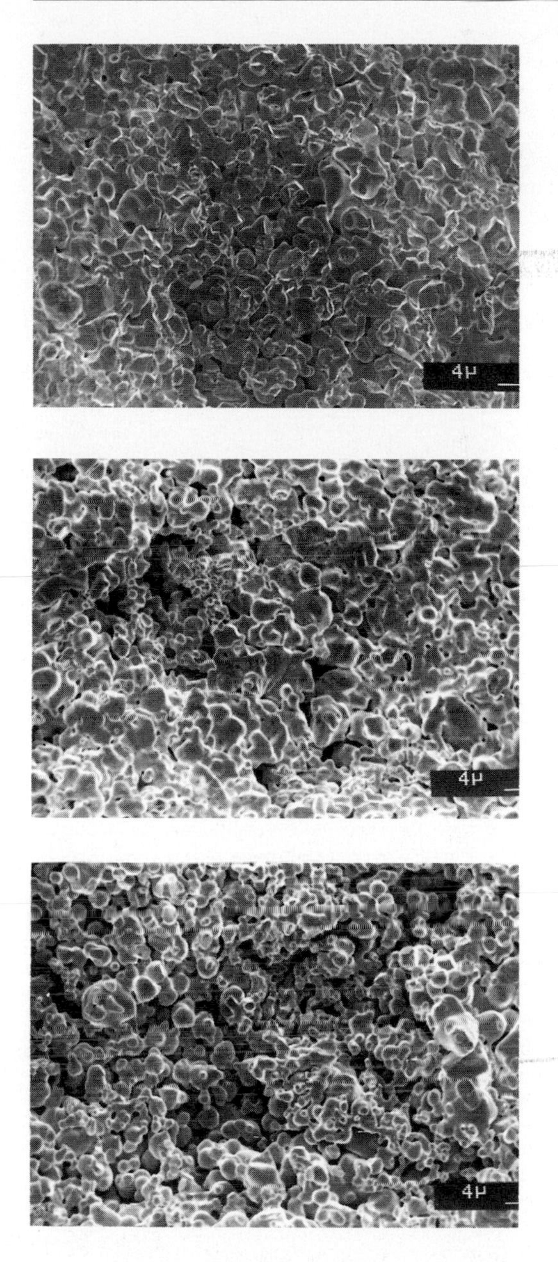

Figure 3.56. The fractal structure of holes found in a sintered metal compact can be used to study the sintering behaviour of the assembly of metal grains [82] (reproduced by permission of VERGLAG SCHMID GMBH and J. Lezanski).

starting point. This would seem to be a good point to take a rest before we start the next chapter, in which we explore scientifically what happens to people undergoing a randomwalk in one and two dimensions. It is hoped that the random meanderings of this chapter will have convinced the reader that fractals can be useful. As we extend our fractal meanderings into diverse dimensional spaces we shall encounter other useful fractals, as well as fascinating mathematical patterns.

References

[1] The profiles in Figure 3.1 are outlines of profiles originally reported by Balasubramanian and co-workers; see R.E. Balasubramanian, P.N. Singh and P. Ramakrishnan, "Effect of Some Particle Characteristics on the Bulk Properties of Powder," *Powder Metall. Int.*, 16, No. 2, (1984), 56-59.
[2] S.P. Mehrota, "Mathematical Modelling of Gas Atomization Process for Metal Powder Production," *Powder Metall. Int.* , 13 (1981) 80-84.
[3] H.C. Nuebing, "Production and Properties of Aluminum Powder for Powder Metallurgy," *Powder Metall. Int.*, 13 (1981) 74-78.
[4] B.H. Kaye, J.E. Leblanc and P. Abbot, "Fractal Description of the Structure of Fresh and Eroded Aluminum Shot Fineparticles," *Part. Charact.*, 2 (1985) 56-61.
[5] F.T. Fong, J.K. Beddow, and A.F. Vetter, "A Technique to Show Effects of a Chemical Reaction Upon Particle Morphology," in Proceedings of the Powder and Bulk Solids Conference, Rosemont, Illinois, 1978., published by the Cahners Exposition Group.
[6] J.K. Beddow and T.P. Meloy, "Advanced Particulate Morphology," CRC Press., Boca Raton, FL, 1980.
[7] M. Matsushista, M. Sano, Y. Hayakawa, H. Honjo and Y. Sawada, "Fractal Structures of Zinc Metal Leaves Grown by Electro Deposition,"*Phys Rev. Lett.*, 53 (1984) 286-289.
[8] The theory of electric storage batteries (called accumulators in UK are discussed in many first-year level physics textbooks). The one that I found particularly useful is by C.G. Wilson "Electricity and Magnetism," English Universities Press, London., 1958.
[9] W. T. Elam, S. A. Wolf, J. Sprague, D. U. Gubser, D. Van Vechten, G. L. Barz, Jr., and P. Meakin, "Fractal Aggregates in Sputter Deposited Niobium Germanium Films," *Phys. Rev. Lett.*, Vol. 54 (1985) 701-703.
[10] D. A. Weitz and M. Oliveria, "Fractal Structures Formed by Kinetic Aggregation of Aqueous Gold Colloids," *Phys. Rev. Lett.*, 52 (1984) 1433-1436.
[11] The fractal agglomerate in Figure 3.10 described by Weitz and Oliveria is also to be found in a review article by T. C. Lubennsky and P.A. Pincus, "Super Polymers, Ultra Weak Solids and Aggregates," *Phys. Today*, October (1984) 44-50.
[12] D.M. Bigg and D.E. Stutz, "Mounded Composites as EMI Shields," *Ind. Res. Dev.*, July (1979) 103-105.
[13] K. Nangrani and S. Gerteisen, "Conductive Composites Blunt EMI Effects," *Res. Dev.*, July (1985) 60-63.
[14] M. Peleg and M.D. Normand, "Characterization of the Ruggedness of Instant Coffee Particle Shape by Natural Fractals," *J. Food Sci.*, 50, (1985) 829-831.
[15] M. Peleg and M.D. Normand, "Mechanical Stability as the Limit to the Fractal Dimension of Solid Particle Silhouettes," *Powder Technol.*, 43(2), 15th July (1985) 187-188.
[16] Personal communications from Dr. Peleg, Dept. of Food Engineering, University of Massachusetts, Amherst, MA, U.S.A., Dr. Riley, University of Bath, UK and Dr. J.C. Williams, Honorary Fellow, Chemical Engineering Dept., University of Bradford, UK.
[17] J.H. Brown, "Unit Operations in Mineral Engineering," International Academic Services, Kingston, Ontario, 1979.
[18] A.M. Gaudin, "Principles of Mineral Dressing," McGraw-Hill, New York, 1939.
[19] E.G. Kelly and D.J. Spottiswood, "Introduction to Mineral Processing," Wiley, New York, 1982.
[20] B.H. Kaye, "Application of Recent Advances in Fineparticle Characterization to Mineral Processing," invited contribution to the Symposium "Challenges in Miner-al Sciences," 23rd Annual Meeting of the Metallurgical Society of CIM, Quebec City, August 19-22, 1984; published in *Part. Charact.*, 2, (1985) 91-97.
[21] B.H. Kaye, "Fractal Description of Fineparticle Systems," in K. Beddow (Ed.), "Particle Characterization in Technology," Vol. 1, CRC Press, Boca Raton, FL, 1984, Ch. 5.

[22] The detailed data generated in the measurement of the roughness of the surfaces were presented in B.H. Kaye, "The Description of Two Dimensional Rugged Boundaries in Fineparticle Science by Means of Fractal Dimensions," first presented at a workshop on fractal dimension, Rosemont, Illinois, May, 1985; also published in *Powder Technol.*, 46 (1986) 245-254.

[23] The profilometer traces in Figure 3.16 were originally described by E. Rabinowiez, "Polishing of Surfaces," *Sci. J.*, 22, January 1970, 45-48.

[24] B.J. West and M. Shlesinger, "Random Model of Impact Phenomenon," *J. Phys. A.*, 127 (1984) 490-508.

[25] S.K. Akhter, "Fineparticle Morphology and the Rheology of Suspensions and PowderSystems," MSc Thesis, Laurentian University, 1982.

[26] This type of viscometer is widely used in teaching laboratories. A discussion of the theory of the instrument can be found in many standard textbooks. see, for example, "Scholarship Physics," by M. Nelkon published by Heinemann Educational Books, London, 1966, 68 and 69.

[27] J.A. Herbst, D.J. Kenneberg and K. Rajamani, in "Proceedings of the International Symposium on Instream Measurement of Particulate Solid Properties, Bergen, Norway, August, 1978," Volume 1.

[28] J.A. Herbst and K.B.S. Sastry (Eds.), "On-stream Characterization and Control of Particulate Processes," report of a workshop held at Asilomar Conference Grounds, Pacific Grove, CA, June 18-23, 1978, The Engineering Foundation, New York, 1978.

[29] For the non-specialist an understandable discussion of the diffraction pattern of regular and random array fineparticles is given in W. Bolton, "Patterns of Physics", McGraw-Hill, London and New York, 1974. This book has rescued many struggling first-year physics students at Laurentian University who find the more usual University first-year text heavy going. The paperback edition of the book has a beautiful picture of the fractal structure of lightning on the front cover.

[30] The various commercial eriometers are continuously developing and it is not possible to give a complete guide to which instrument uses which logic in the breakdown of the energy pattern in a complex diffraction pattern generated by a laser beam passing through an array of rugged fineparticles. The technical literature of the various companies manufacturing eriometers should be consulted if one is interested in an extensive study of this problem. The basic theory for breaking down a diffraction pattern to yield a size distribution of the scattering fineparticles was first worked out by J.H. Talbot; See J.H. Talbot, in M.J. Groves and J.C. Wyatt-Sargent (Eds.), "Particle Size Analysis 1970," Society for Analytical Chemistry, London, 1972, pp. 96-100. A good review of the basic theory was given by B.B. Weiner, "Particle and Drop Sizing using Fraunhofer Diffraction," in H.G. Barth (Eds), "Modern Methods of Particle Size Analysis," Wiley, New York, 1984, Ch.5. Commercially available eriometers include the following instruments: The **Microtrac** eriometer is manufactured by Leeds and Northrup Instruments, Microtrac Products, St. Petersburg, FL, U.S.A. The **CILAS** Eriometer is marketed by CBL Compagne Belge Lasers, Meersstraat 130-B, 9000 Gent, Belgium.

[31] J. Leblanc, "The Shape and Size Characterization of Respirable Dusts," *M.Sc. Thesis*, Laurentian University, 1986.

[32] R. Murphy, "Optical Information Processing Procedures for Determining the Essential Features of Particle Shape," *M.Sc. Thesis*, Laurentian University, 1972.

[33] W.S. Bickel, H.A. Yousif and W.M. Bailey, "Masking of Information in Light Scattering Signals from Complex Scatterers," *Aerosol Sci. Technol.*, 1, (1982) 329-335.

[34] B.H. Kaye, "Fractal Dimension and Signature Waveform Characterization of Fineparticle Shape," *Am. Lab.*, April 1986, 55-63.

[35] D.A. Sutherland "A Theoretical Model of Floc Structure," *J. Colloid Interface Sci.*, 25, (1967), 373-380.

[36] D.M. Sutherland and I. Goodarz-Nia, "Floc Simulation - the Effect of Collision Sequence," *Chem. Eng. Sci.*, 26, pp. 2071-2085.

[37] I. Goodarz-Nia, "Floc Simulation: Effects of Particle Size and Shape," *Chem. Eng. Sci.*, 30, (1975) 407-412.

[38] I. Goodarz-Nia, "Floc Simulation: Effect of Particle Size Distribution," *J. Colloid Interface Sci.*, 52, (1975) 29-40.

[39] H.G.E. Hentschel and J.M. Deutch, "Flory and Type Approximation for the Fractal Dimension of Cluster-Cluster Aggregates," *Phys. Rev. A.*, Vol. 29 (3), March 1984, 1609-1611.

[40] R.C. Ball and T.A. Witten, "Particle Aggregation versus Cluster Aggregates in High Dimensions," *J. Stat. Phys.* 36 (1984) 873-879.

[41] A. G. Flook, "The Use of Dilation Logic on the Quantimet to Achieve Fractal Dimension Characterization of Textured Surfaces," Powder Technol., 21(1978) 295–298.

[42] B.H. Kaye and G.G. Clark, "Characterizing Fractal Structure of Flocculated Suspensions," in preparation.

[43] H. Heywood, "The Origins and Development of Particle Size Analysis," Plenary Lecture given at the Conference on Particle Size Analysis, 1970, in M.J. Groves and J.L. Wyatt-Sargent (Eds.), "Particle Size Analysis 1970," Society for Analytical Chemistry, London, 1972, 1-18.

[44] B.H. Kaye and G.G. Clark "Fractal Description of Extra Terrestrial Fineparticles", *Part. Charact.*, 2 (1985) 143-148.

[45] G. Taubes, "U2 Mission to Catch the Dust of Comets," *Discover* (1983) 74-77.

[46] W.B. Whalley and J.D. Orford, "The Use of the Fractal Dimension to Quantify the Morphology of Irregular Shaped Particles," *Sedimentology*, 30, (1983) 655-668.

[47] W.B. Whalley and J.D. Orford, "Analysis of Scanning Electromicroscopy Images of Sedimentary Particle Form by Fractal Dimension and Fourier Analysis Methods," *Scanning Electron Microsc.* 11, (1982), 639-647.

[48] This discussion of the formation kinetics of fumed silica and the photographs of fumed silica fineparticles are from trade literature from Cabot Corp., Tuscola, IL., 61953, U.S.A.

[49] B.H. Kaye, J.E. Leblanc, G.G. Clark, "A Study of the Physical Significance of Three Dimensional Signature Waveforms," *Part. Charact.*, 1 (1984) 59-65.

[50] V. Timbrell, "Inhalation and Biological Effects of Asbestos," T.T. Mercer, P.E. Morrow and W. Stober (Eds.), "Assessment of Airborne Particles," C.C. Thomas, Springfield, IL, 1972, Ch. 22.

[51] W. Stöber, "Dynamic Shape Factors of Non-Spherical Aerosol Fineparticles." in T.T. Mercer, P.E. Morrow and W. Stober (Eds.), "Assessment of Airborne Particles", Published by Charles C. Thomas, Springfield, IL, 1972, Ch. 14.

[52] W.A. Bloor, "Dangerous Dust," review in Spectrum Notes No. 170, (1980) 12-14. Spectrum Notes is produced by the Science Unit of the London Press Service and published by the Central Office of Information, Hercules Road, London, SE1 7DU, UK.

[53] G.J. Sem, "Aerodynamic Particle Size: Why is it Important," *TSI Quart.*, 10, No. 3, (1984) 3-11. *TSI Quart.*, is a house journal published by TSI Inc., 500 Cardigan Road, P.O. Box 43394, St. Paul, MN, 55164, U.S.A.

[54] V. Timbrell, "An Aerosol Spectrometer and Its Applications," T.T. Mercer, P.E. Morrow and W. Stober, (Eds.)," Assessment of Airborne Particles," Charles C. Thomas, Springfield, IL, 1972, Ch. 15.

[55] A.M. Hollenbach, M. Peleg and R. Rufner, "Effects of Four Anticaking Agents on the Bulk Characteristics of Ground Sugar," *J. Food Sci.*, 47 (1982) 538-544.

[56] P. Kotrappa, "Shape Factors for Aerosols of Coal, Uranium Dioxide and Thorium Dioxide in Respirable Size Range," in T.T. Mercer, P.E. Morrow and W. Stober (Eds.), "Assessment of Airborne Particles," Charles C. Thomas, Springfield, IL, 1972, Ch. 16.

[57] W. Zeller, "Direct Measurement of Aerosol Shape Factors," *Aerosol Sci. Technol.*, Vol. 4, 1985, 45-63.

[58] Photographs provided by and used with kind permission of R.G. Pinnick, Department of the Army, White Sands Missile Range, New Mexico, 88002-5501, U.S.A.

[59] R.G. Pinnick, G. Fernandex, B.D. Hinds, C.W. Bruce, R.W. Schaefer and J.E. Pendelton, "Dust Generated by Vehicular Traffic on Unpaved Roadways, Size and Infra Red Extinction

Characteristics," *Aerosol Sci. Technol.*, 4, (1985) 99-121.
[60] Personal communication of Dr. Stern, World Health Organization, Regional Office for Europe, 8 Scherfigsvej, DK-2100 Copenhagen, Denmark.
[61] C.A. Amann and D.C. Siegla, "Diesel Particulates What They Are and Why," *Aerosol Sci. Technol.*, 1, (1982) 73-101.
[62] B.H. Kaye, "Trends in Fineparticle Characterization," in the proceedings of Particle Size Analysis, N. Stanley-Wood and T. Allen (Eds.), September, 1981, British Institute of Chemistry, London, 1982, 3-15.
[63] D.B. Kittelson and D.F. Dolan, "Diesel Exhaust Aerosols," Particle Technology Laboratory, Mechanical Engineering Department, University of Minnesota, Minneapolis, publication No. 387, 1978.
[64] R.J. Cheng and A.W. Hogan, "Microscopic Study of Lead Iodide-Nucleated Ice Crystals," *Microscope*, 18, No. 4 (1970) 299-302.
[65] Photo provided by Dr. Brownbridge, Kerr-McGee Corporation, used with permission.
[66] J.G. Cabrera and C.J. Hopkins, "The Measurement of Particle Size of Pulverized Fuel Ashes," in T. Allen and N. Stanely-Wood (Eds.), Particle Size Analysis, 1981 , John Wiley, New York, (1982) 127-142.
[67] G.L. Fisher, D.P.Y. Chang, and M. Brummer, "Flyash Collected from Electrostatic Precipitators: Micro Crystalline Structures and the Mystery of the Spheres," *Science*, 192, (1976) 553-555.
[68] G.L. Fisher, B.A. Prentice, D. Silberman, J.M. Ondov, A.H. Biermann, R.C. Ragini and A.R. McFarland, "Physical and Morphological Studies of Size Classified Coal Flyash," *Environ. Sci. Technol.*, 12, (1978) 447-451.
[69] J.A. Small and W.H. Zoller, "Single Particle Analysis of the Ash from the Dickerson Coal Fired Power Plant," in "Methods and Standards for Environmental Measurements," Proceedings of the Eighth IMR Symposium, September 20-24, 1976, Gaithersberg, MD, NBS Special Publication No. 464, National Bureau of Standards, Washington, DC, 1977.
[70] F.R. Kirchner, P.F. Dunn and C.B. Reed, "Toxicologic and Physiochemical Characterization of High Temperature Combustion Emissions," *Aerosol Sci. Technol.*, 2 (1983) 389-400.
[71] Y.S. Cheng, R.L. Carpenter, E.B. Barr and C.H. Hobbs, "Size Distribution of Fineparticle Emissions from a Steam Plant with a Fluidized Bed Coal Combuster," *Aerosol Sci. Technol.*, 4 (1985) 175-189.
[72] "Process for Recycling Vulcanized Rubber," Gould Inc., Cleveland, OH, 1978.
[73] A.E. Martens, Bausch & Lomb, Rochester, NY, personal communication.
[74] R.R. Llinas, "The Cortex of the Cerebellum," *Sci. Am.*, 232, No. 1, January (1975) 56-71.
[75] J.B. Rigaut, P. Berggren and B. Robertson, "Resolution - Dependence of Stereological Estimations: Interpretation, with a New Fractal Concept of Automated Image Analyzer Obtained Results on Lung Sections," *Acta Stereol.*, 2 Supple. 1, (1983) 121-124; Proceedings of the Sixth International Congress on Stereology, Gainesville, FL, October 9-14, 1983.
[76] J.P. Rigaut, "An Empirical Formulation Relating Boundary Lengths to Resolution in Specimens Showing Non-Ideally Fractal Dimensions," *J. Microsc.*, 133, Part 1, (1984) 41-54.
[77] J.P. Rigaut, C. Lantuejoul and F. Deverly, "Relationship Between Variants of Area Density and Quadrant Area - Interpretation by Fractal and Random Models," Proceedings of the Fourth European Symposium on Stereology," *Acta Stereol.*, in press.
[78] J.P. Rigaut, P. Berggren and B. Robertson, "Stereology, Fractals and Semi-Fractals, the Lung Alveolar Structure Studied Through a New Model," Proceedings of the Fourth European Symposium on Stereology, *Acta Stereol.*, in press.
[79] J.P. Rigaut, "Fractals, Semi-Fractals," in G. Cherbit (Ed.), "A Biometri in la Dimension Non Entiere et Ses Applications," Masson, Paris, 1986, in press.
[80] M. Batty, "Fractals–Geometry Between Dimensions," *New Sci.*, April 4, (1985) 31-35.
[81] L. Grossman, "The Most Primitive Objects in the Solar System," *Sci. Am.*, No. 2 (1975) 30.
[82] J. Lezanski, and W. Rutkowski, "Infiltration of a liquid in Sintered Tungsten," PMI, April 1987.

4 Delinquent Coins and Staggering Drunks

4.1 A Capricious Selection of Terms that Describe Random Events (a section which may be skipped over by the reader who is sure he can define randomness)

It will come as no surprise to the North American reader who has pursued the text of this randomly organized study of fractals from page 1 to the present page systematically that I moonlight as a lexicographer. The European reader needs to be told that in North America "**to moonlight**" means to work at another job in the hours outside normal working hours. Like many words to be discussed in this section, "to moonlight" is not listed in ordinary dictionaries. Dr. Johnston, who wrote the first comprehensive dictionary of the English language, defined a **lexicographer** as "a harmless drudge who compiles lists of words." The technical description of a lexicographer is a person who writes dictionaries.

I began my career in lexicography in 1963 when I was invited to revise the Powder Technology section of "Chambers Technical Dictionary" [1]. I was paid one shilling and sixpence (about 20 cents) for every entry I submitted which made it into the dictionary. My love/hate relationship with dictionaries started when I was given a copy of "Chambers Etymological English Dictionary" at the age of twelve [2]. **Etymology** is defined in "Chambers dictionary" as "the science of investigation of the derivation and original meaning of words. It is derived from Greek root words, ETYMOS, true, and LOGOS, a discourse." Using an etymological dictionary, one can discover undreamed of interesting origins of familiar words. Thus, I remember clearly the interest I experienced when I discovered that **scrutiny** comes from "scruta," the Latin word for rags or trash. To scrutinize an area originally meant to investigate a location thoroughly down to a search through the rags. When we say that ideal fractals have the same structure no matter what scale of scrutiny we use to examine their structure, we are using a picturesque way of describing the intensity with which we examine the fractal. The mathematician looking at fractal curves is as intense in his examination as a Roman soldier searching for hidden weapons amongst the rags of his barbarian captives!

The hate part of my attitude to dictionaries comes from the fact that, if you spell a word wrongly at a first attempt, it may be impossible to find the correct spelling hidden in the dictionary. For example, if one spells the word cabbage in the form kabbage, one will never find the correct spelling by starting off in the k section of the dictionary! As a poor speller, I have spent many fruitless hours searching through dictionaries. Practical scientists often become impatient with discussions over the meaning of words, but a failure to be precise when using scientific terms can lead to uncertainty when interpreting experimental data. As fractal geometry evolves, the word fractal dimension is being used with many different implied meanings, as well as those discussed in Section 5.5. When studying random events, the difficult subject is often made more difficult by the sloppy use of language. This section of this chapter has been written in an attempt to clarify what we mean by random events.

Several students complained to me after early lectures on fractals that fractal dimension was not in their dictionaries. Mandelbrot (see chapter 1 reference [18]) tells us in his book that he had to keep inventing words for ideas and concepts that he developed as he fashioned a theoretical structure for the subject which is now known variously as fractal analysis or fractal geometry. As we introduce new terms coined by Mandelbrot, we shall examine their etymology and find that Mandelbrot not only made use of Greek and Latin words, but also that he drew upon colourful words from everyday speech. The term fractal was coined by Mandelbrot in the early 1970s and it may be some time before fractal and fractal dimensions are listed in common usage dictionaries.

A major problem with dictionaries of everyday English is that they list older usages of a word and fail to incorporate new meanings for old terms. At one time, I used to give lectures on the origin and history of scientific terminology to students at the School of Translation at Laurentian University. The students used to come to me for help with specific difficulties they encountered in their translation projects. One student who came for help was translating a passage discussing the use of a holograph of a chess set to create a three-dimensional image. The student looked up the word holograph in "Chambers Etymological Dictionary" and found it defined as follows: "**Holograph**, a document, e.g. a will, wholly in the handwriting of the person from whom it proceeds." The derivation is given as being from the Greek "holos" meaning whole, and "graphein," meaning to write. This is still the legal meaning of the term holograph. However, this legal meaning obviously did not make sense in the passage the student was translating. Long after the dictionary was written, Gabor, a world-famous scientist, had invented a new way of recording three-dimensional images and used the term **optical holography** to describe his system. Gabor, unaware of the existing legal use of the term holograph, coined holograph by comparison with a photograph. Photograph means a drawing made with light (from the Greek "photos," meaning light). A holograph contains all ("holos") the information for a three-dimensional image.

Given the general inadequacies of dictionaries for specialist terminology, it is not surprising that most people fail to look up precise definitions of technical terms. In particular, it is my experience that many students never use dictionaries and attempt to stagger along with knowledge acquired through working with words. This can be dangerous. Considering, for example, the word random, the reader can demonstrate for himself that most people have a very poor intuitive knowledge of the properties of a random variable. If an individual is asked to generate at random a sequence of digits from 0 to 9, the answers given will usually be biased. Thus, in Table 4.1, the results of an attempt by an individual student to generate 100 random digits are summarized. Under the table listing the 100 digits, the frequency of each digit is shown. It can be seen that the student seems to give a low frequency of 1. He also failed to give pairs

	Student	Row Doublets	Table	Row Doublets
	0 3 4 6 8 5 2 7 6 4	0	0 9 2 1 4 5 0 2 3 8	0
	5 8 2 1 3 8 7 9 2 3	0	8 0 5 4 7 9 4 4 5 7	1
	2 4 8 5 7 6 9 8 2 3	0	7 2 8 3 9 8 5 9 3 3	1
	2 8 9 7 5 6 4 9 7 1	0	3 3 5 0 4 0 5 1 1 0	2
	9 5 8 7 6 2 4 9 5 4	0	7 7 1 2 2 1 9 4 3 4	2
	7 6 9 2 3 8 4 3 7 9	0	1 4 9 3 6 6 9 1 7 7	2
	8 2 4 3 5 7 9 0 4 0	0	0 0 6 9 2 2 8 9 7 5	2
	9 8 0 9 6 0 4 5 0 2	0	0 5 1 6 9 2 0 6 9 5	0
	8 4 5 2 1 0 8 9 4 2	0	3 7 9 9 7 4 0 2 3 3	2
	7 4 5 1 9 3 8 7 8 6	0	0 4 1 3 8 2 7 0 7 4	0
Column Doublets	1 1 1 1 0 1 2 1 1 2		1 0 0 0 0 1 2 0 1 1	

Digit	Frequency	
	Student	Table
0	7	13
1	4	9
2	12	11
3	8	12
4	13	10
5	10	9
6	8	5
7	11	12
8	14	6
9	13	13

Table 4.1. Most people have a poor appreciation of the pattern of events which occur in random systems. This is shown by the fact that the student guessing numbers at random in a sequence of 100 digits fails to give any pairs or triplets of the same digit. If the 100 digits between 0 and 9 are assembled in a square matrix, even though the student failed to give doublets when generating the numbers in sequence, the vertical juxtaposition of the digits generates several pairs and a triplet, as shown in the array in (a). A similar table abstracted from a random number table in which the digits are generated by magnifying random noise in an electronic computer is shown for comparison in (b).

and triplets of numbers at a frequency which would normally occur in a run of 100 random digits. Thus, it will be noticed that pairs of digits occur fairly frequently in the vertical direction for the student-generated 100 digits, whereas the student gave none in the sequence as generated "at random."

For comparison purposes, 100 digits taken from a published random number table are shown beside those generated by the student. This short sequence of random numbers is taken from the larger **random number** table shown in Figure 4.1 [3]. A random number table is constructed so that, as one moves along the top of the table, the probability of any one digit between 0 and 9 appearing at any one position is equally probable and independent of what appeared in the proceeding position. Thus, if one were to take a large page of random numbers, all digits should appear in the table with equal probability. The random numbers shown in Figure 4.1 will be used in several experiments described in this book. Many small calculators and minicomputers have a random number generator built into their circuits so that students can generate random

09214	50238	04991	38139	54996	24342	44496	45239	11940	19241
80547	94457	70275	12913	66313	27041	82993	54067	94659	52521
72839	85933	25735	97083	95520	67343	92772	82290	64448	51612
33504	05110	71056	23827	81160	99154	74567	24323	54782	49874
77122	19434	53361	96189	74674	12911	51085	38413	89676	48350
14936	69177	88984	77642	21103	86088	01892	98501	52334	48965
00692	28975	18511	94665	02539	61959	66426	70460	86063	47929
05169	20695	38413	33058	58268	99421	40918	71663	20459	89178
37997	40233	37986	47625	58328	72816	47809	63172	53068	60343
04138	27074	19521	92675	81223	10821	09372	44162	13190	23149
57018	48311	85683	24918	92363	02657	95035	27533	44350	30630
36833	12647	23317	33589	70249	43025	47726	00697	19107	10206
02407	50852	93431	50190	22752	88188	54086	96633	48277	74546
18262	28862	26009	73861	58056	30040	45701	93153	70875	90002
16299	72896	27331	12010	25253	14297	00219	64076	37103	37204
83640	49766	48898	92471	86275	30263	63327	12406	10596	24572
88542	58925	70098	51059	29124	48902	56164	57073	89487	47870
74431	18037	47466	66705	64853	25188	22771	74068	75267	09929
93015	96164	58353	71861	74252	80911	00949	94873	33572	13311
30959	26102	54232	20634	61525	39054	67094	89310	67315	40704
31807	13883	34418	40128	14012	45972	01955	53691	43651	97139
00222	19051	57923	78178	27208	64159	08611	35303	64406	13668
30819	51795	51846	13614	02299	61338	42121	49239	70620	40624
97938	50482	67045	66065	09955	96300	90516	85318	02907	03587
52515	24368	33672	06587	99830	03793	03969	20378	20910	08181
50610	75873	91321	09540	40859	32448	92396	23883	45388	28647
69717	83974	54937	87844	17886	53805	87912	77475	44639	71482
97249	47377	74593	18857	58058	73848	46648	21447	36373	92867
29923	21615	41497	37024	18455	08032	25195	25157	26079	74194
63705	32600	86350	14916	25111	27417	75832	49646	06111	04891
95881	86318	91286	01200	21988	20193	38483	29150	66153	63363
14228	78493	28896	91083	04917	64476	32721	34551	53512	31049
43389	41211	42779	99658	17407	81851	93136	00686	34511	26308
71780	97671	23274	08354	93647	09487	09249	39114	95321	68868
19546	86817	33871	46810	96289	20999	82234	14439	01055	12694
91929	37078	93994	57510	76568	53999	61956	30806	70581	82932
32445	63790	53392	65969	02655	61789	34890	21801	93593	46397
01539	00404	48532	59423	10216	94216	72663	29584	32825	48985
87021	20496	85498	54783	52784	55271	95869	45431	02404	68955
55838	41225	90745	91552	69416	17585	25440	14663	68111	87329
43363	15252	72228	65696	22481	35882	38551	66715	15694	64788
08244	50246	34789	39353	30361	29546	63729	30748	50760	68964
11922	05779	18851	47404	71310	13814	68904	00435	25596	90328
29716	71306	87109	65844	21539	35310	87619	04578	76573	86447
09802	59796	45679	32349	04003	09420	73664	68922	75476	57222
15762	62552	03655	50339	07854	69253	62467	00350	80773	79359
64387	30765	06075	35803	53746	84275	27068	35636	96685	27993
15565	22973	99027	09790	16180	26516	91008	27747	90740	21830
19376	21862	00109	81002	15965	40199	48350	17593	62201	37421
50855	96963	36084	32781	40027	49365	87493	11055	34100	67092

Figure 4.1. A random number table can be used to simulate many different systems in which the variables are changing at random [3]. (Reprinted from page 15 of **A Million Random Digits with 100,000 Normal Deviates** by The RAND Corporation (New York: The Free Press, 1955). Copyright 1955 and 1983 by The RAND Corporation.)

numbers for themselves when undertaking a simulation study. However, the accurate generation of random number tables is a difficult subject and has been the subject of much scholarly discussion. Some of the random number generators in small computers may not generate true random numbers in a long sequence [4]. A discussion of the technical difficulties involved in generating random numbers is beyond the scope of this book. We shall assume that all random number tables used in our exploration of fractal dimensions are as random as modern computer techniques can make them.

The etymology of the word random is not clear. "Chambers Etymological Dictionary" says that it comes from an old French word "randon," meaning to gallop. Another dictionary says that it comes from a word meaning violence and impetuosity. This latter dictionary gives the derivation as being from "randir," to run impetuously. Probably both definitions can be merged when we recognize that in medieval battles, a lot of the fighting involved man-to-man confrontation. Moreover, the warriors involved in this type of fighting were not noted for their sobriety. Certainly in the word random we have the idea of an intoxicated warrior on horseback galloping in all directions during a fight. Until 20–30 years ago, new scientific words were usually coined in the English-speaking world from Latin, Greek or French terms because Latin and Greek were international languages amongst scholars of the Renaissance and French was the international language of culture. However, as the momentum of scientific discovery swung to America, and as the practice of requiring all university entrants to have Greek and Latin studies as part of their background diminished, there has been increasingly a tendency to go to living languages for scientific terms. Thus, the word **laser** is not derived from classical languages, but is an acronym made out of the initial letters of the phrase "light amplification by stimulated emission of radiation." (the term **acronym** means a word made from the "tips" of words in a phrase). If we had gone to Anglo Saxon for a word to describe random events, we may have called random numbers "**berserk numbers**." "Berserk" is Anglo-Saxon for a bear skin. Its modern use to describe crazy, unpredictable behaviour comes from the fact that Norse warriors used to dress in bearskins when they went into battle. Again, in their fighting style, they slashed at all and sundry. Their "aimless" fighting was probably influenced by the fact that they were usually high on mead, an alcoholic drink made from fermented honey, before they went into battle.

If scholars had reached back to Latin instead of French for a word to describe random numbers they might have described them as **capricious numbers**, from the Latin word for goat. Whatever term is used, we see that the essential idea in a random variable is that it is an unpredictable variable that varies in a berserk manner, capriciously changing from value to value.

After many years of working with random variables, I have come to suspect that sometimes it is wiser to define a **random system** as one whose apparently arbitrary behaviour and structural organization have an underlying set of causes not yet perceived by logic. The apparent unpredictability of a "randomly" varying system is not always evidence for a suspension of the laws of cause and effect, but rather a proclamation of our lack of understanding of the physical causes generating the observed "random" behaviour. Part of the fascination of fractal geometry is the discovery of patterns of order in apparently chaotic structure and capricious variables. In this book, we shall use the term **random** to describe patterns of events which appear to behave in a capricious manner. However, we shall find on several occasions that exactly what is meant by "random" is often the key idea as to what has to be reformulated in our thinking before we can understand the pattern of events generated by an experiment.

4.2 Chance, Probability and Error

True "randomness" is often an essential ingredient in a game of chance played by gamblers. The words chance and probability are often used interchangeably in everyday speech. Games of chance have been played for as long as we have had records of man's activities. However, games of chance have also been intimately connected with decision taking when individuals sought to know the "will of the gods." In the autumn of 1984, I visited a Zulu village as part of a holiday after a lecture tour of South Africa. A highlight of the visit was a consultation with the local witch doctor. The witch doctor selected me from amongst a group of tourists and then threw a set of small bones onto the ground. She studied the pattern made by the bones and told me something about myself from the pattern of bones. She told me that I had saved for a long time to come on the tour. Normally this would have been a good guess, since my accent obviously identified me as a northern European. However, since somebody else had paid for my tickets, she was wrong! Nevertheless the witch doctor was a very shrewd character. By casting the bones and looking at the pattern that they made, she was following a very ancient tradition of predicting the future. The word **chance** comes from the Latin word "cadentea," which means "a falling down," particularly of dice. The word **dice** comes from a Latin word which means "that which is given." It is related to the modern Christian name Donna, meaning "a gift from God." Thus, chance is what happens when our lives are predicted by the throw of objects on to the ground and "dice" were so called because, when thrown on the ground, the numbers they displayed were supposed to give information on the future. Even in modern English in North America, the phrase "that's the way the cookie crumbles" or "that is the way the chips fall" indicates that some people believe our life patterns are determined by randomly generated patterns of events.

The word **probability** is derived from the Latin word "probare," to test. It is defined by the dictionary as "that which may be assumed or proved to be likely." Thus, the simplest random variable that we encounter in early studies of science is the tossing of a coin. If we have a coin, which is assumed to be unbiased, we are told that heads and tails of the coin are equally probable. In a study of probability, it is usual to arrange statements of probability so that the total possibility of all different events is 1. Thus, the probability of heads or tails being shown by a tossed coin is said to be 1 and the chance that one particular coin tossed will land with the head showing is said to be 0.5. If one looks at the probability of shaking a six using a cubic die with six faces on it (die is the singular of dice), each number has the probability of 1/6 (approximately 0.166) of being displayed when the die falls on the table. Strictly speaking, probabilities are numerical estimates of the frequency of events occurring by chance as determined by experiment or theory.

Like many students of my generation, my first encounter with the science of probability and statistics was in the discussions of experimental error. One of the most boring laboratory experiments that I ever participated in was to use a device known as a screw caliper to measure the diameter of a piece of copper wire 20 times. The exact value that one obtained for the diameter of the wire at any one attempt depended on many small variables. Thus, if you tightened the driving screw of the caliper too tightly you could actually flatten the wire and change its diameter. The same piece of wire was used by several generations of students, and its real diameter was probably changed by wear, brutal handling and/or the accumulation of dirt at various places along the wire. It had been bent backwards and forwards by many students before I measured it, and all these factors contributed to the assessment of the diameter of the wire being an indeterminate task. Calculating the average of our data set of 20 measurements involved adding up all the measurements and dividing by the number of measurements. The word **average** has an interesting history. It was a term used in Maritime insurance [5]. Medieval merchants who sent off their ships to far places, hoping to make enormous profits, were sometimes disappointed by the fact that, if the ship hit a storm, part of the cargo had to be thrown overboard to lighten the ship so that it could ride out the storm. The "average" was the value of the cargo left on board the ship after the storm, and represented the amount of money to be shared out amongst the merchants, including those whose original contributions to the cargo had been thrown overboard. It is useful to remember that an average is a value which has been calculated after some important details of the experimentally determined information have been thrown overboard [4].

The danger of working with averages, without bearing in mind the physical significance of the data used to calculate the average, is illustrated by the sad fate of the statistician who attempted to walk across a river-bed. He was told the following information. At 2-foot intervals across the river, the depth of the river in feet is 0 (at the bank), 2, 3, 9, 9, 3, 2, and 0 (at the other bank). The statistician quickly calculated the average depth of the river at just over 4 feet. Since he was 6 feet tall, he set off to walk across the river. Unfortunately he could not swim and so he drowned in the statistical variations in the middle of the river!

Because the average does not tell us anything about the range of experimental measurement, students are taught to calculate the standard deviation of the measurements. A technique for estimating the standard deviation will be discussed later in this chapter.

Because the first encounter with statistical reasoning is often associated with experimental error, many students tend to regard statistical thinking as being related to error. Statistical methods of data handling have been defined as salvage methods for inept engineers. It requires a real shift in thinking for the average physics student to start thinking about statistical variables where variation in numerical data is not due to error, but to real random fluctuations in a variable.

The development of theorems for describing the observed patterns of fluctuation in random systems was late to develop in the scientific revolution of the western world. M.G. Kendall has discussed this surprising fact [6]. He pointed out that because the casting of lots was linked, in the mind of the average person, with the pattern of events determined by the Gods or God, no-one looked for patterns generated by the frequent casting of the dice, since any observed pattern could change if the mood of the Gods changed. In one of his articles Kendall notes:

"It might have been supposed that during the several thousand years of dice playing preceding say the year A.D. 1400, some ideas of the permanence of statistical ratios and the rudiments of a frequency theory of probability would have appeared. I know of no evidence to suggest that this was so....I think it very likely that before the Reformation, the feeling that every event, however trivial, happened under divine providence, may have been a severe obstacle to the development of a calculus of chances" [6].

It is generally agreed that the scientific study of the probability patterns of random events was developed under the stimulus of the preoccupation of noblemen, in the 17th and 18 centuries, with gambling for high stakes. Thus, Montroll and Schlesinger [7], at the beginning of a very informative review of the science of randomwalks, make the following observations:

"Since traveling was onerous and expensive, and eating, hunting and wenching generally did not fill the seventeenth century gentleman's day, two possibilities remained to occupy the empty hours - praying and gambling. Many preferred the latter. Hence it is not surprising that efficiency in the estimation of gambling odds kept close to the forefront of the state of the computatorial art and indeed motivated much of it" [7].

Montroll and Schlesinger also note that the first real attempt to study gambling odds, by scientists such as Pascal (1632-1662) and Demoivres, did not lead to the rapid publication of data on probability. They comment:

"the lack of written references, in the probability calculus, is not necessarily indicative of a lack of contemporary interest. Knowledge of chance was so rudimentary that any capacity of gauging gambling odds accurately was worth a great deal of money. Huygens, visiting France in 1657, found intense interest being taken in the doctrine of chances among mathematicians but encountered also a certain coyness about the disclosure of results" [7].

At the same time as there was a growing interest in the mathematics of gambling, there was also a rapid growth of the insurance industry. People who could calculate the life expectancy of people taking out life insurance policies were in great demand. Again, information on the pattern of events in the "random variable" of the "life expectancy" of various occupations was often treated with great secrecy to make sure that the life insurers always made a profit in their gambling upon the life expectancy of their clients.

Studies of such subjects as the frequency with which a pair of dice will show the number 12 is described as a study of stochastic processes. A **stochastic process** is defined as a process which has some element of probability in its structure. An authoritative etymological dictionary gives the origin of the word as being from a Greek word which means "to guess" [8].

Gambling as an activity thrives upon the fact that most individuals have a poor appreciation of the pattern of events generated by stochastic processes. Let us quote Kendall again:

"During the dark ages, gambling was prevalent throughout Europe. Efforts on the part of the Church and State to control the evils associated with it were as ineffectual as they are today. Nothing is more indicative of the persistence of gambling than the continual attempts made to prevent it" [6].

When I worked in Chicago, one of the mathematics professors with whom I used to have lunch occasionally was approached by an organization to calculate the probability of events occurring within a particular gambling game. It was only after some time that the mathematician realized that the organization using him as a consultant was part of an organized crime syndicate aiming to fix the odds of the game so that they could always win by using their knowledge of the probability patterns of the game. It was with some difficulty that he managed to distance himself from his would-be sponsors.

In the 1970s, the federal police in Canada sponsored some work at Laurentian University to study if the odds in a particular pin-ball game were equitable. By Canadian law, the game could only be played legally if there was a "reasonable" chance of winning. The only way to determine the probability of winning with this particular game was to employ students to play the game many times so that the pattern of probabilities could be determined by experiment. The gambling machines were set up in the basement of the University and the students were paid to play the machines **ad nauseum** (a Latin phrase meaning until they were sick of it! Ad nauseum is related to the Latin word for "things of the sea" - nauta; ad nauseum is how you feel when you are sea sick). If one carries out any experiment ad nauseum, to determine the probability patterns of a long sequence of experimental events, the long running sequence of events is described technically as an **ergodic process**. This comes from a Greek word meaning tiresome or boring.

Having defined several of the terms used in a mathematical discussion of random events, we can now start to model stochastic processes that generate fractally structured systems.

4.3 Monte Carlo Technique for Studying Stochastic Processes

The fact that many gamblers have a poor knowledge of the pattern of events in a stochastic process, can be illustrated by the widespread gambling fallacy that, if a number has not appeared in a long run of a throw of a die, then it becomes more probable at the next throw of the die (a **fallacy** is defined as "an apparently genuine but really illogical argument," from the Latin word "fallere," to deceive). Consider, for example, the occurrence of the number 6 on the throw of a die. If one is betting that at each throw of the die the number 6 will appear, then if the winning odds are less than 6 to 1 in an ergodic sequence, a gambler will always lose. If, in any particular set of numbers generated by the sequential throwing of the die, the number 6 fails to appear in a long run, then the gambler is tempted to make a big bet that 6 will appear on the next throw. He bases this decision on the grounds that, since 6 has not been thrown for some time it is more probable at the next throw of the dice.

We can study how frequently the number six will appear in the throwing of a die by carrying out a simulation experiment using the random number table in Figure 4.1. In our simulated experiment, we ignore the digits 0, 7, 8 and 9 as we encounter them in the table, since they do not appear on a six-sided die. Moving along the top column of the random numbers and interpreting the digit in the table as a throw of the die, we would generate the sequence of simulated throws as

2 1 4 5 2 3 4 1 3 1 3 5* 4 6

If our simulated sequence of numbers had been a real game, then by the time the gambler has thrown the die 12 times without a six (the situation when the five marked with an asterisk in the sequence above would be the simulated throw of the die) he would be tempted to "up his bets," because "a six was overdue." He would lose again on the next bet but his friend, who then decided to make a big bet, would be delighted with the "overdue 6" at the 14th simulated throw. The temptation to bet high when an event has failed to occur as anticipated by the gambler is known as "the gambler's ruin."

The fallacy that a random event is more probable if it has not occurred for a long run is a widely held piece of folk logic. A news item which appeared in the Canadian press, August 14th, 1985, illustrates how people feel about the clustering of events. In late July and early August of that year there were some spectacular air disasters. These included the Air India disaster over the Atlantic Ocean near the Irish Sea, the crash of an L1011 aircraft in Texas and the crash of the Air Japan Boeing 747 into the side of a mountain. After the crash of the Japanese flight, the following announcement appeared on the front page of the local newspaper:

"Statistical Catch up"

"The recent spate of airline tragedies is a statistical catch up and should not raise new fears about the safety of flying" says the President of the International Federation of Pilot Associations.

He was using the gambler's ruin fallacy in reverse to say, "don't worry, we are not witnessing a deterioration in air safety, just a statistical catch up."

To show how we can simulate a "pattern" of events which happen by chance in random events, let us simulate the occurrence of accidents in a company with a large number of employees. Let us assume that previous records established that, on average, there was a minor accident once every ten days. To simulate one accident every ten days at random, we can decide that every digit in the random number table in Figure 4.1 represents a working day, and that the number 9 would represent an accident. Starting again at the beginning of the table, we could simulate the occurrence of accidents in the plant over a sequence of days. If we write 0 for an accident-free day and 1 for a day with an accident (that is, when 9 appears in the table) the sequence of events on sequential days would be as follows:

010000000000110000110000000100000

Most people would be happy with this sequence, and think that the accidents would happen according to the average prediction. However, look what would happen if we started to simulate accidents using the row 35 moving from right to left. The generation of the sequence would be as follows:

010000000010000000001110010001 PANIC BUTTON !!!

At the point reached after 30 days in this simulated sequence of events, some one would probably "press the panic button" because accidents should not happen so frequently. If such a sequence of events happened in a real situation, union and management personnel would probably have a crisis meeting to do something about the number of accidents in the plant, even though the accident frequency was compatible with random fluctuations in an average of 1 accident every 10 days. The management, after their panic meeting, would supposedly tighten up all operating procedures, but even if they did nothing, continuing simulation of the accident sequence from the point reached above would show that by random chance there would be 18 accident-free days after the panic button point. Although the sequence of accidents shown above and the subsequent 18 accident-free days are entirely in accordance with random chance, the works committee, which was activated at the "panic button stage," would be convinced that there had been a real problem sequence of accident days which, by their own efforts, they had alleviated and created the run of 18 accident-free days. If a works spokesman had tried to calm the fears of the workers at the point labeled "panic button" in the above sequence by saying that the pattern of five accidents in ten days was a statistical catch up because they had a good run of accident-free days, no-one would have believed him, just as probably no-one believed the pilot who talked about statistical catch up [9]. In a later discussion, we shall learn to give fractal significance to sequences of events such as those discussed in this paragraph (see the discussion of Rosiwald intercepts in Chapter 6).

If people are unprepared for the clustering of events in a randomly generated sequence of events, they are even less prepared for the clustering that can occur in two-dimensional space when random events are generating the observed pattern. Let us imagine that the random number table in Figure 4.1 is a section through a paint film which is 10% by volume pigment. Let us further assume that the pigment fine particles are all the same size. This would mean that we could take the table in Figure 4.1 and make every 9 a pigment fineparticle. In Figure 4.2, the simulation clusters formed by this random positioning of pigment fineparticles are shown.

The first thing to notice is that it is very difficult to describe the structure of these clusters but that they bear a strong resemblance to the carbonblack clusters discussed in Chapter 3. The larger ones appear to have an embryonic fractal structure and hint at the possibility that systems formed by the clustering of random processes in space are probably describable by fractal dimensions. Trying to describe the clusters shown in Figure 4.2 in general terms is difficult because the English language is poor in terms used to describe shape. I have attended lectures where such clusters are described as "fractal animals" or "percolation insects," etc. Indeed, in one lecture I saw so many different fractal insects that I had thought I had wandered into a lecture on entomology rather than fractal dimensions (**entomology** is the formal name for the study of insects).

When we simulate the behaviour of a system which incorporates stochastic behaviour by using a randomness generator to vary the behaviour of the system, we are using a process known as a **Monte Carlo routine**. This term was first used in science in a publication on a stochastic process by Metropolis and Ulam in 1949 [10]. Historically, the term Monte Carlo was used as a code name during the Second World War for top-secret calculations being carried out to predict the flux of neutrons in an atomic bomb. The flow of millions of neutrons, following random paths through a mass of uranium molecules, could only be modelled on a computer and not predicted from theory [11]. Since the paths of the neutrons varied at random, and since the building of the atomic bomb was a big gamble, the calculations were given the code name Monte Carlo because of the fact that Monte Carlo, the capital of the tiny principality of Monaco, was the

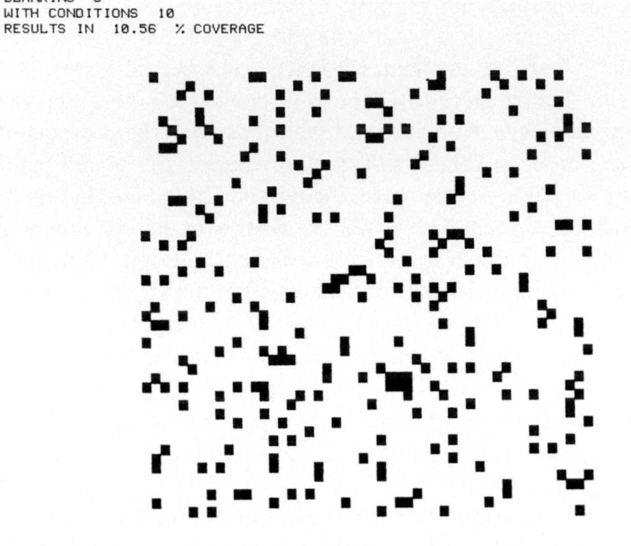

Figure 4.2. To simulate the structure of a paint film containing 10% by volume of a monosized pigment, every digit except 9 in the random number table in Figure 4.1 is considered to be empty space. The digit 9 becomes a pigment fineparticle.

gambling centre of the world. The fact that computers are becoming more and more powerful, and less and less expensive, is making it possible to simulate the behaviour of many systems using Monte Carlo routines, when consideration of such techniques was prohibited by expense as recently as the early 1970s. In this book, we shall employ several Monte Carlo routines to simulate the structure and behaviour of systems which can be described by fractal dimensions.

4.4 Randomwalks in One-Dimensional Space

The term **randomwalk** to describe the movement of an object undergoing Brownian motion and similar types of motion was first used in 1905. In that year Karl Pearson, a professor at London University, wrote a letter to *Nature* (a well known scientific journal) in which he requested help in the solution to the following problem:

"A man starts from a point 0 and walks a distance A in a straight line. He then turns through any angle whatever and walks a distance A in a second straight line. He repeats this process N times. I require the probability that after these N steps he is at a distance between R and $R + dr$ from his starting point at 0" [12, 13].

We shall begin our study of randomwalks by looking at a simple randomwalk in one-dimensional space. Consider the progress of a drunk in one-dimensional space to and from a lamp–post. To create one-dimensional space, we place the drunk in a ditch. Because of the ditch, the drunk can only move towards or away from the lamp–post (also in the ditch!). To keep our discussion simple, we shall assume that the drunk always takes steps of the same size and that he is sufficiently drunk that any movement he makes involves a step either to or from the lamp–post with equal probability, no matter how far from or close to the lamp he is when he takes the next step. We can model the progress of the drunk by using the random number table to simulate his behaviour. Thus, we shall decide that an odd number in the random number table represents a step towards the lamp–post and an even number represents a step away from the lamp–post (for this purpose the digit 0 is treated as if it were an even number). We can plot a graph of this simulated drunk's progress over a period of time, as shown in Figure 4.3(a). In this graph, D denotes the distance of the drunk away from the lamp–post after given total steps of N. After the first 50 steps, we see that the drunk is only 2 steps away from the lamp–post. We also note that in this particular simulated randomwalk, the drunk has visited the lamp–post 15 times.

We can summarize the drunk's progress by recording whether or not he was at the lamp–post. Instead of counting steps, we can record the times when the drunk is at the lamp–post. If we assume that the drunk took one step per second, we can label the horizontal axis "time." For this simple record, we would mark the set of events on a time axis as shown in Figure 4.3(b). The first thing to notice about this set of points on the "time" axis is how they appear to "cluster." As in the case of our simulated accident record described in the previous section, those unfamiliar with stochastic processes are surprised by this clustering of random crossing events.

To increase the reader's interest in developing some mathematical ideas for describing the packing of points on the line, such as the pattern in Figure 4.3(b), it should be pointed out that there are physical systems which will behave in a similar way to our staggering drunk, and that the ideas we develop concerning the clustering of events in stochastic systems will have important application in applied science. For example, an important device for capturing fineparticles in air leaving an industrial process is an **electrostatic precipitator**. The basic construction of such a device is illustrated in Figure 4.3(c). When using this type of device to clean up the air, the central wire is operated at a high voltage with respect to the cylindrical wall. The fineparticles carrying a positive change are repelled by the central wire to be captured by the wall. Let us assume that in order to assess the efficiency of the device, over and above pure chance collection of the dust fineparticles by random diffusion, we wished to study the movements of fineparticles through the device before any voltage is applied to the central wire.

We shall also assume that the flow of gas through the precipitator is turbulent with the individual fineparticles undergoing random movement. Since in our precipitator model we are not concerned with movement around the central wire as compared with movement towards the wall, we can simulate the movement of individual fineparticles by a randomwalk in two rather than in three dimensions. On this basis, we can now view the randomwalk of Figure 4.3(a) as being the track of a fineparticle moving with a turbulent air stream through a simple electrostatic precipitator before a capture voltage is applied to the system. The events on the time axis in Figure 4.3(b) now become

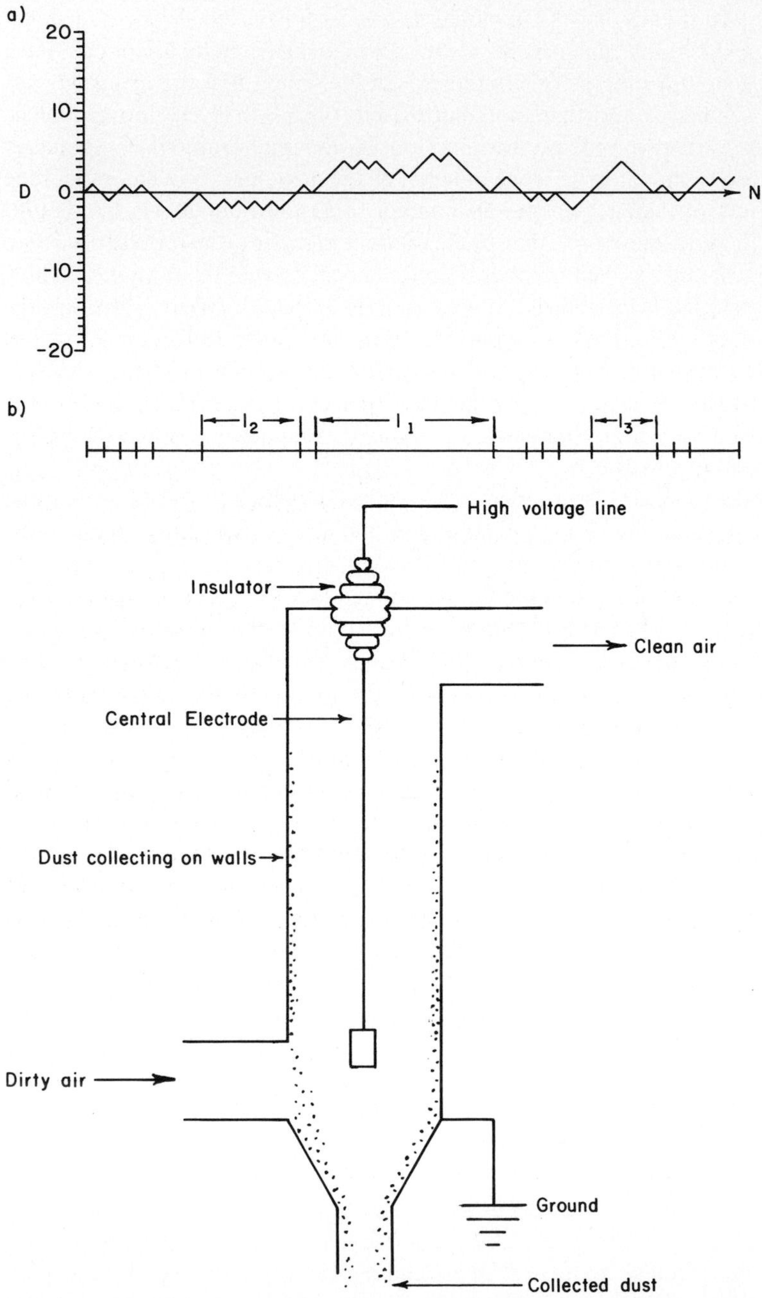

Figure 4.3. The randomwalk of a staggering drunk in one dimensional space generates a cluster of events on the time (number of steps) axis of the graph recording the randomwalk.
(a) Step record of the randomwalk.
(b) Event record of the drunk's visit to the lampost.
(c) Simple electrostatic precipitator.

collisions of the dust fineparticle with a central wire. Using this simple model to study fineparticles in an electrostatic precipitator, we could now simulate the effect of giving the wire an electric voltage with respect to the outer cylinder by altering the probability with which the fineparticle moved to and from the wire. For example, we could change the conditions of the randomwalk so that at any one step the fineparticle was five times as likely to move away from the central wire as it was to move to the central wire. This would simulate the effect of an applied voltage on the wire which would be attempting to drive the fineparticle to the wall against the randomizing effect of the turbulent air flow [14].

In developing our model of electrostatic precipitators, we could decide that when the total distance from the wire equals 20 times D, the size of the individual steps in the model, the fineparticle can be considered as captured by the wall. Using the data from a randomwalk model with and without enhanced probability of moving away from the central wire, we would be able to calculate how long the cylinder would have to be for turbulent diffusion to throw the fineparticle to the cylindrical wall by chance, and compare this distance to the increased capture frequency when the voltage on the central wire made it more probable for the fineparticle to walk away from the central wire towards the wall.

In industry, this type of modelling would probably not be considered a useful study, since very high voltages are used and the performance of the device without an applied voltage would not be of interest to the engineer. However, from an academic point of view, a discussion of this simple model is useful in that it starts to illustrate how randomwalk models can be of interest to the practising scientist.

One has to be very careful when using randomwalk Monte Carlo modelling because of the enormous variations that occur in the behaviour of randomly fluctuating systems, variations which are often surprising to those with little experience of stochastic processes. For example, in Figure 4.4 a second randomwalk carried out using exactly the same algorithm as that used to generate Figure 4.3(a) is shown. In this second randomwalk, we see that the drunk only visits the lamp–post once and that after 50

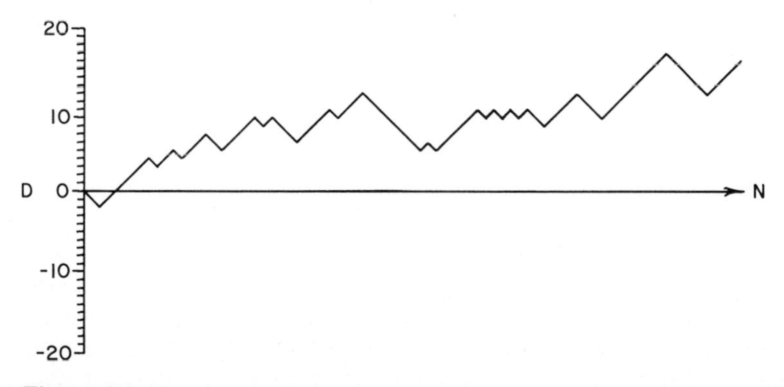

Figure 4.4. Progress made in moving away from the lamp-post by a staggering drunk varies enormously in any one set of steps, as can be seen by comparing the randomwalk of 50 steps shown in this figure with that in Figure 4.3.

strides he has actually managed to move 16 steps away from the lamp–post. If Figure 4.4 had been our first experimental simulation of the drunk's random progress, many students would have been tempted to think that there was something wrong with the random number generator, and that our simulated drunk could "see" where the lamp–post was each time he took a staggering step. In a problem of this kind, before we could estimate the average expected distance moved by a drunk from the lamp–post after 50 steps, we would have to carry out hundreds of simulated 50-step randomwalks to generate the pattern of dispersal expected after 50 steps.

The difference between the two randomwalks in Figures 4.3 and 4.4 also serves to caution the reader about seeing meaning in random patterns. If a randomwalk record of the variation in a physical system is difficult and/or expensive to repeat, one can be tempted to create grand theories on sparse evidence from random patterns. For example, Jacchia has drawn attention to the similarity of the recorded variations in world climate to the type of variation generated in a randomwalk such as those in Figures 4.3(a) and 4.4 [15]. Scientists looking at a graph of variation in temperature plotted against the year, which look like Figure 4.3(a), could claim it as evidence of a stable climatic period, whereas they would interpret a set of data such as that in Figure 4.4 as evidence that the climate of the earth is warming up. In reality, two sets of temperature records as different as those in Figures 4.3(a) and 4.4 are only evidence that the world's climate is fluctuating randomly. The applied scientist should always be sceptical of assigning causative patterns to a perceived "meaning" in a sparse set of unrepeatable data.

4.5 Delinquent Coins and Cantorian Dusts

In Figure 4.5, the record of a randomwalk described by William Feller in 1950 is shown [16]. This randomwalk is a record of the progress of a gambler's gain or loss when playing a simple gambling game of "pitch and toss." Let us assume that this game is played by Henry and Thomas, who each start off the game with a pile of coins. Every time a test coin is spun and it lands "heads" on the table, Henry wins a coin from Thomas. If the test coin comes down "tails" Henry loses a coin to Thomas. The height of the dots above the line in Figure 4.5 represents the number of coins that Henry has won from Thomas. When the dots are below the line, they represent the number of coins Thomas is winning from Henry. To those unfamiliar with stochastic processes, a surprising aspect of this record of win and loss is that in the first 500 pitch and toss activities, only rarely was Thomas in a winning position. For most of the time, Henry's winning dominated the game and he would be described as being lucky or on a winning streak. When presenting the data in Figure 4.5 in his book, Feller indicated that the graph in Figure 4.5 was specially selected from amongst those he had generated in experiments, since many of the graphs of the gambler's progress he had generated in his experiments looked too wild to be believable! His statement regarding the data is worth quoting at length:

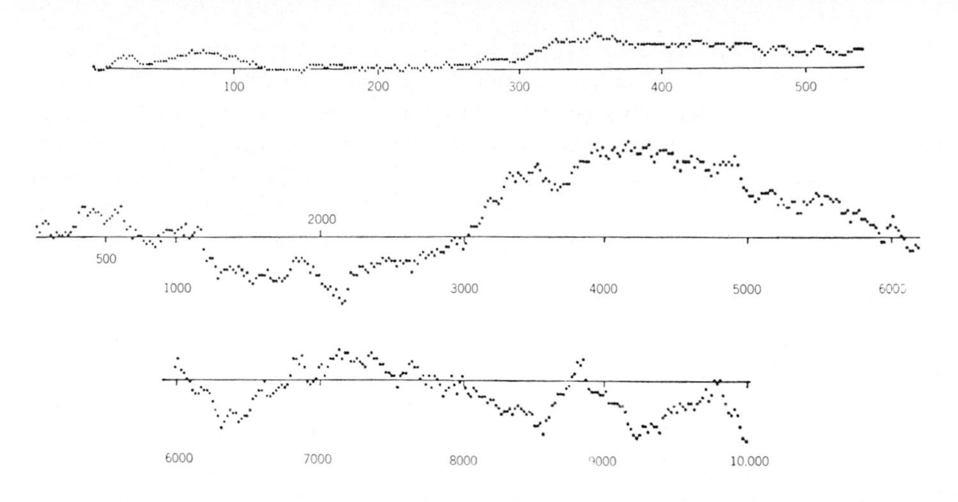

Figure 4.5. A famous record of a randomwalk model of a gambler's gain or loss in a game of "pitch and toss." This randomwalk, described by Feller, inspired Mandelbrot to use fractal dimensions to describe the packing of points on a line [16, 17] (from W. Feller, "Fluctuations in Coin Tossing and Random Walk," in "An Introduction to Probability Theory and Its Applications," Vol. 1, Ch. 3, 1950, reprinted by permission of John Wiley & Sons, Inc. Copyright 1950, John Wiley & Sons, Inc., New York).

"If a modern educator or psychologist were to study the long run of case histories of individual coin tossing games (that I recorded), he would classify the majority of coins as maladjusted. If many coins are tossed N times each, a surprisingly large proportion of them will leave one player in the lead almost all of the time and in very few cases will the lead change sides and fluctuate in the manner that is generally expected of a well behaved coin" [16].

The reader now knows why delinquent coins share equal billing with staggering drunks in the title of this chapter. Again, comments by Feller on the appearance of recorded randomwalks stress how most people have very little experience on which to judge the progress of a randomwalk (and perhaps that many psychologists are not to be trusted as interpreters of observed behaviour!).

The recorded randomwalk in Figure 4.5 played an important role in the birth of fractal geometry. Mandelbrot tells us in his book that in the 1960s he was asked to study noise signals that occurred in data transmission lines. When studying a pattern of noise events in the transmission lines which looked like the events in Figure 4.3(b), he remembered Figure 4.5, which he had studied in a probability textbook [17]. It suddenly struck him that the clustering of the noise signals on the telephone lines he was studying resembled the clustering of cross-over points in the record of the randomwalk winnings versus losses in Figure 4.5. Mandelbrot tells us that the key idea leading to the evolution of fractal dimensions came when he made an intuitive connection between the points of the line, such as those in Figure 4.6, with points on a line in a well known mathematical figure called the **Cantorian triadic set**, named after George Cantor, a German mathematician (1845 and 1918). Cantor developed a special branch of mathematics dealing with the problems of infinitely large numbers. For instance, he studied the problem of deciding how many points there are on a line. Cantor showed that one can answer the question, "how many points are there on a line," without knowing the length of the line. All lines have an infinity of continuous points! The fact that two lines of different lengths are said by mathematicians to contain the same number of points, infinity, seems contrary to common sense. The reason why specialist discussions about the number of points on a line appear to go around infinite circles will be discussed in the last chapter of this book, in a section entitled "The Philosophical Impact of Fractal Geometry." In that section, we shall find that the answer to the question, "How many points are there on a line?," all depends on what you mean by "a point."

To understand how Mandelbrot developed the use of fractal dimensions to describe sets of points on a line using the Cantorian triadic set, let us consider what happens if we were to let our staggering drunk wander back and forth to the lamp–post for an infinite period of time [18]. For an infinitely long walk, the step size of the drunk becomes infinitesimally small compared with the time axis on which his meanderings are recorded. For such a walk, the time axis of the graph summarizing the randomwalk of the drunk would have an infinite number of points. An infinite number of points connected together form a line of dimension 1. Although the set of points representing the number of times that the drunk was at the lamp–post would contain an infinite number of points, it would be a set of disconnected points with a packing density on the line less than that of the infinite number of points forming one-dimensional space on which the crossing points were recorded. Therefore, the crossing points, although infinite in number, would not fully occupy all of the infinite points forming the time axis, and could be allocated a dimension less than 1 to describe the packing density of the points representing visiting events on the time line. Mandelbrot describes the packing of points on a line which are infinite in number but which do not form a continuous line as constituting a **Cantorian dust** with a dimension less than 1.

To understand what Mandelbrot meant by a Cantorian dust, consider the sketch shown in Figure 4.6(a). This diagram is used by Mandelbrot to describe the structure of the Cantorian triadic set of points on a line space. To help visualize the structure of the Cantorian set, Mandelbrot asked us to consider the following theoretical experiment. What would happen if a superhuman blacksmith could cut the bar shown at the top of the diagram into half, and hammer the material in the two halves until they occupy two-thirds of the space of the original bar. The hammering would increase the density in the bars and generate the system shown in Figure 4.6(a). The blacksmith then repeats the process with a whole series of cuts and compressions, in which each time a residual bar is cut in half and hammered until it occupies two-thirds of its original space. The progress of the cut and compress experiment is illustrated by the sketches in Figure 4.6(a). Mandelbrot said that the process was to be carried out until

> "neither the printers press nor the eye could follow the compression of the material into the slugs, which would eventually be infinitely small and have infinite densities."

If one were to draw a line through the bottom set of slugs shown in Figure 4.6(a), the dots on the line represent, at the resolution of the diagram, the infinite number of points in the Cantorian triadic set described by Cantor. In theory the points marking the end and beginning of the infinitely dense slugs are an infinite number of points if we could look at them with infinite resolution. However, they also have measurable spaces between them which are characteristic of the way in which the points pack on the line.

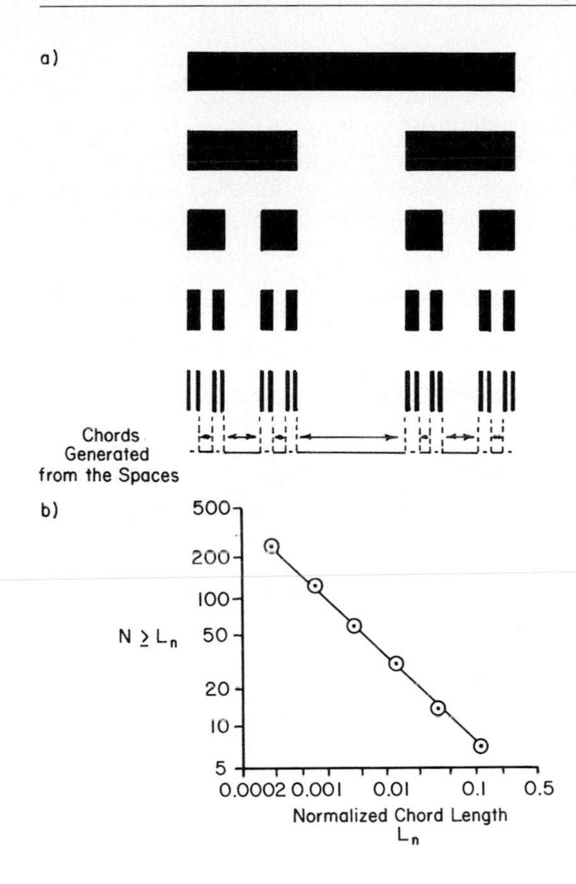

Figure 4.6. Mandelbrot was able to describe the clustering of points on a time line generated by noise signals by comparing them mentally with the pattern of points in line space of a Cantorian triadic set of points.
(a) Construction algorithm for forming a Cantorian triadic set of points of fractal dimension 0.63.
(b) Chord set of spacing between the points of a Cantorian triadic set area characteristic chord population (from B.B. Mandelbrot, "The Fractal Geometry of Nature," W.H. Freeman & Co., San Francisco, 1983, reproduced by permission of B.B. Mandelbrot).

Thus, in Figure 4.6(a) the intervals between the slugs and their frequency are shown. Again, theoretically, the process of recording the length of the gaps between the slugs could be carried out down to the enumeration of an infinite number of infinitely small gaps separating the points on the line. The fact that the points defining the limits of the Cantorian triadic set are infinite in number but fail to pack a line densely enough to produce a line of dimension 1 is a paradox. A **paradox** is defined as a statement that is apparently absurd or self-contradictory but is, or may be, really true.

In the language of set theory, the points forming the Cantorian triadic set of points on a line is known as a subset of points. The mathematical paradox of Cantorian set theory is that on a line there are many subsets of points which contain an infinite number of points but which fail to occupy space efficiently to generate a line of dimension 1. Because counting the number of points on a line leads us into infinite paradoxes we can switch our investigation from "counting points" to looking at the efficiency with which a point set occupies line space. For example, it can be shown from mathematical reasoning that it is useful to allocate a dimension of 0.63 to describe the density with which the Cantorian triadic set in Figure 4.6 occupies space along a line which, if infinitely packed, would have a dimension of 1.

Mandelbrot pointed out that the structure of the Cantorian triadic set is too regular to describe natural events of the real world, but that one could imagine that, if the basic pattern of points in Figure 4.6 were to be randomized with respect to each other, one would have a statistically self-similar set of points related to the Cantorian triadic set. Mandelbrot was able to show that the packing of noise events on the transmission lines he was studying can be described by fractal dimensions of less than 1. In applied science, fractal dimensions of less than 1 occur whenever we study events on a line scan used to probe a fractal dimension in a higher dimensional space, and will be discussed in more detail in Chapter 6.

Mandelbrot created the term Cantorian dust to describe subsets of points on a line. Of all the terms that he coined in his book on fractal geometry, this is the only one that I wish he had not invented. The reason for my lack of enthusiasm for this particular term is that one of the applications of fractal geometry in fineparticle science is the use of the dimensions of Cantorian point set to characterize the structure of dust clouds in air pollution and in the experimental study of dust deposited on a filter. As will be discussed in Chapter 6, one can draw a line across such a deposit and the points on the line formed by its interception with dust fineparticles form a Cantorian dust, which can be related to the size distribution of the deposited dust. In this situation it becomes confusing with the word dust having more than one meaning. However, since the term has become well established, we shall continue to use it.

Mandelbrot showed that a set of chords defined by the points of a Cantorian set would have the probability distribution function given by

$$P(L > 1) = 1^{-D}$$

when P is the probability of L, a given chord length is greater than or equal to l, and D is the fractal dimension of the Cantorian dust.

This equation is the basis of an experimental technique for determining the dimensionality of a Cantorian dust. Thus, if we look at the Cantorian dust in Figure 4.6 we can measure the length distribution of the chords between the points defining the Cantorian dust. The longest chord is L_1, the second longest chord is L_2, the third longest is L_3, and so on. If the probability of chords defined by the Cantorian dust is of the type indicated by the above equation, it can be shown that a plot of the number of chords equal to or less than a given length L on log-log graph paper will give a straight line of slope $-D$, where D is the fractal dimension of the Cantorian dust.

In the Figure 4.7 the chord distribution of the randomwalk in Figure 4.3(a) is plotted. It can be seen that this generates a linear dataline of slope m. It is dangerous, however, to interpret the slope of the dataline in Figure 4.7 as an accurate estimate of the fractal dimension of the points on the line representing the drunk's presence at the lamp–post because of the extreme variability of the chord distribution generated in a 50-step randomwalk. Thus, the data in Figure 4.4, which should generate the same dataline, gave us only one chord. One would have to have a much longer randomwalk than 50 steps before one could place any confidence in the estimate of the fractal dimension of the Cantorian dust to represent the time events when the drunk was at the lamp–post.

The fact that an apparently random variable, in this case the staggering drunk in one-dimensional space, can generate a set of chords definable by a regular mathematical function which display the type of dataline shown in Figure 4.7, comes as a surprise to

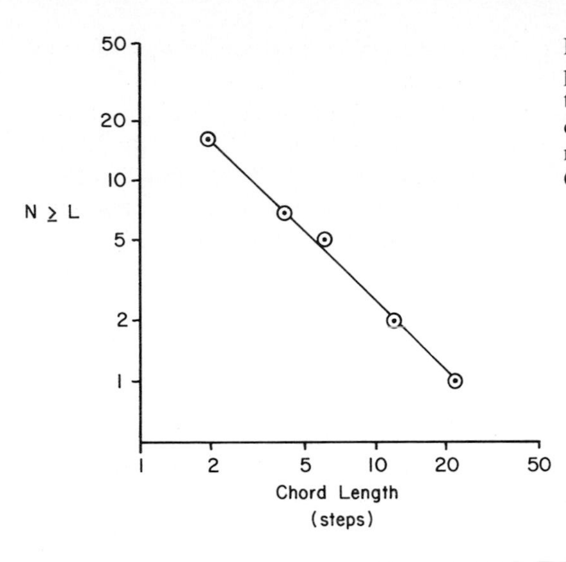

Figure 4.7. The chord distribution of periods between visits to the lamp-post of the drunk's walk in Figure 4.3 demonstrates the logic involved in measuring the fractal dimension of a Cantorian dust in one-dimensional space.

many students. The Greeks' view of the universe was that before, the gods organized everything, matter was in a complete state of disorder which they called "**chaos**." From one point of view, we can call the dataline in Figure 4.7 one of the surprising patterns of chaos. Discovering patterns in chaotic systems is an important branch of modern physics [19]. The opposite of chaos in Greek mythology was "cosmos." The theory of how the universe came into being is known as **cosmology** (one wonders how many people know that cosmetics are technically described as devices designed to bring order out of chaos in the decorated human face!). In our exploration of fractal dimensions, we shall discover several surprising patterns in apparently chaotic systems [20]. Whenever we chance upon a pattern of order in apparently random systems, we shall hope to indicate why that particular pattern has arisen in that particular system. However, sometimes we shall not be able to explain the reason for a given pattern. In such situations, one can only accept the fact that the pattern has been discovered and hope that further understanding of the system will eventually give a reasonable explanation of the origin of the observed pattern.

The two randomwalks shown in Figures 4.3(a) and 4.4 are examples of what mathematicians call a "**Markov chain**," named after the Russian mathematician A.A. Markov (1856-1922). A mathematical dictionary describes a Markov process as "a stochastic process in which the future is determined by the present and is independent of the past." Thus, in our drunk's staggering progress, the way in which he moves at any one time, which we can call the present, is independent of the past. His next step is determined by probability and not by past meanderings. However, the zig-zag lines in Figures 4.3 and 4.4 are records of his past movement. His location at a particular instant in time is not independent of past movements. A record of the result of a **Markovian process** is known as a Markovian chain.

We shall discover that whenever a Markovian chain of events exists, there will probably be a fractal pattern embedded in the pattern of events. Thus, in the case of the staggering drunk, his progress is a Markovian chain, and the intersections of his historic track with the zero displacement line generate a Cantorian dust which has a fractal dimension.

A mineral processing engineer and/or a metallurgist looking at the Markovian chains in Figures 4.3(a) and 4.4 without reading the labels on the axes may have a Rorschach reaction that he is looking at the spread of a crack in a rock or a piece of metal. If he now wishes to define the structures of the crack, he can draw a line through the crack and characterize the crack structure by the dimensionality of its Cantorian dust. These two engineers have the option that they do not need to wait for the crack to cross an arbitrary reference line such as that of zero displacement-they can draw their line through the average progress of the crack to define its structure (we shall discuss the fractal description of structures in Chapter 10).

If a geographer was to be presented with the Markovian chain in Figure 4.3 without any indication of what was plotted on the graph, his Rorschach reaction would probably be that he was looking at a section through a land mass to describe the topography (the surface structure) of the region. Thus, the Markovian chain below the central reference line would look like the depth contours of a lake, and the central part of the chain would look like a range of mountains with two or three river valleys being shown in the topographical section. Mandelbrot tells us that because he recognized the similarity between the Markovian chain representing a randomwalk and a section through a land mass, he went on to create algorithms which could be used to create realistic looking landscapes and geographical entities by randomwalks in two-and three-dimensional space (see reference 17, p. 240).

4.6 The Devil's Staircase and Crystal Structure

If one generates a graph of the distribution of the slugs in the compressed segment of the bar used to illustrate the structure of the Cantorian triadic set in Figure 4.6(a), one generates what is known as the **Devil's staircase.** This graph is shown in Figure 4.8. Let us assume that the original bar had a mass of 1. Since the compression of the separated slugs used to visualize the Cantorian set does not alter the amount of mass present in the original bar, the sum of the mass in the individual slugs will equal the mass present in the original bar. Therefore, by the time that one moves from one end to the other of the sequence of slugs defining the Cantorian triadic set, one must reach a total of mass 1. En route to this total, there must be an infinite series of small increases in mass as one encounters each small slug of infinite density. Therefore, the number of steps in the Devil's staircase is infinite. It can be shown that the infinite jumpy curve defining the Devil's staircase has a finite length of 2 and a dimension of 1. Mandelbrot stresses in his book that the curve outlining the Devil's staircase is continuous, but non-differentiable.

A photograph of the Devil's staircase should be in every elementary book on calculus to warn the teacher that, when teaching differential calculus, he should discuss with the students that many curves encountered in nature have no differential function.

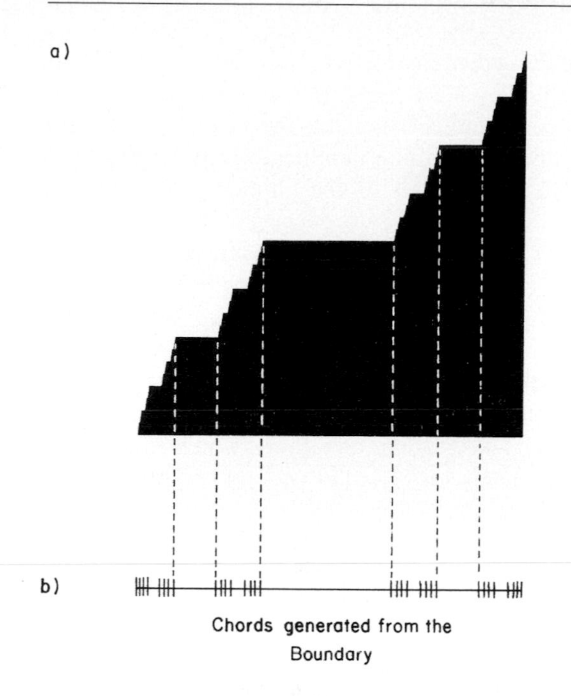

Figure 4.8. A graph of the mass distribution in the dense slugs used to visualize the Cantorian triadic set generates the Devil's staircase. This staircase increases in height "over infinitely many, infinitely small highly clustered jumps corresponding to the mass of slugs," defining the Cantorian set in Figure 4.7 (reference 17, p. 82).
(a) Devil's staircase.
(b) Projecting edge patterns discerned in a crystal boundary can generate a Cantorian dust the fractal dimension of which is characteristic of the crystal structure (from B.B. Mandelbrot, "The Fractal Geometry of Nature," W.H. Freeman & Co., San Francisco, 1983, reproduced by permission of B.B. Mandelbrot).

My Rorschach reaction to the Devil's staircase was that it reminded me of the type of sketch used in many elementary physics textbooks to explain why a wire stretched beyond its elastic limit changes from shiny to dull. The dullness of the strained wire comes from the fact that the crystal planes in the overstretched wire slip relative to each other to produce a rough surface which does not reflect light as well as the undisturbed surfaces which existed before deformation.

If one encountered a crystal boundary in metallurgy or in geology that looked like the Devil's staircase, one could generate a fractal dust on the axis of a graph of edge structure by projecting the edges on to the abcissa to form a Cantorian dust as illustrated in Figure 4.8(b). The dimension of the Cantorian dust formed in this way could then be related to the rugged, non-differential boundary of the crystal being examined.

4.7 Pin-ball Machines and Some Random Thoughts on the Philosophical Significance of Fractal Dimensions

The Markovian chains in Figures 4.3(a) and 4.4 can be regarded as the first steps in modelling another problem of interest to the fineparticle specialist using a Monte Carlo routine. Let us consider that the zero point on the time axis of the randomwalk in Figure

4.3(a) is the outlet of a chimney stack emitting dust fineparticles into the air. The stream of fineparticles coming out of a real chimney stack is known as a chimney **plume**, because from a distance it looks like a plume (feather) stuck into the surface of the earth. Let us assume that the fineparticles will be blown along the time axis of the plume by a steady wind as they disperse from the central axis of the plume by a turbulence generated randomwalk.

To model the dispersion of fineparticles in a chimney plume, we could carry out many randomwalks such as those in Figures 4.3(a) and 4.4 and superimpose them to find the distribution of fineparticles down-wind of the chimney stack at any given distance from the chimney stack. However, we can also model the random dispersion of the fineparticles using **an analogue computer**. In a digital computer, the variables are represented by numbers and the behaviour of the system is calculated using numbers. In an analogue computer, the system to be studied is modelled using something which mimics the physical behaviour of the system to be studied, with variables that are easier to handle. In the days before the development of cheap electronic computers, many difficult problems in engineering had to be solved using analogue computers. Thus, the flow of water through a network of pipes was modelled by measuring the flow of electricity through a network of wires in which the electrical resistance of the wires was made similar to the flow resistance of the pipes through which the water would flow. The use of analogue computers in applied science has almost been eclipsed by the development of mini-, macro- and super-digital computers. However, using an analogue computer to study randomwalks has the advantage that the analogue computer can make the randomwalk very visible.

The analogue computer that we shall use to study the dispersion of fineparticles in a chimney plume is shown in Figure 4.9. A set of pins are arranged in a symmetrical pattern on an inclined board, at the base of which is a set of compartments. Steel balls are dropped down through the array of pins one at a time and allowed to accumulate in the compartments as shown in the diagram. This type of board was originally developed by Sir Francis Galton (1822-1911) who was interested in the possible links between heredity and intelligence. He was a first cousin of Darwin, who developed the theory of evolution. In 1889, Galton wrote a book entitled "Natural Inheritance," in which he described the system shown in Figure 4.9. He called the system a Quincunx board, because of the fact that the number 5 (which in Latin is "quinque") as shown on a die ⁙ can be seen in the pattern of nails [21]. The pattern of balls accumulating at the base of the Quincunx board can be seen to be a bell-shaped curve. In terms of our chimney plume, the distribution of the balls in the containers at the bottom of the Quincunx board would describe the concentration of the fineparticles across the plane through the plume at the end of the board. Remember that we have allowed the balls to accumulate at the base, whereas in a real chimney plume the individual fineparticles would be undergoing many turbulent diffusion steps and all of the fineparticles would be present at the same time in the cross-section of the plume shown by the line AB across the board [22, 23].

The "**bell curve**", defined by the accumulated balls in our simple analogue computer, occurs in many systems studied in applied science. Thus, Warren Weaver in a book called "Lady Luck" tells us that this type of curve can describe:

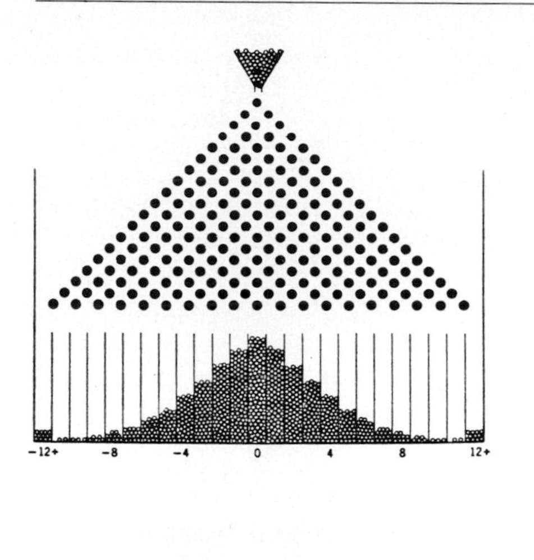

Figure 4.9. The dispersal of fineparticles in a turbulently diffusing chimney plume can be modelled with a simple analogue computer using nails and steel balls (from PR0BABILITY AND EXPERIMENTAL ERRORS IN SCIENCE, L.G. Parratt, Dover Publications, Inc., NY, 1971).

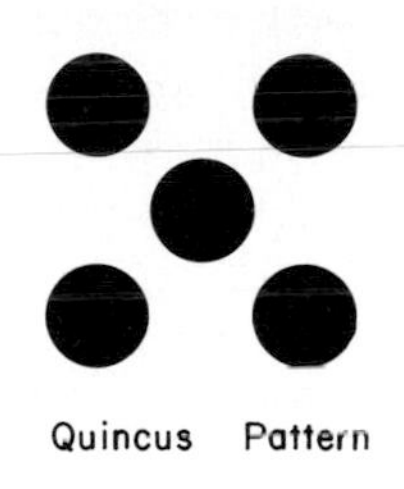

1. The results of I.Q. tests taken by many students.
2. Deviations from the point of aim in artillery fire and bombing.
3. Heights of army recruits.
4. The lengths of leaves on a tobacco plant.

This type of curve occurs so often in science that it is often called the "**normal**" (that is, the "usual") **probability distribution**. The normal probability curve is also called the **Gaussian probability curve**, in honour of the mathematician Johann Carl Friedrich Gauss (1777-1855). Another name for the bell curve is the **arithmetic probability curve**. In this book, the preferred term is the Gaussian probability curve.

Let us now consider why Galton would be interested in this type of analogue computer. Let us imagine a single ball moving through the array of pins. As it met the first pin, there would be a probability of it moving to left or right. The mechanics of the movement to the left and right would depend on many minor factors, such as the angle of approach of the ball to the pin, the energy of the ball and the cleanliness of the ball and the pin. However, after it has bounced to the left or right at the first pin, it now has another independent encounter with a pin at the next layer of the board, which will decide whether it will go to the left or the right with equal probability. In this way, any individual ball works its way down through the pins as shown in Figure 4.10, where the overall Markovian chain of one of the balls moving down through the pins is illustrated. From one point of view, the final position of a ball at the bottom of the Quincunx board in Figure 4.10 can be regarded as having been caused by, in this case, 19 different

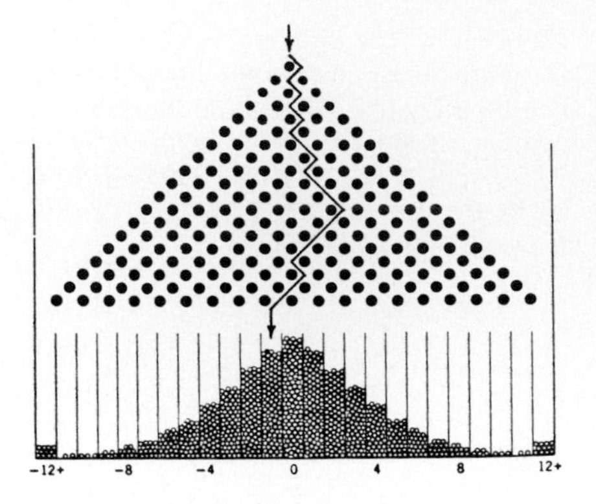

Figure 4.10. The Markovian chain representing the recorded trajectory of a ball down to a given position at the bottom of the Quincunx board can be regarded as having been caused by 19 causes represented by the 19 layers or lines of pins as indicated in the diagram (from PROBABILITY AND EXPERIMENTAL ERRORS IN SCIENCE, L.G. Parratt, Dover Publications, Inc., NY, 1971).

collisions or causes as it moved en route to the bottom of the board. Each horizontal line of pins can be regarded as a set of causes. From this perspective, the final position of the ball is caused by the chaotic interaction of the nineteen causes acting on the ball. Hence we can state that the Quincunx board illustrates that when many small causes interact chaotically, that is at random, to produce a final magnitude of the physical variable, that variable will often be distributed according to the Gaussian distribution function. Stated in another way, a system created by the random interaction of many equally strong variables will generate a set of observed values that will define a bell curve when plotted as a graph. Although not usually stated in this way, the bell curve is actually one of those surprising patterns which occur in apparently randomly fluctuating systems. One of the theories of inherited intelligence that Galton wished to investigate was that the intelligence (whatever that is) of a group of descendants of two people would have levels of intelligence distributed according to the Gaussian probability distribution. Thus, the Quincunx board was built to model the distribution of intelligence levels that would be produced by "random chance." The resemblance of the Quincunx board to pin-ball machines in amusement arcades is obvious. If one were to randomize the positions of the pins to a small extent, it would probably not affect the probability of arrival of any particular ball in a given slot at the bottom of the pin-ball machine. The structured pattern of the Quincunx board is an illustration of how one can often restrict the amount of randomness in a physical model of a stochastic process because redundant randomness does not alter the statistics of the process. A randomly moving ball through a regular array of pins is sufficient randomness to generate a randomly distributed variable. It is also obvious that one can deliberately alter the positioning of the pins to "fix" the probabilities of arrival at the base of the board. This non-random variation of the pins is used to fix the odds in a pin-ball machine used for gambling purposes. There is really no such thing as the **laws of chance**. Chance does not obey any laws. What are called the laws of chance are surprising patterns in chaos which may well surprise us again if we seek to repeat our observation of a stochastic process. To demonstrate the non-causative nature of the so-called laws of chance, consider the nature of the distribution of marks obtained by fifty students in a class test. The factors determining how well these students did in a given test would be a combination of some of the following causes:

1. The native intelligence of the child.
2. The interest of the child in the subject.
3. The capability of the teacher of conveying the concepts under test.
4. The freshness of the student (watching late-night television can be a negative cause in the performance of students!).
5. The obscurity or clarity of the test questions.
6. The emotional stability of the student at the time of this test.

These many small causes operating amongst a large number of children will produce a mark per student which will generate a bell curve when plotted as a graph. However, this distribution function only describes the current level of achievement of the group of students and does not predict their individual performances. Any or all of the above causes could change and shift the position of individual students in the class rating, or produce a considerable change in all of them. Thus, I once went from the bottom of a school class of 82 students to the top of the class in successive terms, but I do not think that my intelligence changed during the holidays between terms.

Some educators appear to have been so mesmerized by the bell curve that they have stopped using their critical judgement when evaluating the performance of students. I had a personal experience of the inappropriate use of the bell curve to grade students one semester when I was asked to teach a course on the Impact of Science and Technology on Society for another department of the University. I had 16 students in the class. The class was an elective and anyone who anticipated that they were not going to obtain good grades dropped out very early on in the semester. These dropouts cut off the bottom end of any bell curve that one might like to have imagined as being descriptive of the students. When I came to hand in the grades, I allocated 5 As. I was told by the departmental head that in a class of that size there could only be one A and that he would shift the grades with a bell curve. The net consequence was that the average student grade was shifted down by about 10%. I vigorously protested about the inappropriate use of bell curves, pointing out that if the students knew at the beginning that there was only going to be one A, this would be counter–productive as an incentive to good performance in the class.

I also pointed out that if they persisted in the use of the bell curve, they would never be able to demonstrate any improvement in student quality in their collegiate activities. Since these arguments failed to shift the attitude of the head of the department, I was driven to suggesting the immodest probability that perhaps I had been a very good teacher, and that the students deserved better grades because they had been well taught. This final argument also fell on stony ground. The head of the department was bent on being procrustean. Procrustes was a person in Greek mythology who operated a kind of hotel on the outskirts of Athens. When you visited his establishment, you were offered free hospitality for the night but it was not mentioned that to avail yourself of this hospitality you had to fit the available bed exactly. If you were a little longer than the bed, then parts of your anatomy were chopped off to ensure a good fit. On the other hand, if you were too short you were stretched on the rack until you fitted the bed. (His name in Greek meant "the stretcher," similar to the type of name we give to describe modern individuals as the bouncer, the basher or the enforcer who removes rowdy customers from a pub or club). It is my opinion that teachers who use bell curves to grade students should be given a free week's holiday at a motel run by a modern day Procrusteas. I offered such a vacation to the head of the department who changed the grades of my students.

At several points in our discussion of data interpretation we shall have occasion to use the adjective procrustean. A **procrustean** device is one that achieves uniformity by violent treatment. Many graph papers with unusually structured scales on the ordinate and abcissa can be procrustean in their display of datapoints. In some fractal dimension studies, investigations start off with a preconceived notion of what fractal dimensions should fit a given natural boundary. Consequently, they sometimes pummel their data procrusteanly to fit the desired fractal dimension. The wise applied scientist will adopt the strategy that he will let a fractally structured system manifest its inherent fractal dimensions and then seek to understand the physical significance of differences between anticipated and measured fractal dimensions, rather than force their data to fit preconceived ideas.

The diagram shown in Figure 4.9 is based upon a discussion of experimental error in a book by Lyman Parratt [22]. He uses the data in Figure 4.9 to discuss the experimentally established fact that errors in physical measurement are often Gaussianly distributed. Unfortunately, in a discussion of accuracy and precision, the term **error** is often used when a better term would be uncertainty. Again, all too often in everyday speech, **accuracy** and **precision** are used as interchangeable terms when they have different distinct meanings. The precision of a numerical estimate of physical quantity indicates the variation that would occur if the measurement were repeated. It is an estimate of the limitations of the measurement procedure. Consider, for example, the simple experiment discussed earlier in which 20 students attempted to measure the diameter of a wire using a screw caliper. If the average of the measurements were to be quoted as 1.02 mm, the precision of this statement is taken to be that we know that the average diameter of the wire lies between the limits 1.03 and 1.01, since the precision of the method is quoted to two decimal places. The statement 1.02 is a very precise statement. However, if the caliper were calibrated in tenths of inches instead of in millimetres and the student had misread the scale, the quoted value could be inaccurate although precise, because the scale had been wrongly interpreted by the student during the investigation. Quoting inches instead of millimetres is an error, but the spread of values from 1.01 to 1.03 is not an error, it is a statement of the uncertainty or lack of precision in the measurements. If the 20 measurements generated by 20 students measuring the diameter of the wire were to be tabulated, in all probability the spread of uncertainty would be describable by a bell curve. The uncertainty represented by the bell curve would probably be due to two causes. First, each of the small measurements would have a small range of imprecision but it may be, as already discussed, that the use of the same piece of wire by many students had probably created a real range of diameters at different points on the wire, so that some of the uncertainty in the quoted value of the diameter could represent a real variation in the diameter of the wire. Thus, the spread of measured values reported by the 20 students could vary owing to at least two causes:

1. the lack of precision in the measurement;
2. the fact that there was no direct answer to the question, "What is the diameter of the wire?," because the diameter itself varied in space.

The word "error" comes from a Latin word which means to wander away from a position. The term "error" should be reserved for uncertainty arising from the investigator's inability to estimate precisely and accurately the value of a given quantity. The term "uncertainty" should be used to describe a spread of values which may be dependent on a real variation in the quantity being measured. One of the important processes in applied science is the inspection of a spread of values generated in an experiment and attempting to distinguish between error and uncertainty. One should always clearly quantify and clearly separate the lack of precision and certainty in a reported measurement which is due to error and that which is due to real fluctuation in the quantity being estimated precisely. New students often learn statistical methods of handling experimental data in a "cook book" style, that is, they learn how to calculate averages and standard deviations but often remain essentially uncomfortable with statistical reasoning. Parratt states that because scientists often lack a fundamental understanding of the patterns of probability, they are often "spiritually uncomfortable" in their own fields and science as portrayed by the scientist cannot fit comfortably in the society of other human activities and knowledge. When measuring the fractal dimensions of a system, it becomes very important to distinguish between inherent variations in a variable and experimental error in the measured data. Thus, when constructing a polygon of side λ to a rugged fineparticle profile to estimate its perimeter, the precision with which we can measure a given polygon may be high. However, because the actual polygon structure of side length λ can vary considerably with the starting point used to construct the polygon, the range of estimates of perimeters at a given value of λ can be high. This variation in perimeter estimates is not error or lack of precision, it is actually information on the ruggedness of the profile. Particularly for coarse resolution exploration of a rugged boundary, the variation in perimeter estimates at a given stride length of exploration is related to the fractal dimension of the boundary.

After attempting to teach several generations of students the art of the statistical description of physical systems and variations in the estimates of fractal dimensions of a given system, it is my conclusion that the students are uncomfortable with some of the surprising patterns of chaos because they fail to distinguish between descriptive and deterministic laws of science. Philosophers argue over the reasons why western science blossomed so dramatically in the 17th century, but it is generally agreed that one important element in the development of modern science came when man stopped believing that the events of the world were the manifestations of the capricious acts of many gods. When man started to believe that the universe was the manifestation of the rational acts of a single God, he searched for order and found it (or, as some atheists would say, he thought he "found it"). Thus people like Galileo, Kepler and Newton were deeply religious men who sought to seek order in the universe as a way of understanding the ways of God. Kepler was driven to develop his theory of the elliptical orbit of the planets because, when he measured their orbits exactly and checked the assumption that they were circles, he found that diameters at right-angles differed by 2%. This is a quantity most students would be happy with! Kepler tells us that his motivation for looking for alternative geometric structures for the planetary orbits, leading to the discovery of the elliptical orbits to describe the motion of the planets, was that "God did not make bad circles."

The progress of science over the last three centuries has changed the philosophical perspectives against which students are taught the structure of the universe. Implicitly and explicitly, they are taught that everything that one sees in this world is a manifestation of a cause and effect relationship. It is drilled into them that the three laws of motion set out by Newton govern this world and that every effect must have a cause. Consequently, the student is reasonably happy with Newton's laws of motion that connect mass and velocity with force. This type of law is deterministic in that, if we hit a ball with a given force, and if we know the conditions of flight, then the flight of the ball is determined by its environment. When the student looks at the distribution of fineparticles across a chimney plume and he is unable to predict the position of any particular fineparticle after it undergoes a Markovian chain of motion, he becomes uneasy with the lack of determinism in the description of the movement of an individual fineparticle. He has been drilled to seek cause and effect relationships and he is uncomfortable with bell curves because, when he uses "probable description," he must abandon "cause and effect" determinism.

The legal world has great difficulty in living with probabilities, since it too is structured on a simple alternative of "guilty" or "not guilty." A trial lawyer once came to ask me, "Does asbestos cause cancer?" When I said to him "yes and no," that some kinds of asbestos, in some kinds of individuals, in some kinds of situations, could cause cancer, there was a look of bafflement on his face. He was frustrated that there was no simple relationship of cause and effect between asbestos fibre and cancer [24].

We could avoid the philosophical unease that tends to accompany an introduction to a study of the patterns of probability if we insisted on distinguishing between determinative and descriptive relationships. When a physical variable is found to be describable by a Gaussian distribution, this relationship does not govern the behaviour of the system, but only describes the outcome of observed behaviour. For example, the bell curve only describes student grades and does not determine them in the absence of procrustean minded teachers.

Many physics students have difficulty with a subject known as **quantum mechanics**, because they arrive at the subject thoroughly indoctrinated with the predictive laws of mechanics. In quantum mechanics, absolute certainty slips away and the students find themselves having to deal with probable relationships, which at the present level of knowledge have no logical explanation. For example, the pattern of events created by millions of photons passing through a narrow slit leads to a distribution of energy which we describe as a diffraction pattern. The students are baffled when they encounter the measured energy distribution in a diffraction pattern, because they feel they ought to be able to understand this type of pattern just as they understand the movement of a baseball when hit by a bat. In actual fact, the relationships of quantum mechanics are not causative relationships from predictive laws; they are observed patterns of behaviour which often defy intuitive understanding.

Scientists themselves are divided as to the absolute meaning of the descriptive relationships of quantum mechanics. Some scientists see an analogy between the relationships of quantum mechanics and the surprising patterns of chaos in stochastic processes, and they take the relationships of quantum mechanics to be evidence that the universe as a whole is one stochastic soup.

Other scientists look at the relationships describing the movement of photons through a slit as being a limited description of reality with which they must work until they are able to gain a deeper predictive understanding of the movement of individual photons; to this group of scientists stochastic description is a temporary expedient. Those who wish to adopt this latter perspective, regarding the philosophical implications of quantum mechanics, can take heart from the fact that this was the position of Einstein. Einstein refused to believe that the universe was a stochastic soup and made the famous comment that, "God does not play dice with the Universe." His view of the Universe was that underneath observed patterns of behaviour, there were still unknown causes to be discovered in a rationally constructed universe. Lincoln Barnett in a biography of Einstein reports:

"Einstein more than once expressed the hope that the statistical methods of quantum physics would prove a temporary expedient."

Someone who believes that the universe is ruled by "laws of chance" is reverting to the pre-renaissance fatalism that whatever happens is the capricious will of the Gods, whoever they are.

Some interesting randomwalk data which illustrate the difference between descriptive relationships and predictive laws have been reported by Henderson [27]. He studied groups of people walking around Sydney in Australia. One group of people he studied were students walking on a footpath outside a library at the University of Sydney. His data are shown in Figure 4.11. It can be seen that the distribution of velocities of the 693 students approximates to a bell curve, and that the distributions of the speeds of the various students could be described using the Gaussian distribution function. However, the relationship does not describe the things which are causing the individual students to move at a given speed. Furthermore, if one were now to assume that the students were being chased along the path by a hungry tiger, their new distribution function may still be Gaussian, but it is a safe bet that the average speed of motion would be much higher than for the data in Figure 4.11. Thus, the new distribution of speeds would differ from

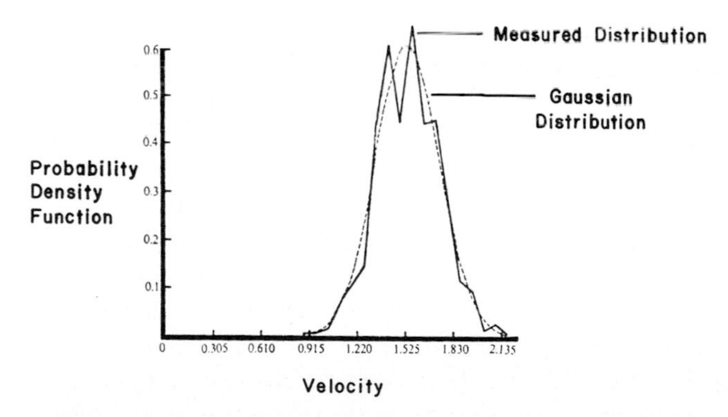

Figure 4.11. Bell curves are useful descriptive relationships as illustrated by the data of Henderson who studied students walking about at random. (reproduced courtesy of Dr. L.F. Henderson and reprinted by permission from Nature, Vol. 229, pp. 381-383. Copyright (c) 1971 Macmillan Journals Limited).

those observed by Henderson because a new cause was operating in the environment of the students. In fact, when observing the motion of the students, the detection of a sudden increase in speed may be one way of detecting roaming tigers!

Fractal dimensions have many things in common with the Gaussian distribution. First of all, the way in which a structure which can be described by fractal dimension is generated can often be modelled by a randomwalk process. Boundaries of systems that can be described by fractals tend to be generated by the random interaction of many relatively equal causes. Like the Gaussian distribution, a fractal dimension describing a system is only a descriptive relationship and not a deterministic one. A shift in the interaction of the underlying causes creating the fractal structure will change the magnitude of a manifest fractal dimension (see, for example, the discussion of the effect of electrostatic forces on the structure of simulated soot agglomerates in Section 5.3).

The utility of a fractal dimension used to characterize a physical structure lies in its descriptive power and in the clues that the magnitude of the fractal dimension provide to the investigator concerning the type of interacting causes which are generating the observed phenomena.

Gaussian distributions and other probability distributions are sometimes used when, although we can understand underlying causes generating individual items within a population, for the purposes of a given study a detailed study of causes of individual items is not appropriate to the problem being studied. Thus, if we were providing uniforms for the army, we would need to know the probability distribution of sizes so that we could provide the right number of uniforms for each size group of soldiers. If one were a custom designer of individual uniforms, one could investigate each individual soldier to determine the size and height of the individual to design a tailor-made uniform for each soldier. However, data on each soldier would not be of interest to the many manufacturers of uniforms.

In the same way, fractal dimensions are sometimes useful as summaries of data which otherwise would swamp the investigator. For example, when we come to study the structure of a porous body, we shall find that we can gather all sorts of information on the number and size of the various holes in the porous body; however we shall also discover that in some situations the "holiness" of a porous body can be described by a fractal dimension between two and three. There is already some indication that this type of fractal dimension is a useful summary of the structure of the body which enables us to discuss the structure of a porous body with someone else without overwhelming them with zillions of data on millions of holes.

In Figure 4.12, some interesting profiles which illustrate how a fractal boundary can be generated by the random interaction of many causes are presented. Profile A is a profilometer trace of the surface of an alumina as provided by the manufacturer. A profilometer is a device for quantifying the structure of rough surfaces. A needle riding at the end of a moveable arm is moved across a surface and the up and down movements of the needle are magnified and recorded. The maximum difference between the height and depth of the troughs shown in the profilometer trace of the initial surface is 13 μm, but the depth of troughs between adjacent mountains is less than this and none is much larger than 2-3 μm, which is approximately six times the wavelength of visible light. This type of finish is not as highly polished as a mirror but certainly looks "shiny."

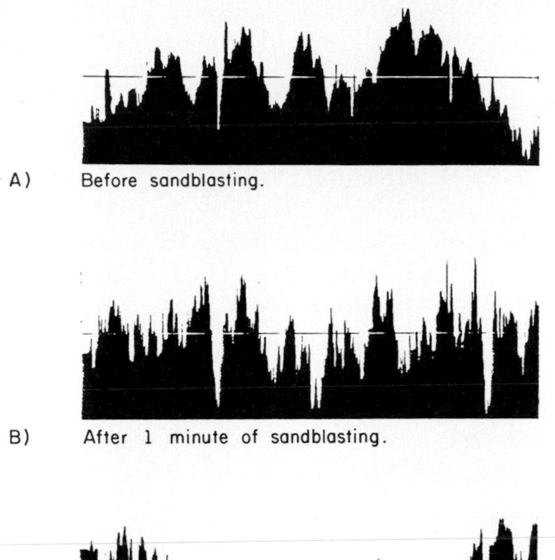

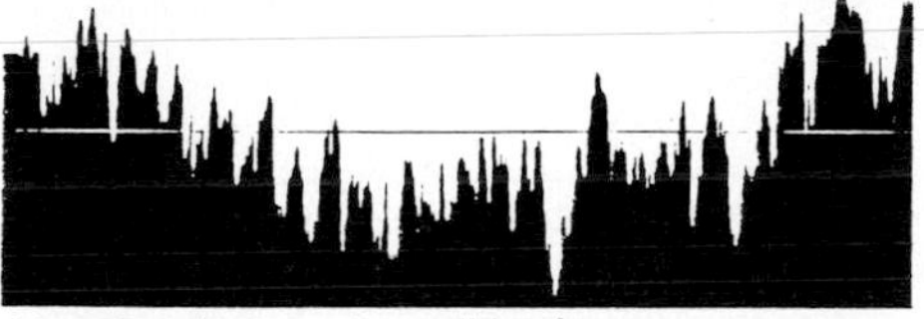

Figure 4.12. The roughness of a surface can be increased by sandblasting. The bombardment by many sharp sand grains gouging out pieces from the surface generates a boundary formed by many small interacting causes and manifests a fractal structure (from R. Trottier, MSc. Thesis, "Fractal Description of Fineparticle Systems," Laurentian University, 1986; reproduced by permission of R. Trottier).

In Figure 4.13, the fractal dimension of the boundary of the profilometer of the "as produced" alumina sheet as measured by erosion dilation logic is shown as line A. It appears to have two regions of fractal structure. At coarse resolution it appears to have a fractal dimension of 1.28 and at high resolution 1.18.

The same surface, after being sandblasted for 1 min. is shown in Figure 4.12(b). Note that the linear scale describing height of the mountains and troughs on the surface, as determined by the profilometer, is different from that in Figure 4.12(a). To explore the type of ruggedness, as distinct from the absolute roughness of the surface, the magnification of the graphs in Figures 4.12(a), (b) and (c) have been adjusted so that they have the same distance between maximum depth and maximum height of the troughs and mountains on the rough surface.

The Richardson plot for the dilation logic exploration of the profilometer traces of the 1 min. sandblasted surface is shown in Figure 4.13, line B. It can be seen that at coarse resolution the roughness has increased to an average value of 1.35, but that at high resolution the second region has increased in roughness from 1.18 to 1.25. Thus, not only has the absolute roughness changed but also the fractal dimension of the surface has increased. The roughness created by the sandblasting can be regarded as a random interaction of many causes. Each small fineparticle of sharp sand hitting the surface at high velocity gouges out a small piece of metal to create a roughness element not present in the surface before the fineparticle hit the surface. The profilometer trace after 1 min. is a graphic record of the combined effect of the gouging actions of thousands of small

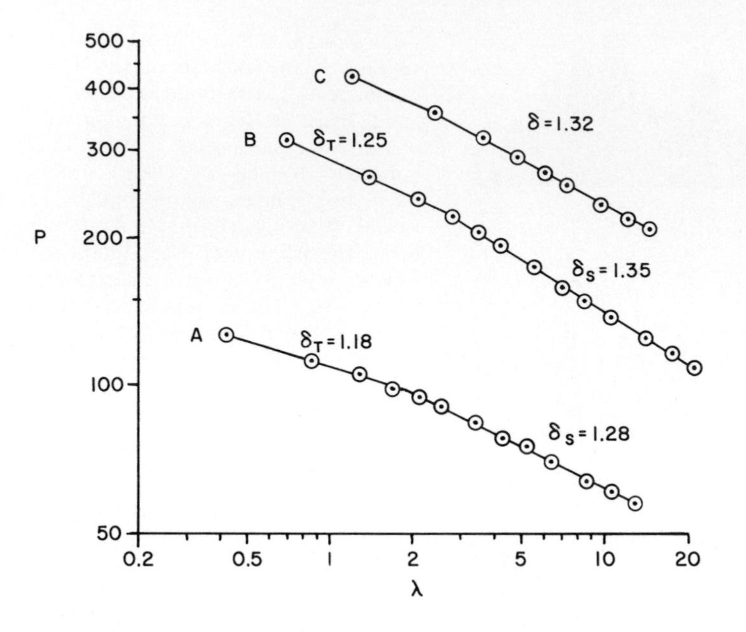

Figure 4.13. The fractal dimensions of the rough boundaries in Figure 4.12 can be quantified using dilation logic [26, 27].

fineparticles hitting the surface at random. Thus, the change in roughness from 1.28 to 1.35 at coarse resolution has been caused by the random interaction of many fineparticles undergoing a randomwalk from the nozzle of the sandblasting equipment, to impact on and gouge out the surface of the alumina.

In Figure 4.12(c), the profilometer trace of a surface sandblasted for 5 min. is shown. Again, the absolute roughness has increased. However, the fractal dimension of the surface has actually decreased slightly and there are no longer two regions of structure. This probably corresponds to the fact that, during the sandblasting action, after a certain roughness has been reached the actual fractal dimension of the profile will remain approximately constant, since any sharp peak that develops at any instant during the sandblasting is likely to be removed by a subsequent gouging action of another fineparticle hitting the surface. Thus, the continued erosion of the surface by the sandblasting produces an increase in absolute roughness but the structure of that roughness, as described by its fractal dimension, remains approximately the same because of the pattern of events in the stochastic process (sandblasting) producing the roughness [25, 26].

In Figure 4.14 shows the profilometer trace and fractal characterization of a piece of freshly shattered rock. Studies are under way to link the fractal description of the freshly produced boundary of shattered rock to the health hazard of freshly shattered dust and to the crack propagation through the rock which caused the generation of the fragments.

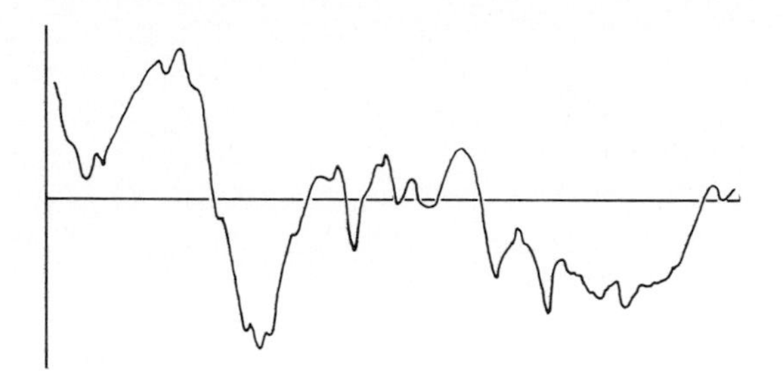

Figure 4.14. The profilometer trace of a freshly shattered rock fragment manifests fractal structure which may be related to the crack structure producing failure of the rock matrix which lead to the generation of the fragment [24].

4.8 Plumes with Fractal Boundaries

The way in which fineparticles disperse in a "smoke plume" emitted by an industrial chimney stack is an important aspect of pollution control engineering. When modelling the dispersal of a smoke plume, engineers usually specify the angle a dispersing plume makes with the horizontal line drawn from the top of the chimney stack parallel to the ground. They also specify the cone angle that the widening plume makes as it moves from the chimney stack. At any distance from the chimney stack, they assume that the density of smoke within the widening plume can be described by a Gaussian distribution. A complex detail often omitted from the computer modelling of the smoke plume dispersal is the fact that the edges of the plume are not straight. They are often curly because of the turbulent interaction of the hot smoke plume with the colder surrounding air. Thus, a real chimney plume often has ragged edges of the type shown in Figure 4.15.

The largest chimney stack in the world is in Sudbury, Ontario, at the processing plant of International Nickel Company Ltd. The plume from this chimney stack is eminently visible from Laurentian University. One of the laboratory studies undertaken by students who take a course on fractal geometry at Laurentian University is to photograph the plume and measure the fractal dimension of the boundaries of the plume. On some days the turbulence of each side of the plume is symmetrical, but on the day the photograph shown in Figure 4.15 was taken the lower boundary was almost straight and the upper boundary more turbulent. The fractal dimension of the boundary of the plume, as measured by the structured walk technique, is summarized for the two boundaries of the plume in Figure 4.15. For each boundary, the perimeter estimates of the rugged boundary are normalized with respect to a line passing through the middle of the variations in the rugged boundary as illustrated in Figure 4.15. Eventually, it is hoped that such studies can link the turbulence of the boundary of the smoke plume and the weather conditions, and help to further our understanding of the physical forces effective in dispersing a chimney plume.

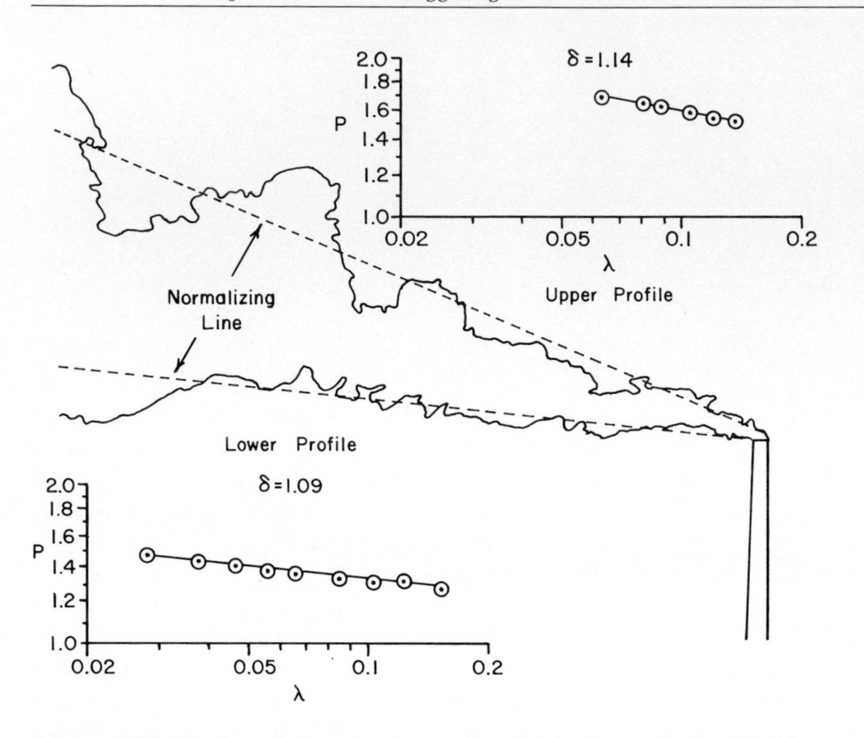

Figure 4.15. The boundaries of a plume of smoke leaving an industrial chimney stack interact turbulently with the surrounding air. This interaction produces a boundary with a fractal structure. Depending on the weather conditions, the two sides of the plume may have the same or different fractal structures.

4.9 Gaussian Graph Paper, Fractal Distributions and Elephants in the Face Powder

In a discussion of fractal systems, it is sometimes useful to present data describable by a Gaussian distribution in a different format to that of the familiar bell curve. For example, let us consider that the number of balls in the slot at the bottom of the Quincunx board represents the heights of recruits to the Munchkin army in the Land of Oz. If we now proceeded to calculate the number of soldiers less than or equal to a given height, we would obtain a set of data which when graphed would have the appearance shown in Figure 4.16(a) This type of curve shown is described by mathematics as an **Ogive curve**. Ogive is an Arabic word used to describe a curved arch of the type found in many mosques. The name Ogive is given to the curve because if one combines it with its mirror image it looks like an Arabian arch.

The Ogive data curve in Figure 4.16(a) is described as a cumulative distribution function and this particular curve is a cumulative undersize distribution function. Because the Gaussian distribution curve occurs so frequently in natural systems, mathematicians have devised a special type of graph paper in which the percentage axis is distorted mathematically, so that the Ogive curve appears as a straight line on the mathematically stretched graph paper. This type of graph paper is variously known as **probability graph paper, Gaussian probability graph** paper or a**rithmetic probability paper**. Engineers, with their penchant for short terms, describe the paper as a probit paper and refer to the mathematically described probability scale as a **probit scale**. We shall use the term Gaussian probability paper.

In Figure 4.16(b), the size distribution of Munchkin soldiers is summarized on Gaussian probability paper. In Chapter 3, we considered the fractal boundaries of a set of rock fineparticles left after valuable ore had been recovered from the crushed material. At that time we did not discuss the range of the magnitude of fractal boundaries present in the rock tailings. The boundary fractal was simply quoted alongside each profile. The variations in the fractal structure of the boundaries arise from many small causes interacting at random. It is therefore reasonable to assume that the distribution of fractals manifest by the different rock tailing fineparticles may be Gaussianly distributed. In Figure 4.17, the calculated distribution for the occurrence of

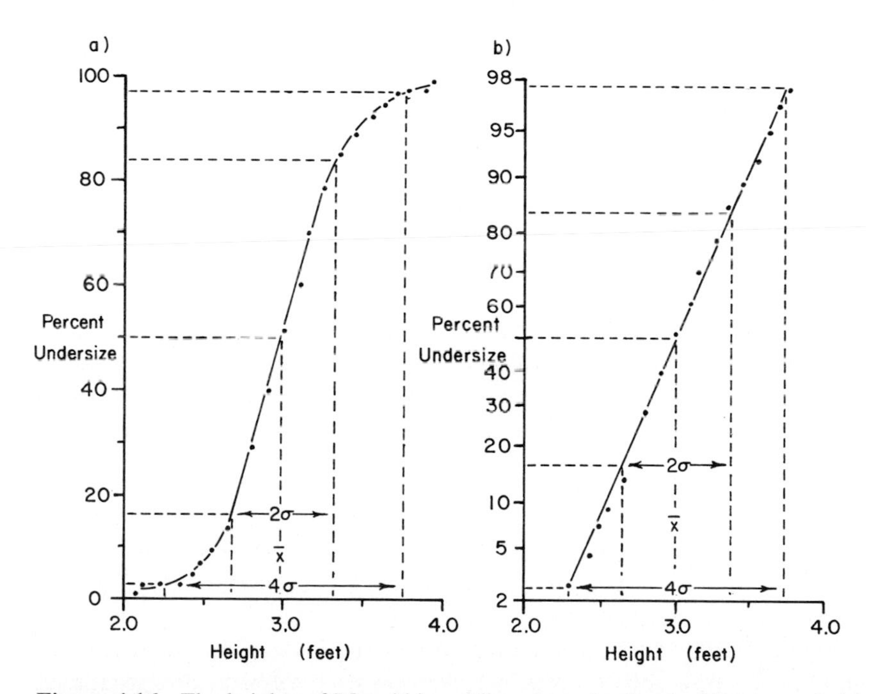

Figure 4.16. The heights of Munchkin soldiers from the Land of Oz are probably Gaussianly distributed.
(a) Undersize distribution of the soldiers plotted as an Ogive curve.
(b) The undersize distribution of the Munchkin soldiers generates a straight line when plotted on Gaussian probability paper.

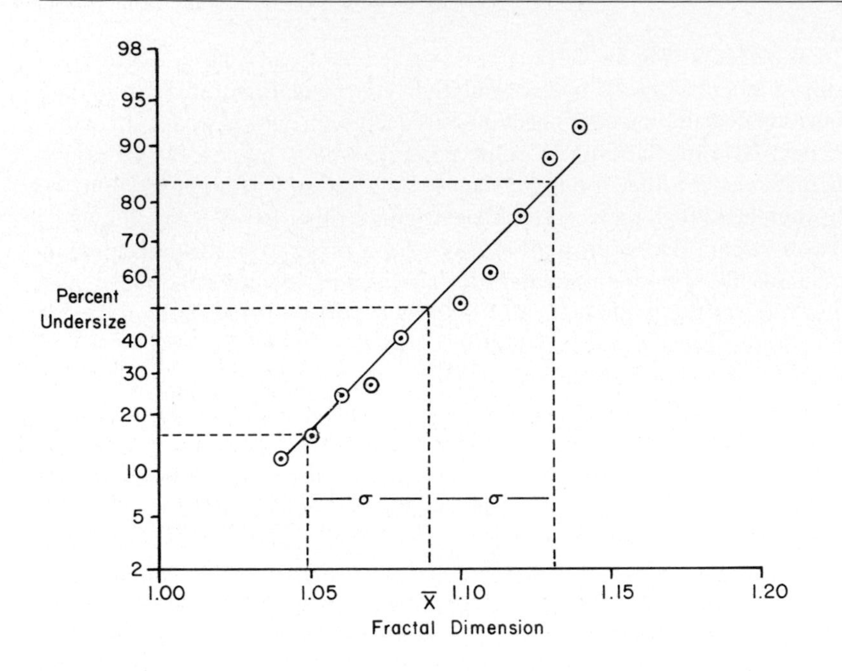

Figure 4.17. The range of fractal boundaries present in rock tailing fragments can be described on Gaussian distribution probability paper.

a fractal dimension in a rock tailing fragment of a value less than or equal to the stated fractal dimension on the ordinate of the graph is shown. Note that the abscissa is the probability scale of the type shown in Figure 4.16(b). The fact that the fractal information in Figure 4.17, plotted as a cumulative occurrence of a fractal equal to or smaller than a stated value, gives a straight line on Gaussian probability paper is evidence that the range of fractal dimensions in a population of shattered rocks is probably describable by a Gaussian distribution function. The average fractal dimension and the range of fractals present in this type of distribution may help engineers to understand how the rock fragments are formed. The reader is warned, however, that interpreting the scatter of points about an apparent data line on Gaussian probability paper is a difficult problem that should not be attempted by the inexperienced student. Scatter at the "tails" of the straight line through the datapoints on Gaussian probability graph paper should not be treated like scatter on ordinary graph paper. The central points on this type of graph paper contain more information than the data in the "tails" of the graph. The scatter in the tails data represents rare events.

Not all fractal variations in a population of fineparticles are describable by the Gaussian distribution. In a later chapter, we shall have occasion to discuss fractal variations in a set of fineparticles which are describable by a very different probability distribution.

In our discussion of the structure of Richardson plots data relationships in Chapter 2, the reader was warned of the perils of extrapolation in the graphical treatment of data.

Extrapolation on Gaussian probability graph paper beyond the outer datapoints also leads to strange paradoxes. Thus, on the probability scale of Gaussian probability paper, one can never plot 0 or 100% percent. The probability scale theoretically extends to infinity at both ends of the scale, so that we can display data from infinitely rare events. Extrapolation of datapoint information into the realm of rare events is sometimes useful, but any range of extrapolation should always be checked against physical reality.

To illustrate the danger of over-zealous extrapolation on Gaussian probability paper, when teaching the use of probability graph paper to my students I warn them against the reality of the "elephant in the face powder compact probability" paradox. It just so happens that the size distribution for some types of powder used in cosmetic technology is describable by a Gaussian distribution. If the size distribution of the grains of face powder is plotted on Gaussian probability paper, one can extrapolate the curve into the region of the probability of large fineparticles to discover that there is a very small, but finite, chance that one could discover an elephant in a face powder container!

In reality, in the case of face powder manufacturing technology, fractionation techniques are used to ensure that no fineparticles larger than 30 μm exist in the face powder, since anything as large as that would generate a tactile impression on the face (this is a fancy way of saying that large grains of face powder would scratch the face). Therefore, although extrapolation of the probability distribution of the face powder grain distribution would indicate that there is a finite probability of having grains larger than 30 μm, grains larger than this size are removed from the face powder by a physical process which makes the extrapolated region beyond 30 μm forbidden territory for real data (in other words, all elephants are sieved out of the face powder at the cosmetics factory!).

The students enjoy the nonsense of the elephant in the compact, but I have come across equally strange predictions in fineparticle technology when scientists interpreting implications of extrapolated data have failed to check theories against reality. If we look at the fractal distribution in Figure 4.17, any real fractal population of grain boundaries cannot be extrapolated beyond the fractal dimension of 1, nor can fractal dimensions for a two-dimensional boundary go above 2. Extrapolation predication of any data relationship describing fractal distributions of boundaries on Gaussian probability paper should always be checked against physical reality.

As mentioned earlier, when discussing the limited information left in the average value of a set of data, a numerical estimate of the range of values present in a set of data is often provided by calculating a quantity known as the standard deviation of the data. The mathematical procedure for calculating this quantity is set out in standard texts on statistical methods, but it is worth noting that an estimate of the standard deviation of a set of data can be calculated quickly from the dataline generated on Gaussian probability paper. It can be shown that if data are Gaussianly distributed, then 95% of all data points are within 4 standard deviations of the mean value of the data. If we mark out this range between 2.5% and 97.5% on the probit scale and divide the range of values of the variable between their limits by 4, we have a good estimate of the standard deviation. Thus, for the data on the Munchkin soldiers, we can state that the average height of the soldiers is $\overline{x} \pm \sigma$ where $\overline{x}$ is the mean height of the soldiers and the standard deviation is σ.

References

[1] T.C. Collocott (Ed.), "Chambers Dictionary of Science and Technology," Chambers., Edinburgh., 1971.
[2] A.M. Macdonald (Ed.), "Chambers Etymological English Dictionary," Chambers., Edinburgh, 1912.
[3] Rand Corp., "A Million Random Digits with 100 000 Normal Deviates," Free Press Publishers, Glencoe, IL, 1952.
[4] See, for example, the discussion of techniques for generating random numbers given in Yu. A. Shreider (Ed.), "The Monte Carlo Method. The Method of Statistical Trials," Pergamon Press, Oxford, 1966.
[5] See, for example, the discussion of the word "average" in M.J. Moroney, "Facts from Figures," Penguin Books, Harmondsworth, 3rd ed., England, third ed., 1956.
[6] M.G. Kendall, *Biometrica*, 43 (1956) 1.
[7] E.W. Montroll and M.F. Schlesinger, "On the Wonderful World of Random Walks," J.L. Lebowitz and E.W. Montroll (Eds.), in "Non-Equilibrium Phenomena," Part 2 of "Stochastics in Hydrodynamics," North Holland Physics Publishing, Amsterdam, 1984, Ch. 2.
[8] E. Klein, "A Comprehensive Etymological Dictionary of the English Language," Elsevier, Amsterdam, 1971.
[9] E. Fairstein, "Do Accidents Occur in Threes?" (Letter to the Editor), *Am. Sci.*, 72 (1984) 232.
[10] K. Metropolis and S. Ulam, "The Monte Carlo Method," *J. Am. Stat. Assoc.*, 44 (1949) 335-341.
[11] McCracken, "The Monte Carlo Method," readings from *Scientific American* on mathematical thinking in behavioural sciences, Freeman, San Francisco, 1968.
[12] K. Pearson, *Nature* (London) 72, (1905) 294.
[13] G.H. Weiss and R.J. Rubin, "Random Walks Theory and Selected Applications," in I. Prigogine and S.A. Wright (Eds.), "Advances of Chemical Physics," Wiley, New York, 52 (1983), 363-505.
[14] For an introductory article discussing electrostatics and electrostatic precipitators see A. Ross-Innes, "Static Electricity an Ancient Enigma," *New Scientist,* May 6, 1982.
[15] L.G. Jacchia, "Some Thoughts About Randomness," *Sky Telescope*, December (1975) 371-374.
[16] W. Feller, "Fluctuations in Coin Tossing and Random Walk," in "An Introduction to Probability Theory and Its Applications," Vol. 1, Wiley, New York, 1950, Ch. 3.
[17] B.B. Mandelbrot, "The Fractal Geometry of Nature," W.H. Freeman, 1980, p. 241; an updated and augmented edition of the original book, "Fractals, Form Chance and Dimension," published in 1977.
[18] M. Gardner, in the Mathematical Games section of *Scientific American*, "The Rambling Random Walk and its Gambling Equivalent," *Sci. Am.*, May (1969) 110-120; "Random Walks by Semi-Drunk Bugs and Others on the Square and on the Cube," *Sci. Am.*, June (1969) 122.
[19] J. McDermott, "Fractals Will Help to Make Order Out of Chaos," *Smithsonian* (Journal of the Smithsonian Institution, Washington, DC), 14, No. 9 (1983) 110-117; see also D. Campbell and H. Rose (Eds.), "Order in Chaos" North Holland Physics, Amsterdam, 1983.
[20] "The Mathematics of Mayhem," *Economist*, September 8 (1984) 87-89.
[21] W. Weaver, "Lady Luck," Anchor Books, Science Study Series, New York, 1963.
[22] L.J. Parratt, "Probability and Experimental Errors in Science," Dover, New York, 1971.
[23] A device called the Hexstat, which is essentially the same as the Quincunx Board, is manufactured and described by H. Ruchlis and E. Marcus in the Teachers Manual; Harcourt Brace and World Inc., New York, 1965. In this device the movement of a set of small steel balls passing through a scattering device simulates the generation of the widely used probability distributions.

[24] See discussion of the lawyers dilemma in occupational hygiene in the forthcoming book by B.H. Kaye, "Quality Air in Enclosed Spaces."
[25] B.H. Kaye, "Fractal Geometry and the Characterization of Rock Fragments," in the proceedings of "Fragmentation, Form and Flow in Fractal Media," Neve Ilan, Israel, January 6th-9th, 1986, R. Englman and Z. Jaeger (Eds.), *Israel Physical Society*, 1986.
[26] R. Trottier, M.Sc. Thesis, "Fractal Description of Fineparticle Systems," Laurentian University, 1986.
[27] L.F. Henderson, "Statistics of Crowd Fluids," *Nature* (London) 229 (1971) 381-383.

5 Fractal Systems Generated by Randomwalks in Two-Dimensional Space

5.1 Randomwalks on a Rectangular Lattice in Two-Dimensional Space

Mathematically, the simplest way to let our staggering drunk emerge from his one-dimensional ditch to stagger about in two-dimensional space is to confine his movements to explorations of a two dimensional lattice. Thus, we allow our liberated drunk to move around the restricted space shown in Figure 5.1. A **lattice** is defined in the dictionary as "a network of crossed laths or bars named after the type of grid used to protect windows." The word comes from the French word for a thin stick. The French mathematician Descartes (1596-1650) was one of the first to study the geometry of points in space. He invented the system of specifying the position, that is the address, of a point in space by means of a lattice such as that shown in Figure 5.1.

The two reference lines *Y* and *X* are described as a co-ordinate system. The word co-ordinate comes from a Latin word meaning "a set of instructions." Thus the co-ordinate system in Figure 5.1 enables us to interpret a set of instructions such as "5 steps in the *X* direction and 6 steps in the *Y* direction" to locate a point in space.

Mathematicians refer to the horizontal reference line in this type of system as the **abscissa** and the vertical axis as the **ordinate**. At the time that Descartes developed his co-ordinate system, scientists were still using Latin as an international language. Descartes used to sign his scientific papers with the Latin form of his name – Renatus Cartesius. In honour of Descartes, mathematicians use the Latin form of his name to describe the co-ordinate system shown in Figure 5.1(a) as a **Cartesian co-ordinate system**.

To set up a Monte Carlo routine for modelling the staggering of the drunk in two-dimensional space, we assume that the drunk's stride is equal to the spacing of the lattice. This stride size is designated by the Greek letter λ in Figure 5.1(a). If we now consider the drunk to start off at an *XY* address of (5, 6) (the *X* co-ordinate is usually written first when giving the address), we know that the drunk is starting off at the point on the lattice in Figure 5.1(a) marked by the asterisk. In a simple system for generating a randomwalk, we assume that the drunk can move from any given point in any one of four directions shown by the numbers 1, 2, 3 and 4 in Figure 5.1(a). We can now select digits from 1 to 4 from a random number table to decide the direction in which the drunk will step. Thus, if we found 3 in the random number table, the first step would take our

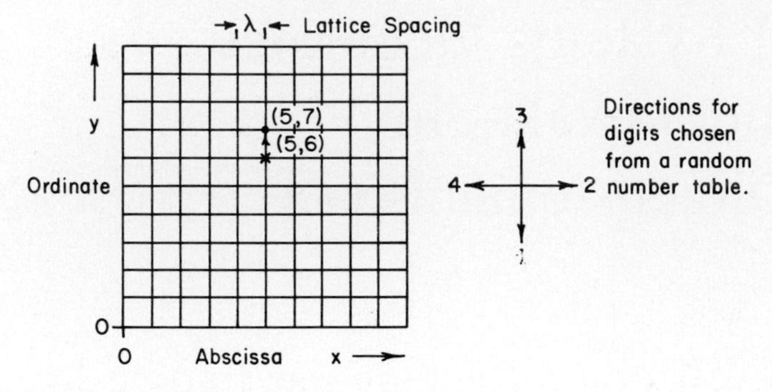

a) Cartesian coordinate system.

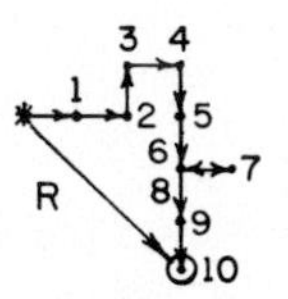

b) 10 step random walk and displacement vector .

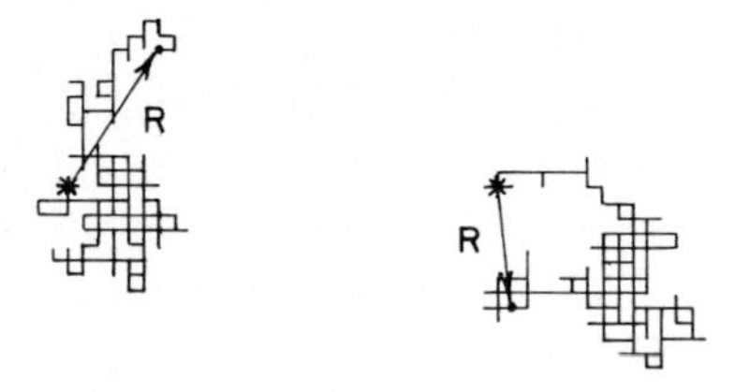

c) 250 step random walks and displacement vectors.

Figure 5.1. The simplest way to let a drunk explore two-dimensional space is to constrain his movements to possible positions in a square lattice. To specify the drunk's position within the lattice at any one time, we digitize the two sides of the lattice to form an address for any point within the lattice. Thus the point marked with an asterisk requires 5 steps in the X direction combined with 6 steps in the Y direction and has the address (x,y).

drunk to the point with the XY address (5, 7). Having reached this new position, he now steps off again in any of the four directions with equal probability. Again, his step direction is chosen from a random number table. In Figure 5.1(b), we show the position of the drunk after 10 random steps chosen in this way. The record of his movement is a Markovian chain and his net displacement is the arrow marked R.

It should be noticed that there are two arrows marking the 7th and 8th steps, which indicate that the 8th step took the drunk back to the position he was at after the 6th step, since in this simple type of randomwalk we allow our drunk to retrace his path or cross over an earlier part of his walk. This type of walk is described as a **non-self-avoiding randomwalk**. Two other similar randomwalks are shown in Figure 5.1(c).

There have been many scientific studies of this type of randomwalk since, although studying the meanderings of a drunk appears to be a trivial problem, it is the same as the scientific study of the random movement of electrons in the lattice of atoms forming a crystal, a problem of great scientific interest.

A detailed study of the theoretical results of the average net displacement of many electrons meandering around crystals and amorphous semiconductors is beyond the scope of this introductory text on fractal structures. The interested reader can find an interesting introduction to this type of problem in a review article by Weiss [1].

Scientists studying the movement of electrons and photons in various materials and modelling the growth of polymer molecules in polymerizing suspensions have developed many different types of two-dimensional randomwalks to take into account the various properties of real solids [2, 3].

When modelling the growing structure of large polymer molecular chains created by the random collision of constituent smaller subunits, scientists study the properties of **self-avoiding randomwalks**. In this type of randomwalk, the drunk is not allowed to retrace his steps or to cross his path if his meanderings take him to a point he has already visited. Scientists have been able to develop many theorems concerning the progress of "non-self-avoiding randomwalks," but solving the theoretical structure of self-avoiding randomwalks is much more difficult. In essence, most of what we know about self-avoiding randomwalks has been generated by using powerful computers to model the progress of this type of randomwalk [2, 3, 4].

Scientists who are interested in the movement of packets of energy in a solid sometimes modify the simple randomwalk system of Figure 5.1 to allow for random absorption of energy moving along a lattice. For this type of study, a packet of energy, such as a photon meandering about a space lattice, can sometimes be considered as being absorbed at random. In the terminology of our model of scientific problems based on a study of the progress of a staggering drunk, we can introduce trap doors at various points on our lattice so that the drunk has a random chance of falling through a trap door. Thus, in Figure 5.2, the points marked by the small squares represent absorption traps for our meandering drunk. By simulating many randomwalks using a computer, we can investigate how long it takes for an average drunk to be absorbed by the system or, alternatively, look at the probability of escaping from the lattice without falling into a trap.

Scientists interested in looking at a bundle of energy moving across a solid are also interested in how long it takes a drunk to get from one side of the lattice to the other. In such cases, it is necessary to introduce an extra variable to the probability of movement in which we allow our drunk to have a rest at any time during his movement through the lattice. When we look at randomwalks in four-dimensional space, we shall briefly discuss this type of randomwalk (see Chapter 6).

So far in our discussion of the randomly moving drunk we have assumed that there is no bias in his movement. However, there are many scientific problems in which we have random motions superimposed upon a general velocity in a given direction. For example, consider the movement of a stream of dust fineparticles towards a filter. This type of system is illustrated in Figure 5.3. It can be assumed that the individual fineparticles of dust are undergoing random motions but that the general flow of air

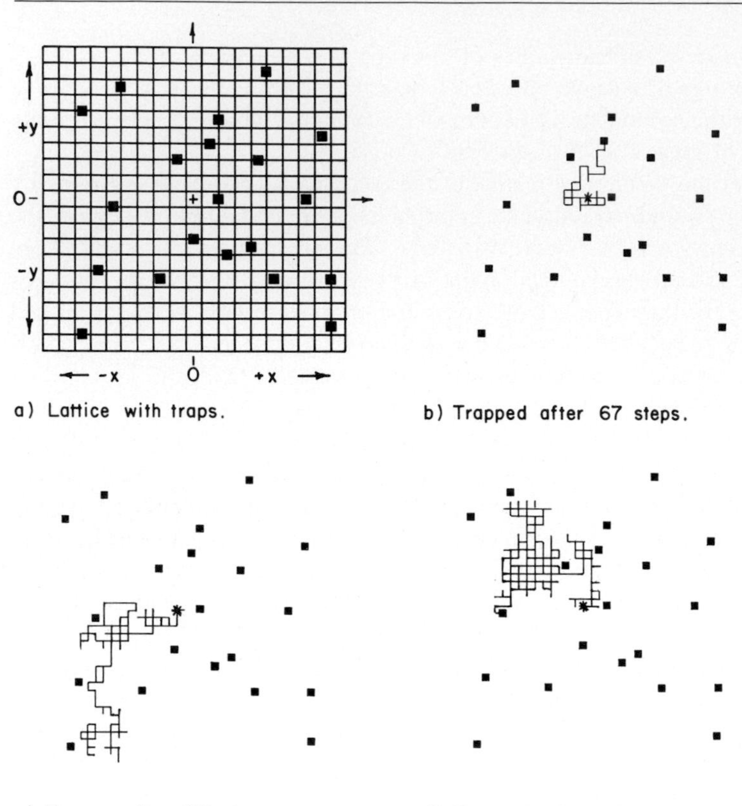

Figure 5.2. Some scientific problems can be described as randomwalks through lattice space in which there are randomly distributed traps.

towards the filter is imposing a general motion upon the general meandering movement of the fineparticles towards the filter. In Section 5.3, we shall discuss the type of deposition pattern that the dust fineparticles generate as they arrive at the surface of the filter by following this type of randomwalk towards the filter. In this section, we shall content ourselves with introducing the basic concepts used when modelling this type of **drifting** or **biased randomwalk**. Thus, if we imagine that we have a lattice placed in a pipe as shown in Figure 5.3(a) to help us specify the position and progress of the randomwalk, we can designate the probability of moving in the general direction of flow by the numbers 1, 2 and 3, whereas the probability of moving against the flow would be designated by the number 4. The probability of moving at right-angles to the general flow in the two possible directions at right-angles to the flow would correspond to the digits 5 and 6.

When modelling the path of any given fineparticle, we must consider that it is already in a random position across the pipe when we start to study its movement. Thus in the Y direction, which represents travel across the pipe, we would first set up an address at the beginning of the walk to specify the location at which the fineparticle enters the pipe.

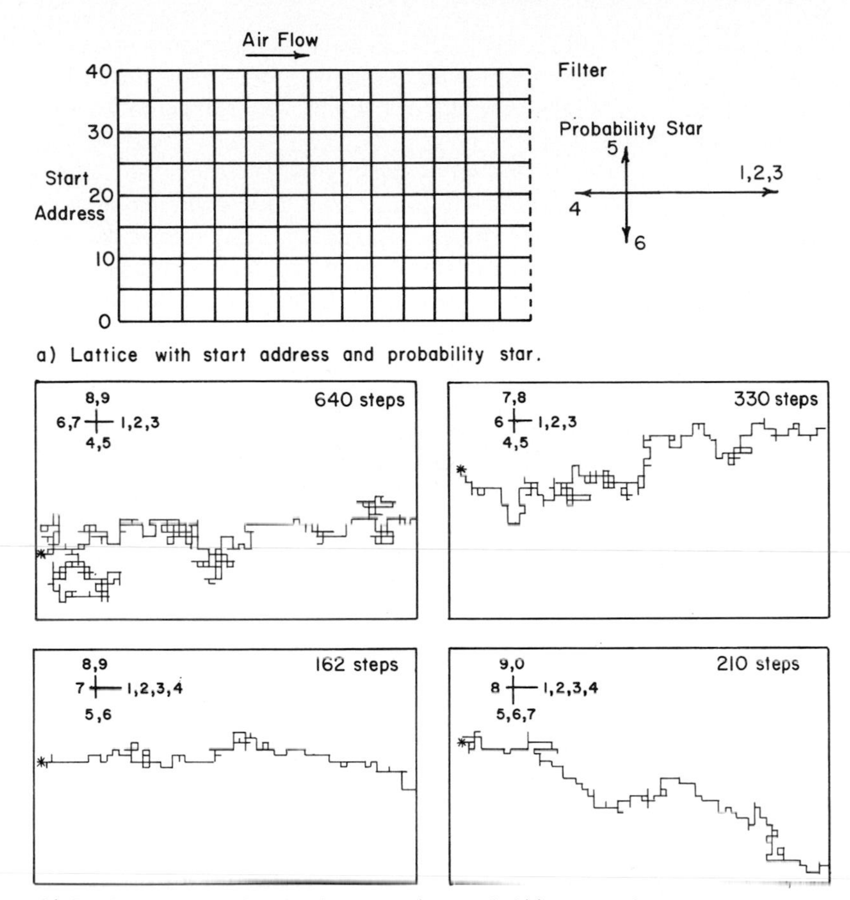

Figure 5.3. The progress of a dust fineparticle in the turbulent stream of air moving towards a filter can be modelled using randomwalk theory.

Thus, if we allocate 40 positions across the pipe, we then select a number between 0 and 40 to designate the entry point of the fineparticle into the pipe with the air flow.

Once we have allowed a fineparticle to enter the tube at random across the diameter, we can now allow it to walk randomly in the tube leading to the filter at the end of the tube by selecting random numbers between 1 and 6 for successive steps. If a step takes a fineparticle back and out of the tube, we decide not to follow the progress of that fineparticle any further, and introduce the next fineparticle at random across the tube.

In a real tube, viscosity effects near the wall, where the velocity is effectively zero, will usually ensure that a fineparticle reaching the wall will be captured by the wall. Thus, in our model, we can estimate wall losses when sampling industrial dust streams by considering that the fineparticle is captured when it hits the wall. Assessing wall losses is an important problem when characterizing the dust present in a smoke plume or in a dusty working environment. In Figure 5.3(b), several dust trajectories down the tube leading to the filter as simulated on a computer are shown. Obviously, the chance of reaching the wall depends on the relative width of the pipe with respect to the step size of our simulated randomwalk.

We can simulate the effect of increasing the gas velocity by increasing the number of digits which will move the fineparticle forward rather than back, forth and around. Note that in reality, the movement of a gas stream carrying dust fineparticles towards the filter in a tube is a three-dimensional problem. However, we can assume that movement around the centre of the tube is redundant, since our main interest is studying wall deposition or filter deposition. Therefore, our three-dimensional problems can be considered as having degenerated to two dimensions for the purpose of computer simulation studies.

When we study the build up of fineparticles on the filter, the simplest situation that we would encounter is a system in which the concentration of fineparticles is very low. In such systems, individual fineparticles would be deposited at random on the two-dimensional plane and never build up a sufficient deposit to cover the surface of the filter. Because this is an important problem in the assessment of air pollution from industrial smoke, we shall study this type of deposition in some detail in Section 5.2. We shall show that the clusters of fineparticles forming on the filter surface by random sequential arrival have a fractal structure. If we consider the other extreme case, where the concentration of fineparticles in suspension is very high, then we can show that the build up of a deposit on the surface of the filter results in a thick deposit which also has a fractal structure. The fractal dimension of the deep dust deposit depends upon the deposition kinetics and the forces operative inside the filter system (see Chapter 6).

The interaction between a whole bunch of drunks meandering around a lattice space over a period of time turns out to be yet another important scientific problem. Consider, for example, the system shown in Figure 5.4(a). We can assume that in the centre of the lattice is a very attractive young lady. The drunks, even in their state of inebriation, will recognize this as a location that attracts them. In this simulation of a randomwalk, we shall represent our drunk by a black square of the same size as a subunit of the lattice. Computer scientists who generate pictures by filling sub-squares of a lattice call the subunits a pixel, which is a contraction of "picture element." Let us assume that the pixel representing the first drunk is allowed to enter the lattice at random. To use our Monte Carlo routine for allowing the drunk to wander around, we must extend the address of the point of entry not only by specifying the X and Y co-ordinates but also by indicating whether the drunk enters from the north, south, east or west. Thus, we would select X and Y at random from random number tables, and then a digit between 1 to 4 to indicate north, south, east or west. Once entered into the lattice, the drunk is now allowed to proceed with equal probability in all directions, so that each step is now taken by selecting a random number between 1 and 4. We can permit the randomwalk to be self-intersecting, since we are only interested in the final position reached by the amorous (or lecherous) drunk. If he wanders out of the lattice we consider him annihilated and let another drunk enter the lattice. However, if the drunk manages to reach the lady at the centre we assume that all his further travel ceases and that he holds on to the lady as tightly as possible.

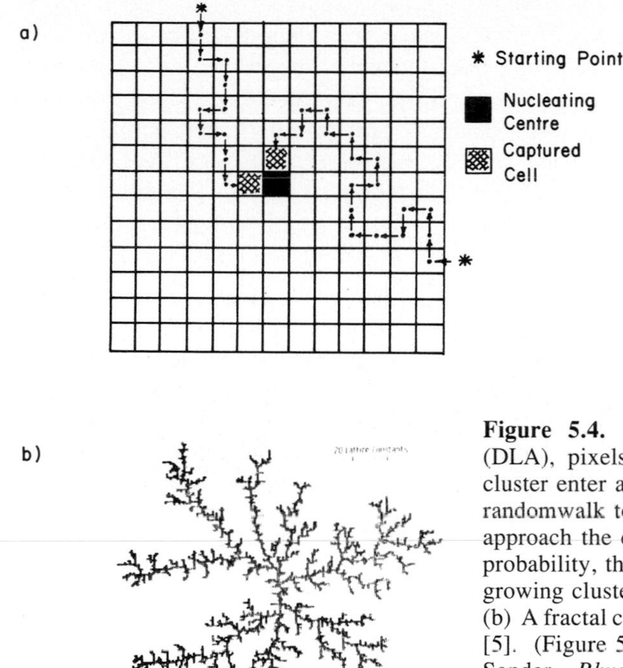

Figure 5.4. In diffusion-limited aggregation (DLA), pixels contributing to the growth of a cluster enter a lattice at random and undertake a randomwalk towards a nucleating centre; if they approach the central cluster with a valid contact probability, they are captured to form part of the growing cluster.
(b) A fractal cluster grown by Whitten and Sander [5]. (Figure 5.4(b) from T.A. Whitten and L.M. Sander, *Phys. Rev. Lett.*, Vol. 47, (1981); reproduced by permission of L. Sander and the American Physical Society.)

Next, we let another drunk enter the lattice. This second drunk may also reach the lady and cease his wandering, or it may be that he encounters the first drunk before reaching the lady. In this event he clings to the first drunk for support and we have an embryonic cluster of drunks who have meandered through the lattice.

In scientific terms, the pixel at the centre of the lattice which we designate as having an attractive force around which the arriving pixels gather is described as the **nucleating centre** or the **seed of the cluster**. The growth of fumed silica fineparticles discussed in Section 3.7 is a scientific problem which can be simulated using this randomwalk Monte Carlo method. We assume that the constituent unit spheres formed in the flame cluster together by randomly occurring collisions in the turbulence of the flame. Initially, growth of a simulated cluster around the nucleation centre will be fairly slow but eventually a cluster can grow to span the limits of the lattice space. This type of nucleated growth is referred to as **diffusion-limited aggregation**, often referred to in short as a DLA process. The study of this type of nucleation – cluster growth using Monte Carlo simulation techniques was pioneered by Whitten and Sander [5]. A famous Whitten and Sander fractal, grown on a computer screen using this technique, is shown in Figure 5.4(b). When simulating the growing of this type of cluster, it is assumed that a wandering pixel (unit square) joins the cluster if it touches the cluster **orthogonally** (a geometric term meaning at right-angles or, simply "square on"). Permitted encounters which are considered to cause adhesion to the growing cluster by orthogonal collision are illustrated in Figure 5.4(a). In one sense, the subunits joining the growing agglomerates can be considered as having been thrown at the growing cluster by the computer. The word ballistic comes from the Greek word meaning to throw, and the type of growth modelled by Whitten and Sander is called **ballistic accretion.**

The Whitten and Sander cluster has a fractal dimension of 1.7. The fractal dimensions of similar clusters can be increased or decreased by varying the permitted motion of the cluster, the size range of individual wandering units and the probability of sticking to a growing cluster when the wandering unit touches the growing cluster. In Section 5.6, various scientific problems modelled using slightly different diffusion limited aggregation Monte Carlo routines will be reviewed. The industrially important aspects of the resultant fractal structures simulated by the technique will be discussed briefly.

So far in this section, all of our randomwalks have been pursued in two dimensional Euclidean space. A very interesting set of randomwalk problems which take place in **fractal space** have been studied in recent years by many different groups of scientists [2, 3, 6 – 10]. To understand what we mean by fractal space, consider the Whitten and Sander fractal system shown in Figure 5.5 [7]. Imagine now that, instead of this representing a cluster of small square subunits joined together to form a cluster, the

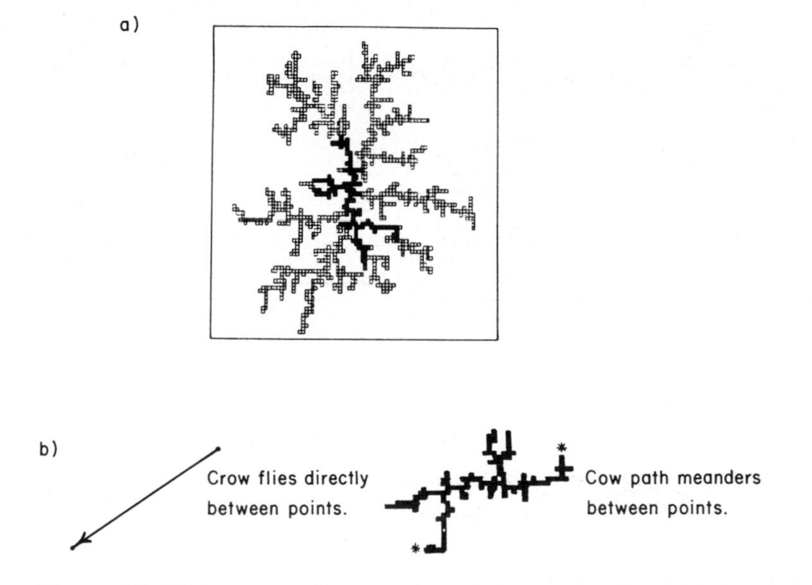

Figure 5.5. The tortuous passages existing in a porous body can define a fractal space in which gas molecules moving at random can be considered to undergo a randomwalk in fractal space. The fractal space defined by the square ABCD has been created by truncating the structure of a Whitten and Sander's type of fractal system, constructed by Stanley and Meakin [8].
(a) Fractal space created from a fractal aggregate (from P. Meakin and H.E. Stanley, "Spectral Dimension for the Diffusion-Limited Aggregation Model of Colloid Growth," *Phys. Rev. Lett.*, October (1983); reproducd by permission of American Physical Society and H.E. Stanley).
(b) Crow distances and cow distances within the fractal space are important parameters to be considered when characterizing the properties of fractal space (the crow space is the direct distance between two points which would be the path of a crow flying from one tree to another; the cow's distance indicates the path followed by a meandering cow!).

subunits outline tortuous paths possible through a two-dimensional slab of material, outlined by the dotted square A,B,C,D. If we now had a gas molecule undergoing random motion inside such a porous body, it would have to stagger around the fractal passages in order to find its way through the maze of passages. This is a very important type of problem in science and technology. For example, if we considered the system enclosed by the square ABCD as representing possible paths through a porous membrane, then we could look at the movement of gas molecules from AB to DC to see how fast they could pass through the membrane. Molecules with high velocity will move through the random passages more quickly than a more sluggish, massive molecule. Such differential motion of different sized molecules forms the basis of the separation of mixtures of gases by porous membranes. For example, this is the physical basis of one method of manufacturing enriched uranium, using a gaseous mixture of uranium isotopes, for use in nuclear reactors and atomic bombs. Again, in powder metallurgy, as a compact of metal powder is consolidated by sintering, the vapour molecules of metal must find their way out of the maze of pathways as the body collapses during the sintering process to form a denser body. The fractally tortuous passages in Figure 5.5 could represent drainage channels in consolidated tailing ponds, passages in a plastic bodies used to deliver controlled drugs, escape routes for radioactive gas molecules through cracked granite at radioactive burial sites or through the walls of cracked nuclear reactors, or radon gas percolation through soil into the basement of a house. It is not surprising that the study of the fractal structure of porous bodies is a growing area of materials science research.

In Chapter 7, we will briefly discuss an extension of a model such as that in Figure 5.5 to three-dimensional space, and indicate that if one is interested in the time dimension of the movement of randomly moving molecules through the fractal maze one will find it useful to discuss the fourth-dimensional fractal space, which scientists have described as the **fracton dimension** of a system [2].

5.2 The Use of Polar Co-ordinates to Describe Random Progress in Two-Dimensional Space

Cartesian co-ordinates are not always the most useful mathematical technique for describing the random motion of an object in space. As an alternative procedure, one can use what is known as the **polar co-ordinate specification of position.**

The basic concepts of the polar co-ordinate system for specifying a point in space are illustrated in Figure 5.6. In this co-ordinate system, we specify the position of a staggering drunk by measuring the distance from the lamp–post to the drunk by means of a straight piece of string. Then, if we know the angle made by the piece of string with a given reference line and its length, we have a complete specification of the position of the drunk. Thus, in polar co-ordinates, we measure the distance from a reference point called the pole of the system (hence the name) and the angle made by the straight piece of string to the position in space with respect to a reference direction. Thus, in Figure 5.6 the point P_1 is specified by the values R_1 and θ_1. The address of the point P_1 in polar co-ordinates is written in the form (R_1,θ_1). Similarly, the point P_2 is specified by the address (R_2,θ_2). The reference pole used to set up the polar co-ordinates is referred to as the **axis** of the system. We can always convert polar co-ordinates to Cartesian co-ordinates, and vice versa, by using Pythagoras' theorem for a right angled triangle. The conversion is illustrated in Figure 5.6.

The first step in using a Monte Carlo routine for modelling the staggering of the drunk in two-dimensional space using polar co-ordinates is to set up a correlation between possible directions in space and a random number table. Thus, for the model shown in Figure 5.7(a), we would permit eight different directions in space, with each direction being specified by the digit indicated in the figure. In the simplest model of the randomwalk, we would restrict our drunk to equi-sized steps of size λ. To simulate the progress of a randomwalk, we would start off at the lamp–post and permit the drunk to take a step in the direction specified by selecting one to eight from a random number table. Thus, if we were to start off at the top left-hand corner of the random number table shown in Figure 4.1, as we move down the table the first step would be in the direction 8. As shown in Figure 5.7(b) this first step would take the drunk to point B. At point B the random number selection procedure would direct us to move in direction 7 to reach point C. The third step would be taken in direction 3 and would bring the drunk back to point B. The fourth step would again be in direction 7, so that after five steps we are still at point C. Step 6, however, is in direction 1, which takes us to point D.

The next permissible direction from the random number table is direction 3. Thus, after six steps we are at position E. We can now draw the dotted line shown in Figure 5.7(b) to generate the polar co-ordinates of the resultant displacement after the Markovian chain of events represented by ABCBCDE. The address of the point E is specified in polar co-ordinates by the address $P_f(x_f,y_f)$ where the subscript f denotes final. The arrow drawn from the axis of the system to the final point illustrates the magnitude and direction of the resulting displacement. In the language of physics and mathematics, a line drawn in space which carries information concerning the magnitude and direction of a variable quantity is known as a **vector quantity**. The word vector comes from the Latin word "vehere" meaning to carry; thus a vehicle carries people and we make a vector quantity, such as shown in the graph, carry extra information compared with a number, which usually carries only one piece of information.

It can be shown that as time goes on the drunk has less and less probability of returning to the lamp–post and that the probable position after taking N steps of length λ completely at random is given by the equation

$$\lambda\sqrt{N} = E$$

where E is the expected distance from the lamp–post.

The decision to use either polar or Cartesian co-ordinate systems to model a given scientific problem depends upon the structure of that problem. Thus, when studying the behaviour of electrons in a crystal, the lattice randomwalk is the appropriate system to use since its physical structure is closer to the system being studied. On the other hand, polar co-ordinates are very useful in studying a whole class of problems concerning the

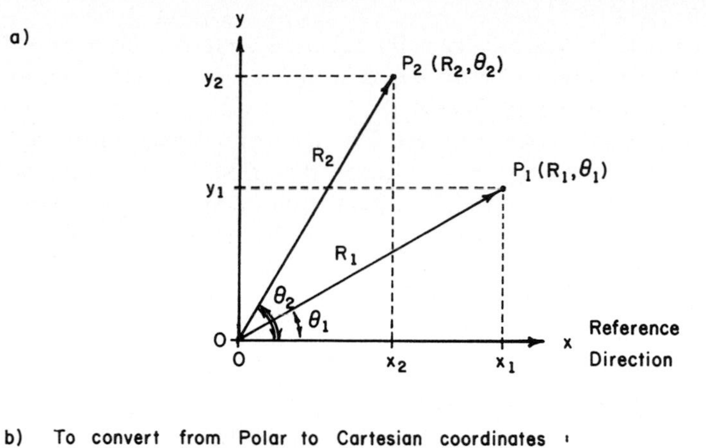

b) To convert from Polar to Cartesian coordinates :

Pythagoras' Theorem : $R^2 = x^2 + y^2$

$x = R \cos\theta$

$y = R \sin\theta$

Figure 5.6. Polar co-ordinates represent a geometric system for specifying a point in space and are sometimes more convenient than Cartesian coordinates.
(a) The point in space P_1 is specified by the length of the line R_1 drawn in the direction θ_1, the angle being measured with respect to the reference direction shown.
(b) Conversion of a polar co-ordinate address of a point to the Cartesian address using the theorem of a right angled triangle developed by Pythagoras.

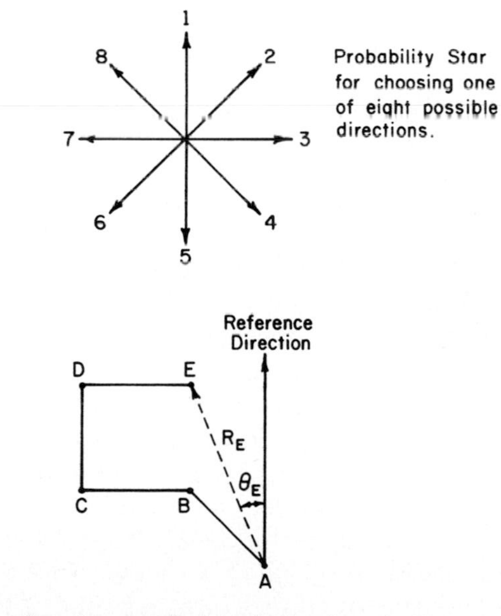

(R_E, θ_E) specifies the magnitude and direction of the Resultant Displacement Vector.

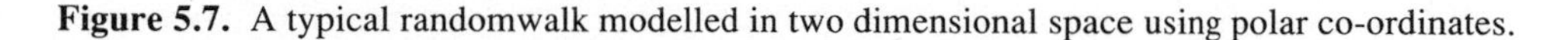

Figure 5.7. A typical randomwalk modelled in two dimensional space using polar co-ordinates.

diffusion of gas molecules and the movement of colloidal fineparticles in a suspension. The rapid irregular motion of colloidal fineparticles in a suspension is known as Brownian motion [11 – 17].

The term **Brownian motion** to describe colloidal fineparticle movement in suspension, is used to honour the Scottish botanist Robert Brown, who first observed this type of motion in 1827. When Brown looked at pollen grains from a flower suspended in water through a microscope, he discovered that they were in rapidly irregular motion. Initially he thought that this type of motion was unique to living cells such as pollen. Further experiments demonstrated that fineparticles of other substances, such as bits of wood and chips of glass and granite, moved in the same way. We now know that this irregular, unpredictable motion is due to irregular bombardment of the fineparticles with molecules, which pushes and shoves the fineparticles and changes their direction depending upon the balance of molecular bombardment at any one time. It is interesting that Einstein predicted Brownian motion from a mathematical theory without knowing that it had actually been discovered a century earlier [18]. To model Brownian motion in polar co-ordinates, we increase the complexity of the model used to generate the path in Figure 5.7 by permitting a whole range of step sizes. A randomwalk of this type is shown in Figure 5.8. Note that when tracking Brownian motion in a real system, we really have no justification for joining up the several positions with the zig-zagging lines shown in Figure 5.8. It has been shown many times that if we decrease the period of observation when studying the motion of a colloidal fineparticle, the intermediate motion between any two points on this high-resolution graph of the movement looks just as ragged and unpredictable as the large-scale plot at coarse resolution. Hence the joining of two points by the displays of Brownian motion is not a representation of reality. We do not know where a fineparticle has been between two observed positions.

In mathematical language, it can be stated that the Brownian motion track is self-similar at all magnifications. This means that if we see a track without any labels on it, we cannot know the magnification at which the track was monitored. The track representing Brownian motion has no differential function. Thus Brownian motion satisfies all the theoretical requirements for treating it as a fractal system.

Mandelbrot has told us that the study of Brownian motion was an important stimulus in his development of fractal geometry. Note that the fractal dimension of a Brownian motion track is 2, because if a colloidal fineparticle is given long enough, it will zig-zag back and forth and visit every point in a plain. Again, this fact brings out the important difference between the fractal dimension of a short track and the fractal dimension of the system from an ergodic point of view. If we look at a short-term track between the points A and B in Figure 5.8, we would deduce a fractal dimension for this zig-zag boundary but we would underestimate the fractal dimension of an infinitely long Brownian motion. Remember that the ideal fractal dimension is a measure of the capacity of a curve or boundary to fill space and not the actual filling of space by a short segment of any particular line or boundary, although we can use the concepts of fractal geometry to describe short range ragged systems if we record the resolution limits of our observations.

Theoretically, any one molecule or colloidal fineparticle moving around for an infinite period will cover the entire area of a plain, but if we consider many molecules

STEP	LENGTH R	ANGLE θ	STEP	LENGTH R	ANGLE θ
1	15	111	26	7	348
2	2	42	27	2	267
3	5	9	28	20	100
4	4	140	29	16	26
5	6	324	30	17	253
6	17	10	31	9	278
7	18	86	32	4	123
8	17	31	33	16	293
9	9	302	34	1	258
10	1	279	35	13	249
11	10	139	36	10	285
12	14	95	37	16	215
13	5	106	38	13	85
14	9	132	39	19	11
15	9	301	40	6	200
16	3	79	41	5	201
17	9	240	42	14	132
18	13	258	43	14	267
19	17	75	44	7	238
20	20	239	45	8	67
21	14	280	46	14	315
22	14	84	47	5	35
23	20	67	48	16	252
24	1	335	49	3	132
25	9	21			

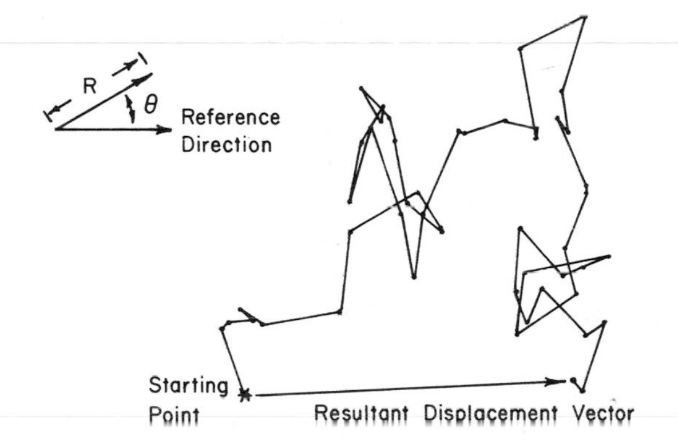

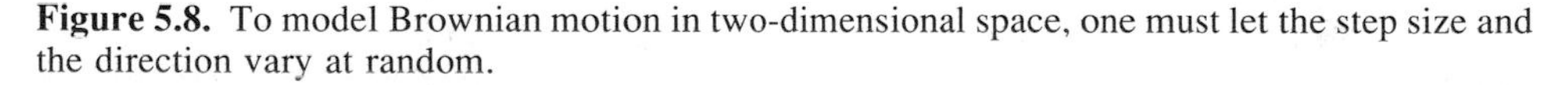

Figure 5.8. To model Brownian motion in two-dimensional space, one must let the step size and the direction vary at random.

leaving the same original axis after a given time they will spread out so that the probability of being a certain distance from the starting point can be described mathematically. Thus, in Figure 5.9(a), we show the effect of superimposing 20 randomwalks of 49 steps each. The terminal point of each of the randomwalks indicated by the asterisks cannot be taken as representing the location of the moving fineparticles after a given period of time. Figure 5.9(a) also shows the expected average displacement from the axis given by the equation quoted earlier in this section. It can be seen that the dots representing the location of the molecules after 49 steps do not lie exactly on the line but are clustered around the expected distance, forming a diffuse boundary.

In scientific terms, the fineparticles are stated to have diffused away from the axis. The word **diffuse** originally meant to pour out all around, to spread in all directions. Thus we are hoping that this book will help to diffuse information on fractals throughout the scientific community. The word comes from two Latin words, "dif" meaning

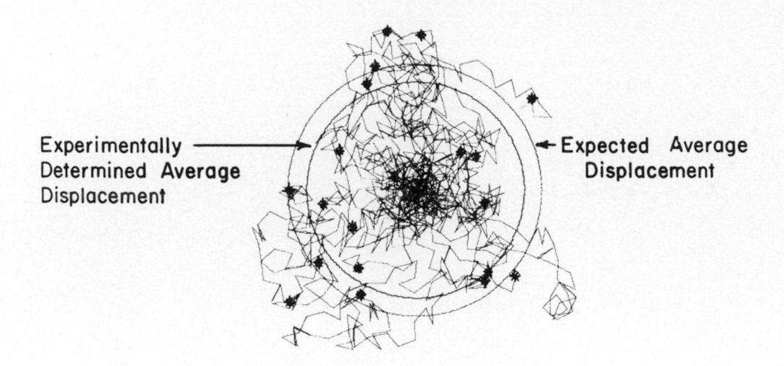

a) 20 random walks of 49 steps each, simulating Brownian Motion.

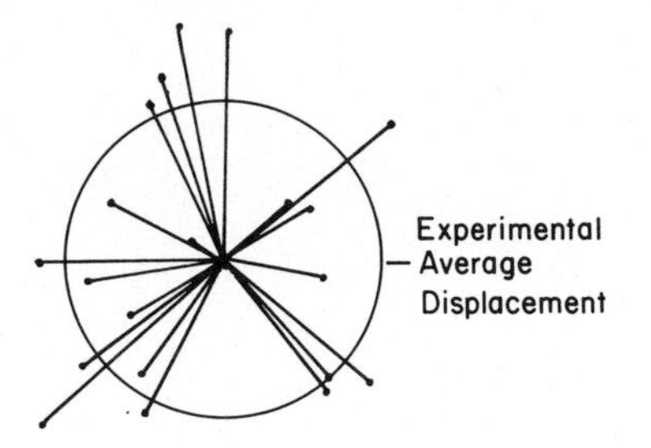

b) Scatter of the end points about the experimental average displacement.

Figure 5.9. If we consider the movement of many colloidal fineparticles from a central point, their locations after a given number of steps form a diffuse boundary around the expected average distance from the point from which the colloidal fineparticles are dispersed.
(a) Diffuse boundary created by 20 fineparticles diffusing from a central point with 49 random steps.
(b) Final positions of the 20 fineparticles about the expected diffusion limit.

asunder, and "funder," meaning to pour out. As time increases, the average distance of the diffusing molecules increases and the boundary delineated by the molecules about the average displacement becomes diffuse.

In Figure 5.9(b) the data in Figure 5.9(a) are simplified to show the location of the individual fineparticles with respect to the centre of the circle. The distance of an individual's location from the centre is taken to be the displacement vector for the walk as shown in Figure 5.9(b).

A quantitative description of dispersing items about an expected distance can form an important element of many scientific problems. For example, the axis of the diagram of Figure 5.9(a) could be the nozzle of a spray-generating system used to disperse pesticide-containing droplets. As the droplets move towards the ground they may be subjected to random turbulence which may cause them to undergo random motion en route to the ground. We can ignore the movement in the direction towards the ground, except to point out that this variable controls how long the droplets are diffusing before they hit the ground. Movement away from the nozzle can then be modelled by a randomwalk, using polar co-ordinates to look at the dispersion of droplets as they hit the ground. The model that we use to look at the dispersion of an individual droplet involves the modelling of 49 steps taken at random. We can regard these 49 steps as small causes interacting at random to create the final position of the droplet. Therefore, it is reasonable to anticipate that the distribution of droplets around the expected average dispersion will be Gaussianly distributed (see the discussion of the Gaussian distribution in Sections 4.7 and 4.9).

In Figure 5.10(b), the distribution of distances, α, of droplets from the expected dispersion for the data in Figure 5.10(a) is plotted on Gaussian probability paper. The fact that a straight line can be drawn through the data points confirms that the dispersion of the droplets can be described by the Gaussian function.

When studying a problem such as the dispersion of pesticide droplets, we are concerned not only with the distance of the droplets from the nozzle but also with the uniformity of dispersion around the nozzle. If we look at the point defined on the perimeter of the average dispersal circle by the line drawn from the location points at

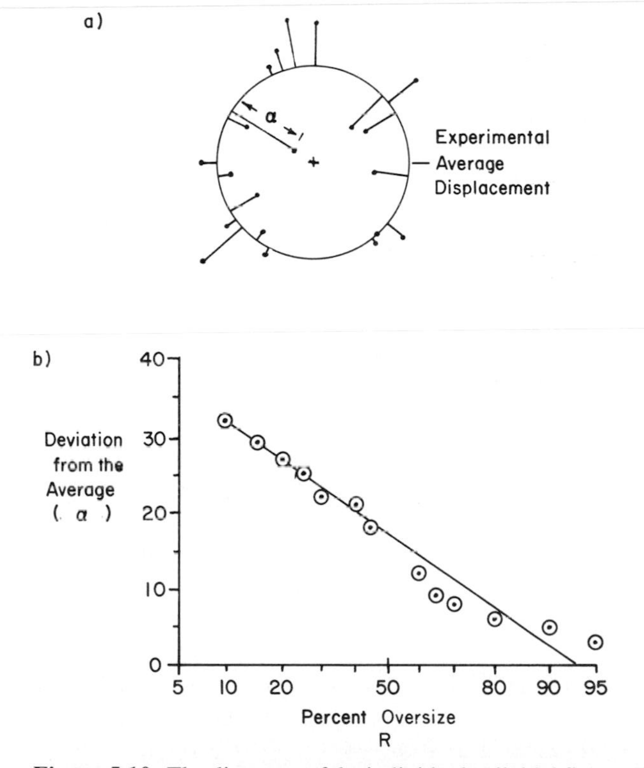

Figure 5.10. The distances of the individual colloidal fineparticles from the average displacement of the group of fineparticles is Gaussianly distributed.
(a) Simplified representation of the data in Figure 5.9; it represents the distance along the radial distance to the circle of average dispersion.
(b) The values of α plotted on Gaussian probability paper as a cumulative distribution function.

right-angles to the circle, we generate a set of intercepts on the dispersion circle, which are shown separately in Figure 5.11(a). The points on the line represent a Cantorian set with dimensionality less than 1. The chord distribution between the points on the circle of average dispersion forms a set linked to the dimensionality of the set of points. The distribution of the chord lengths around the circle of average dispersion for the data in Figure 5.11(a) is plotted in Figure 5.11(b). The slope of this line can be used to characterize the dispersion of the droplets around the nozzle perimeter.

The model of droplet dispersion that we have discussed in the previous paragraph is a simple one which can be expanded to take into account variations in the environmental conditions operative when crop spraying is in progress. For example, if a wind is blowing when the droplets are generated, the drift of the droplets due to the wind can be built into the randomwalk simulation by giving an increased probability of movement in the direction of the wind. Again, in industrial spray technology a great deal of interest is centred around the possibility of increasing the efficiency of spray dispersal and deposition by giving an electrostatic charge to the individual droplets. In this situation a high voltage is applied to the nozzle to repel the droplets from the nozzle and to generate a charge on each droplet. Therefore, when modelling by a Monte Carlo routine

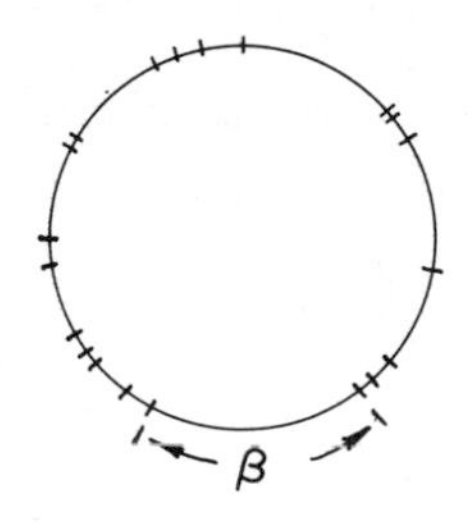

a) Cantorian Dust formed by the intersection of displacement vectors with the average displacement circle.

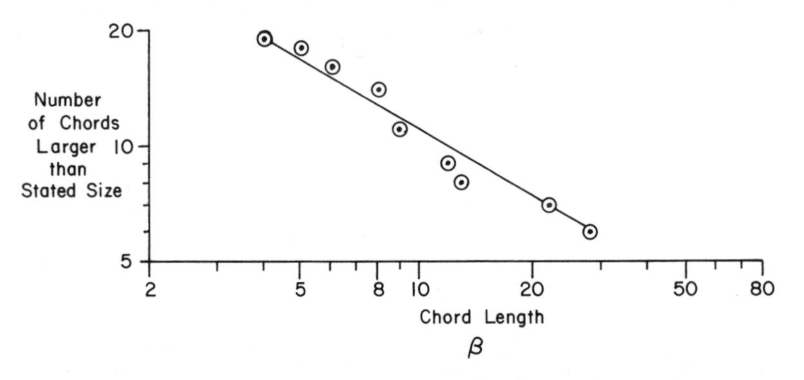

b) Distribution of the chord lengths formed on the circle of average displacement, plotted on log-log graph paper.

Figure 5.11. The points on the circle of average displacement created by projecting the location of the individual randomly dispersed colloidal fineparticles along a radius create a Cantorian dust of dimension less than 1. The chord length between the elements of the Cantorian dust can be related to the dimensionality of the dust.

the dispersal of electrostatically charged droplets, one must now increase the probability of movement in the direction along the radius away from the nozzle because of mutual and electrostatic repulsion, the droplets will not be allowed to come closer to each other than a specified distance. Also, the droplets will not be able to move backwards because they are being repelled from the nozzle by electrostatic forces. The result is a special case of a self-avoiding randomwalk model. One could simulate the dispersal using different simulated charges on the droplets and compare the predicted pattern with those observed from the experimental study of deposited droplets generated in real spray studies.

Students encountering randomwalk theory for the first time sometimes find it surprising that any individual drunk over a period of time has a tendency to move away from the lamp–post. An alternative method of looking at the probability of movement away from the lamp–post can sometimes be helpful when trying to grasp the physical significance of the diffusion of a horde of drunks about the expected distance. In Figure 5.12, consider the small circle labeled A drawn on the circle of radius R_1 around the starting point for all of the randomwalks. The area of the small circle outside the circle drawn around the starting point is a measure of the probability that the drunk will move away from the lamp–post, whereas the area of the small circle inside the circle around the axis is a measure of the probability that the drunk will move back to the lamp–post. The ratio of these two sub-areas of the possible directions from a point on the circle at a given distance from the lamp–post is such that the drunk is always more likely to move

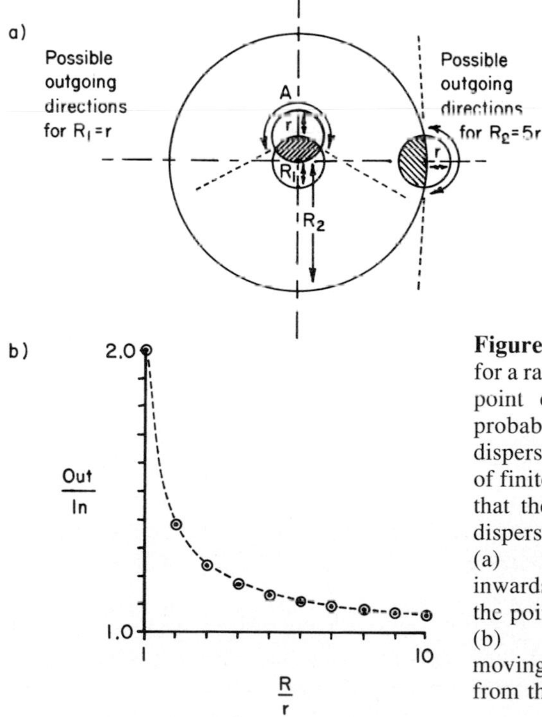

Figure 5.12. The probability of moving outwards for a randomly moving fineparticle starting from a point on a plane is always greater than the probability of moving inwards towards the dispersal point for all average displacement circles of finite value. This can be interpreted as the fact that the randomly moving elements undergoing dispersion move in a curved probability space.
(a) Illustration of the probabilities of moving inwards or outwards at any given distance from the point of dispersal.
(b) A summary of the probability changes of moving inwards or outwards at any given distance from the dispersal point.

away from the lamp–post than towards it. Only when the distance from the lamp–post becomes infinite do we have equal possibilities of moving to or from the lamp–post. A circle of infinite radius is a straight line and if we consider a line of drunks in space, they would tend to diffuse equally away from their starting position to either the east or west of their starting point.

An alternate way of describing the fact that the drunk has a higher probability of moving away from the lamp–post than he has of moving towards it can be appreciated by considering an alternative way of modelling the movement of the drunk. Let us imagine that we have a ball which is capable of moving at random on a two-dimensional surface. To create a two-dimensional space on which the ball would have a higher probability of moving away from the initial centre would be equivalent to slightly curving the space on which the ball moved. Thus, a map of the probability of movement in two-dimensional space would be like a shallow upside-down saucer with the slight curvature of the saucer giving the ball a higher probability of moving downhill rather than to the top of the saucer. Thus, although the **mapping space** (the graph paper on which we plot the movement of the drunk) in two-dimensional space is flat, one can conveniently describe the **probability space** for the movements of the drunk as being curved. In more advanced physics, one sometimes hears a statement that a certain type of space is curved. An intuitive interpretation of what we mean by a **curved space** is very difficult. It is helpful to realize that in some ways talking about curved space is like saying that the drunk undergoing diffusional movement away from the lamp–post is operating in curved probability space.

Now that we have established the basic concepts and vocabulary of randomwalk modelling in two-dimensional space, we can now proceed to look at specific systems which generate fractally structured systems in two-dimensional space.

5.3 Randomwalk Modelling of Fractal Deposits in Two-Dimensional Space

In Section 4.7, we discussed how the pin-ball machine can be used to model some randomwalk systems. Figure 5.13 shows a special type of pin-ball machine built by Parman to illustrate the randomwalk basis of an important scientific process known as chromatography used by analytical chemists to separate the ingredients of complex mixtures of chemicals [19]. The pioneer work which led to the development of the sophisticated chromatographic systems to be found in the modern analytical chemistry laboratory was carried out by the Russian scientist Tswett (1872-1920), who was interested in the nature of plant pigments.

In his initial experiment carried out in 1906, Tswett poured an extract of plant pigments suspended in a hydrocarbon fluid into the top of a column of powdered limestone held in a glass tube. He then continued to pour more hydrocarbon fluid into

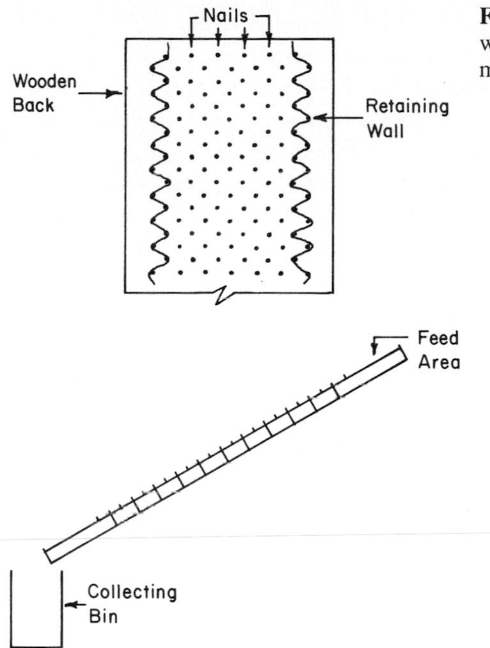

Figure 5.13. A special-purpose pin-ball machine which can be used to model the fractionation of molecules in a chromatographic column [19].

the tube. The different pigments in the original suspension had different affinities for the surface of the limestone and therefore the pigment molecules moved down the column at different speeds, depending on how easily they were adsorbed and desorbed from the surface of the limestone. As a consequence, the different pigments present in the original mixture formed different coloured bands down the column of the limestone powder. Because of the rainbow appearance of these separated pigments down the limestone column, Tswett called his fractionation method **chromatography**, which comes from the Greek words for colour and writing [19].

Modern procedures built on the pioneering work of Tswett are still referred to as chromatography, even though there may be no colours visible in the modern chromatogram. Modern procedures usually pass liquid through the column until the molecular species to be fractionated emerge one by one from the bottom of the column. The emerging fluid is monitored to see what is coming out from the column at different times. The record of the arrival of the various substances fractionated in the **chromatograph** column over a period of time is known as a chromatogram. The technical term for the process of washing the molecules out of the column is **elution**, from the Latin word "elurere," meaning to wash away. The liquid used to wash the molecules to be separated is described as the **eluent** or mobile phase.

Parman built the equipment shown in Figure 5.13 to help students visualize the absorption-desorption process going on in a chromatographic column [20]. Parman poured a mixture of two different sized steel balls down his special pin-ball machine. As the balls rolled down through the pins they collided with the pins, at a frequency determined by the diameter of the balls relative to the spacing of the pins. Thus, the smaller balls tend to participate in fewer collisions than the larger ones and migrate at a faster speed down the pin-ball machine. Hence a mixture of the balls put in at the top of the column is separated by size at the base of the column, as shown in Figure 5.14. Therefore, from one important point of view, separations achieved in chromatography involve randomwalk theory, with the variation in the permitted steps of the randomwalk involving the pore size distribution between the fineparticles forming the powder beds through which the molecules to be fractionated are moving with the added complication that the residence times of the randomly walking molecules vary depending upon the affinity between the molecule and the surface and the thermal energy of the molecules (this randomwalk model is a special type of chromatography called **hydrodynamic chromatography**, which is discussed in detail in the next chapter).

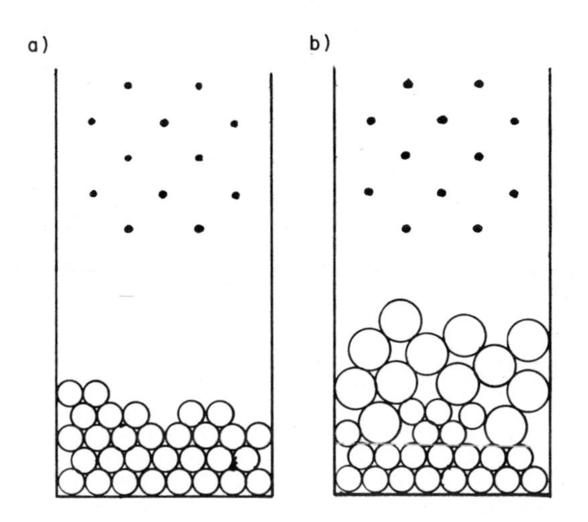

Figure 5.14. The pin-ball machine in Figure 5.13 can be used to fractionate spheres of different sizes into two groups. (a) Mono-size spheres passing down the pinball machine form a uniform deposit across the base of the machine. (b) Two different sizes of steel balls can be separated into two distinct groups by exploiting the differential movement of the two sized balls down the pin array.

In **thin-layer** or **paper chromatography**, the molecules to be fractionated are separated in a fluid moving through a thin layer of paper or adsorbent. In Figure 5.15 we show the separation fronts created by placing a drop of ink at the centre of a circular filter-paper. After the original drop of ink has been placed on the paper, further drops of water are added to the centre of the paper. The water moves out through the filter-paper by capillary action towards the perimeter, where it drips away or evaporates. The moving water attempts to carry the pigment molecules with it, but the molecules are alternatively adsorbed by the fibres of the paper and have to wait until chance variations in the thermal energy of the water-molecule system enable the molecules to break free to travel with the water a short distance before they are adsorbed into another fibre in the paper structure.

In Figure 5.16(a), the simulated structure of a thin layer of paper made from long fibres is shown (for a description of how this simulated paper was constructed, see Chapter 7). A simulated randomwalk of two different molecules across the fibrous mat, with the individual molecules undergoing random motions in between adsorptions on to and off the fibres, is shown in Figure 5.16(b). In the Monte Carlo routine used to

Figure 5.15. In one form of paper chromatography, a drop of solution containing the molecules to be fractionated is applied to the centre of a filter-paper. This drop is then eluted to the perimeter of the filter-paper by subsequent addition of drops of the eluent (water). The migration fronts created by the pigments in the ink are obviously fractal fronts.

generate the data summarized in Figure 5.16(b), when the molecules are adsorbed on a fibre they are allowed to reside at that position for a time fraction which varies at random and from molecule type to molecule type [21].

The appearance of the migrating front in Figure 5.15 illustrates that the front created in paper chromatography has a fractal structure and it can be appreciated from the features of the model shown in Figure 5.16 that the fractal dimension of the migration boundary in Figure 5.15 depends on the molecular mobility of the dye molecules and

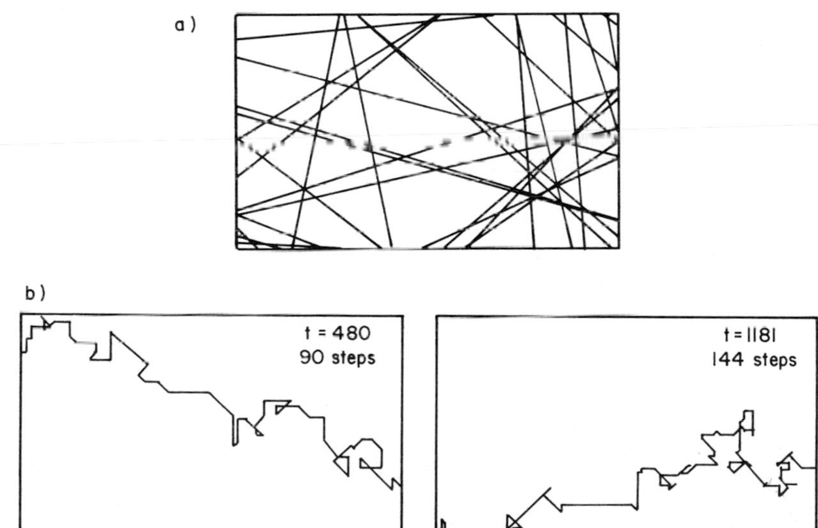

Figure 5.16. The migration of pigment molecules in paper chromatography can be modelled by a Monte Carlo routine by which one assumes the pigment molecules are undergoing a randomwalk involving adsorption on and desorption from the fibres of the paper.

(a) Simulated structure of a paper made from long cellulose fibres.

(b) Simulated randomwalk of three pigment molecules for a fixed period of time.

the packing density of the fibres in the paper structure. An alternative way of stating this latter variable is that the fractal of the migrating boundary will be related to the pore structure of the paper, which can be described in terms of a Sierpinski fractal as described in Chapter 7. The significance of the multiple fronts visible in Figure 5.15 will be discussed in more detail in Chapter 7.

In traditional paper chromatography, the aim has been to minimize the ruggedness of the migration boundary in order to increase the sharpness of the separation of different molecular species present in the material being fractionated. In column chromatography, attempts to improve the resolution of the technique have focused on providing uniformly sized fineparticles as the column material to create a uniform pore structure in the column [22]. Similarly, in paper chromatography, uniform pore-structured paper gives better resolution in the fractionation of the molecules. Experiments are under way at Laurentian University to characterize the pore structure of paper from the fractal front of a migrating ink blot [21].

Equipment similar to that used by Parman to model chromatography has been used by Nowick and Mader to make two-dimensional models of the crystal structure of metal alloys [23]. In Figure 5.17, the assembly table used by Nowick and Mader to create two-dimensional arrays of spheres is shown. The system could be used to produce either two-dimensional arrays of spheres of the same size, or mixtures of different sized spheres. Note the irregular sided container to the assembly plane to randomize any effects due to the wall restricting the random packing in the plane. The feed hoppers are equipped with notched disks to regulate the flow rate of spheres into the assembly area. The hoppers are also equipped with closely spaced pins to avoid any rapid motion of the spheres out of the hoppers into the notched feed discs. Spheres are fed down the

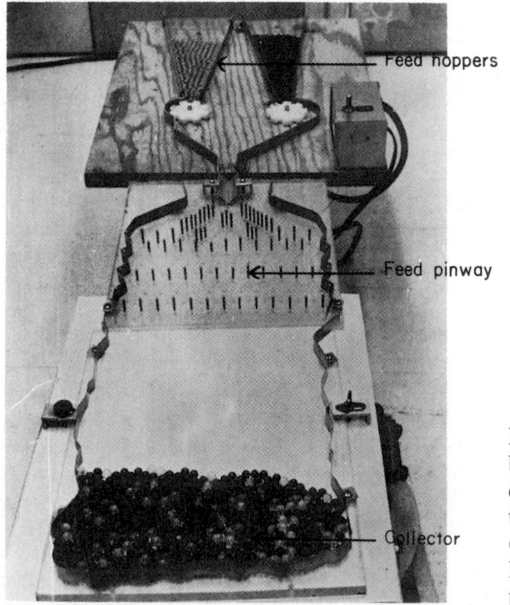

Figure 5.17. Nowick and Mader used a pinball type machine to produce a two-dimensional array of spheres in their studies of the simulated structure of metal alloy systems. (Copyright 1965 by International Business Machines Corporation; reprinted with permission.)

feed pin array to the assembly area. The rate at which they move through this array can be controlled by altering the elevation of the feed assembly. The whole table can also be vibrated gently to change the pattern of the assembled spheres in the base collector.

The basic concept utilized by Nowick and Mader in their experiments is that if one were to photograph the two-dimensional array of spheres in the base collector, then one could use a reduced negative of the array to generate an optical diffraction pattern by shining a laser through the negative. This would simulate the diffraction of X-rays passed through a thin film of metal atoms in a real alloy. When using the negative of the sphere array to diffract light, the ratio of the wavelength of the light used to create the optical diffraction pattern to the size of the images of the spheres in the negative has to be the same as the ratio of the size of metal atoms in a thin film of the metal to the wavelength of the X-rays used to generate an X-ray diffraction pattern.

To test the validity of their model system, Nowick and Mader first assembled a regular array of spheres to simulate a perfect thin film of pure metal. In Figure 5.18, a picture of the array is shown alongside the optical diffraction pattern of the array. This

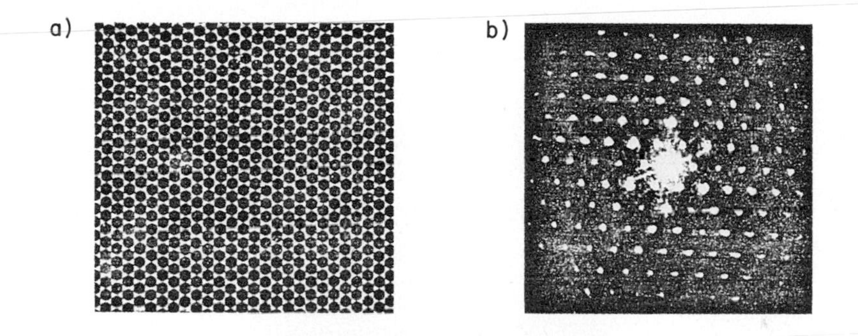

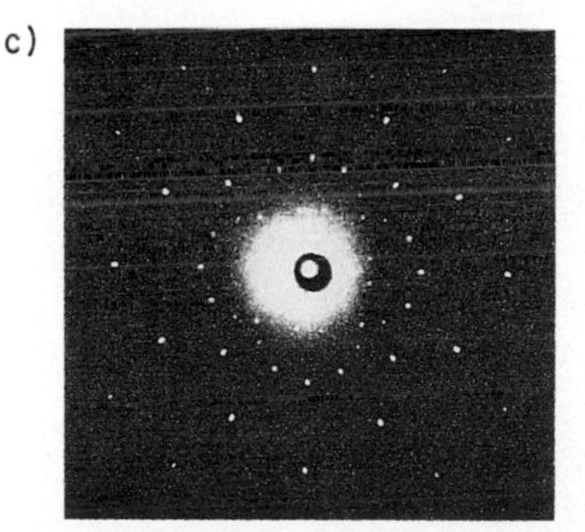

Figure 5.18. The diffraction pattern from the photographic negative of a random array of mono-sized spheres is similar to that of the X-ray diffraction pattern produced by a crystalline substance, wherever a randomly assembled system models amorphous material.
(a) The regular array of spheres.
(b) Optical diffraction pattern formed by a photographic negative of the array in (a).
(c) An X-ray diffraction pattern of a crystalline substance.
(Copyright 1965 by International Business Machines Corporation; reprinted with permission.)

type of diffraction pattern is very spotty. It is known that this type of diffraction pattern corresponds to diffraction by a regular structure of the type found in a pure crystal. Thus in Figure 5.18(c), the X-ray diffraction pattern of a pure crystal generated by X-rays is shown [24].

In Figure 5.19, a random array of mono-sized spheres, assembled by allowing the spheres to run down the feed array system as fast as possible, is shown. Note that the surface of this random array is a fractal boundary. Note also the almost complete absence of regular packing of the spheres within the array, except for very small islands of regularity. This type of structure is similar to that achieved by rapidly cooling molten metal before the atoms have time to arrange themselves in a regular lattice. It is described as the amorphous state. The optical diffraction pattern, generated using a negative photograph of the array of Figure 5.19(a), is shown in Figure 5.19(b). The almost continuous rings and the collapse of the overall pattern towards the centre of the rings is characteristic of the structure of amorphous materials, as shown by the X-ray of an amorphous material shown in Figure 5.19(c) [24].

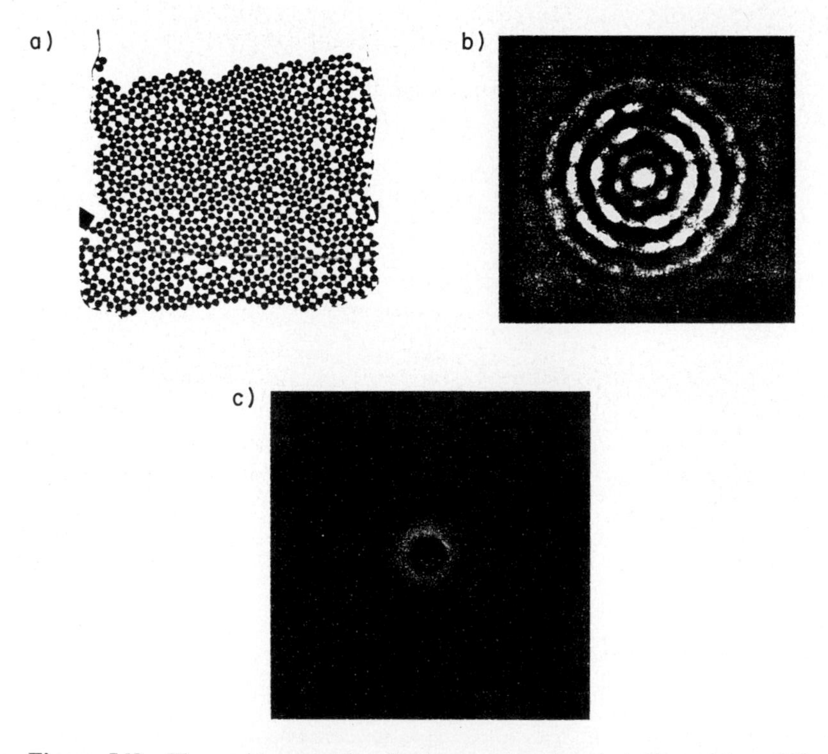

Figure 5.19. The rapid assembly of spheres using the pin-ball machine of Nowick and Mader simulates amorphous structure.
(a) Randomly assembled array of spheres.
(b) Optical diffraction pattern of a negative of the sphere assembly shown in (a).
(c) X-ray diffraction pattern of an amorphous substance.
(Copyright 1965 by International Business Machines Corporation; reprinted with permission.)

Metallurgists use a technique known as **annealing** to change the properties of materials. A dictionary defines the process of annealing as,

"heat treatment of material such as metal or glass to remove strains resulting from previous operations and to make it less brittle and tougher. The process commonly consists in holding a material at a pre-determined temperature for a pre-determined time, then cooling it at a rate and to a temperature that will produce the desired properties."

The word anneal is interesting in that it is one of the few scientific and technical terms which is based on an old English root word. This root word is "aelan," meaning to burn; annealed material is subjected to controlled burning to alter its properties.

We now know that what happens during an annealing process is that the atoms within a material are given enough thermal energy for them to be able to arrange themselves into larger islands of regularity which we call crystallinity. Thus, the annealing temperature has to be high enough for the atoms to be able to move into a better packing

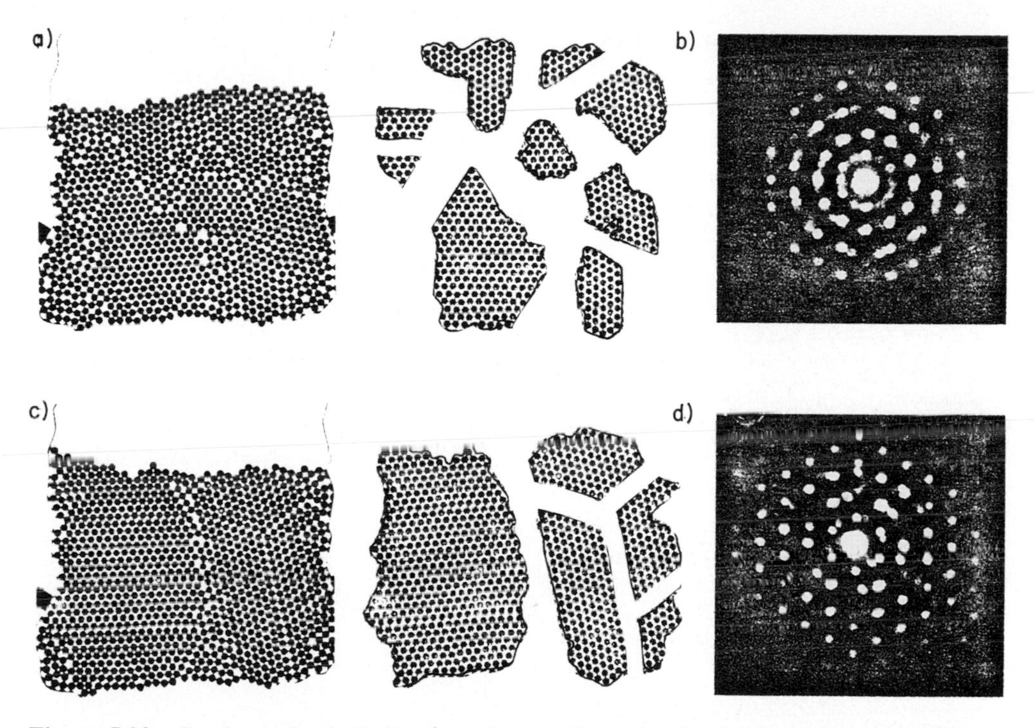

Figure 5.20. Gentle mechanical vibration of a two-dimensional array of mono sized spheres simulates the process of annealing used in metallurgy.
(a) Array of spheres produced by slow assembly techniques shows several islands of regularity.
(b) The optical diffraction pattern produced by a photographic negative of the array in (a) shows intermediate structure between a crystalline and an amorphous substance.
(c) The array in Figure (a) can be annealed by gentle vibration to reduce the number of spheres in the random array.
(d) The optical diffraction pattern of a negative of the array in (c) shows progress towards the diffraction pattern of a single crystal produced by the annealing of the array in (a).
(Copyright 1965 by International Business Machines Corporation; reprinted with permission.)

structure, but not high enough for them to move towards the irregular structure present in the material just before it melts. On the other hand, one does not always want the islands of crystallinity within a material being annealed to grow too large. It is for this reason that the time at which a material is held at the annealing temperature is critical to the success of the annealing process. The mechanical simulation of an annealing process for a two-dimensional array of the simulated amorphous material shown in Figure 5.18 is the application of vibration to the array. This gives the spheres modelling the behaviour of the atoms enough energy to enable themselves to rearrange their packing structure to form a regular array without giving them enough energy to bounce all over the board, which would simulate the evaporation from the surface of the material and the melting of the structure. In Figure 5.20, a two-dimensional array of mono-sized spheres assembled slowly by Nowick and Mader is shown. This slow growth rate enables some islands of regularity to grow. The array can be broken up into several islands of regularity as illustrated. Not surprisingly, the optical diffraction of the array in Figure 5.20(a) shows a structure intermediate between that of Figures 5.18(b) and 5.19(b), as shown in Figure 5.20(b). Nowick and Mader subjected the array in Figure 5.20(a) to gentle vibration to produce the array in Figure 5.20(c). It can be shown that this simulated annealing process caused growth of the major island present in Figure 5.20(a) to produce essentially a four crystal assembly. By comparing the two figures, it can be seen that the major crystals in Figure 5.20(a) grew until they met at a boundary which became the grain boundary limiting the two islands of regularity. The optical diffraction pattern of Figure 5.20(c), shown in Figure 5.20(d), illustrates the evolution of the pattern in Figure 5.20(b) towards that of a complete irregular crystal shown in Figure 5.18(b).

Fractal geometry may find an application in describing the progress of an annealing process by looking at the grain boundaries in materials subjected to various amounts of annealing and treating the grain boundaries as if they were growing fractal structures. Thus, in Figure 5.21, a set of pictures presented by Nowick and Mader to show the progress of an annealing process are given. The array in Figure 5.21(a) was produced at a slow rate of deposition and a simplified, abstracted grain boundary fractal from the

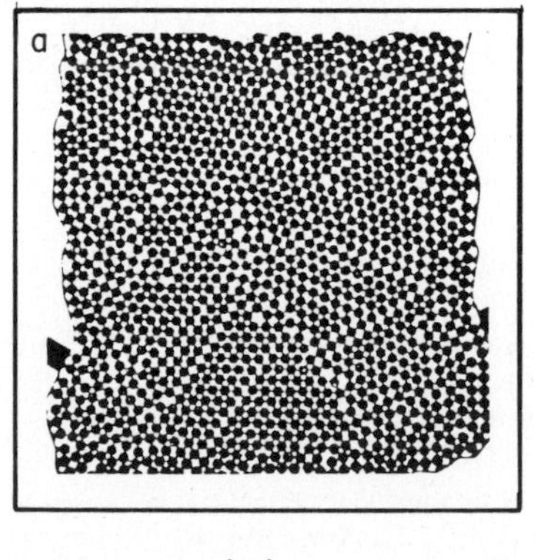

(a) No vibration (b)

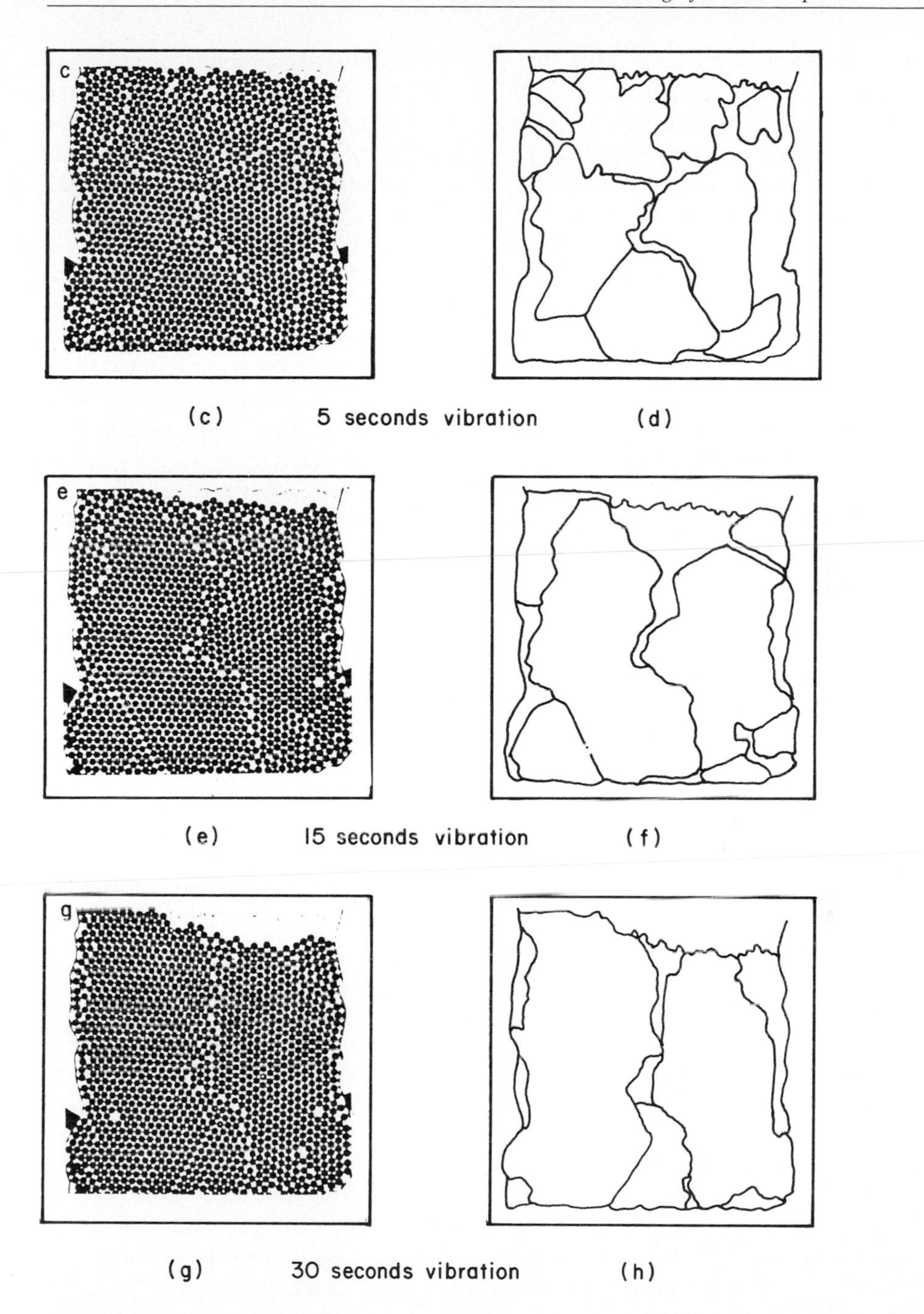

Figure 5.21. The progress of the annealing process was illustrated by data published by Nowick and Mader [23]. The boundaries between the islands of regularity appear to be fractals which change in complexity as the system is annealed. (Copyright 1965 by International Business Machines Corporation; reprinted with permission.)

diagram is shown in Figure 5.21(b). Figure 5.21(c) shows the structure of the initial array after being subjected to 5 s of vibration. The abstracted fractal boundary fractal, shown in Figure 5.21(d), is obviously less complex than the boundary fractal of the initially deposited array. In Figure 5.21(e), the structure achieved after 15 s of annealing is shown. Again, the boundary delineating islands of irregularity is much simpler than that in Figure 5.21(f). After 30 s of annealing, we are left with some island voids and a single fractal boundary separating two large crystals with some minor dislocations in the regular packing as shown by the dotted lines. When discussing the structure of cracks in Chapter 8, we shall discuss techniques for characterizing fractal structures such as those abstracted from the simulated arrays as summarized in Figure 5.21. We have discussed the sphere arrays produced by Nowick and Mader in some detail since they will prove to be very important when discussing the fractal figures produced by driving oil out of sandstone with a driving fluid such as water, a physical system discussed in detail in Chapter 9.

As mentioned earlier, the prime objective of the Nowick and Mader experiments was to simulate the structure of thin films of **alloys**. An alloy is defined as a mixture of two or more metals. The word comes from the Latin term "alligare," meaning to bind together. This word has also given us the word "ligament" for string-like parts of the body that hold the body together. At the molecular level, an alloy consists of atoms of different sizes. The presence of different sized atoms in a crystal lattice causes distortion of the lattice a phenomenon that generates the physical properties of an alloy. Nowick and Mader used three different sizes of spheres in various mixtures to simulate the crystalline structure of alloys. In Figure 5.22, some of the mixtures of different size spheres described by Nowick and Mader are shown. It can be seen that the presence of one sized atoms in a predominant array of other atoms caused distortions of the majority atom arrangements. Nowick and Mader were able to relate the diffraction patterns of simulated alloy mixtures to the properties of those alloys.

From the randomwalk-fractal perspective of this book, the major interest of the systems in Figure 5.22 is the apparent clustering of a randomly dispersed system. The similarity between the structure of the dispersed species in Figure 5.22 and the clustering of the fineparticles deposited on a filter will become apparent when we discuss this problem in detail in Chapter 7.

In Chapter 6, when we discuss the applications of fractal logic to a description of powder mixing processes, the structures of the system shown in Figure 5.22 will be interpreted in terms of problems encountered in the blending of dry powders. For example, in such a system the small percentages of smaller or larger spheres in the four pictures in Figure 5.22 could represent grains of a drug compound dispersed in starch, the mixture being used as a raw feed material for a tableting machine. Obviously, in such cases one is now very interested in the amount of dispersed species in any subsection of the overall system. This problem will be discussed in Chapters 6 and 7.

Today, because of the increasing power and falling costs of computer time, the simulation studies carried out by Nowick and Mader could probably be generated more efficiently using computer graphics technology. Voss has carried out some very elegant and beautiful computer graphics simulation studies of the growth of dendritic crystals during electrodeposition using a two dimensional randomwalk model [25, 26].

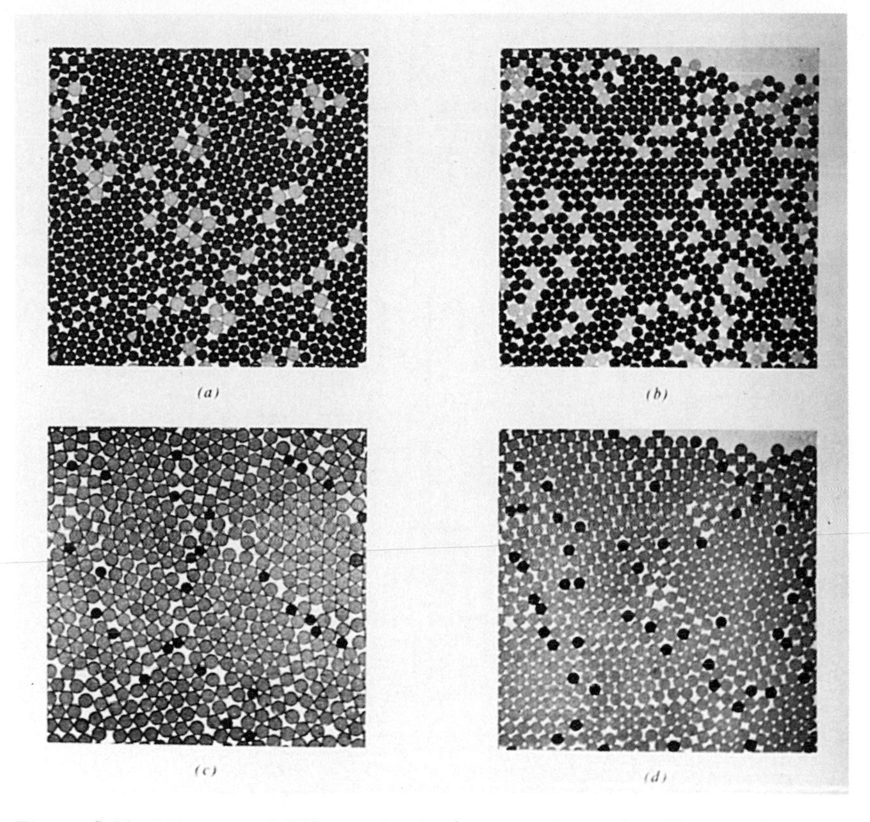

Figure 5.22. Mixtures of different sized spheres can be used to illustrate the structure of metal alloys. (Copyright 1965 by International Business Machines Corporation; reprinted with permission.)

The basic concepts of the logic used by Voss in his model of electrolytic deposition can be understood from the simplified version shown in Figure 5.23. At the beginning of the experiment, it is assumed that a fraction of the pixel squares of the lattice are occupied by ions of the electrolyte. Thus, if we were depositing copper from a copper sulphate solution, the black pixels in Figure 5.21 would represent copper (II) ions, Cu^{2+}, where Cu is the chemical symbol for copper. The word **ion** means wanderer or traveller. When copper sulphate is placed in a solution it splits into two ions denoted by Cu^{2+} and SO_4^{2-}. The formula SO_4^{2-} indicates that the sulphate ion is made up of one sulphur atom and four oxygen atoms, with two negative charges. The base of the lattice square would represent the deposition electrode.

It can be seen that the fraction *f* of pixel squares occupied by copper ions represents the concentration of the electrolyte. Occupied pixels away from the deposition surface are referred to as mobile pixels. In the computer programming, each of the mobile pixels is examined in random order, then one of its neighbourhood sites is selected by chance as a possible next position. Thus, if we looked at the pixel with the address (4, 6) marked

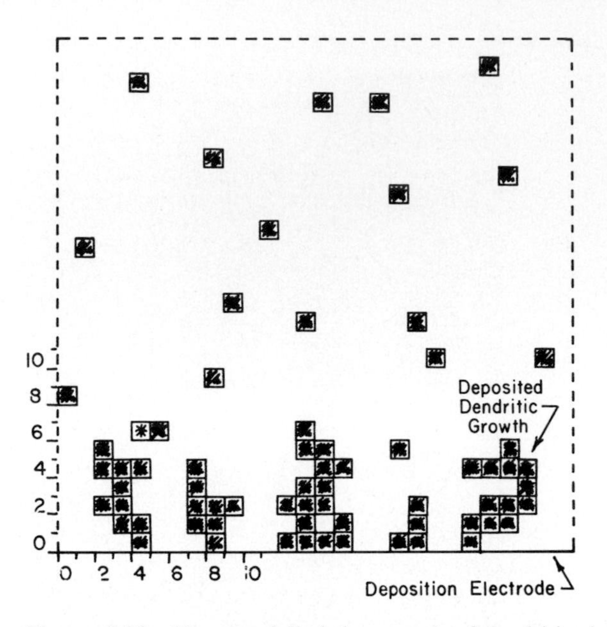

Figure 5.23. Voss modelled the growth of dendritic deposits formed by electrolysis using a randomwalk model in which many individual pixels are given an opportunity to grow to dendritic structures formed at the electrode [25, 26].

with an asterisk in Figure 5.23, we would then look at the nine possible pixel sites to which the asterisked pixel could move. If the selected new address is unoccupied, the pixel moves to this new site. If it is occupied, the pixel remains fixed. Thus, if our asterisked pixel was instructed by chance selection to move to the pixel site given by the address (5, 6), the computer would discover that this site is occupied and the asterisked pixel would stay in its present location.

If, on the other hand, the asterisked pixel was instructed to move to the address (3, 5) the computer would check and find that this was empty. It would also discover that a move to this square brought the asterisk pixel adjacent to the growing dendritic crystal which started to grow from the deposition surface at the address (4, 0). In this situation, the pixel would be considered to have joined the dendritic growth. If, however, it moved to the pixel address (3, 6), this diagonal encounter with the tip of the dendritic tree would not be considered as a sufficiently strong encounter for the pixel to join the tree. It would be still considered to be a free agent, able to wander around the lattice space (that is, the simulated solution). As the dendrites built up from the deposition surface, fresh pixels were allowed to come into the lattice space to maintain the fraction of occupied pixels above the dendritic growth at the constant value of f.

Voss also incorporated into his Monte Carlo routine a probability factor governing the possibility of permanent adhesion to the dendritic growth, or subsequent movement away back into the free space above the deposition tree after the initial orthogonal contact of a wandering pixel with a deposition tree. When modelling electrolytic deposition, Voss pointed out that altering the probability of permanent adhesion corresponds to changes in the current flow in the electrolytic cell. Thus, if zero current is flowing through the cell, there is zero possibility of adhesion to the growing deposit and the copper ion can be regarded as having bounced off the deposit, when it encounters the electrode pixels or deposited tree already formed. On the other hand, at high current densities, the probability of sticking approaches certainty. In mathematical terms, S, the probability of adhesion, becomes 1.0 at high current densities. In Figure 5.24, the appearances of simulated deposition structures at a volume fraction of occupied pixel space 0.05 for several probabilities of adhesion are shown. It can be seen that at high current densities, corresponding to certainty of adhesion, the deposit consists of highly dendritic crystals of high fractal dimension. At lower current densities the deposit becomes, in Voss's terms, "more 'moss like' rather than 'fern like'." The simulated systems in Figure 5.24 illustrate the fact that if one wishes to form a dense, strong coating during electrolytic deposition, one should try to use as low a current density as possible. In Chapter 3, we discussed the fact that one of the mechanisms causing failure in an automobile lead storage battery was the shorting out of the plates owing to dendritic crystals bridging the gap between the plates. The data in Figure 5.24 suggest that one can prolong the life of a lead storage battery by operating at low current densities, both when charging and discharging the system. In Figures 5.25 and 5.26 the effect of changing the concentration of the electrolyte is demonstrated. It can be seen that the quality of the coating is improved by increasing the concentration of the electrolyte. Notice that all of the simulated deposited coatings in Figures 5.24, 5.25 and 5.26 under low current conditions form fractal surfaces, the ruggedness of which could be characterized using some of the techniques discussed in Chapter 3.

In Figure 5.27 (see Plate 2 at the beginning of the book), a colour version of one of the experiments carried out by Voss is shown. In this photograph, the deposits formed early in the simulation are coloured red, those deposited later in the experiment are coloured yellow and the latest stages of deposition are coloured green. It can be seen

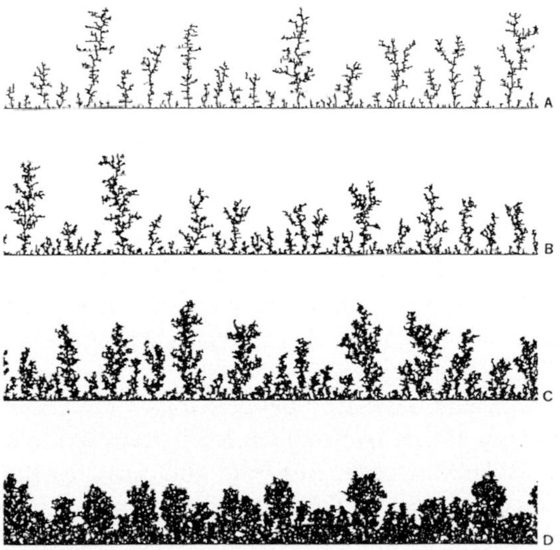

A

B

C

D

Figure 5.24. The structure of simulated deposits grown at a given electrolyte concentration varies according to the probability of adhesion, which is proportional to the current density; f = 0.05 for all deposits.
(a) Sticking probability $S = 1$.
(b) $S = 0.02$.
(c) $S = 0.05$.
(d) $S = 0.01$.
(Reproduced by permission of Richard F. Voss/IBM Research).

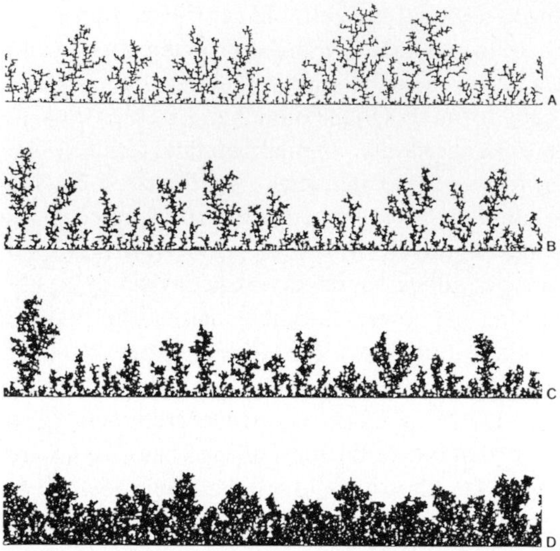

Figure 5.25. The complexity of simulated electrolytic deposits increases with increased electrolyte concentration. For the deposits shown the electrolyte concentration $f = 0.20$
(a) Sticking probability $S = 1$
(b) $S = 0.02$
(c) $S = 0.05$
(d) $S = 0.01$
(Reproduced by permission of Richard F. Voss/IBM Research).

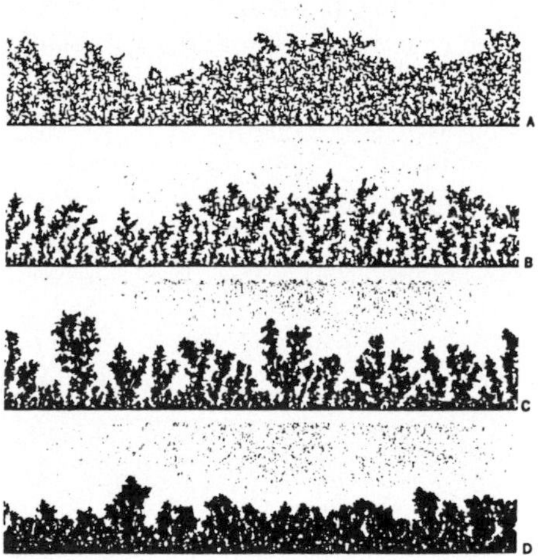

Figure 5.26. At ion concentration $f = 0.40$, the boundary becomes a virtually continuous deposit with a fractal boundary.
(a) $S = 1$
(b) $S = 0.02$
(c) $S = 0.05$
(d) $S = 0.01$
(Reproduced by permission of Richard F. Voss/IBM Research).

that at simulated high current densities, the randomly moving pixels find it very hard to penetrate the growing deposit, that is, they find it hard to avoid the seaweed-like tentacles of the growing dendrite which appear to mop up the pixels rather than to let them move down into the lower branches of the deposit. On the other hand, for the system simulated at low probability of sticking a uniform front which grows relatively uniformly over a period of time is formed. The similarity between the dendrites in Figure 5.27(a) and the frost patterns growing on a window is obvious. It would seem reasonable to assume that some kind of preferential capture of water molecules by the growing tips of the frost crystals operates during the growth of a frost crystal. If one looks carefully at the mossy-type deposit in Figure 5.27(b), it can be seen that this system appears to be porous. In some technologies, the porous nature of an electrically deposited metal is important (see the discussion of sponge fractals in Chapter 9).

When crystals grow on a rod dipped into a molten system, or if one looks at the build-up of frost coatings on a wire, the highly dendritic structures simulated by Voss are no longer applicable to the growth of the deposit. To explore possible mechanisms underlying different types of surface deposits in other types of deposition processes, Vicsek added two more variables to the Monte Carlo routine developed by Voss [27].

The two extra phenomena that Vicsek added to his model were the effects of surface tension and local evaporation-deposition on the structure of a surface deposit. To understand the basic physics of these two factors, consider the system shown in Figure 5.28. Figure 5.28(a) shows various sized regular clusters of pixel squares representing clusters of atoms. If a pixel such as the open-textured square is joined to each of the clusters as shown, there is always a chance that variations in thermal energy will enable the atom represented by the open-structured pixel to escape from the cluster. In a liquid, the interatomic molecular attractions hold any cluster of atoms into the tightest possible packing, to minimize the energy of the cluster. The force holding atomic or molecular clusters together in the liquid state is referred to as **surface tension**. Historically, the name developed because a drop of liquid such as mercury when placed on a surface does not wet this surface but appears to be surrounded by a skin that pulls the drop into a

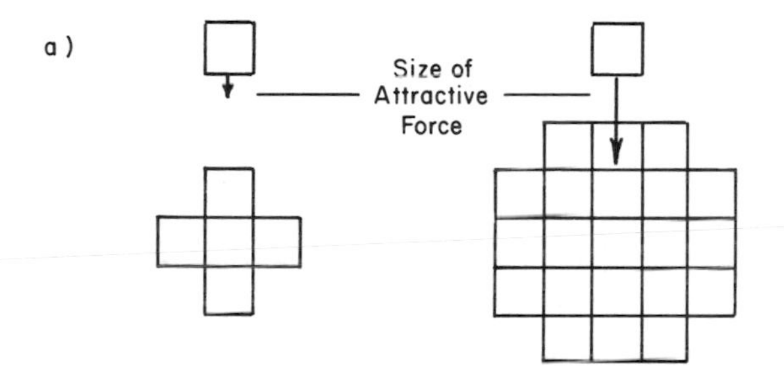

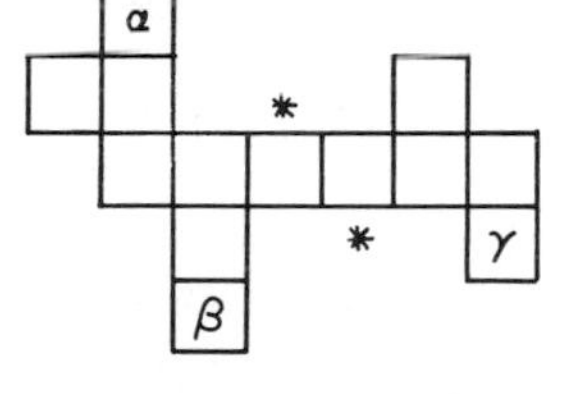

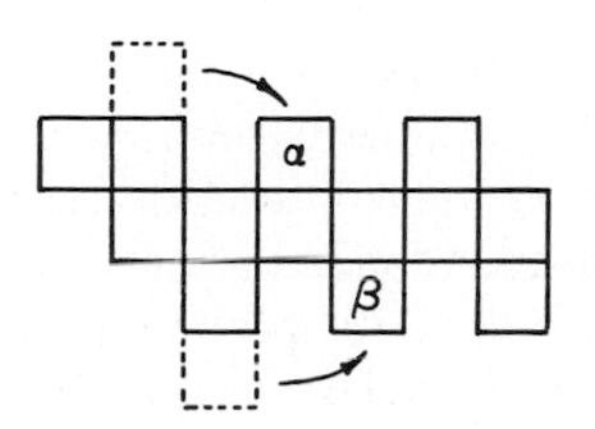

Figure 5.28. Vicsek extended the deposition model due to Voss by adding two factors which took into account the different forces encountered by an arriving pixel in the neighbourhood of a dendrite containing different numbers of occupied sites.
(a) Illustration of the force experienced by a simulated molecule arriving at various sized clusters.
(b) Thermal agitation may cause an arriving pixcl to break free from the dendrite momentarily, only to be recaptured by a surface set of pixels with higher attractive forces at the points marked by a asterisks.

spherical shape. Early scientists thought this meant that the surface was in tension, as if it were surrounded by a rubber membrane. For this reason, the force holding the drop together was described as surface tension and the term has persisted into modern science. A drop wets a surface when the pull of the molecules in the surface is a greater force on the molecules of the liquid than the internal forces pulling the drop together. At the molecular level, when only a few molecules or atoms form a cluster, differences in the size of the cluster result in very different forces on an outer molecule, such as those represented by the open-structured pixels in Figure 5.28(a).

New students of physics are often baffled to find that in a mixture of drop sizes in a spray, smaller drops evaporate to feed the size of the larger droplets [28]. Traditional science explains this growth of the large drop at the expense of the smaller ones as being due to the fact that the **vapour pressure** over a small drop is higher than that over a large drop. From the perspective of molecular attraction and dynamics, this is equivalent to saying that molecules can escape more effectively from the surface of a small drop because there are fewer molecules pulling it back when it attempts to leave. If we now apply this logic to a dendritic crystal of the type forming in the simulated systems of Voss, one can see that at the tip of the dendrite branches there is less force due to the other molecules to hold an impinging molecule on the tip of the dendrites. Therefore, it is not surprising that in a dendrite system in which the intermolecular attractions are relatively weak it is hard for tips to grow on the dendrites and the tip impinging molecules have a high probability of leaving the crystal again than if they impinge on a position such as those marked with asterisks in Figure 5.28(b). In a formal physics statement, it is said that the surface tension at points such as α, β and γ is less than that at the points marked with asterisks. When Vicsek built these two phenomenon into his Monte Carlo model, he obtained less dendritic types of structures as illustrated in Figure 5.29. If a molecule attempts to leave a point such as a, it can either move off by chance to the left or the right. If it moves to the right, there is much more intermolecular attraction on the molecule than if it moves to the left. This means that there is a high chance that if it attempts to move to the right, there is a high probability that its escape will be foiled and it will end up being attached to the pixel area denoted by the asterisks in Figure 5.28(b). In the language of computer science, this is equivalent to stating that one allows the position of an impinging pixel to be relaxed to one of the nearest or next nearest neighbour sites which is most occupied by molecules. It should be noted that these two mechanisms correspond to reality in the behaviour of highly fragmented powders. It is well known that if one attempts to store a finely divided powder made of needle-like crystals, the movement of atoms or molecules in the vapour phase (that is, molecular movement to and from the surface of the crystals) results in the loss of the sharp needles and self-sintering of the powder mass to form a cohesive set of crystals. Such a type of degradation of the powdered nature of the material is often blamed on moisture since this can also achieve the same result, but the consolidation can also be described in terms of molecular attraction using the model in Figure 5.28(b). This difference between freshly shattered powders and stored powders also accounts for the fact that when studying silicosis, the freshly shattered quartz crystals appear to have a higher chemical activity than aged quartz powders in the lungs. It is also the reason why stored explosive powders in stock piles of ammunition lose their effectiveness.

Figure 5.29. Vicsek was able to model the crystallization of a material under different sets of conditions by introducing factors related to the surface tension of the material being crystallized and the chance of redeposition along the surface of the deposit due to local evaporation - deposition mechanism.
A = factor related to the surface tension of the material.
B = factor related to the probability of relocation on an adjacent site of a deposited pixel.
(Tama's Vicsek, *Phy. Rev. Lett.*, 53(1984) 2281 published with permission of T.Vicsek and the American Physical Society.)

For the systems generated by Vicsek and reproduced in Figure 5.29, the factor *A* is related to the force of attraction holding impinging molecules to the already deposited surface. Thus *A* is related to the surface tension of the material. The factor *B* is related to the probability of relocation on an adjacent site, again due to molecular attraction. It can be seen that as the factor *A* increases, the dendritic structure of the deposit tends to decrease and that a nearly regular pattern develops. Vicsek has shown that this pattern actually has a characteristic wavelength which can be related to the forces and kinetics of deposition [27, 29].

A model very similar to that used by Voss has been used by Sapoval et al. to study the interdiffusion of metal atoms when two dissimilar metals are placed in close contact [30]. Thus, if one places two pieces of metal such as gold and lead in close contact, random motion of the atoms at the surface will cause intersurface penetration of the atoms. As Sapoval et al. pointed out, a good contact between two media is often realized after heating to increase the energy of the individual atoms to increase the mobility of the atoms and to permit diffusion across the metal interface to occur in a reasonably short time. Sapoval et al. showed that the diffusing atoms create a fractal boundary and that the structure of this fractal boundary is important when considering the properties of metal junctions.

Another important system in technology in which fractal systems are generated by basic units undergoing a randomwalk in two-dimensional space is the pattern of deposition formed by dust fineparticles arriving at the surface of a filter. In Section 5.1 we discussed how the movement of a dust stream to a filter can be considered to be a two-dimensional problem, since movement around the axis of the pipe does not affect the efficiency of deposition on the wall or the base filter. In the same way, if we now switch our attention to the deposited fineparticles on the filter, the problem collapses to a two dimensional one when we are not concerned with what happens above the filter. Looking at the structure of the deposit on the filter, we are only concerned with the (x,y) address on the filter surface at which they are deposited. One of the earliest procedures for measuring the amount of dust deposited in a filter involved measuring the percentage of the area of filter covered by the dust by optical reflectance [31]. Scientists who developed these procedures were well aware of the problem that late-arriving dust could be deposited on top of the dust fineparticles collected earlier in a sampling experiment. They derived equations to allow for the loss of coverage of the filter due to overlapping dust fineparticles [31]. Based on experiments, specialists recommend that in order to avoid overlapping dust when sampling an industrial dust, the proportion of a filter surface area covered by dust should be less than 5%. Unfortunately, it is not easy to judge by eye what constitutes five percent coverage. Inadvertently health and safety workers have sometimes worked with much higher coverages when monitoring dusty atmospheres, thus causing errors in their estimates of dust hazards. In recent years, there has been growing concern for considering the actual physical structure of dust in a working environment. In particular, regulations with regard to air pollution and industrial dusts are concerned with permitting different levels of respirable and non-respirable dusts. In such a situation, the characterization of the size of dust fineparticles deposited on a filter becomes more difficult since if a cluster of dust fineparticles is present on the filter it is very important to decide whether the cluster is formed by overlapping smaller units, which will constitute a respirable dust hazard, or is formed in the air before deposition on the filter and is of such a size that the agglomerate would not constitute a respirable hazard. The wrong classification of a cluster on a filter could result in gross errors in the estimates of respirable dust in the air. This problem is not always appreciated because some members of the scientific community are sometimes unaware of the clustering behaviour of random events, in the same way that any people are often unaware of the clustering of random events in one-dimensional space. Many scientists do not anticipate how frequently clusters can be formed by chance collision on the filter-paper.

An elementary model for demonstrating the surprising clustering that occurs on a filter surface by random chance is illustrated by a simple Monte Carlo routine summarized in Figure 5.30. If we consider the structure of a random number table, we can let each digit in the table represent a dust fineparticle when interpreted according to certain guidelines. Thus, to simulate a 5% deposit of a mono-sized dust fineparticle (the recommended level in industrial hygiene studies), we could let the area occupied by the digit 8 in the table become a square dust fineparticle, representing 1/100th of the area of the filter, if the neighbouring digit to the right when moving horizontally along the row of digits is an odd number. On average, the next neighbour to the right will be an even number 50% of the time. Since the digit 8 occupies one tenth of the space of the random number table, using half of them will generate a coverage which is 5% of the total area.

In Figures 5.30(a) and (b), a typical section of a random number table and its transformation to represent a 5% random coverage of a filter-paper by mono sized dust fineparticles are shown. In Figure 5.30(c), the worst clustering which occurred within

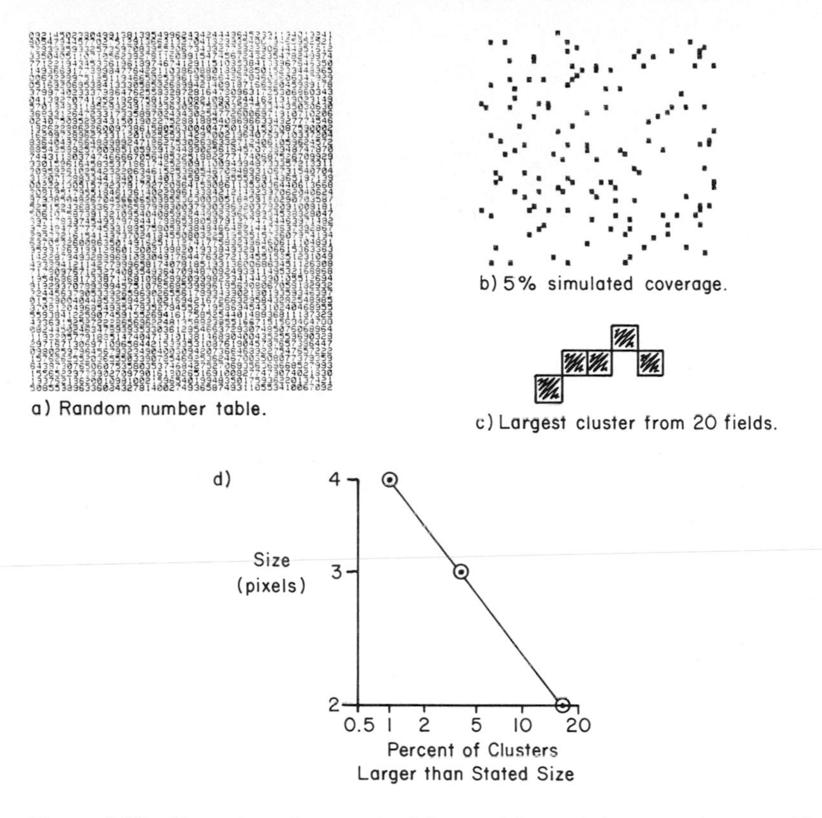

Figure 5.30. Clustering of mono-sized fineparticles arriving at random on a filter surface can be simulated by transforming random number tables according to certain rules. Thus, to simulate a 5% coverage of the filter every 8 in the random number table in (a) is transformed into a square pixel representing a fineparticle if its right-hand neighbour is an odd number.

20 simulated transformations is shown. The data in Figure 5.30 were generated during a class session concerned with industrial dust assessment. The students needed extensive persuasion to convince them that the high clustering in Figure 5.30(c) was not due to a faulty random number table!

The fractal structure of the simulated clusters in Figure 5.30(c) is obvious. If we redraw one of the clusters with slightly overlapping circles, we can see that the cluster resembles those of some of the diesel exhaust discussed in Chapter 4, Section 4.5 (see also the discussion of the structure of diesel smoke in section 5.4). In Figure 5.30(d), the size distribution of the clusters which arose during the simulation experiment summarized in Figures 5.30(b) and (c) fields of view of the size illustrated in Figure 5.30(b).

In Figures 5.31 and 5.32, data similar to that in Figure 5.30 are presented for 10% and 20% simulated densities of coverage.

The graph paper used in Figures 5.30(d), 5.31(c) and 5.32(c) to summarize the size distribution of the simulated clusters, formed by random events, is different from the

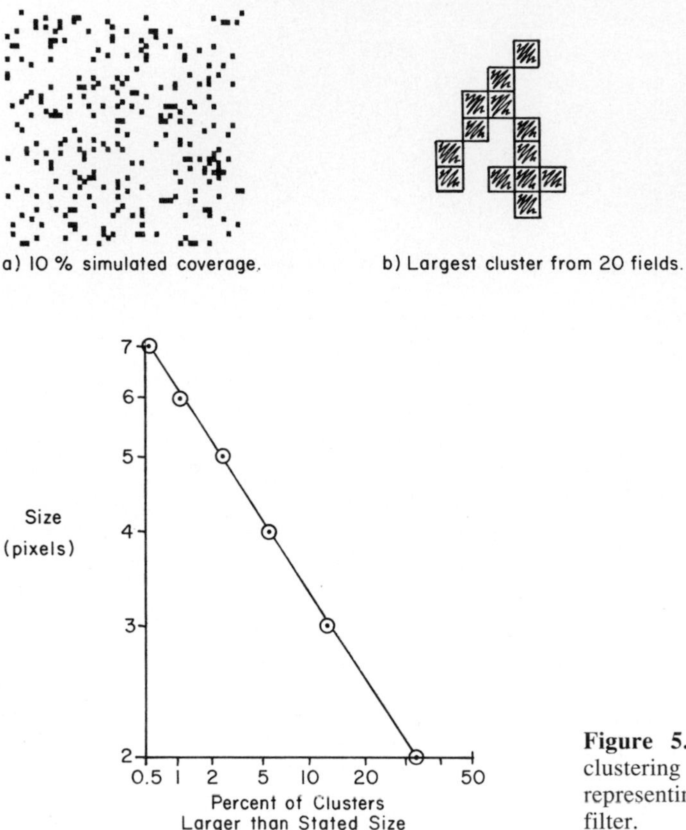

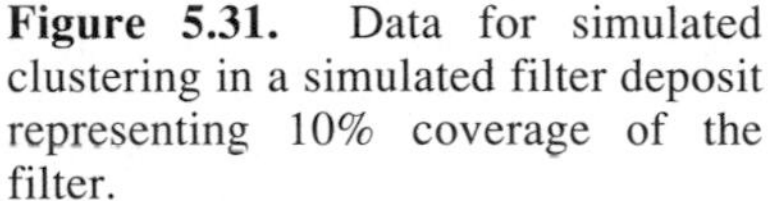

Figure 5.31. Data for simulated clustering in a simulated filter deposit representing 10% coverage of the filter.

Gaussian probability paper discussed in Section 4.9. On the graph paper used in these figures, the probability scale along the abcissa is the same as that on Gaussian probability paper, but the ordinate is a logarithmic scale. It can be shown that systems which are describable by linear relationships on such graph papers require the interaction of many causes but with the proviso that rare events (large clusters in this situation) require favourable interaction of the causes. Thus, fluctuations of the actual number of 7s in a sub-area of the table would obey the Gaussian probability distribution but, the clustering to form a large cluster is a relatively rare event. Therefore, to form a large cluster, one needs the favourable interaction of the interacting causes. In mathematical terms, the size distribution of clusters such as those simulated in the data in Figures 5.30 – 5.32 is described as a **log-normal distribution** and the special graph paper used in the presentation of the data is known as **log probability** graph paper. Log-normal distributions occur frequently in nature and in everyday life. Thus the salaries of workers in a bank will often be log-normally distributed, with a high salary requiring favourable interaction of causes or even the occurrence of a rare cause added to the other causes in the situation. Thus, for example, the highest paid member of a company may either be the son of the chief shareholder or may have married the boss's daughter; rare events,

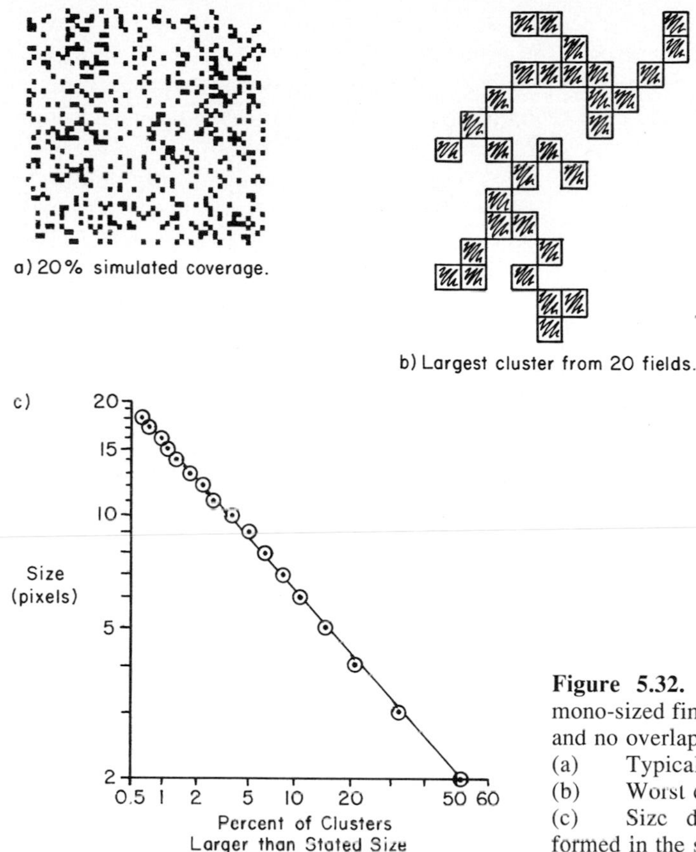

Figure 5.32. Simulated clustering of mono-sized fineparticles at 20% coverage and no overlap permitted.
(a) Typical simulated field of view.
(b) Worst clustering in 20 trials.
(c) Size distributions of clustering formed in the simulation study.

which do not contribute to the average salaries of the ordinary workers. In Chapter 4, we discussed how difficult it is to be sure that a time sequence of accidents is consistent with a given average accident rate. In the same way, clustering of events in two-dimensional space can sometimes cause people to panic because they do not understand how often events will cluster in two dimensional space by random chance. Thus, if the simulated deposits in Figures 5.30(b) and (c) are re-interpreted as showing death rates from cancer, where the chance of cancer deaths for every ten households in a city in a year was 5%, then it would be hard to convince people that the cancer cluster in Figure 5.30(c) occurred by random chance. Such a clustering would probably stimulate citizens' action for investigation into the effluent from a local factory. Such an investigation might show a causative relationship for the cluster, but it might not! The specialist who studies patterns of disease in a community is called an epidemiologist. Studying the spread of epidemics in society and of cancer in the body can be helped by concepts from fractal geometry (see the discussion of epidemics and cancer metastasis in reference 34).

It is interesting that many studies of airborne dust have reported that the size distribution of the dust can be described by a log-normal distribution function. What is not always clearly stated is whether this distribution is being formed on the filter or actually existed in free space before filtration [21, 32]. Obviously, a series of experiments should be carried out when sampling a new dust to establish at what concentration of coverage of the filter paper the size distribution function remains independent of the coverage [21].

In Figure 5.33, a simulated cluster which illustrates the build-up of dust clusters by overlapping deposition of individual fineparticles in a more sophisticated Monte Carlo routine than that used to generate the data in Figures 5.30 and 5.32 is shown. In this Monte Carlo routine, the x and y axes of the deposition space were split up into 80 digital addresses. Then various sized circles were allowed to arrive at the filter surface with the (x,y) address of the centre of the circle being selected at random from random number tables. In this Monte Carlo routine, overlap of arriving profiles is permitted. The probability of a certain size circle being deposited on the simulated filter space was allowed to vary according to a postulated size distribution of aerosol fineparticles arriving at the filter. The reader is reminded that the models presented here are relatively simple and they suffer from some limitations. One of the important limitations of this Monte Carlo routine is that the boundaries of the lattice can be regarded as intersecting clusters. Thus, when we see a cluster such as that labeled in Figure 5.32(b), we really do not know how big the cluster is since the boundary may have cut through the edge of the middle of a cluster. This boundary effect can distort the measured size

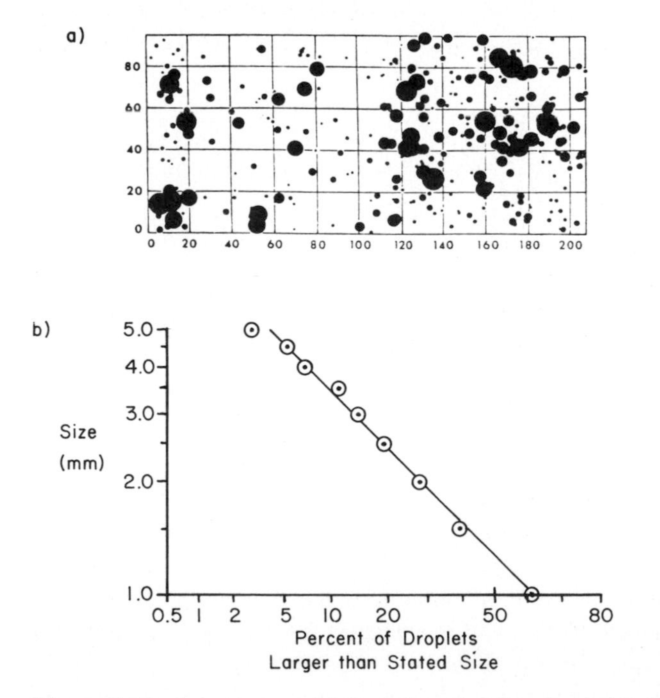

Figure 5.33. In a more sophisticated routine for simulating the size of clusters deposited on a filter-paper, one can allow for the overlap of various sized dust fineparticles arriving at the filter paper [31]. (Reproduced by permission of the British Occupational Hygiene Society.)

distribution of clusters [21]. Therefore, deciding which clusters are to be counted when assessing dust samples is an important problem in the automated image analysis of dust samples [33]. When sizing a powder as viewed through a microscope, it is usual practice **not** to record information on clusters touching a boundary of the field of view.

In Section 5.1, we discussed the possibility of a drunk escaping from a lattice in which there were traps to capture him. A scientific problem which is related to this problem is that of finding a minimum path around randomly distributed traps. Stanley has discussed a problem of this type using the diagram shown in Figure 5.34 [34, 35].

Stanley discusses the problem in terms of a mine-infested battle field which has been littered with mines at random by drunken enemy troops. It is assumed that two generals on the same side are at locations A and B and a messenger must take an important dispatch from A to B. The straight-line distance between A and B is described by Stanley as the Pythagorean distance or, using earlier terminology, the crow's flight distance between the generals. To estimate the time of travel needed to deliver the dispatch, the soldier needs to know the minimum path he can travel between the generals through the mine field. This is the type of problem that one meets in leisure activity magazines. Finding the path by eye is a relatively simple task for a human brain. However, we must remember that the human brain has enormous processing capacity. Thus a human can readily find the path A to B traced out with a pencil. We might make some false starts going along the pathways shown by the dotted lines but we would quickly home in on the minimum path. However, finding the minimum path through the minefield can be a difficult task for a computer. Although we may never have to carry dispatches through a randomly distributed minefield, Stanley tells us that his figure was

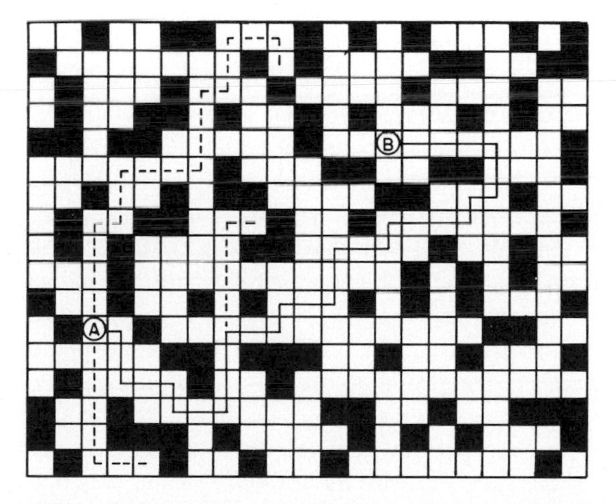

Figure 5.34. Finding the minimum path through a semi-occupied maze can be discussed in terms of finding a minimum route through a randomly structured minefield. This is a randomwalk problem which mimics the transfer of electrical signals in heart tissue [34, 36] (from H.E. Stanley and N. Ostrowsky, (Editors), "On Growth and Form; Fractal and Non Fractal Patterns in Physics," Martinus Nijhoff Publishers, Dordrecht, Holland, p. 29. "Copyright (c) 1985 Martinus Nijhoff Publishers, Dordrecht, Holland." Reproduced by permission of Martinus Nijhoff Publishers and H.E. Stanley).

adapted from a scientific paper by Ritzenberg and Cohen who used it to model the spread of electrical excitations in heart tissue, and that could be the start of many interesting simulations of random walks related to the life sciences [35, 36].

No doubt many readers will be tempted to simulate some of the interesting problems discussed in this section using home computers. When attempting to carry out this type of simulation, the reader should bear in mind the comments made in Section 4.1 on the problems of generating truly random numbers in a computer. Thus, Dietrich Stauffer has the following comments to make on random numbers used in Monte Carlo routines.

"Random number generation is an art not a science. I recommend that you program in a simple efficient way whatever runs on your computer. If the results are suspicious compare them with those obtained using a different random number generator and the rest should be left to the experts" [37].

5.4 Pigmented Coatings and Percolating Systems

My scientific interest in the structure of paints began in 1962 when I went to work for the Whiting and Industrial Powders Research Association [38] (this organization is known by the acronym WIPRIC, pronounced Whip Reck).

Whiting is a name used in Great Britain for a type of white pigment manufactured from naturally occurring pure deposits of calcium carbonates. For reasons that will be discussed later in this section, it is used in paint technology as an extender pigment. It is also used as a filler in plastics and many other industrial applications. Since Whiting is also the name of a small fish, the reader can be forgiven if he thought that I went to work on problems associated with the fishing industry (it is rumoured that manufacturers of fishing nets used to visit WIPRIC on a regular basis until they were educated as to the meaning of the term in the pigment industry).

The level of sophistication of my interest in paints increased considerably when I went to work at the Illinois Institute of Technology Research Institute (IITRI) in Chicago.

At the time that I joined IITRI in 1963, they had a major contract from NASA to look at the design of paints to be used on spacecraft. The problem was that traditional paints degenerated in outer space because of the bombardment of the pigment fineparticles. This degeneration was caused by the fact that without the protection of an atmosphere such as that which protects paint on earth, energetic radiation constantly attacked the spacecraft. Also, it was important to be able to design the paint so that its reflectance of light energy was as high as possible without increasing the weight of the paint. Every calorie turned away from the outside of the spacecraft by the reflecting paint was one calorie less that had to be removed from the interior of the spacecraft by the onboard air conditioning system. This in turn meant less work for the cooling system, which saved fuel. Saved fuel means a saving of weight in the spacecraft over a long journey. This in turn would save fuel in the rocket boosting the space capsule into space.

In common with many other problems tackled by NASA at that time, traditional industries had to deal with much more sophisticated systems than was the norm in the

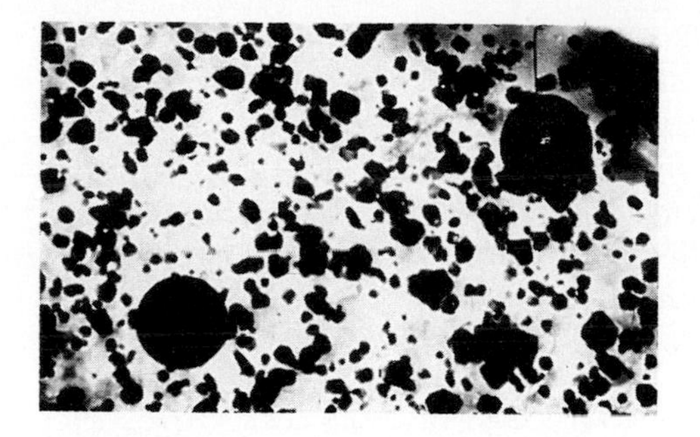

Figure 5.35. The structure of a typical coloured paint film is illustrated by this electron micrograph of a section through a dried paint film. The irregularly shaped crystals are titanium dioxide white pigment fineparticles. The space between the pigments is filled with an organic binder substance. The two large circles represent organic colour pigment. Electron micrograph provided and used with permission of Dr. DeWitt of PPG Paint Industries, Akron, Ohio.

existing industry. White paint used to cover the side of a house is not too demanding on the scientific structure of the paint, but reflecting paint for spacecraft represented a whole new level of sophistication [39]. To understand the problem that I was asked to work on in 1964, we need to understand the basic way in which a white paint is made and how light behaves inside a paint film. In Figure 5.35, an electron microscope picture of a section through a coloured paint is shown. The irregular crystals represent titanium dioxide crystals. The two circular pigments are organic dye pigments. Light entering the paint film is diffracted back and forth and around by the titanium dioxide pigments. A large crystal of titanium dioxide is transparent which means that it diffracts light without absorbing it. Powdered titanium dioxide looks very white because light entering the heap of powder is scattered back efficiently by multiple diffraction within the paint film. This is the same behaviour that one can observe with ice. A large ice cube without bubbles is clear with a slightly bluish tinge. If it is crushed with a hammer, it becomes white because now the many tiny crystals work together to diffract light back out of the crushed material to create a sensation of whiteness. Freshly formed snow in which the ice crystals are tiny is the whitest of white snow.

It should be noted that maximizing the reflective power of paint is not the same as making the paint look whiter than white. It is rumoured that, in some of the earlier experiments directed at improving the paint properties for spacecraft, NASA asked the paint industry to make the paint whiter. The paint industry responded by putting tiny blue granules, of the type used to make washing powders produce whiter than white shirts, in the paint. Unfortunately, this lowered the performance of the paint with regard to the total reflection of radiant energy in outer space. The tiny blue granules are ultraviolet-absorbing crystals which fluoresce, that is, give out light, in the visible wavelengths range. However, they absorb more ultraviolet radiation energy than they give out as visible light. Therefore, when considering the overall energy balance for the paint coating, adding the "whiteners" that made the paint whiter than white caused a deterioration in the overall energy-reflective power of the paint.

The fact that shirts washed in such detergents cause fluorescence of the coated material can be seen from the fact that white fabric coated with such whiteners glow in the dark for a short time after the lights are turned off. I remember going to watch a choir, enrobed in tuxedos and white dress shirts, give a concert. After the singing all the lights were dimmed and the choir was supposed to walk off the stage in darkness. However because of the fluorescence of the white shirts, what should have been a mysterious dark and solemn conclusion became a hilarious spectacle as the audience was treated to a fantastic ballet of glowing blue triangles floating up and down and across the stage.

Another reason why the visible "whiteness" of the paint does not necessarily correspond to the reflective power of the paint is that often one is as much interested in the opacity of the paint film as in the reflective power. Thus, as one paints over a surface that has bright red letters on the surface, one does not wish to see the red letters through the covering paint. To improve the opacity of white paint, the industry often adds a small amount of carbonblack. Far from causing the paint to look black, the tiny dispersed black pigments act like radiation sinks, to reduce the energy level of light being back scattered from the underlying surface to an external observer. This is similar to the fact that just as only small amounts of organic pigment will cause strong colour effects in a paint film, only a small amount of carbonblack is required to increase the opacity of the paint surface. Increasing the opacity of white paint in this way appears to make it "whitier" to the observer even though it reflects less energy.

It can be shown from the theory of optics that titanium dioxide is effective as a white pigment in paint because of its high refractive index with respect to the **binder**, which is the material holding the paint together. In Figure 5.35, the binder is the white spaces holding the pigment fineparticles. **The refractive index** of a material is a measure of how well a material diffracts light. A diamond has a high refractive index compared with ordinary window glass. Powdered diamond would make a good paint pigment if diamonds were more abundant. Leaded glass, used to make "crystal" tableware, has a higher refractive index than window glass, which is why "crystal" tableware looks so brilliant.

Although a paint film contains relatively few colour pigments, they have a strong effect on the behaviour of the paint since light is continually being bounced on to and off the colour pigment by multiple diffraction by the surrounding titanium dioxide crystals. On the other hand, calcite crystals, which are the basic crystals present in the whiting pigment referred to earlier in this section, has a refractive index very close to that of the binder substance. Therefore, the whiting pigments are almost invisible in a paint film compared with the crystals of titanium dioxide.

Calculations show that the refractive index of air is as far below that of the paint binder as the refractive index of titanium dioxide is above the binder. Therefore, from a light-scattering point of view, thousands of tiny bubbles would be as effective in reflecting energy from a coating on a spacecraft as a coating containing traditional white pigments. The reflective power of many tiny bubbles is demonstrated by the whiteness of freshly generated shaving foam from an aerosol dispensing can. Unfortunately, a rigid white foam, although it would reflect energy efficiently, would not be very tough. It would be eroded quickly by the bombardment of micro-meteorites encountered in outer space. Early, experiments aimed at marketing a "pigmentless" paint, in which the whiteness is generated by air bubbles, have not been particularly successful because of the lack of toughness of the coating.

It is the general opinion in the paint industry that to make a pigment efficient from a light-scattering point of view, one should aim to have a pigment size about the same as the wavelength of light being scattered by the pigments. This would mean in general terms that the pigment fineparticles should be approximately 0.5 μm in diameter (the wavelength of yellow sodium light is 0.55 μm). It was this folklore in the paint industry which led to my involvement in the project to design efficient paints for spacecraft, since it was reasoned that we needed to aim for an optimum pigment size to achieve maximum scattering power in the spacecraft paint. I refer to the idea that the pigment size has an optimum performance with respect to light-scattering properties as folklore, to focus attention on what I consider to be an equally important aspect of the performance of a paint film, the prevention of clustering of pigment fineparticles within a dried paint film. It is no good having a pigment of mono-sized 0.5 μm spheres if they are clustered together in the paint film when one makes the paint. Furthermore, in my opinion rugged, elongated crystals are likely to be more efficient in scattering light than spherical fineparticles of the same size. This is not to deny that the size of the pigment is important. In fact, my personal opinion is that the finer the better, down to about 0.3 μm, but not because of any special light-scattering properties of individual fineparticles. The finer the pigment, for a given pigment loading the more scattering surfaces exist in the pigmented film (note: the white paint for spacecraft would not contain the organic colour pigments in Figure 5.35).

When I was asked to look at a section through a paint film of the type shown in Figure 5.35, I pointed out the difficulty of deciding if the rather large rugged pigment scatterers were individual fineparticles created by the pigment manufacturing process, or were clusters formed by chance clustering in the paint film. I then suggested that perhaps the way to improve the reflecting power of the paint did not lie in a concern for the pigment size alone but in a study of the frequency of random clustering of pigment fineparticles. If one could interfere with clustering formation in the paint drying process, or could add something during the manufacturing of the pigments, then one might improve the reflectance properties of the paint.

When I suggested that clustering of the pigment inside a pigmented film might be an important phenomenon, we found that there was a surprising lack of information on how frequently clusters could occur by random chance at a given pigment loading. A useful measure of the pigment loading of a paint film is the pigment volume concentration (PVC) of the pigment. The PVC of a paint film is the amount of pigment in a dried paint film, expressed as a volume fraction of the pigmented coating. Thus a 0.2 PVC would indicate that 20% of the volume of a pigmented coating represented actual pigment.

To generate information on the clustering of pigments that occurred by random chance in a paint film at various PVC levels, a series of experiments to model the clustering of mono-sized pigments in simulated paint space were carried out [40].

To simulate the random positioning of pigment in a paint film, unit squares (pixels) of a lattice were occupied at random by picking the x and y co-ordinates of the occupying pixels from random number tables. When a position was selected, if that

Figure 5.36. Simulated cluster of mono-sized pigment fineparticles at various pigment loadings.
(a) 0.2 PVC.
(b) 0.3 PVC.
(c) 0.45 PVC.
(d) 0.5 PVC.
(Reproduced by permission of National Aeronautics and Space Administration.)

position was already occupied by a pigment pixel, then that selection was ignored and the next pixel position selected. When looking at the growth of clusters in a pigment, one is only interested in the loss of scattering centre individual pigment fineparticles and not on the strength of the agglomerates formed by the clustering and therefore for this type of cluster all contacts, diagonal and orthogonal, are considered to contribute to the growth of a cluster and to constitute a loss of an individual scattering centre. In Figure 5.36, the appearance of simulated pigment clusters at 0.2, 0.3, 0.45 and 0.5 PVC are shown.

In Figure 5.37, data on the loss of scattering centres due to the growth of clustering are summarized. The number of scattering centres per unit area of the simulated paint film at various PVC values is plotted. It can be seen that after a PVC of 0.17 has been reached, the absolute number of scattering centres decreases. This corresponds to the fact that above this PVC an arriving pixel in the simulation study has a greater chance

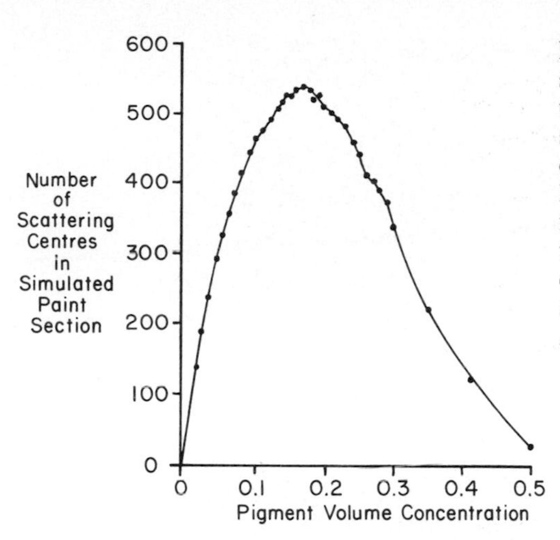

Figure 5.37. Simulated clustering of monosized pigment in paint films demonstrates that there is a maxmimum number of independent scattering centres at a pigment volume concentration of 0.17. Above this pigment loading, the chance that an individual pigment fineparticle will participate in a cluster is such that the number of independent scattering centres decreases even when more pigment is added (reproduced by permission of National Aeronautics and Space Administration).

of joining a cluster than it has of forming an independent scattering centre. It is interesting that a survey of the paint literature indicates that this value of 0.17 for the maximum scattering power of a paint pigment loading, as predicted by the Monte Carlo routine, is similar to the experimentally determined peak in the scattering power-PVC relationships of paints discussed by several investigators [41, 42]. Note that if the pigment volume concentration is increased until all pigment and extender fineparticles touch each other, the mixtures of spheres assembled by Nowick and Mader become models of a densely packed paint film.

As mentioned earlier, whiting is used in the paint and plastics industry as an extender pigment. Extender pigments were cheaper by weight than the organic binder of the paint or the plastic of a molded object and therefore whiting was added in large quantities to paint and plastics to lower the cost of the final product. Many other powders were also used as extender pigments, such as crushed chalk, powdered slate and a divided form of silica known as silica flour. Although the extender pigment was originally used to lower the costs of the paint, scientists began to discover that whiting and other calcium carbonate type fillers actually improved the scattering power of titanium dioxide pigments for a given titanium dioxide loading. As mentioned earlier, because the refractive index of calcite (the main constituent of whiting) is close to that of the binder of the paint film, whiting pigment is virtually invisible in the paint film. In the early work on the optical properties of paint film structures, this led to the conclusion that it could not be involved in improving the scattering power of the paint film. If, however, we now look at the problem of preventing clustering in a paint film, it appears reasonable to assume that the invisible extender pigments in the paint film increase the scattering power of the expensive titanium dioxide by competing for space in the paint film and preventing the formation of inefficient large clusters of titanium dioxide fineparticles. To investigate this possible role of extender pigments in paint films, experiments were carried out under the NASA project to simulate the role of extender pigments as "cluster preventors." Thus, in Figure 5.38, two diagrams from the text of

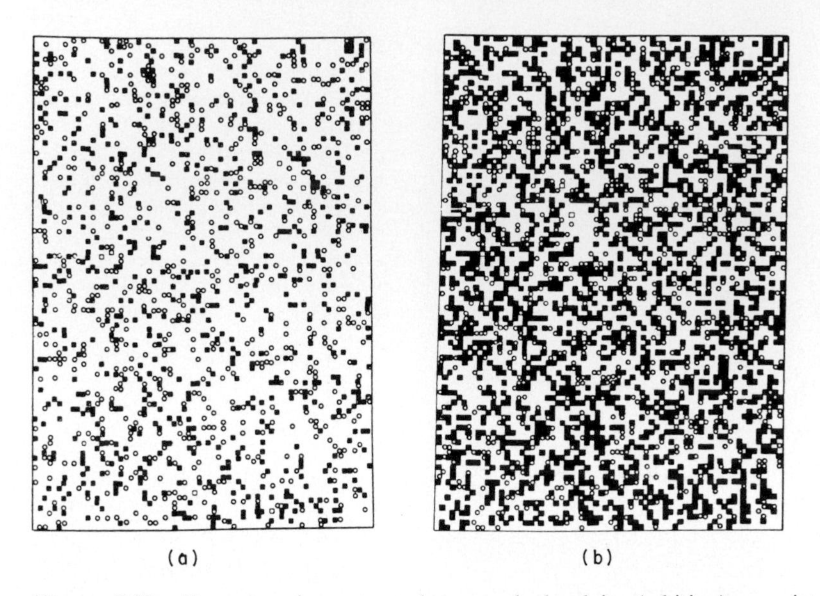

Figure 5.38. Extender pigments such as crushed calcite (whiting) are virtually invisible to incoming photons because their refractive index is close to that of the paint binder. However, they improve the light scattering properties of titanium dioxide pigmented paints by mechanically competing for space and preventing the formation of inefficient clusters of light-scattering pigments.
(a) Simulated structure of a paint film containing scattering pigment and extender pigment at concentrations of 0.1 PVC for each component.
(b) Simulated structure with 0.3 PVC high efficiency pigment and 0.1 PVC extender pigment. (Reproduced by permission of National Aeronautics and Space Administration.)

the NASA report are shown. These and similar experiments showed that extender pigment was efficient at preventing the build-up of large clusters of pigment and increased the number of scattering centres at any given PVC. Thus, historically, industrial paint technologists had actually increased the efficiency of an expensive pigment (titanium dioxide) accidentally by adding extender pigment for economic reasons not relevant to the optical performance of the paint (see also the discussion in Chapter 9 of the role of fractal powder pigments in reinforcing the strength of rubber).

Having discovered a possible role for invisible extender pigments, it seemed to our group working on the design of paint for spacecraft that we could increase the efficiency of the extender material by arranging for each pigment fineparticle to carry its own extender with it into the paint film, so that the "blocking" effect of the extender was immediate and less dependent on random chance. At that time, I was also working with a different group at ITTRI on the production of microencapsulated flavour droplets for use in chewing gum. The idea was to trap flavour droplets inside a gelatine envelope so that the flavour did not evaporate before the chewer chewed his gum. We started to look for possible encapsulation material for pigments to be used in the spacecraft paint, since if we could microencapsulate the titanium dioxide pigment with extender material this would achieve the maximum prevention of clustering. However, we ran out of money before we completed the project. As a last resort, we suggested that partial encapsulation might be useful. Thus, if we could coat larger pigment fineparticles with a coat which held on to the pigment tenaciously as it was fractured in a grinding process, one would achieve partially coated pigments which could be effective in increasing the scattering power of clustered pigment fineparticles by allowing photons to sneak through the less optically tight structure of the cluster. The type of clustering that one would build with partially encapsulated pigments can be appreciated from Figures 5.39 and 5.40. In Figure 5.39(a) it can be seen that in adjacent clusters, the two simulated pigment fineparticles with a partial coating of extender material in two-dimensional space (in this case it is assumed that one out of four of the simulated faces is covered with extender material) show that in seven out of the sixteen possible combinations photons could sneak through the cluster structure because of the presence of the extender material. If we try to look at possible clusters of three partially encapsulated pigment fineparticles in two-dimensional space, we find that the combination of possible clusters increases rapidly and that many of the cluster combinations have "sneak through" paths for photons. As the clusters are allowed to grow, the number of possible configurations increases very rapidly to the point at which it becomes difficult to predict the probability of the pigment fineparticles being opened up by the presence of the extender layer, but a few typical complex clusters generated in the original study are shown in Figure 5.40. It was at this point in our study of the potential utility of partially encapsulated pigments that our research came to an end. I remain convinced, however, that microencapsulation is a good way to improve the scattering power of expensive pigments.

When studying the clustering developed within a pigmented coating as demonstrated by the data in Figure 5.35, the effect of the shape of the pigment fineparticles on the less

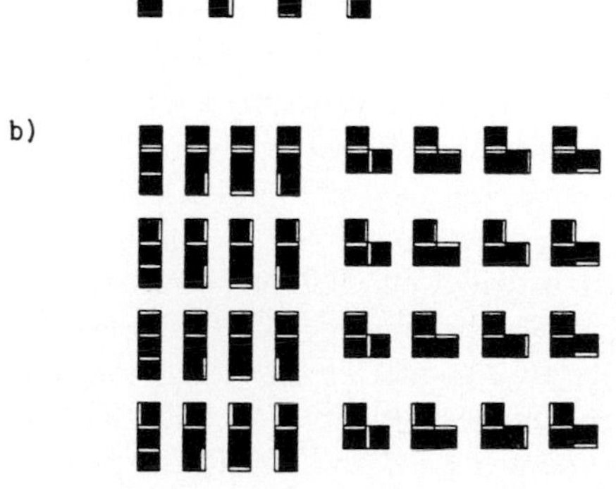

a)

b)

Figure 5.39. Partially encapsulated pigments carry their own extender pigment. Chance positioning of the partially encapsulated pigments in a paint film often permits pathways for photons to "sneak through" pigment clusters.
(a) Possible configurations of two partially encapsulated pigment fineparticles.
(b) Possible configurations of three partially encapsulated pigment fineparticles.
(Reproduced by permission of National Aeronautics and Space Administration.)

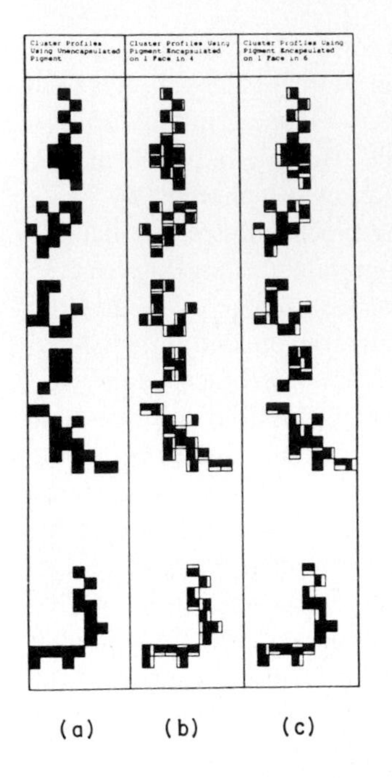

Figure 5.40. The complexity of possible pigment clusters increases considerably as one considers larger clusters.
(a) Representative clusters of uncoated pigment fineparticles.
(b) Representative clusters of fineparticles containing extender coating on one out of four facets.
(c) Representative cluster structure with a higher incidence of coated pigment facets.
(Reproduced by permission of National Aeronautics and Space Administration.)

optically tight structure of the clusters was questioned. Accordingly, a second study was carried out to simulate the clustering of pigment fineparticles having an aspect ratio of 2:1, where the pigment fineparticles could now have different orientations after their address had been found within the lattice. In Figure 5.41, the build up of clusters of these elongated fineparticles at various concentrations of pigment are shown. It can be seen that fairly large fractally structured agglomerates build up in the pigment by chance clustering, as generated in a simulated study of the film containing elongated fineparticles. In Figures 5.36 – 5.41, it can be seen that chance positioning can build large fractally structured clusters, although we do not know of the concepts of fractal geometry when we cancel out the experiments. The question of whether or not the crystals of titanium dioxide dispersed in the paint film in Figure 5.35, where there are already fractally structured units of smaller subunits, cannot be readily answered by studying the electron micrograph shown (see, however, the discussion of the titanium dioxide pigment in Section 3.7.6). In some situations, we know that the pigment or filter in a paint film is fractally structured before being dispensed in the film. Consider for example, the dispersion of carbonblack dispersed in plastic, shown in Figure 5.42.

It can be seen from the summary of Trottier's data presented in Figure 5.42(b) that the distribution of structural fractal dimensions in the populations of carbonblack pigment can be described using the log-normal distribution function. Again, the problems created by chance clustering in a pigmented coating when characterizing the structure of a pigmented material can be illustrated by considering the structure of the dispersion shown in Figure 5.42(a). An experienced operator would suspect that the junctions within a composite profile marked by the arrows shown on Figure 5.42(a)

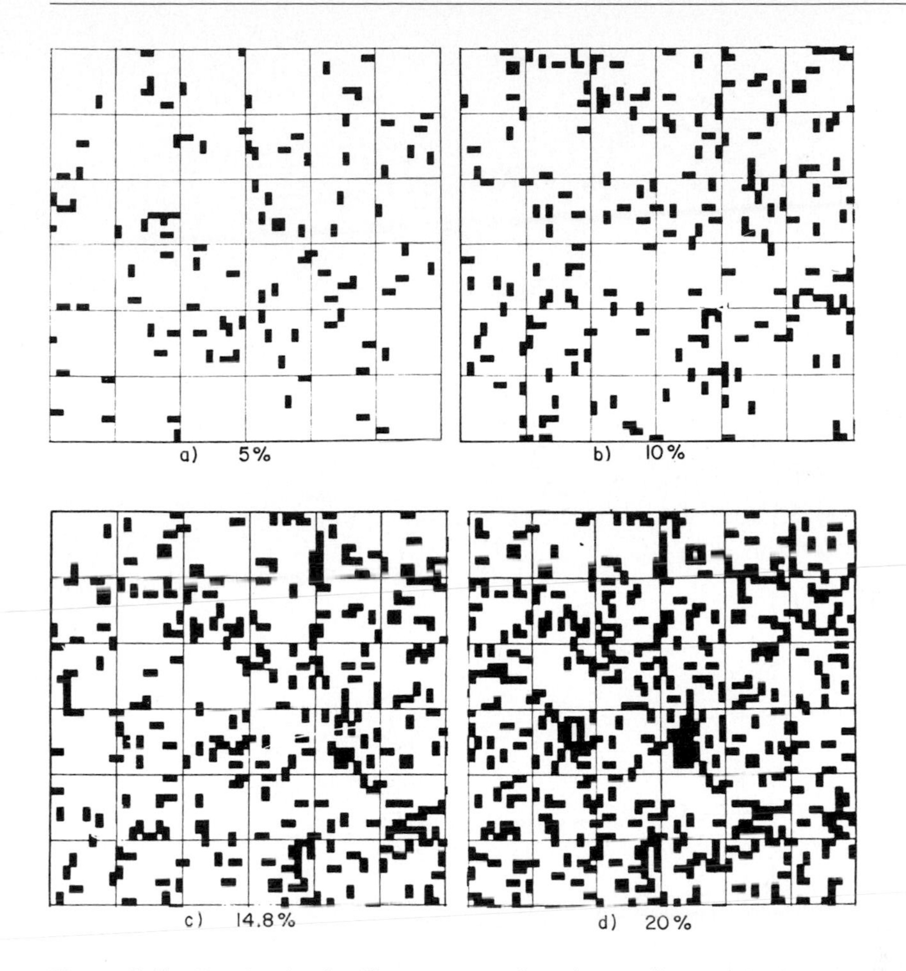

Figure 5.41. Simulated paint film structure when pigment fineparticles have an aspect ratio of 2 (reproduced by permission of National Aeronautics and Space Administration).

were formed by the chance touching of pigment fineparticles in the pigmented coatings. The severe dispersion techniques used to create a coating such as that in Figure 5.43 would probably have sheared off any real connections in the carbonblack pigment as tenuous as those marked by the arrows in Figure 5.42(a) (see the discussion of the Sierpinski fractal of this carbonblack system in Chapter 6). A possible technique for deciding how many clusters may have been formed by random chance is discussed in Section 5.6 when discussing the processing of photographs of diesel exhaust fumes by erosion-dilation logic.

The percentage of the field of view covered by carbonblack is 7.8%, equivalent to a PVC of 0.078. The carbonblack pigments obviously have a fractal structure and Remi Trottier has measured the structural fractal of all of the carbonblack agglomerates shown in the photograph [43].

a) Section through a toner bead. Arrows denote potential weak points.

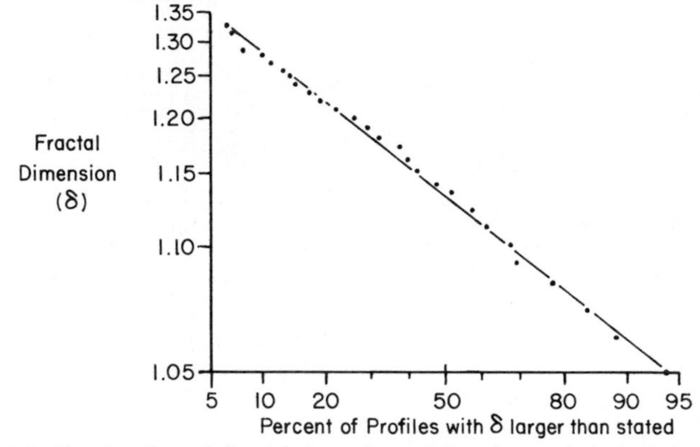

b) Distribution function of fractal dimensions of the dispersed pigments.

Figure 5.42. Toner, the dry powdered ink used in xerography, consists of carbonblack pigments dispersed in a plastic matrix. This electron micrograph shows a section through a toner bead at a magnification of a 100 000. The structural fractals of the dispersed pigment can be described by a log-normal distribution function [43] (diagram used by permission of D. Alliet, Xerox).

5.5 Mathematical Description of Fractal Clusters

In the Preface, we stated that one of the aims of this book was to introduce biologists and pre-university students to the fascinating world of fractals. If this book succeeds in tempting such readers to explore the advanced scientific literature on fractal systems, they will soon encounter mathematical descriptions of fractal clusters such as those found in our simulated paint films, employing concepts such as "radius of gyration" and "Fourier transforms." In this section, we shall attempt to domesticate these concepts for the average reader. Readers with a scientific background may wish to skip this section since they could well find it boring. On the other hand, readers who are only interested

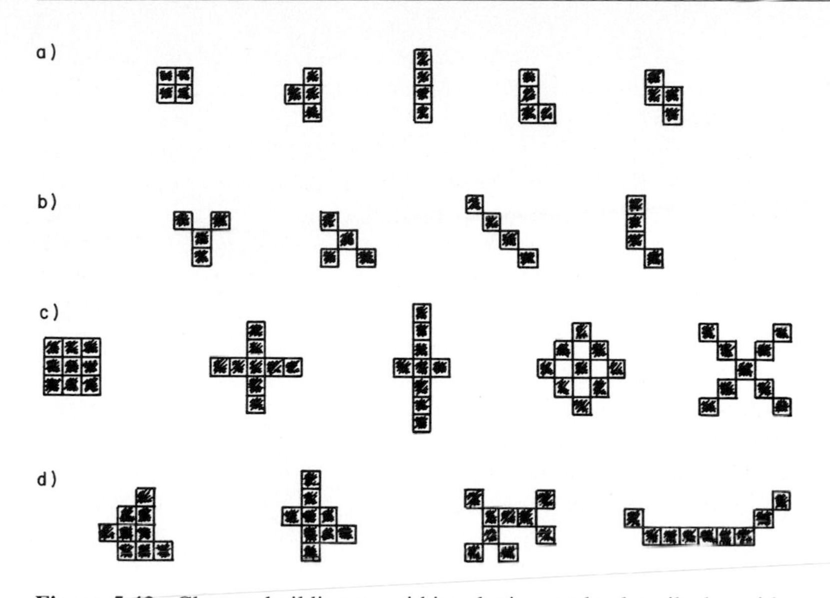

Figure 5.43. Clusters building up within a lattice can be described as either animals or insects depending on the strength on the bond between adjacent elements of the cluster. Animals are built up from orthogonal cluster bonds, whereas insects may contain one or more diagonal links in their structure. Orthodox scientists studying the dynamics of lattice occupancy in statistical physics may consider such insects as unorthodox, that is, non-permitted creatures in their world of lattice structures. Animals and insects may also be divided into two groups, those possessing symmetry and those which are asymmetric.
(a) "Animals" constructed from four pixels.
(b) "Insects" constructed from four pixels.
(c) "Symmetrical animals and insects" from more than four pixels.
(d) "Asymmetrical animals and insects" from more than four pixels.

in gaining a general knowledge of fractal structures can also skip over this section and procced at maximum speed with their randomwalk through fractal dimensions without bothering about the ultimate utility of such ancillary concepts.

When we look at the clusters simulated in Figure 5.40, it is difficult to describe the structure with everyday words. Scientists concerned with the build-up of such clusters in fractally structured systems have developed a vocabulary to use when studying clusters similar to those in Figure 5.40, by describing them as **lattice animals**. However, the clusters in Figure 5.40 are not lattice animals but **lattice insects**. To understand what is meant by these two terms, consider the systems shown in Figure 5.43. If we consider four closed packed pixels, they can be arranged into two sets of patterns shown in Figure 5.43(a). These types of clusters, which are only linked by strong orthogonal joints, are lattice animals. On the other hand, clusters that contain diagonal joints form the sets of patterns labeled "unorthodox" lattice insects in Figure 5.43(b). It will be recalled that when we looked at two-dimensional randomwalks and the growth of clusters, we only permitted orthogonal junctions to form strong bonds. In some problems, this corresponds to physical reality in that diagonal junctions are obviously very weak. However, in some fractal systems, diagonal junctions may correspond to physical reality; for example, in the pigment clusters that we have been discussing in this section the cluster is supported by the paint binder and as far as photons are concerned lattice insects constitute a pigment cluster which bars the path of the photon on its way as it tries to rattle around the paint film. When we come to look at the capture of fineparticles in a filter, we shall find that some fineparticles build tree-like structures by attaching to each other by means of electrostatic forces and form clusters which are more like lattice insects than lattice animals. The term lattice animal has been used widely in the terminology of fractal geometry, but the term fractal insect is relatively new. The word **insect** comes from two Latin words which mean "cut in two." Thus, many insects such as an ant appears to be almost cut into two parts; a structure which allows a fair amount of articulation in the movement of the insect. The terminology "unorthodox" is used in describing the "insects" in Figure 5.43 to warn the reader that such articulated clusters are not normally considered to be permitted entities in many of the published discussions of fractal clusters. The reader should check carefully when looking at any simulation studies of fractal clusters to see if the scientist permits both insects and animals to exist in the occupied lattice space.

When we start to look at clusters with more than two components, the permitted structures, whether they be animals or insects, start to become, more complicated. Thus, in Figures 5.43(c) and (d) a selection of some of the possible animals and insects are shown. One can separate these clusters into symmetrical and asymmetrical clusters, but this is obviously a limited way of describing the various complex structures that are possible.

Scientists have adopted a technique from a study of the structure of rotating masses to describe the structure of fractal animals and insects in a mathematical way. This technique depends upon treating each pixel of the cluster as if it were a thin tile of mass M. They then calculate the **radius of gyration** of the cluster [44]. There is no painless method for describing the radius of gyration to readers who do not have a background in mathematics. However, I shall attempt to illustrate the basic concept of the radius of gyration, and its physical significance by discussing the data in Figure 5.44. To understand what is meant by the radius of gyration we have to first describe what is meant by the **centre of gravity** of a cluster. Thus, if we made a cardboard model of a cluster made out of tiles, as shown in Figure 5.44(a), we know that if we were to try to balance the assembly on a pin we would have to place the pin at the point P. This balance point for the model of the cluster is described as the centre of gravity. We also know from experience that the centre of gravity of each individual tile is the centre point shown in each tile. Mathematically, we calculate the centre of gravity for an ensemble of tiles such as that shown in Figure 5.44(a) by measuring the distance of the centres of gravities of the components from the centre of gravity of the cluster. Mathematically, the centre of gravity is defined as the point about which the sum of the products of the distance of the centre of each tile times the mass of the tile for all of the tiles is zero. For a simple symmetric cluster such as 5.44(a), the position of the overall centre of gravity of the cluster is the centre of gravity of the central tile. Calculation of the centre of gravity for an asymmetric cluster is not a simple process but can be readily achieved with the aid of a computer. Sometimes, the centre of gravity of a cluster does not lie within the

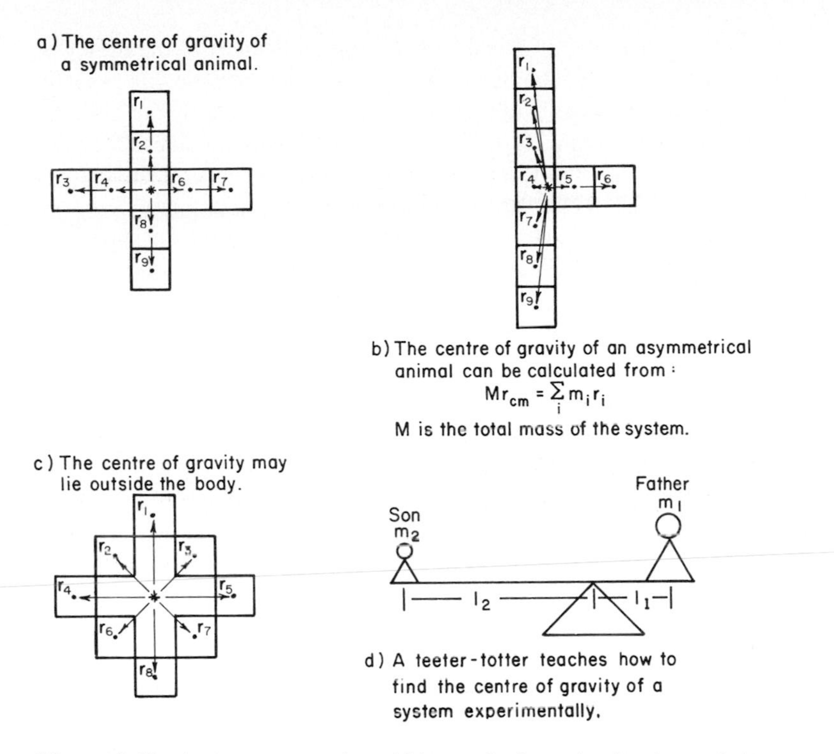

Figure 5.44. An important point within any lattice animal or insect is its centre of gravity. The centre of gravity is defined as the point about which the moment of the individual elements of a cluster is zero.

physical structure of the cluster. Thus, for the cluster in Figure 5.44(c), the centre of gravity lies in the free space inside the cluster. In fact, the distances from the centre of gravity of the cluster to the centres of gravity of the individual tiles now have to be regarded as thin strong wires of negligible weight if we are to be able to consider attempting to balance the cluster on the centre of gravity of the point P.

Perhaps the easiest way of understanding the calculation of the centre of gravity of a system is to consider the tactic employed by a father who wishes to be able to operate the machine found in playgrounds known variously as a teeter-totter or a see-saw (I grew up in England where it was called a see-saw; I was baffled when I went to live in Chicago and heard that my children had been playing on a teeter-totter, whatever that was!). This machine consists of a long, thin, strong plank placed on a pivot, as shown in Figure 5.44(d). Mathematicians describe the pivot point of the teeter-totter as a fulcrum. This sounds very technical, until one realizes that fulcrum is the Latin word for bedpost! Perhaps the scientific use of this term recalls the fact that the first teeter-totter was improvised by a Roman father using a piece of wood and a bedpost.

A father who takes his son to the playground soon learns that to initiate and maintain a pleasurable rocking motion of the teeter-totter, he needs to adjust his initial length l_1 to a smaller distance to balance out the effect of the smaller weight of his son on the other side. The effect of the force exerted at a distance l_1 by the weight of the father is given mathematically by the product of the distance from the fulcrum times the weight of the father. This quantity is known as the **moment of the force** exerted by the father's weight. The term **moment** comes from the Latin word "Mover," which means to move. Thus, the moment of the force is the ability of the force to move the arm of the teeter-totter or any similar system in mechanics. The teeter-totter operates satisfactorily when the moment of the son's weight equals the moment exerted by the father's weight. In this situation, the pivot is at the centre of gravity of the teeter-totter system of "plank + father and son." We all know that in such a situation it is possible to achieve a delicate balance which is known as **equilibrium**. Equilibrium seems to be a very fancy word until we realize it comes from two Latin words, "aequus," meaning equal, and "libra," which is a balance used to weigh objects (this word and its meaning are well known to the general public from the astrological sign Libra).

Thus, when we have the teeter-totter in equilibrium, everything is just balanced. When we calculate the centre of gravity of a system such as the clusters in Figure 5.44(a), (b) and (c), we are adjusting the position of the pivot point until the moments of the weights of each of the tiles balance out so that the pivot at the centre of gravity is like the fulcrum of our teeter-totter. Mathematically, the procedure for finding the centre of gravity of a complex cluster involves making a reasonable guess at the location of the centre of gravity, calculating the moment of all the individual forces about that point, and adjusting our estimate until the combination of all the moments of the individual tile weights about the adjusted centre of gravity is zero.

When we look at the cluster a in Figure 5.45, we can consider it as an aerial photograph of another machine found in the children's playground, the roundabout or carousel as it is known in the UK and USA, respectively. On such a machine, children sit at the four extremities of the cross, which is then made to spin around the centre of gravity. When we start to consider rotational movement about a pivot, we need to use the concept of **moment of inertia**. The term **inertia** is defined in a dictionary as "a property of matter by which it tends to remain at rest" (inertia comes from a Latin word meaning idle or hard to move). Thus, inertia of an object is that property which resists attempts to move it. If we were to try and move the cross around the pivot, the inertia of the system would resist movement and we would have to exert a force around the pivot to overcome the inertia of the body. The quantity known as the moment of inertia of a system, such as the cross cluster α can be shown to be mathematically equal to the sum of the product of the mass of the individual tiles times the square of the distance of the centre of the gravity of each tile from the pivot. Mathematicians have found it convenient to take this sum and make it equal to the product of a quantity denoted by I:

$$I = \sum md^2 \quad \text{(for all tiles)}$$

where $\sum$ (Greek sigma) represents "the sum of" and indicates that we have carried out the operation of first multiplying the mass of each tile m by d^2, where d is the distance of the centre of gravity of the tile from the centre of rotation, followed by the second operation of adding up all the separate md^2 terms. The moment of inertia of a body is not a fixed quantity like its weight. The value of I depends upon the position of the pivot

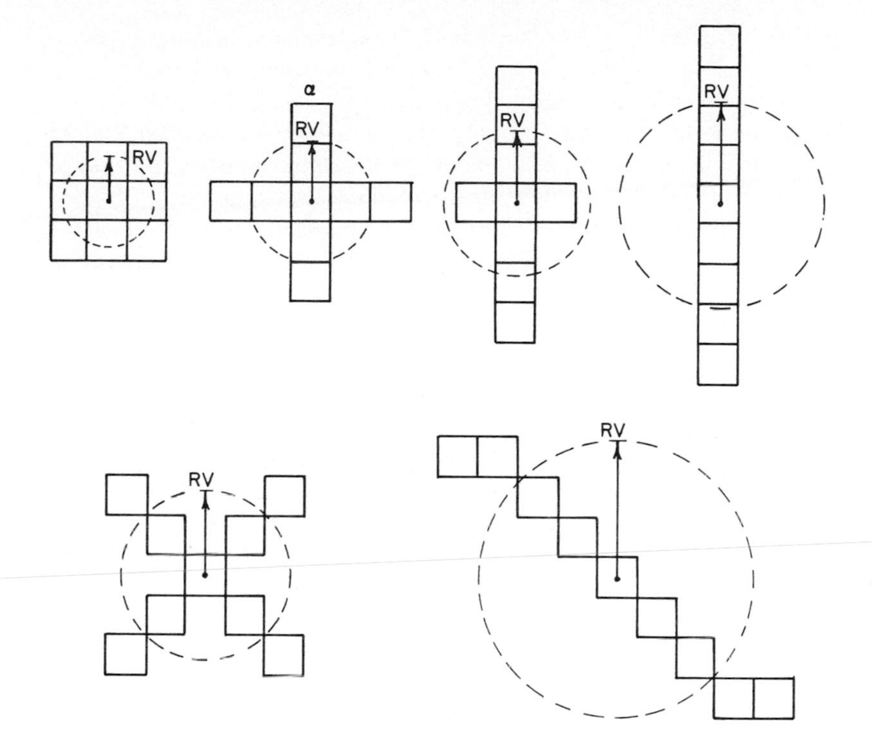

Figure 5.45. The radius of gyration, which is a measure of the distribution of a mass within a cluster, can be used to describe the structure of a cluster. *RV* equals the radius of gyration about an axis coincident with the centre of gravity of the cluster.

around which motion is occurring. Again, mathematicians have found it convenient to develop a mathematical concept called the radius of gyration:

$$I = \sum md^2 = MK^2$$

where M is now the total masses of the body (in this case the sum of the mass of the constituent tiles) and K is the radius of gyration of the rotating body. The radius of gyration is a measure of the distribution of mass around a pivot. In a discussion of the structure of fractal clusters, scientists often calculate the radius of gyration of the cluster as a measure of the spatial distribution of the elements in the cluster. The radius of gyration can be related to the fractal structure of large complex agglomerates [44].

In Figure 5.45, several lattice animals and insects are shown together with their calculated centres of gravity and radii of gyration. In general terms, it can be seen that for clusters containing the same number of tiles the more the individual tiles are spread out from the centre of gravity the higher is the radius of gyration of the cluster.

Many of the scientists who have looked at fractal clusters have used a mathematical technique known as **Fourier transformation** in two dimensions to calculate a fractal dimension of the cluster. Except for students specializing in advanced physics and mathematics, Fourier transformation tends to be a mysterious mathematical transformation encountered in a discussion of the structure of complex musical notes. In this encounter, the student learns that the quality of a note depends upon the mixture of frequencies embodied in the note and that this frequency mixture can be unscrambled using the mathematical techniques developed by the French mathematician Fourier (1768-1830). Unfortunately, this first acquaintance with Fourier transforms seems to fix in the mind of students that the mathematical technique only applies to vibrating systems in one dimension. In fact, the technique is applicable in two and three dimensions and can be used to transform many different types of mathematical descriptions of various systems. For example, what is known as the Fraunhofer diffraction pattern in optical theory is a Fourier transform in a two-dimensional plane of the structure of the object creating the diffraction pattern. Fraunhofer diffraction patterns are essentially those in which the diffracting objects are at least several multiples of several times smaller than the wavelength of light used to generate the pattern, and the pattern is observed at a distance which can be considered to be infinitely distant from the diffracting object. This type of optical diffraction pattern is named after the German scientist Joseph Fraunhofer (1787-1826). In Figure 5.46, the optically generated two-dimensional Fourier transforms of the lattice animals and insects in Figure 5.45 are shown. The fact that the two-dimensional Fourier transform of a fractal object is related to the fractal dimension has already been demonstrated when we looked at the diffraction pattern generated by Koch triadic islands of different orders and fineparticles of different ruggedness, as reported in Figures 3.19, 3.20 and 3.22.

If we know the Cartesian co-ordinates of each of the tile elements of a fractal cluster, we would usually have in the memory of the computer enough information to use, create or study a profile. Computer programs are used to calculate the two-dimensional Fourier transform of clusters and hence evaluate the fractal dimension of the cluster. It is because this is the preferred technique used by many scientists who have carried out studies into the growth of clusters by diffusion limited aggregation discussed briefly in Section 5.1 and discussed extensively in Section 5.7 that I have attempted to explain the concepts involved in a Fourier transformation of the structure of a cluster [45 – 48].

5.6 Percolating Pathways And Scaling Properties

If we take the simulated paint film of 0.5 PVC shown in Figure 5.36(d), we can amuse ourselves by considering it to be a maze through which we must find our way from one side to the other of the paint film. Our success at penetrating the maze will depend on how large a gap we require to be able to go through elements of the maze. In Figure 5.47(a), we see the different paths through the maze of 0.5 PVC when we assume that we are allowed to progress through diagonal escape routes as shown by the progress through the small insert in the top left-hand corner showing the actual structure of the maze. The other paths in the diagram have been shown as tracks uncluttered by the actual structure of the maze. It can be seen that we are able to make our way completely

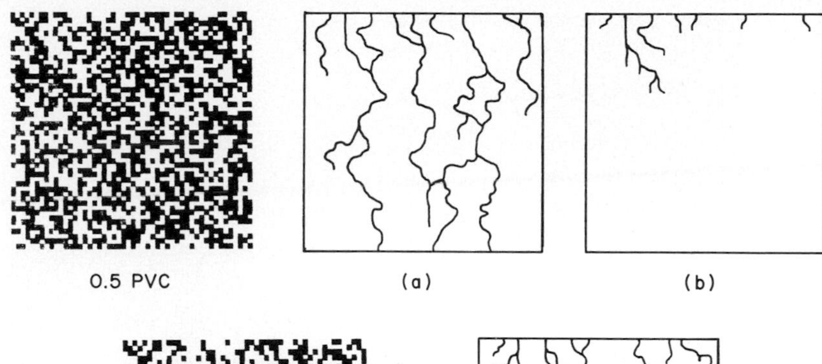

Figure 5.47. The critical occupancy level of a two-dimensional lattice at which it is possible to find continuous (that is, "percolating") pathways through the lattice via the occupied pixels depends on which is considered to be a permitted pathway.
(a) Percolating pathways exist at 0.5 PVC (50% occupancy) if diagonal and orthogonal junctions are allowed to carry the pathways.
(b) At 50% occupancy percolating paths do not occur if one is only allowed to pass through orthogonal pixel junctions.
(c) At 60% occupancy percolating paths exist through the body. It can be shown theoretically that the critical occupancy for percolating paths via orthogonal junctions is 59.28% [49].

through the maze even if we find that many of the paths end in dead ends. At first sight, it would not seem reasonable to allow us to squeeze through diagonal contacts, but if the white squares were copper blocks we could assume that the contact at the corners of a diagonal linkage is sufficiently large to allow electric current to flow and the pathway through the maze would represent the flow of electric current. In Chapter 9, we shall consider how cracks progress through a stressed lump of material. A diagonal contact between simulated growing crack elements may be sufficient for the crack to propagate through the system. Therefore, the white squares in our simulated paint film may be used to simulate crack growth in a stressed rock, with the path showing when a crack traversing the field of view can break up a piece of material. However, if we were to put water on the top of the maze it is reasonable to assume that only orthogonal junctions would form continuous pipes for the water to flow through the porous body. If we use the 0.5 PVC model in Figure 5.36(d) and only consider escape routes with orthogonal junctions, we find that we are not able to go through the maze as illustrated by the tracks reproduced in Figure 5.47(b).

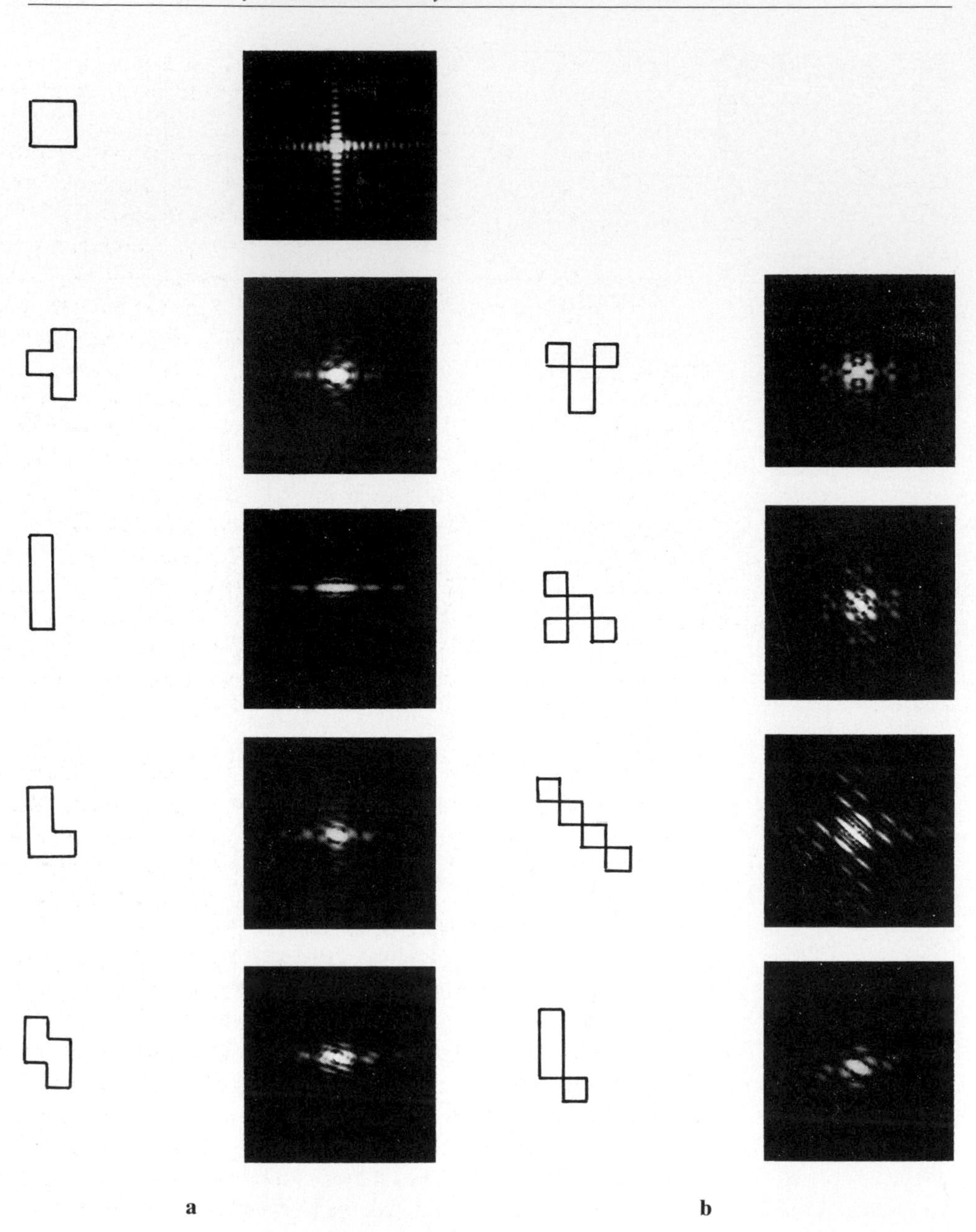

Figure 5.46. The Fraunhofer diffraction pattern of a lattice animal or insect is a two-dimensional Fourier transform of the cluster profiles. Optical diffraction patterns are generated using a photographic negative and laser light and are shown here beside two profiles used to generate them.

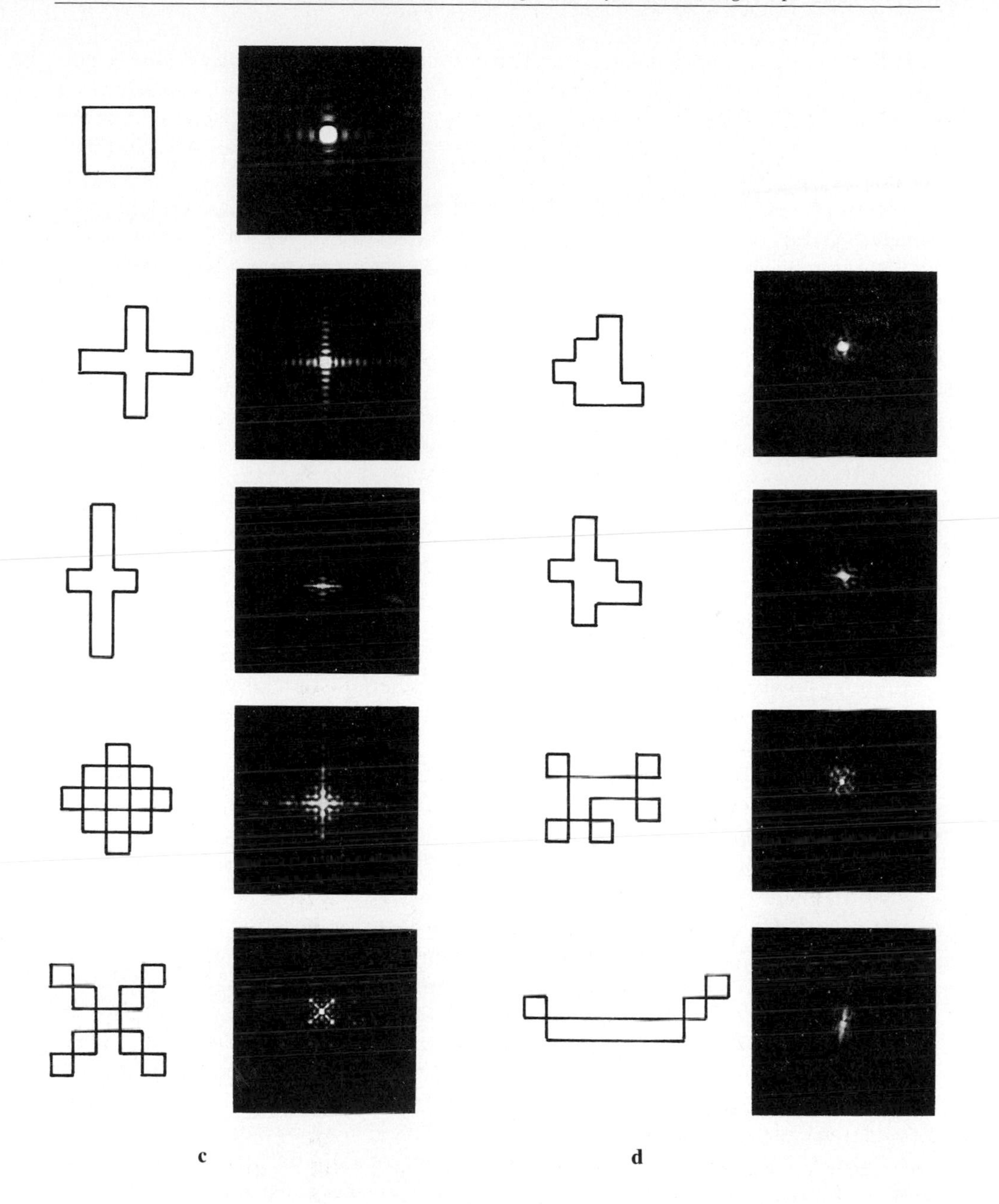
c
d

In Figure 5.47(c), the simulated structure for 0.4 PVC is shown. If we now regard this as a porous body focusing on the white areas of the structure, this constitutes a model of a porous body which is 0.6 porous. If we were now to put water on top of the porous body, the continuous paths shown in Figure 5.47(c) would exist. We could then say that water percolates through such a body.

The word percolate comes from two Latin words "per," meaning through, and "colare," to flow or to strain. A **colander** is defined in the dictionary as a vessel having small holes in the bottom to use as a strainer or a separator. This word also comes from the Latin "colare." The colander is a coarse sieve used in cooking for such tasks as separating peas from water; a coffee percolator works by allowing the hot water to percolate down through the coffee grains. We shall look at both systems again in Chapter 6 and discover that both serve as models for studying filters and filtration. A study of the build-up of clusters in a lattice in two and three dimensions until one builds up an infinite cluster, which is said to percolate the lattice, is an important branch of statistical physics known as **percolation theory.** When I first met the term percolation physics in discussions of fractal geometry, I was baffled by the fact that percolation theory is the preferred name amongst "statistical physicists" for the study of the build-up of clusters in a lattice before a percolating path exists. It would seem that a more logical name would have been "the study of the statistics of cluster formation at pre-percolation levels." However, the overall study of the structure, size and location of clusters in two or three dimensions and higher dimensions is widely accepted as constituting "percolation physics." Unfortunately, scientific terminology is not always strictly logical, it evolves as "common usage" and we have to learn to live with minor illogicalities such as this.

It now appears that when I was studying the growth of pigment clusters in simulated paint films I was studying percolation physics without knowing that the subject existed. In our original study for NASA, we realized that a study of the build-up of clusters until they formed continuous paths through the pigment could be important in the design of conductive paints bearing metallic pigments. In our original report in 1966, we looked at the size distribution and span length of clusters with a view to applying the experience we gained in cluster dynamics to predict a pigment volume concentration at which useful continuous paths would exist through the body when using metal pigments. Using today's terminology, we sought to understand when the growth of clusters provided percolation paths via the metal pigments for electrical conductance.

My personal difficulties when starting to read the percolation literature written by "statistical physicists" in fractal geometry came from two factors. Firstly, as I have already stated, the word percolation was so firmly linked in my own mind with "through paths" that I could not understand the preoccupation of some of the workers in percolation with the growth of clusters in pre-percolation situations. Secondly, I knew from my own simulation studies that complete paths existed at 0.5 PVC in my system and at first I could not understand why the theories of statistical physics predicted that one had to go up to a theoretical value of 0.59 PVC before percolation paths existed. I discovered later that the difficulty could be resolved by the fact that if only orthogonal junctions are permitted as contributing permissible pathways through the system, then 0.59 PVC is required before percolating paths exist. However, the statement that only orthogonal junctions were considered valid pathways in statistical physics literature is often either tucked away in the small print of scientific papers or is assumed to be known from previous scientific publications. This is why in my earlier discussion of different legal junctions existing in a cluster I have stressed the fact that one should always check to see if any particular writer is permitting diagonal junctions as well as orthogonal junctions in his growing clusters.

It has been established by extensive theoretical considerations that if only orthogonal junctions are permitted, then the lattice has to be occupied up to a critical concentration of 0.5928 PVC in my terminology or in the terminology used by percolation specialists, the lattice must be occupied by pixels forming 59.28% of the area of the lattice before percolation paths exist in two dimensional space. In Figure 5.48 (see Plate 3 at the beginning of the book), the physical structure of the critical occupancy density is illustrated by some elegant computer simulation studies carried out by R.F. Voss et al. of the IBM Thomas J. Watson Research Center [54]. These diagrams were also reproduced in an article by Orbach [49], which is very clearly written and was historically the article that helped me clarify my ideas on percolation physics.

Percolation studies in statistical physics pre-date the appearance of Mandelbrot's book on fractal geometry. When Mandelbrot's book was published, many workers in the field quickly recognized the relevance of fractal geometry to their systems. As a consequence, there has been a rapid generation of scientific publications dealing with percolation physics and fractal geometry (see the discussion of the integration of the two areas of theory presented by Mandelbrot [50]).

This literature has grown so rapidly that whole books have been written on the subject [51, 52]. It is not possible within the scope of this book to provide more than an introduction to the subject along with the presentation of key ideas and concepts. From the introduction given here, it is hoped that the interested reader may graduate to the

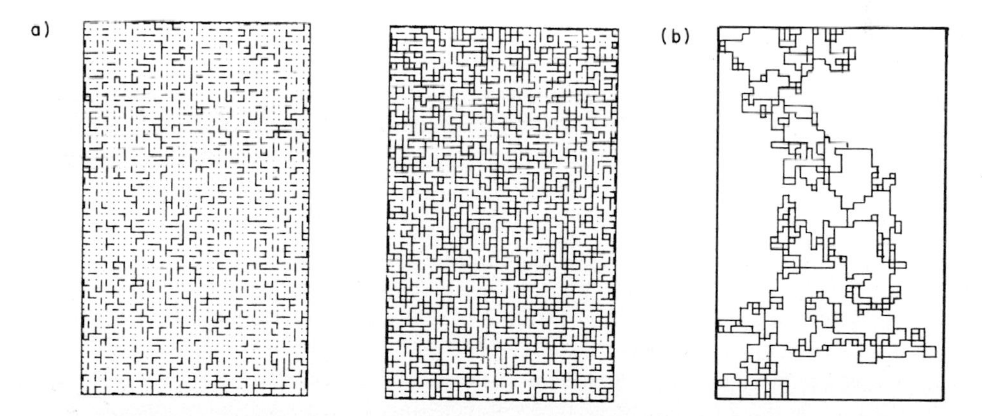

Figure 5.49. In bond percolation, one can imagine that copper sticks occupy at random the bonds between knots in a string mesh, that is assuming the original strings are non-conductive material such as plastic.

(a) A lattice occupied with 0.2 of the bonds being copper.

(b) 0.5 occupancy creates percolating paths through the random network.

specialist literature to discover for himself where fractal geometry is helping in advanced theoretical physics.

When discussing percolation theory, it is useful to define three main branches of percolation theory. The first is known as **site percolation**. All of the work in this chapter so far has been concerned with site occupancy on a rectangular lattice, where we consider pixels of defined area occupying space on a rectangular lattice and is part of percolation theory known as site percolation. Although most of the material discussed in this book will be concerned with square pixels on a square lattice, one can use other geometric lattice elements such as hexagonal to look at different types of site percolation (see the brief discussion of hexagonal lattice work as diffusion-limited aggregation canceled out by Meakin discussed in Section 5.7). The second type of percolation theory is concerned with a build-up of a matrix of bonds of the type shown in Figure 5.49. A brief discussion of the importance of bond percolation will be given later in this section. The third type of percolation theory is concerned with the flow of fluid through porous networks and is usefully described as **invasion percolation**. Discussion of invasion percolation in this book is delayed until Chapter 8, in which we discuss the creation of fractal fronts when a liquid invades a porous body.

Percolation studies have been found useful in studying so-called critical phenomena [53] and have also been applied to the formation of colloidal gels, the structure of branched polymers and of the dependence of ferromagnetism on temperature [34].

The simplest property of systems such as those shown in Figure 5.48 is the average density of the lattice. If we consider making a reference point inside of an occupied lattice such as that shown in Figure 5.50, then if we draw a search area of side length L as shown in 5.50 (a) (ii) then if we consider an occupied tile of the lattice as having a mass M we can calculate the mass of tiles in the search area by counting the occupied squares. We can then obtain a value for the density of material in the search area by dividing the mass of tiles in the area by the magnitude of the search area. Thus we have the relationship:

$$\sigma = \frac{nM}{L^2}$$

where σ is the average density of the lattice area L and n is the number of tiles of mass M present inside the circle.

If we were to consider a very large area of a network, we know that this quantity would have an average value which for a porous body would be related to the porosity of the structure by

$$\frac{M(1 - \varepsilon)}{A}$$

where ε is the porosity (that is, the percentage of voids) in the lattice and M is the mass the lattice would have if completely occupied.

Studies of a lattice such as those in Figure 5.50 have shown that from a certain range of magnitude of search area the mass within the search area is given by the relationship

$$M = \delta B$$

where B is a constant which varies according to what is known as the lacunarity of the system (defined and discussed later in this section) and δ is a fractal dimension of the

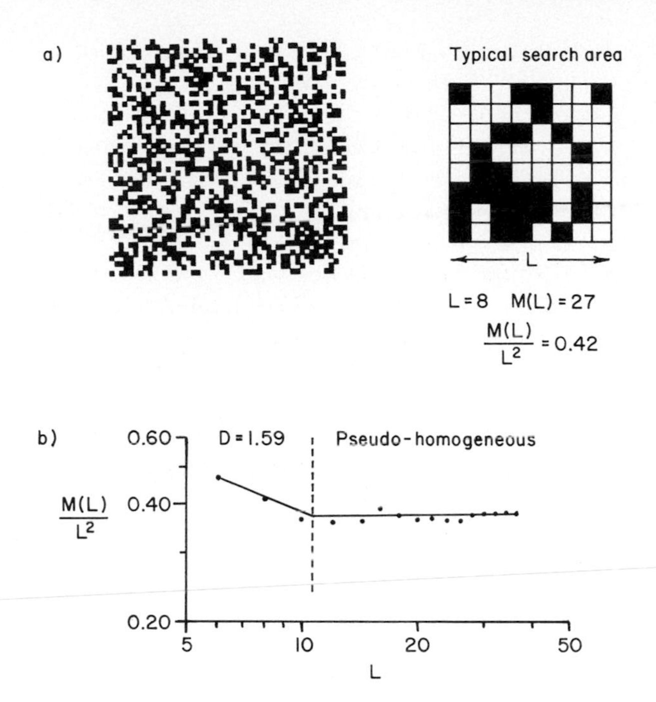

Figure 5.50. The structure of a randomly occupied lattice manifests fractal structure over a certain range of inspections. The lower limit for fractal structure is the size of the unit lattice. The upper limit for fractal structure is the length of the largest cluster present below the percolation limit; at occupancy levels above the percolation limit the size of the largest "lake" is the limit of size manifesting fractal structure. The upper length limit defining the manifestation of fractal structure is known as the percolation correlation length of the system.
(a) A partially occupied lattice used to generate the data on the average density of the lattice used to construct the data curve in (b). This curve demonstrates that the fractal structure of the lattice over a range of inspection sizes is the fractal dimension of the structure of the lattice. The Euclidean dimension of the lattice space is 2.

structure of the occupied area of the lattice. It can be shown that over a certain range of values, the equation that we have written above has to be written in the form

$$M = \lambda^{(D-\delta)}$$

where D is the Euclidean dimension of the system and λ is the side length of the search area. This relationship tells us that if we plot a graph of the logarithm of the average density versus the logarithm of a characteristic length of the search area, we will obtain a relationship of slope $(D - \delta)$ over the range of inspections, for which the occupied lattice manifests fractal structure. The fractal dimension defined in this way is found to be 1.9 [54].

The range over which the density of occupation of the lattice can be described by fractal geometry and can be illustrated from the data of Voss shown in Figure 5.48. Using the same terminology as used in earlier discussion of lattice occupation, we denote λ as the side length of the unit pixel. Obviously, the concept of average density of the lattice has no meaning for the inspection area of side length smaller than λ, so that this represents the lowest magnitude of the search area that can be used to explore the fractal structure of the density of material in the lattice. Voss has shown that, at occupancy levels before a complete percolation path exists, the size of the largest cluster constitutes the upper limit of the length of the search area used to investigate the density of occupation before the average density of the lattice becomes a constant quantity. Thus, in Figure 5.48(a), the length of the largest cluster shown in white is the magnitude of the upper limit for a search square at which the average density of occupation will be fractally distributed. This upper limit on the search area which will result in a manifestation of fractal structure is known as the **percolation correlation length,** usually denoted by the Greek letter η. In formal terms it is stated that the lattice manifests fractal structure for scales of inspections given by the relationship

$$\eta \geq R \geq \lambda$$

where $\geq$ means "greater than or equal to", R is the diameter of the search circle and η is the percolation correlation length for a probability of finding a pixel in a given large area equal to P.

In Figure 5.48(b), which is a slightly higher lattice occupation level than that in Figure 5.48(a), the white cluster now crosses the entire field of view and can be regarded as infinite in structure. For this system, the percolation correlation length is of the order of the next largest cluster coloured bright yellow in the figure. The structure of the occupied lattice is now fractal for search areas governed by the relationship

$$\eta = R$$

where R is now of the order of a quarter of the grid size.

When we come to the structure in Figure 5.48(d), we note that it looks far more homogeneous than the other structures. This **pseudo-homogeneous** structure is quantified by the fact that for such a system the percolation correlation length is now much smaller than the grid size and is approximately equal to the size of the largest hole in the structure.

Orbach warned that data on the average density of an occupied lattice within the range of resolution inspections where one can expect to discover a fractal dimension vary considerably because of the growth holes that occur in the lattice. Therefore, to establish the magnitude of the fractal dimension at an acceptable level of confidence requires the collection of a high level of data [49]. The study of the structure of a powder mixture, such as that of an active pharmaceutical powder such as acetylsalicilic acid (aspirin) mixed with starch (called the excipient by the pharmacist) to facilitate the manufacture of tablets of a useful size, is an important industrial problem. The fractal dimension of the mixture at scales of scrutiny below the percolation correlation length is a measure of the inhomogenity of the powder mixture. The magnitude of the percolation length is related to the minimum size of a tablet which will have a reasonable chance of containing a specified dose of the drug without unacceptable fluctuations in dosage from tablet to tablet [68].

Mandelbrot has coined a word to describe the "holiness" of an occupied fractal lattice: **lacunarity**. The origin of this name can be appreciated by looking at the system shown in Figure 5.51 (see Plate 4 at the beginning of the book). This is a picture of a randomly occupied rectangular lattice generated by Voss. The lattice has 3000 square units to a side. The occupied spaces of the lattice represent 59.9% of the available lattice. When we look at this picture without knowing what it represents, it could be an aerial photograph of Northern Ontario (which has 250 000 lakes) or Minnesota (which is known as the land of lakes) with the coloured areas representing villages and towns and the dark blue areas representing lakes. Lacunarity was coined from the Latin word "lacona," which means gap and has given us the modern English word lake. Thus, the lacunarity of a structure is the "gappiness" of the structure or its "lakiness." When discussing percolating systems, Mandelbrot also coined the word **succolating** for fractals which nearly include systems which percolate. The word literally means almost flowing through. We shall not use this term in this book but it is given here for the sake of completeness.

It is interesting that my Rorschach reaction when I first met the field of view shown in Figure 5.51 was that I thought it was a section through a porous body. The picture serves to illustrate how the random packing of fineparticles in a ceramic body before it is fired leaves gross holes. Most people would not suspect that such gaps would be left in the structure of the porous body by random chance. Using Mandelbrot's terminology, the lacunarity of a raw ceramic body is surprising.

Another important concept used in the quantification of the structure of a randomly occupied lattice is the **backbone of a cluster**. The backbone of a cluster is isolated by removing from the structure of a cluster all occupied sites which are said to be dangling or form dead ends. Thus, if we look at the very simple pathway through a porous body created by letting a small lattice be occupied at random by various pixels until a complete pathway exists as shown in Figure 5.52, the backbone of the percolating path can be shown in the simplified diagram illustrated. The physical importance of such a system can be illustrated by a problem in a subject known as hydrometallurgy or **microbiological mining**. In **hydrometallurgy**, which literally means water mining, one tries to use biological agents or chemical agents in a fluid to extract valuable minerals from the ground without actually having to excavate the ore out of the ground or by processing heaps of crushed ore. In the in situ mining process, one attempts to crack the layer of rock carrying the valuable mineral and then one lets water carrying bacteria seep into the cracks. The bacteria are especially chosen so that they will digest

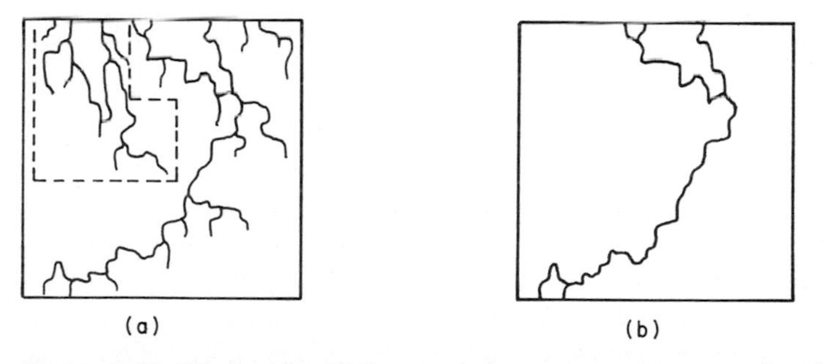

Figure 5.52. The backbone of a percolating tortuous system can be of importance when studying percolating systems.
(a) A tortuous percolating path linking two surfaces of a porous body.
(b) The backbone of the percolating path can be constructed by deleting all "dangling ends" and "dead end" subsidiary pathways from the main percolating path.

(by a combination of biological and chemical processes) the valuable minerals exposed on the surface of the crack. Presumably, since the bacteria are self-propelled, they can penetrate into every nook and cranny of the crack, where they can happily feed on the mineral. However, when the time comes to be able to flush out the mineral dissolved by the bacterial action, the flow of liquid can only be introduced through the backbone of the percolating path. Thus, bacteria which have flourished in a dead end illustrated by the system enclosed in the dotted rectangle may not be able to deliver the digested minerals to the flow of liquid used to clean out the crack after biological activity is finished. The only hope is that during the biological digestion of the mineral some chemical or physical agent can carry the valuable material out of the recesses of the crack. For example, one type of bacteria being used in the mining industry can turn sulphides into sulphuric acid, which can then react with other constituents of the crack face to produce a valuable mineral solution. It is possible that the heat of reaction and/ or bubbles formed by biological activity may help to carry valuable solutions out of the inner recesses into the stream of liquid being used to collect the material generated by the bacteria. There may also be a need to provide additional nutrient fluid into the cracks to improve the efficiency of the process, and again the backbone provides the nutrient flow paths. The mining research group at Laurentian University is currently looking at the optimum structure of cracks created in mineral-bearing layers with the hope that further research in explosives technology may indicate how to produce desirable cracks for the hydrometallurgical extraction of valuable minerals [67].

The backbone of a percolating structure has been the object of many studies [34, 51]. In Figure 5.53 (see Plate 5 at the beginning of the book), some beautiful pictures illustrating the difference between ramified clusters and their backbones are shown. These were generated on a computer by Voss. In these pictures, the changing structures of the clusters as the occupancy of the lattice progresses through a range of occupancy levels up to the occupancy level for complete percolating paths to exist are shown. These pictures were generated for a lattice of 600 x 400 pixel elements. In each figure the largest cluster is shown in white, the second largest is bright yellow, the third is dull yellow, with successively smaller clusters and shapes varying from orange to red to light blue. The remaining smaller clusters are all shown in dark blue. In Figure 5.53(a), the density of occupation is 0.580 (remember that the percolation threshold for pathways through onthogonal junctions is 0.5928) and for this situation all of the clusters are finite in extent. The percolation correlation length below which inspection of the lattice generates a fractally structured average density is approximately the average size of the coloured clusters. In Figure 5.53(b), the back-bones of the cluster in Figure 5.53(a) are shown. It can be seen that the backbones appear significantly smaller than the original clusters. Furthermore, they have a fair amount of open space inside of them. Figures 5.53(c) and (d) are the percolation clusters and their backbones created at an occupancy level of 0.593, which is close enough to the critical percolation occupation level that one can regard the percolation correlation in these figures to be infinity. We note that the largest cluster in Figure 5.53(c) crosses the lattice completely in the horizontal direction and almost makes a connection in the vertical direction. The backbones in Figure 5.53(d) are considerably smaller than the clusters in Figure 5.53(c). In Figure 5.53(e), the lattice occupation is 0.610. Now the largest cluster is infinite in extent and the percolation correlation length is approximately the length of the largest hole. At this level, the backbone of the cluster also spans the entire field, indicating that if these were pathways in a porous powder compact there would be a multitude of possible pathways for the fluid to move through the system. Voss et al. have applied this type of study to look at the real clusters occurring in thin gold films at a density of coverage close to that corresponding to the percolation threshold in a occupied lattice model [56].

Scientists studying the build-up of clusters within a lattice being occupied by random conversion of pixel spaces have found that the sizes of the clusters can be described by what is know as a **scaling function**. The statement that a particular property of a system can be described by a scaling function, or alternatively that a particular physical property is scaling, occurs frequently in the literature on fractal geometry. The meaning of this term can be appreciated by looking at the data summarized in Figure 5.54. If we

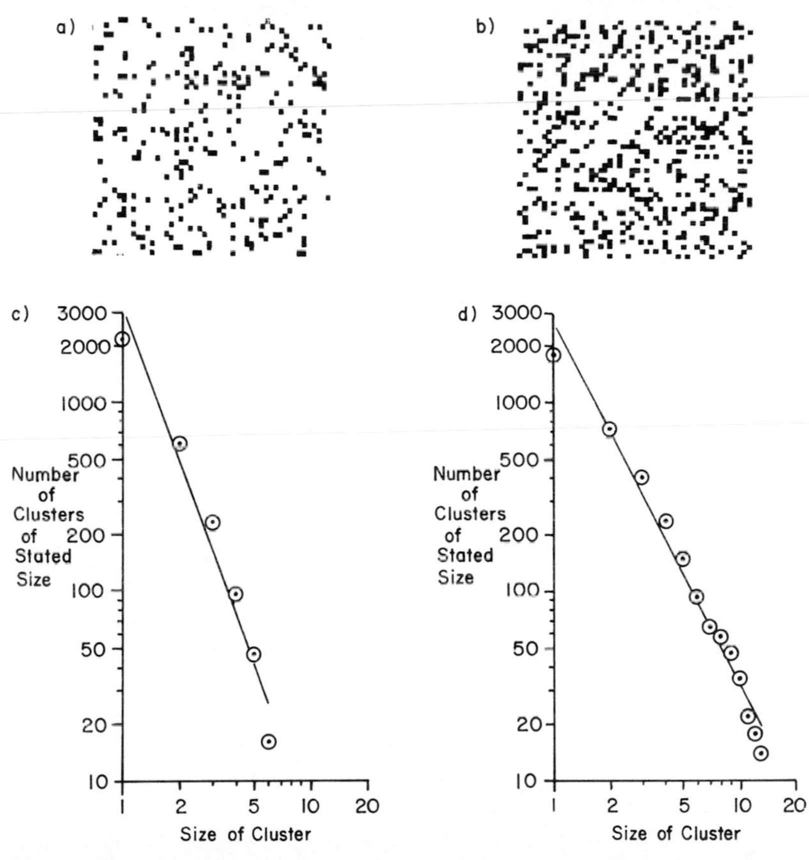

Figure 5.54. The cluster size distribution frequency of growing lattice insects in an occupied lattice is said to be described by a scaling function.
(a), (b) Lattice insects manifest at 0.1 and 0.2 random occupancy of the lattice.
(c), (d) For both fields of view in (a) and (b) the number of clusters versus the size of the cluster forms a linear relationship on log-log graph paper, showing that the cluster frequency is a scaling function.

look at the clusters created in the lattice space at an occupation density of 0.1, we obtain a system such as that shown in Figure 5.54(a). When looking at the cluster structures in the diagram, we permit diagonal linkages since this was carried out in a study aimed at looking at the growth of cracks in a stressed body. For such a situation, the cluster size versus the cluster frequency can be plotted on log-log graph paper as shown in Figure 5.54(c). The linear relationship that can be drawn through the data points in Figure 5.54(c) justify the statement that the frequency of the occurrence of a cluster of given size scales with respect to the size of the cluster. In mathematical terms, this means that one can write the following equation:

$$N(S \geq s) = s^m$$

where N is the number of clusters of size S which is greater than or equal to the stated size s, and m is the slope of the dataline. In Figure 5.54(d), we show the scaling of the frequency of a cluster of a given size against the size of the cluster for a lattice occupied at random at a 0.2 occupancy level. Although this is a simple example of a scaling phenomenon, it helps to clarify the concept involved in describing a system which can be described by a scaling function.

The various aspects of percolation theory outlined in the foregoing paragraphs can be illustrated for a problem in computer technology by the data reported by Kapitulnik and Deutscher [57]. They studied the growth of lead films deposited on a layer of germanium. The way in which the lead was deposited led to the formation of a structure very similar to that of the partially occupied lattice simulated by Voss et al. Thus, in Figure 5.55(a) the structure of an infinite cluster of deposited lead material described by Kapitulnik and Deutscher is shown. It can be seen that the non-conducting islands in the path of the percolating (infinite) cluster are themselves fractally structured. The interest in this particular study was on the conducting path for electricity that existed in the deposited lead, so that the backbone of the percolating cluster was the important structure from an experimental point of view. In Figure 5.55(b), the backbone of the area of a cluster shown in Figure 5.55(a) is shown. In Figure 5.55(c), the average density of lead in a search area of size length L is plotted for both the backbone and the infinite percolating cluster (note that the size of the individual crystallites of lead in Figure 5.56 is approximately $L = 10$). It can be seen that the average density of lead on the germanium surface can be described by a scaling function with a slope which can be interpreted from theory as generating a fractal dimension of 1.90 for the infinite cluster and a fractal dimension of 1.60 for the structure of the backbone (note the fractal dimension is not a simple function of the slope of a curve, but needs to be interpreted in geometric terms as set out by Kapitulnik and Deutscher [57]. Note that the scaling function for the infinite cluster breaks down at small cluster size.)

In Figure 5.56(a), the structure of the lead film at a surface occupancy rate much higher than that required to create a percolating cluster is shown. In Figure 5.56(b), the average density for an increasing search area of side length L is shown. It can be seen that after $L = 40$, the average density remains approximately constant showing that $L = 40$ is the percolation correlation for the film in Figure 5.56(a). In Figure 5.56(c), the number of clusters of size S, n_S is plotted against S, the size of the cluster for a film, the occupancy level of which is well below that required to establish a percolating path. It can be seen that the cluster frequency is a scaling function. Again, it can be shown

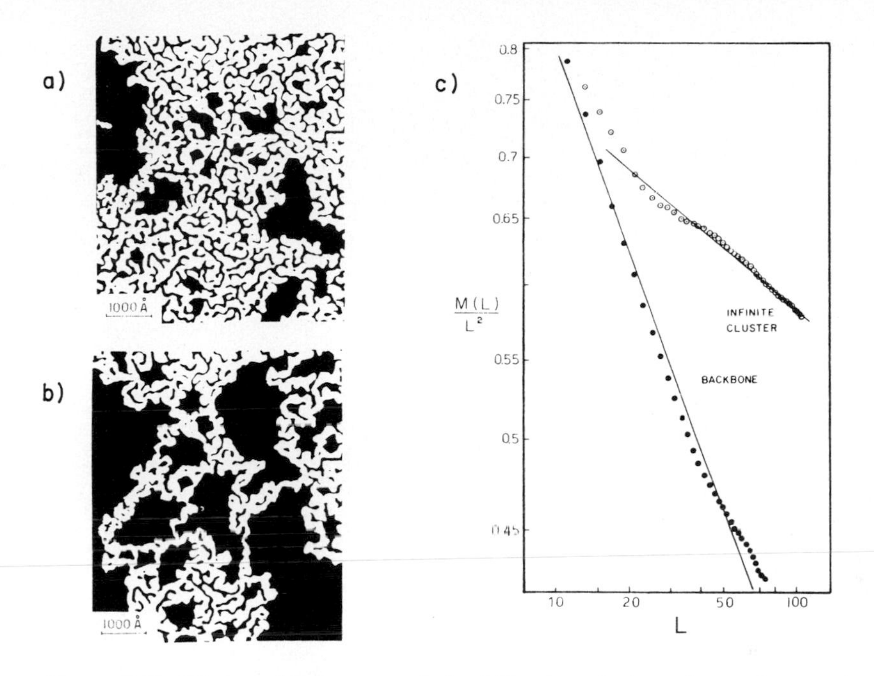

Figure 5.55. Work on the structure of a lead film deposited on a layer of germanium can be used to illustrate the various fractal concepts used to describe percolating thin films.
(a) Part of the percolating infinite cluster of lead on a germanium layer. The black areas are the uncoated residual germanium areas and the white convoluted structures are the deposited lead films.
(b) The backbone of the cluster in (a) carries the electrical current across the surface of the germanium.
(c) The average density of areas of the percolating cluster and its backbone can be described by scaling functions which can be interpreted in terms of fractal dimensions of the structure of the lead films.
(From A. Kapitulnik and G. Deutscher, "Percolation Characteristics in Discontinuous Thick Films of Lead," *Phys. Rev. Lett.*, Vol. 49, November (1982); reproduced by permission of American Physical Society and G. Deutscher.)

that the slope of the line in Figure 5.56(c) is in good agreement with the theories of statistical physics. It should be noted that the data for mass - number population is non-linear for small clusters. If one considers the build up of clusters at very low concentrations of occupancy, it can be anticipated that the build-up of a large cluster is a rare event. However, as one approaches the percolation threshold, the gaps between the growing clusters shrink and it becomes increasingly difficult for small clusters to maintain their identity so that it is not surprising to see that the frequency of the occurrence of small clusters falls away from the levels predicted by the scaling function describing the structure of the large clusters.

As already mentioned briefly, the other main type of percolating structure studied by scientists is known as **bond percolation**. In bond percolation, we take a lattice mesh and let the strings between the knots of the mesh become the objects of interest. For

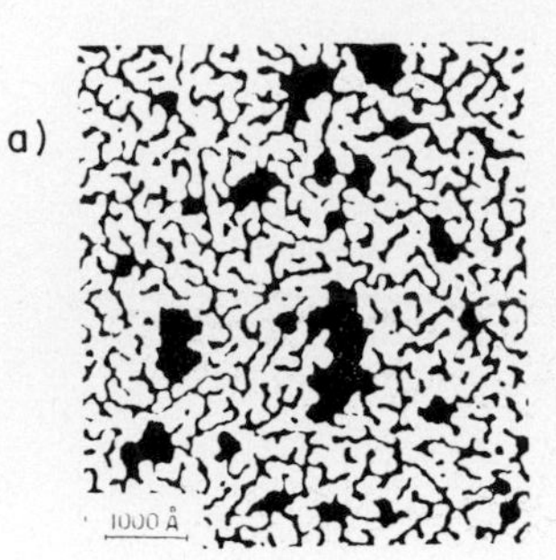

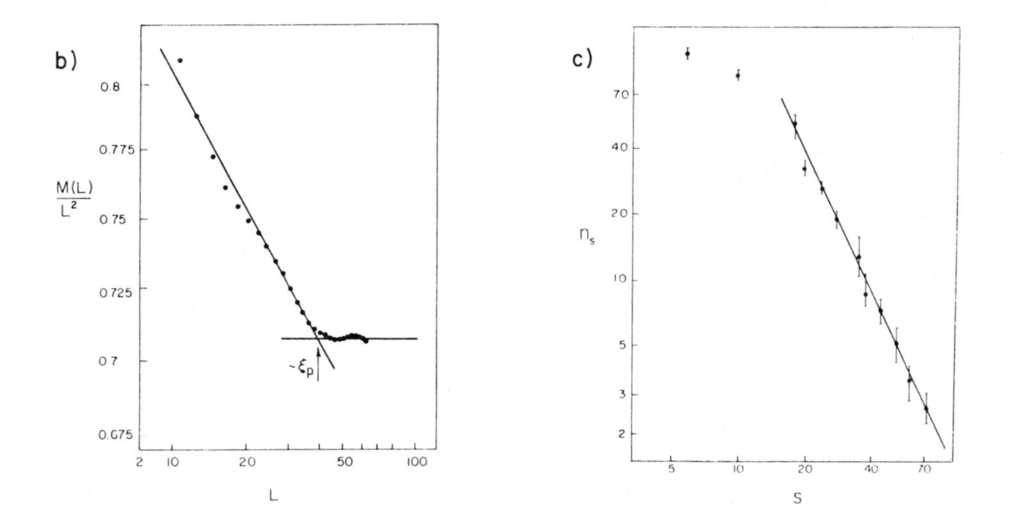

Figure 5.56. The lead film data reported by Kapitulnik and Deutscher can also be used to demonstrate properties of randomly occupied lattice below and above the percolating occupancy level.

(a) Appearance of a lead film at an occupancy level well above that required to establish a percolating pathway.

(b) The average density of the system in (a) as a function of the search square of side length *L*. After a value of *L*= the percolation correlation length is reached, and the average density becomes a constant value. In the terminology of percolation physics, it is said that the average density becomes a Euclidean function after the percolating correlation length has been reached.

(c) The number of clusters of a given size for a lead film of occupancy density lower than that required to establish a percolating path is a scaling function for clusters greater than 20 units in size.

(From A. Kapitulnik and G. Deutscher, "Percolation Characteristics in Discontinuous Thick Films of Lead," *Phys. Rev. Lett.*, Vol. 49, November (1982); reproduced by permission of American Physical Society and G. Deutscher.)

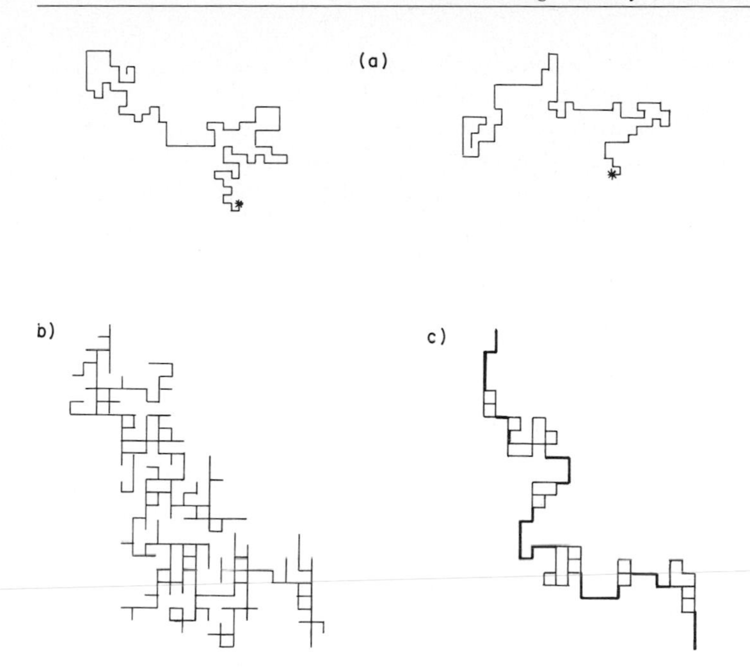

Figure 5.57. Self-avoiding randomwalks on a bond lattice generates clusters that are similar to the growth of polymer molecules or can be used to study the properties of networks of electrical clusters.
(a) Self-avoiding randomwalks terminating in impenetrable loops.
(b) A bond cluster generated by Stanley and co-workers.
(c) backbone of the bond cluster in (b).
(Diagrams (b) and (c) are from H. E. Stanley and N. Ostrowsky, (Editors), "On Growth and Form; Fractal and Non Fractal Patterns In Physics," Martinus Nijhoff Publishers, Dordrecht, Holland, p. 46. "Copyright (c) 1985 Martinus Nijhoff Publishers, Dordrecht, Holland." Reproduced by permission of Martinus Nijhoff Publishers and H.E. Stanley.)

example, Mandelbrot introduces the concepts of bond percolation by considering an array of short sticks placed in a square lattice, where the sticks are either copper (electrically conducting) or PVC (electrically non-conducting). One then studies the number of copper sticks that are needed to replace PVC sticks on the mesh before a conducting path is established. One way of simulating the build-up of a conducting mesh of copper sticks on the lattice is to assume that the total mesh is originally made of PVC, and that we glue copper sticks in a random manner by choosing the address of a knot in a mesh at random and then choose a direction for the copper stick at random from the four possible directions, in which the stick can lie on the network and touch the knot. Two systems of different occupancy levels generated this way are shown in Figure 5.49. After the bonds start to be occupied at a relatively high level, the incoming bond may not be able to go in all four directions and in such situations the next unoccupied direction for that knot address is chosen. Mandelbrot points out that this type of bonded network should percolate at an occupancy of 0.5. Indeed, the system in Figure 5.49(b) (i) percolates along the routes shown in Figure 5.49(b) (ii).

Bond percolation has been studied extensively because it is the basis of Monte Carlo techniques for studying electrical resistance networks and the structure of polymer molecules [58, 59]. This latter subject has been studied extensively by Stanley and co-workers [34] and other investigators. Bond percolation is also similar to the structure of networks created by self-avoiding randomwalks in which only four directions are possible and the step size is unity. Thus, in Figure 5.57(a), a bond cluster generated by a self-avoiding randomwalk of this type is shown. It will be noted that this particular randomwalk ends when the next bond cannot be added, because of a dead end loop facing it at the terminous of the walk. This corresponds to physical reality in polymer physics, in the sense that if the chain of events were a randomly growing polymer molecule the inability to penetrate a loop facing the end of the branch corresponds to the fact that the molecule of the polymer cannot squeeze in between other molecules already in the chain. In Figure 5.57(b), a bond cluster generated by Stanley and co-workers along with its backbone are shown. Stanley has an interesting story to tell about this particular problem. He tells us that in 1977, he made a presentation on bond percolation at a meeting of physicists in Toronto, Canada. In his lecture, he discussed the difficulty of generating the backbone of a bond cluster in a computer simulation model. After the meeting an undergraduate student in the audience named Robert Pike came up to Professor Stanley and proposed a solution to the problem. This student went on to work with Professor Stanley [60].

Stanley differentiates between three types of bond linkages in a cluster. These are:

(a) **Dangling ends.** If the cluster were conducting rods of copper it would carry no current.

(b) **Red bonds.** These are single links in the cluster. This name comes from the fact that if the network was a network of electrical resistors, then when electrical currents were passed through the clusters the single linkage elements would glow red because they would be carrying most of the current.

(c) **Blue links.** These represent multiple network linkages that would not carry as much electrical current as the red ones, so that they would glow gently rather than be bright red.

In Figure 5.57(c), the backbone of the cluster in (b) is shown. The red bonds are shown by the heavy black lines and the blue bonds by the lighter black lines. For an introduction into the extensive literature on polymer chains, one should start with Professor Stanley's review lecture in Reference 34. The percolating path in Figure 5.57(c) can also be regarded as the pathways that would exist in a porous membrane used to deliver drugs over a continuous period. This type of continuous dissolving of a drug to give a steady dose over an extended period of time has been used to develop travel sickness "stick on" plasters used by astronauts to avoid the space equivalent of sea sickness. The medication is randomly dispersed in a porous plastic and the moisture of the skin dissolves the chemical out of the plastic membrane so that it can pass through the skin structure. This type of delivery system is known as transdermal delivery of medication. The pathways available for the delivery of the medication are usual fractal structures [61, 67].

5.7 The Fractal Structure of Clusters Generated by Diffusion-Limited Aggregation (DLA)

In Section 5.1, we discussed a simple model for the simulation of the growth of a cluster by discussing the probability of drunks clustering around an attractive young lady. At that time, we discussed the basic Whitten and Sander technique for simulating the growth of clusters in two-dimensional space, and a typical cluster was presented in Figure 5.4(b).

The simulation carried out by Whitten and Sander is described technically as an "on-lattice simulation of growth by ballistic addition." This terminology implies that the randomly moving primary unit throws itself at the cluster from a random direction and that it approaches the immediate spatial environment of the growing cluster via a randomwalk of the same basic structure as Brownian motion.

Meakin has pointed out that the original study of agglomerate growth by Whitten and Sander stimulated extensive studies of aggregation kinetics, and the various models generated by various workers have found extensive application in various areas of applied science [62]. Again, as in percolation physics, whole books have been written on "fractal geometry and the simulation of aggregation processes" [4]. All that we can do in this section is indicate the directions in which research is progressing and briefly review the applications of fractal geometry in various applied scientific investigations. It should be noted that simple ballistic bombardment of a growing agglomerate without a randomwalk approach to the agglomerate results in a very different type of structure. Sander has pointed out that in this type of ballistic aggregation the subunits which contributed to the growth of the cluster rain down on the cluster, producing a relatively uniform fern like growth from points on the original deposition surface [63].

Dr. Saunder has continued to extend his model of aggregation with various co-workers. For example Richter, Sander and Cheng have defined several modifications of the original model aimed at the computer simulations of soot aggregation. Several of their simulated soot fineparticles are shown in Figure 5.58 [64].

In Figure 5.58(a), the square pixel element has been replaced by a circle, to make the model more realistic in its similarity to real soot. However, it still retains a squarish geometry which, as we will discuss later, hints at the reality that it has grown on a square lattice even though its pixels are circles. In Figure 5.58(b), the circle size used to simulate the growth of the cluster have a diameter three times λ, the lattice spacing over which the units meander to join the cluster. In geometric terms, this has the effect of varying the angle of which the arriving subunit can stick to the growing agglomerate. As a consequence, the simulated agglomerate loses its angular nature and starts to approach the appearance of real soot. An important consideration when simulating the growth is to decide on the probability with which the arriving unit sticks at the surface of the growing agglomerate. In the original Whitten and Sander agglomerate and in the two agglomerates shown in Figure 5.58(a) and (b), it is assumed that as they arrive they have a probability of sticking equal to 1, that is, they are certain to stick when they encounter the growing cluster. In Figure 5.58(c), the probability of sticking using the

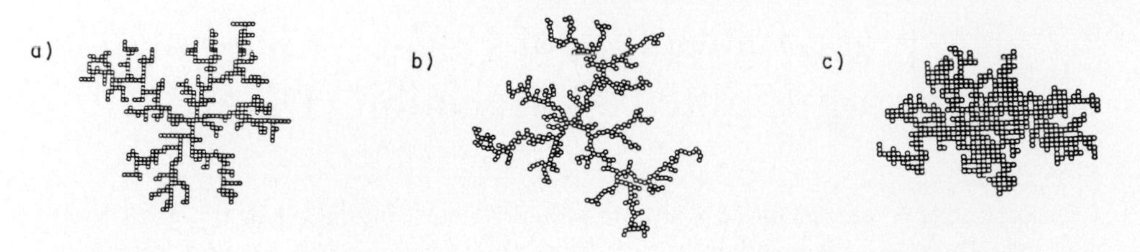

Figure 5.58. Richter, Sander and Cheng have carried out extensive simulations of soot agglomerate using a modified version of the original algorithm put forward by Whitten and Sander.
(a) Five hundred unit agglomerates grown with circular pixels on a square lattice (500 units).
(b) The angular nature of the simulated soot fineparticle is softened by using a pixel circle three times larger than the basic lattice on which the agglomerate is permitted to grow.
(c) If the probability of the permanent attachment of an arriving subunit is modified from 1 to 0.1, a denser type growth results because the arriving units have a greater chance of penetrating past the external fingers of the agglomerate (500 pixel units).
(From R. Richter, L.M. Sander and Z. Cheng, "Computer Simulations of Soot Aggregation," *J. Colloid Interface Sci.*, 100, July, (1984); reproduced by permission of Academic Press, Florida and R. Richter.)

simulation model in Figure 5.58(a) is reduced to 0.1. It can be seen that in this situation a more dense and compact agglomerate is formed. At first sight, it is surprising that the reduction in the probability of sticking results in a denser agglomerate but this will seem reasonable if you consider the fact that after touching the extremity of the growing agglomerate the more mobile (less sticky) unit has a greater chance of penetrating into the space dominated by the filaments of the agglomerate, so that they can join the growing agglomerate further down inside its bush-like structure. In Chapter 3, we discussed the work of Elam and co-workers in which a growing crystal on a surface appeared to have a branch-like structure in the centre, with a thickening of the outer leaf of the structure. If we look at the two simulated clusters in Figure 5.58(a) and (c), we can see that it is reasonable to assume that what happened during the growth of the real crystal in the experiment by Elam and co-workers happened after a certain stage of growth, when the forces of adhesion changed and the probability of sticking for the arriving metal atoms changed to give a higher density structure at the extremity of the fingers of the growing agglomerate. Richter, Sander and Cheng also described a very filimentous type of aggregate produced when one adds a consideration of electrostatic forces to the build-up of the simulated agglomerate. The basic computer system can be used to simulate the three-dimensional growth of agglomerates (see the discussion of the projection of this agglomerate into two dimensions given by Kaye and Clark) [79].

Voss has looked at the growth of agglomerates when more than one subunit is arriving at the overall structure of the agglomerate at the same time. In his model, it is assumed that the lattice on which the agglomerate is being grown has a fraction F of the sites occupied at random and that each of these growth unit pixels are moving around at random. One then starts with a nucleating seed at the centre of the lattice. Then, in each step of the simulation process each of the mobile pixels is examined in random order. One of its neighbouring sites is selected by chance as a possible next position.

If unoccupied, the pixel moves to his new site. If, however, the chosen new location is already occupied, then that particular mobile pixel loses its turn at moving. If the new address for the pixel borders on the cluster orthogonally, then the mobile fineparticle and any of its neighbours which become linked by the new address move and also become part of the cluster. A cluster grown by this process is shown in Figure 5.59(a). The cluster consists of 5000 units grown in an initial cloud, equivalent to an occupation fraction of 0.05. A depletion layer with a few mobile pixels is clearly seen outside the cluster boundary. It can be seen that this type of cluster is very similar to those obtained by Whitten and Sander using a one by one sequential addition of pixel units.

The process used by Voss [25] to grow the system shown in Figure 5.59(a) is known as multi-particle diffusive aggregation (MPDA). In Figure 5.59(b), a 10 000 unit agglomerate grown in a environment containing a fractional occupancy of 0.25 is shown. This type of cluster is much more compact than the surrounding depletion layer, which is very narrow. Obviously MPDA illustrates how a cluster will grow in a cloud of subunits such as those that would exist in the tip of a flame where there are unburnt carbon molecules condensing to spheres, which then turbulently collide to form the soot fineparticles. Voss has also pointed out that in some systems the cluster itself will move,

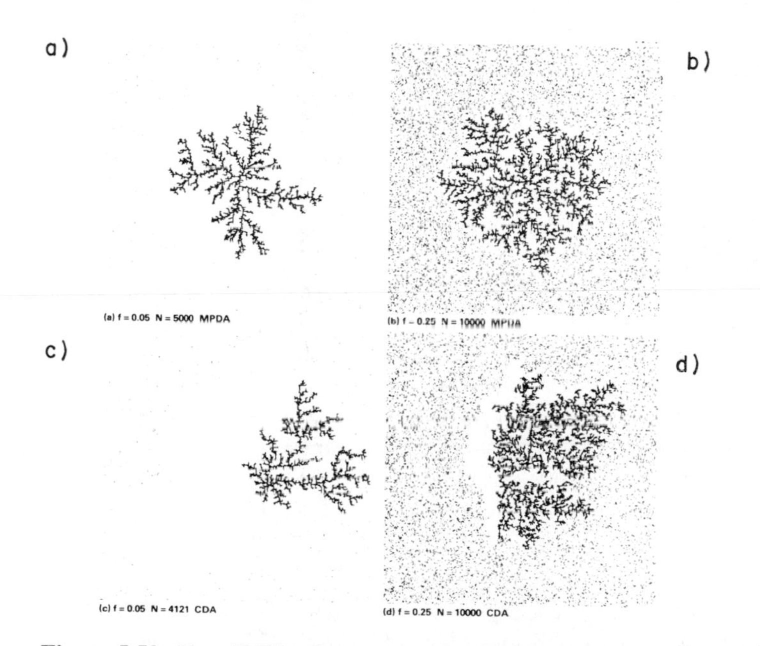

Figure 5.59. Voss [25] has grown complex agglomerates by multi-particle diffusive aggregation (MPDA) and cluster diffusion through a cloud of subunits.
(a) MPDA through a cloud of occupied space $S = 0.05$; the cluster contains 5000 subunits.
(b) Growth of a cluster by MPDA in a fractionally occupied space $S = 0.25$; the cluster contains 10 000 units.
(c) DA aggregate generated by the cluster moving through a cloud of $S = 0.05$ subunits; the cluster contains 4121 subunits.
(d) Cluster produced by CDA for $S = 0.25$ when $N = 10\ 000$.
(Reproduced by permission of Richard F. Voss/IBM Research.)

and clusters grown by simulating cluster movement through a cloud of subunits are shown in Figures 5.59(c) and (d). In this model, the cloud of unit pixels remains fixed while the growing cluster undergoes a randomwalk. At each stage of the randomwalk of the cluster, it moves rigidly and without rotation to a lattice site in a random direction. Any pixels encountered by the moving cluster become attached at each step of the randomwalk. Thus, as the cluster moves it grows by consuming the pixels in its path. Growth stops when the extremities of the cluster reach the end of the initial lattice. In Figure 5.59(c), the cluster contains 4121 subunits growing in a cloud of pixels with an occupancy rate of 0.05. The depleted area created by the sweeping up of the subunits by the cluster's randomwalk can be clearly seen. This type of growth model is known as cluster diffusion aggregation (CDA). In Figure 5.59(d), a CDA cluster of 10 000 units grown in a lattice cloud of concentration $F = 0.25$ is shown. As Voss pointed out, this type of cluster is more compact than that grown by a moving cluster in a more dilute cloud. The MPDA clusters are more symmetric about their origin, while the CDA clusters show much more variation in structure. In Figure 5.60 (see Plate 6 at the beginning of the book), colour-coded versions of the agglomerates in Figure 5.59 are shown. The early growth stages are coloured red, the intermediate stages yellow and the late arrivals green. It can be seen that, as expected the growth of the CDA agglomerates is lop-sided with growth predominating on the leading edge of the cluster as it undergoes its randomwalk. The relevance of the studies carried out by Voss to the growth of snowflakes and a flame-produced pigment and the structure of diesel exhaust is graphically represented by the diagrams in Figures 5.59 and 5.60. The depletions zones around the growing clusters are clearly visible.

Meakin has carried out extensive simulation of agglomerate growth under various simulated conditions [62]. In particular, he has used growth lattice structures other than square lattice. Thus, in Figure 5.61(a), computer-simulated growth on a hexagonal lattice is shown. The original paper [62] should be consulted for details of the permitted steps in the randomwalk approach and the "sticking" rules. A fractal cluster grown in the computer using this lattice and randomwalk addition is shown in Figure 5.61(b).

The striking difference between this cluster and those grown by Sander and co-workers is the overall morphology of the cluster. The morphology is obviously conditioned by the lattice structure and the sticking rules used to generate the cluster. This reinforces the fact that the fractal dimension of a cluster does not describe its overall morphology but its texture and structure. The Sander model always retains its basic rectangular structure, mirroring in its overall outline the basic pixel lattice structure used in its generation [66 – 68].

The difference between the Whitten and Sander and hexagonal Meakin "fractal" crystals illustrates another important physical phenomenon epitaxial growth. When molecules of one metal are vaporized on to the surface of a crystal of a metal having a different crystal structure to that of the arriving molecular species then, in the initial stages of growth, the arriving molecules of metal vapour crystallize in the crystal pattern of the receiving surface. The Greek root words "epi" and "taxes" mean "upon" and "arrangement." Epitaxial growth is one in which molecules arriving on an alien crystal structure arrange themselves according to the alien pattern rather than in their pattern. Epitaxial growth is an important phenomena in the physics of thin films and hence in

a

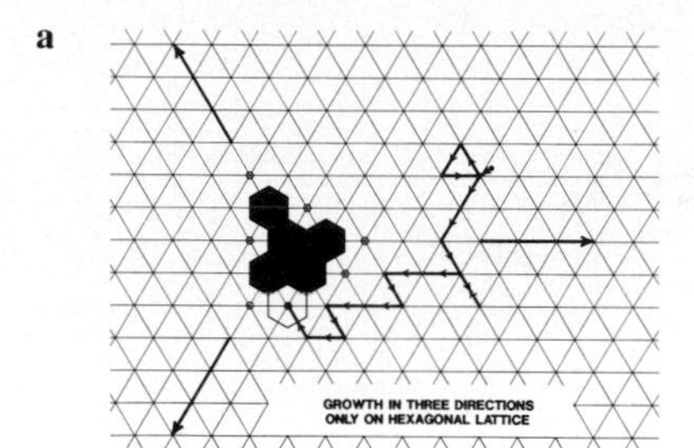

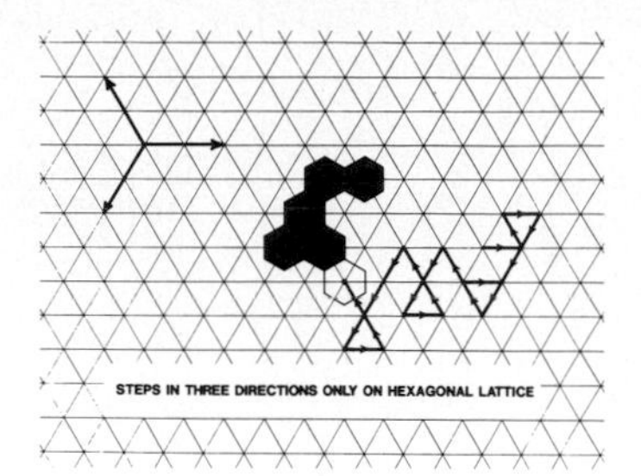

b

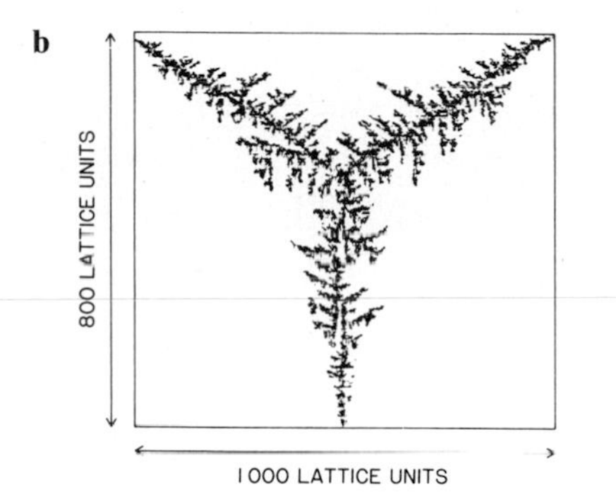

Figure 5.61. The lattice structure and the pixel shape influences the overall shape of the simulated fractal structure, as illustrated by this simulated cluster grown by Meakin on a hexagonal lattice [65].
(a) Lattice structure and pixel movement.
(b) 40 000 cluster on a 1000 x 800 lattice with permitted sticking in three directions only.
(From P. Meakin, "Some Recent Advances in the Simulation of Diffusion Limited Aggregation and Related Processes," *Phy. Rev. Lett.*, A33, 3371 (1986); reproduced by permission of American Physical Society and P. Meakin.)

the generation of computer logic systems. Thus, one can model epitaxial growth on the computer by looking at different crystal lattices as basic structures on which atoms of other metals are deposited in simulated growth.

Meakin has also carried out extensive simulation studies of the different ways in which a fume-type agglomerate can grow in an atmosphere of primary contributory units and colliding embryonic clusters. The term embryonic in biology is used to describe the earlier stages of an organism before it has developed characteristic organs. It comes from Greek root words "en," meaning in, and "bryen," meaning to swell. The biological use of this term obviously comes from the fact that, as a seed starts to grow, the first observable change is that it swells as it begins to sprout. In this book, we shall use the term embryonic cluster to denote a simple cluster containing only a few obvious subunits, but which already begins to show the way in which it will grow. Thus, the six-unit cluster in Figure 5.61(a) is an embryonic cluster which already exhibits rudimentary fractal characteristics. In Chapter 6, we shall show how specialists in aerosol physics are interested in studying embryonic clusters which could pose health hazards from diesel exhausts.

In his computer-simulated fume fineparticles, Meakin has considered the agglomeration procedure in a randomly moving cloud in which embryonic clusters collide and grow, until all of the unitary pixels have joined a major cluster. He also has considered in some of his studies the possibility that some of the clusters can re-adjust themselves within the cluster agglomerate by breaking bonds and rotating to adhere to another portion of the growing cluster. Thus, in Figure 5.62 two of his simulated clusters are

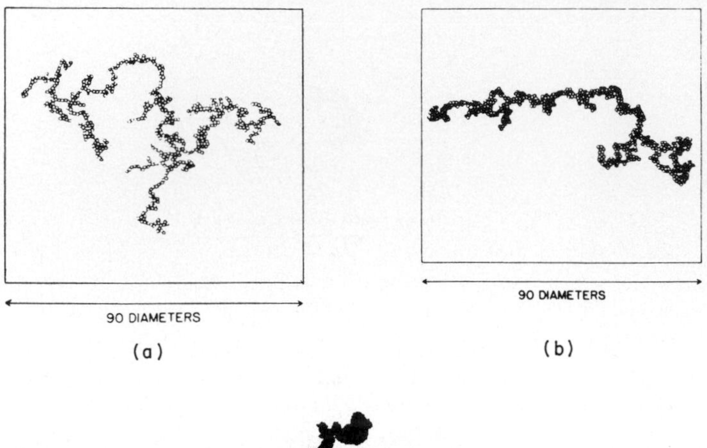

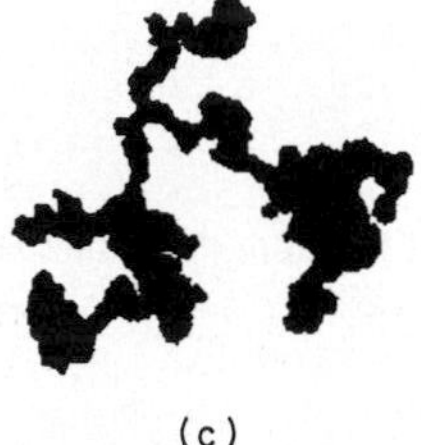

Figure 5.62. By comparing the structure of simulated fumed fineparticles with those encountered in a real situation, one can gain an understanding of the possible dynamics of the formation of the real fineparticle.
(a) Two-dimensional clusters grown by Meakin by mechanical cluster-cluster aggregation using ballistic addition. No reorganization after initial contact between clusters permitted. (This cluster contains 512 primary units) (from P. Meakin and R. Jullien, "The Effects of Random Bond Breaking on Diffusion Limited Cluster-Cluster Aggregation," J. Physique, 46, 1543 (1985); reproduced by permission of P. Meakin.)
(b) 512-unit clusters formed by ballistic cluster-cluster aggregation with one readjusting rotation after initial contact.
(c) Fineparticles present in automobile exhaust photographed by R.J. Cheng [69]. (Magnification 300 000). (From R.J. Cheng, "Microscope," 18, No. 4 (1970); reproduced by permission of R.J. Cheng.)

shown, together with a real fumed fineparticle collected from the smoke coming out of an automobile exhaust. The similarity between the real exhaust fineparticle in Figure 5.62(c) and the simulated cluster in Figure 5.62(a) is obvious. This demonstrates the utility of the simulation study being carried out by Meakin, Voss and other workers in that by varying the rules in which one simulates the growth of agglomerates, one can compare real agglomerates with simulated ones and gain information on the probable mechanisms of formation. Thus, we see that the comparison of profiles in Figures 5.62(a) and (c) suggest that the automotive exhaust material grew in two stages. The final overall agglomerate was probably created in the exhaust system of the car. The primary clusters, which were formed in the original combustion process, collided in the turbulent flow after the primary combustion was complete.

When one examines the structure of the agglomerate in Figure 5.62(c), one immediately questions how many clusters have collided to form the basic structure of the cluster linked by the chain-like elements of the agglomerate. Visual examination would suggest a possible collision of five or six primary clusters at various stages of growth with subsequent growth of branch chains of primary units. In Chapter 2, we discussed the use of erosion logic on an image analyser to break down the structure of an agglomerate of primary fineparticles. In Figure 5.63, the erosion of the agglomerate in Figure 5.62(c) is shown. By the fifth erosion, the agglomerate is clearly broken down into seven primary clusters; however, the decision as to the number of constituent clusters is obviously subjective and must be based on experience [71 – 73]. Figure 5.63 shows the sequential erosion of the automobile emission fineparticle in Figure 5.62, and enables one to estimate the number of clusters which have collided to form the overall agglomerate. From the situation at erosion 3 (approximately), one can postulate that seven clusters joined in the formation of the overall agglomerate. However, one might suspect that the subunit in the lower left of the third erosion may itself have been formed by the collision of three clusters formed in an early stage of condensation of the soot.

It is obvious from an examination of the third stage erosion residual clusters that the fractal dimension of these clusters is much smaller than that of the overall agglomerate. If one were to carry out a dilation logic evaluation of the fractal dimension, then the dilation stage at which there is an abrupt change in the fractal dimension would indicate a stage of operation corresponding to the change from exploring the structure of the overall agglomerate to that of studying the fractal dimension of the subsidiary units. This information could then be used in the reverse operation of erosion to decide when to make a reasonable estimate of the number of clusters participating in the overall agglomerate. The use of erosion dilation image processing and fractal dimension descriptions is opening up new vistas in the area of the characterization of fumed respirable dust hazards, such as those generated from welding fumes and diesel exhaust. We can expect more detailed studies of this kind of analysis of the structure of fumed agglomerate respirable dust in the coming years [71].

So far in our discussion of diffusion-limited aggregation, we have considered only simulated growth. In some very interesting experiments, Hurd and Schaefer have studied a physical system in which they could observe the growth of clusters in two-dimensional space [74 – 78]. They studied the aggregation of silica micro-spheres confined to two-dimensional space, at an air/water interface. The individual spheres were 0.3 μm in diameter. To quote from one of Dr. Hurd's papers [74]:

> "a suspension of silica micro-spheres with a methanol agent was dispersed on to a flat surface of salty water (1.0 normal calcium chloride solution) with a micrometer-driven syringe. Care was taken to avoid turbulence in the sample during spreading since the system is unstable with respect to aggregation in the presence of high electrolyte concentration. Optical observations and photomicrographs were made as the sample proceeded to aggregate. Single spheres could be seen sticking to each other and to developing clusters but by the time a reasonable size of a few hundred units were achieved (1 hour) there were very few single spheres left and the clustering of clusters was evident."

In Figure 5.64, the photomicrograph of a system of growing clusters is shown. Hurd reports that the initial clusters that he showed had much lower fractal dimensions than those of the simulated Whitten and Sander agglomerates. He suggested that this was due

ORIGINAL PROFILE

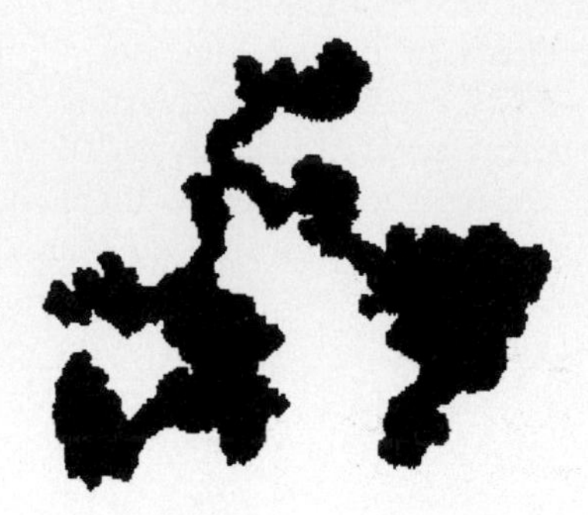

1 EROSION

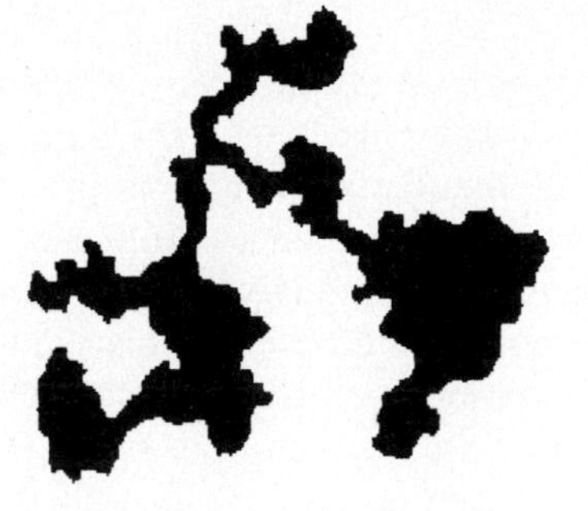

2 EROSIONS

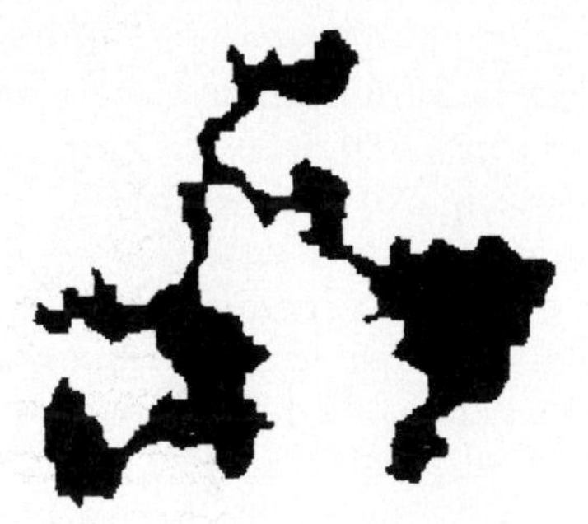

3 EROSIONS

4 EROSIONS

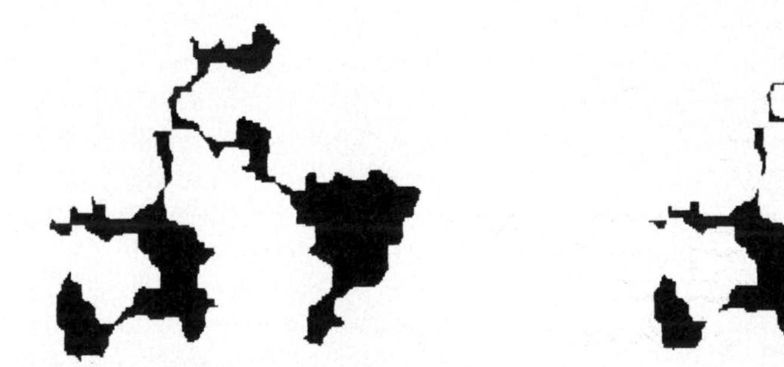

Figure 5.63. The erosion of the agglomerate in Figure 5.62(c). [Reference 31, Chapter 3]. (Reproduced by permission of J. Leblanc.)

to the presence of forces in the two-dimensional space, which interfered with the ideal DLA growth of the clusters. He then simulated various types of cluster growth, assuming different types of surface forces operative within the two-dimensional space at the interface of the liquid. In Figure 5.64(b), several simulated colloidal cluster growths for various types of forces built into the growth models are shown. Again, it is obvious that one can simulate the different types of clusters and compare them with those observed in the real case to gain an understanding of what type of forces are dominant in any given type of colloidal aggregation.

It would be interesting to simulate this type of two-dimensional aggregation using a mixture of white plastic beads with weakly magnetic black plastic beads. One would incorporate a small amount of ferrite powder into the black beads during the manufacturing process to make them stick to each other if they were to encounter each other when moving randomly in a two dimensional space filled with inert white beads. One could vary the type of motion in a shallow dish constituting the two dimensional space holding the mixture of beads. One could vary the forces causing the black beads to adhere to each other in their motion by varying the strength of the magnetic forces between the black beads by varying the amount of ferrite powder put into the black beads. [Note: the clusters of beads present in the simulated two-layer system of Nowick and Mader [23] discussed earlier are obviously similar to the clusters of Figure 5.64(a), and a similar technique to that of Nowick and Mader could be used to assemble a two-

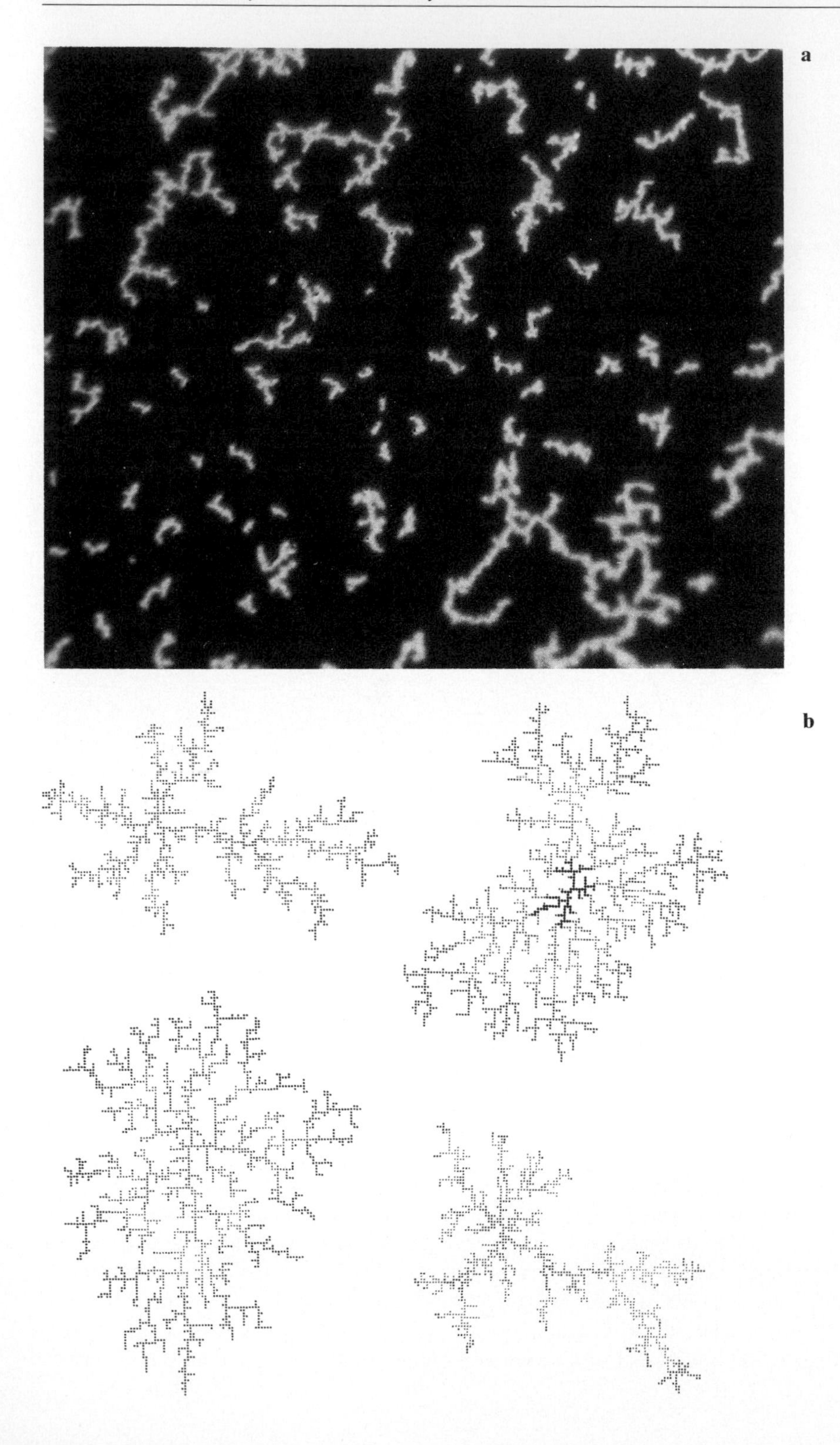
a
b

dimensional array of black and white beads in which the black beads had attraction for each other with subsequent simulated turbulence of the two-dimensional domain to study growth of clusters. Such a simulation of two-dimensional space may not be of interest to working scientists but it would certainly help the novice in the field of cluster aggregation to visualize the dynamics of cluster aggregation].

In some branches of science, the simulation of the structure of agglomerates containing very large numbers of individual units are of interest, but in other situations the scientist is interested in the range of fractal structures present within the population of clusters. Thus, if we were to look at the field of view shown in Figure 5.64(a) as representing real soot fineparticles coming out of a diesel engine, then a prediction of the potential hazard of such soot fineparticles to the lung would require a knowledge of the size distribution of the individual fineparticles and a knowledge of the range of fractal dimensions present in the population of fineparticles. This aspect of the potential hazard from fractally structured respirable dust is discussed briefly in the next chapter, where we expand our fractal horizons from two-dimensional to three- or more dimensional space.

References

[1] G. H. Weiss, "Random Walks and Their Applications," *Am. Sci.*, 71, January-February (1983) 65-71.

[2] For an interesting introduction to the many aspects of randomwalk modelling of physical phenomena, see R. Orbach, "Dynamics of Fractal Networks," *Science*, 231 (1986) 814-819.

[3] H. E. Stanley, "Applications of Fractal Concepts to Polymer Statistics and to Anomalous Transport in Randomly Porous Media, " *J. Stat. Phys.*, 36 (1984) 843-859.

[4] F. Family and D.P. Landau (Eds.), "Kinetics of Aggregation and Gelation," Elsevier-North Holland, Amsterdam, 1984.

[5] T.A. Whitten and L.M. Sander, *Phys. Rev. Lett.*, 47 (1981) 1400.

[6] R. Rammal, G. Toulouse and J. Vannimenus, "Self Avoiding Walks on Fractal Spaces: Exact Results and Flory Approximation," *J. Phys.* (Paris), 45 (1984) 389-394.

[7] Raham and S. Havlin "Exact Fractals with Adjustable Fractal and Fracton Dimensionalities."

[8] P. Meakin and H.E. Stanley, "Spectral Dimensions for the Diffusion Limited Aggregation Model of Colloid Growth." *Phys. Rev. Lett.*, 51 (16), October 17 (1983) 1457-1460

◀ **Figure 5.64.** Hurd and Schaefer studied the kinetics of aggregation of silica microspheres in two-dimensional space using microspheres on the surface of a liquid.
(a) A photomicrograph showing the aggregation of the individual silica microspheres into fractally structured clusters (from A.J. Hurd and D.W. Schaefer, "Diffusion Limited Aggregation in Two Dimensions," *Phys. Rev. Lett.*, Vol. 54, Number 10, March (1985), p. 1043-1046. Reproduced by permission of A.J. Hurd and the American Physical Society).
(b) Simulated colloidal clusters in two-dimensional space created by building various types of forces operative in the cluster formation into the computer simulation model (reproduced by A.J. Hurd, Sandia National Laboratories, Albuquerque, New Mexico).

[9] H.E. Stanley and A. Coniglio, "Flow in Porous Media: the Backbone Fractal as the Percolation Threshold," *Phys. Rev.* B, 29 (1984) 522-524.

[10] P. Meakin and H.E. Stanley, "Spectral Dimension for the Diffusion-Limited Aggregation Model of Colloid Growth," *Phys. Rev. Lett.*, 51 (1983) 1457-1459.

[11] B. H. Lavenda, "Brownian Motion," *Sci. Am.*, 252, No. 2 (1985) 70-85.

[12] R. Hersh and R.J. Griego, "Brownian Motion and Potential Theory," *Sci. Am.*, March (1969) 69-74.

[13] A discussion of the modelling of Brownian motion and the progress of diffusion in the colloidal suspension is given by S. Walfram "Computer Software in Science and Mathematics," *Sci. Am.*, 251 (3), September (1984) 188-203. This article discusses the modelling of self-avoiding randomwalks and Brownian motion.

[14] A readable elementary discussion of Brownian motion is to be found in G. Gamow, "1, 2, 3 Infinity," Bantam Books, New York (1965), (paperback).

[15] J.G. Kemeny, H. Mirkil, J.L. Snell and G.L. Thompson, "Finite Markov Chains in Finite Mathematical Structures," Prentice Hall, Englewood Cliffs, NJ, 1959.

[16] J. Perrin, "Atoms," Van Nostrand, New York, 1916.

[17] E. Nelson, "Dynamical Theories of Brownian Motion," Princeton University Press, Princeton, 1967.

[18] A. Einstein, "Investigations on the Theory of the Brownian Movement," edited by R. Furth and translated by A.D. Cowper, Dover., 1926.

[19] See entry on "Chromatography" in J.R. Newman (Ed.), "Harper Encyclopedia of Science," revised ed., Harper and Row, New York, 1967.

[20] C.L. Strong, "A Simple Laser Interferometer, an Inexpensive Infrared Viewer, and Simulated Chromatograms," "Amateur Scientist," *Sci. Am.*, 226 (2) February (1972) 106-111.

[21] G. Clark, "Fractal Geometry Applications in Filtration Science," *M.Sc. Thesis,* Laurentian University, in preparation.

[22] B.H. Kaye "Shape Characterization in Fineparticle Science and Technology,". Elsevier, Amsterdam, in preparation.

[23] A.S. Nowick and S.A. Mader, "A Hard Sphere Model to Simulate Alloy Thin Films," *IBM J.*, September/November (1965) 358-374.

[24] X-ray pictures taken from W. Bolton, "Patterns in Physics," McGraw-Hill, London, New York, 1974.

[25] R. F. Voss, "Multiparticle Fractal Aggregation," *J. Stat. Phys.*, 36 (1984) 861-872.

[26] R.F. Voss, "On 2-D Percolation Clusters and On Multi Particle Fractal Aggregation," in F. Family and D.P. Landau (Eds.), "Kinetics of Aggregation and Gelation," Elsevier, Amsterdam, 1984, pp. 8-9.

[27] T. Vicsek, "Pattern Formation in Diffusion Limited Aggregation," *Phys. Rev. Lett.*, 53 (1984) 2281-2284.

[28] For a discussion of the evaporation behaviour of a mixture of liquid droplets of different size, see a standard undergraduate textbook of physics; a text I have found useful is D.H. Fender, "General Physics and Sound," English University Press, London, 1957.

[29] Work similar to that of Vicsek and that of Voss was carried out independently by Kertesz and co-workers; reference 34, p. 249.

[30] B. Sapoval, M. Rosso and J.F. Gouyet, "Fractal Nature of a Diffusion Front and Relation to Percolation," *J. Phys. Lett.* (Paris)., 46, February 15 (1985) L149-L156.

[31] J.R. Hodkinson, "Some Observations on Particle Overlap Error in Dust Measurement," *Ann. Occup. Hyg.*, 6 (1963) 131-142.

[32] P.C. Reist, "Introduction to Aerosol Science," Macmillian, New York, 1984.

[33] B.H. Kaye, "Direct Characterization of Fineparticles," Wiley, New York, 1981.

[34] For the serious student of fractals who wishes to move into the more advanced problems of fractal geometry applied to material science, a very useful book is H.E. Stanely and N. Ostrowsky (Eds.), "Growth and Form; Fractal and Non-Fractal Patterns in Physics," Martinus Nijhoff, Boston, 1986. This book is based on the proceedings of a NATO Advanced Studies Institute course along with scientific papers presented as part of the meetings held

at the Institute Etudes, Scientifics de Cargese in Corsica, France, June 26-July 6, 1985. The 11 introductory lectures presented at this meeting are very useful for anyone wishing to pursue a particular topic in depth.

[35] H.E. Stanley, "Form and Introduction to Self Similarity in Fractal Behaviour," reference 34, pp. 21-53.

[36] A.L. Ritzenberg and R.J. Cohen, *Phys. Rev. B*, 30 (1984) 4038.

[37] D. Stauffer, "Percolation and Cluster Size Distribution," reference 34, p.79.

[38] In the UK jointly sponsored research organizations which are supported by indus-trial contributing members and matching government funds have been organized by many different industries. The Whiting Research Association of Quarry Owners produced natural pigment known as "whiting" from quarried calcium carbonate. Whiting is used in many industrial systems. In 1962 the scope of the organization was widened to accept general powder-using industries since manufacturers of whiting had accumulated a great deal of know-how concerning the properties of powdered substances. The inclusion of powder manufacturers in the organization seemed a natural extension of activities. I joined the organization in 1962 to help establish the Powders Division. I only stayed a year before leaving for Chicago. The organization flourished for several years until it was disbanded in the early 1970s with the whiting group reverting back to its own special interests.

[39] G.A. Zerlaut, "Utilization of Pigmented Coatings for the Control of Equilibrium Skin Temperatures of Space Vehicles," in "Proceedings of the Aerospace Finishing Symposium, Fort Worth, Texas, December 8-9, 1959."

[40] These experiments are fully described in B.H. Kaye, M. Jackson and G.A. Zerlaut, "Investigation of Light Scattering in Highly Reflecting Coatings, Volume 3, Monte Carlo and Other Statistical Investigations," Report No. 11TR1 - U6003-19, pro-vided under the term of a contract undertaken for the National Aeronautic and Space Administration Office of Advanced Research and Technology, Washington, DC, Contract Number NASA-65(07). This work was carried out in the period May 1, 1963 to September 30, 1966. The important features of this investigation were also reported in G.A. Zerlaut and B.H. Kaye, "Summary of Investigations of Light Scattering in Highly Reflecting Pigmented Coating," NASA contract report, NASA Cr-844, NASA Administration, Washington, DC, July 1967. This report is more accessible to the scientific investigator than the original IITRI report.

[41] F.B. Steig, Jr., *Off. Dig. Fed. Paint Varnish Prod. Clubs*, 25, (1957) 439.

[42] W.G. Armstrong and W.H. Madson, *Ind. Eng. Chem.*, 39 (1947) 944.

[43] R. Trottier, "Fractal Description of Fineparticle Systems," *MSc Thesis*, Laurentian University, 1987.

[44] A detailed discussion of the calculation of the moment of inertia and the radius of gyration can be found in most university first year level physics textbooks. A discussion of the radius of gyration of fracturally structured clusters can be found in several articles in reference 34.

[45] A technical discussion with photographs of how different features of a fineparticle profile affect the two-dimensional Fourier transform of the image of the fineparticle both by computer techniques and by optical diffraction can be found in A.G. Naylor and C.D. Wright, "Shape Analysis of Particle Profiles Used in the Fourier Transform," in M.J. Groves (Ed.), "Particle Size Analysis," Heyden, London, 1978, pp. 110-119. This book is the proceedings on a Conference on Particle Size Analysis organized by the Analytical Division of the Chemical Society and held at the University of Salford, September 12-15, 1977.

[46] One of the more readable introductions to the general theory of Fourier transform is to be found in R.C. Jennison, "Fourier Transforms and Convolutions for the Experimentalist," Pergamon Press, Oxford, 1961, 120 p. The mathematically gen-erated Fourier transforms in Figure 5.46 were generated by V. Roze of Tracor Inc., Toronto, Ontario and are used by permission of Tracor Inc.

[47] For a general article on the generation of Fourier transforms in two dimensions using optical diffraction techniques, see J. Hecht, "Processing Signals the Optical Way," *High Technol.*, October (1983) 55-61.

[48] An introductory article on Fourier transforms in one dimensional space and the fractal

dimensions of irregular fineparticles is presented in some printed lecture notes prepared by B.H. Kaye entitled "Harmonious Rocks, Infinite Coastlines and Fineparticle Science" (40 typed pages and 35 diagrams). Copies of these lecture notes are available at the cost of printing plus a small charge for postage and handling, from Dr. Kaye, Physics Department, Laurentian University, Sudbury, Ontario, Canada, P3E 2C6. It is hoped eventually that these lecture notes will form a chapter in a planned book entitled "Delightful Instruments and Exciting Moments in Applied Science."

[49] R. Orbach, "Dynamics of Fractal Networks," *Science*, 231, February 21 (1986) 814-819.

[50] B.B. Mandelbrot, "The Fractal Geometry of Nature," Freeman, San Francisco, 1983, pp. 126.

[51] D. Stauffer, "Introduction To Percolation Theory," Taylor and Francis, London, 1985.

[52] G. Deutscher, R. Zallen and J. Adler (Eds.), "Percolation Structure and Processes," *Ann. Isr. Phys. Soc.*, 5, Adam Hilger, Bristol, 1983.

[53] K.G. Wilson, "Problems in Physics With Many Scales of Length," *Sci. Am.*, 241 (1979) 158-179.

[54] R.F. Voss, R.B. Laibowitz and E.I. Allessandrine, *Phys. Rev. Lett.*, 49 (1982) 1441.

[55] The material in Figure 5.52 appeared on the front cover of *Science*, February 21, 1986.

[56] R.F. Voss, R.B. Laibowitz and E.I. Allessandrini, "Fractal (Scaling) Clusters in Thin Gold Films Near the Percolation Threshold," *Phys. Rev. Lett.*, 49, (1982) 1441-1444.

[57] A. Kapitulnik and G. Deutscher, "Percolation Characteristics in Discontinuous Thick Films of Lead," *Phys. Rev. Lett.*, 49, (1982) 1444-1448.

[58] C.M. Guttman, "Monte Carlo Studies of Two Measures of Polymer Chain Size as a Function of Temperature," *J. Stat. Phys.*, 36, (1984) 717-733.

[59] L. De, A. Arcangelis, S. Redner and A. Coniglio, "Anomalous Voltage Distribution of Random Resistance Networks and a New Model for the Backbone at the Percolation Threshold," *Phys. Rev. B*, 31 (1985) 4725-4727.

[60] R. Pike and H.E. Stanley, "Order Propagation Near the Percolation Threshold," *J. Phys. A.*, 14 (1981) L169-L177.

[61] A. Bunde, S. Havlin and R. Nossal, H.E. Stanley and G.H. Weiss, "On Controlled Diffusion-Limited Drug Release from a Leaky Matrix," *J. Chem. Phys.*, 83 (1985) 5905-5913.

[62] P. Meakin, "Some Recent Advances in the Simulation of Diffusion Limited Aggregation and Related Processes," Reprint supplied by Dr. Meakin, Central Research and Development Department E.I. Dupont de Nemours and Company, Experimental Station, Wilmington, Delaware, USA, May 1986.

[63] L.M. Sander, "Theory of Ballistic Aggregation and Deposition," Reprint provided by Dr. Sander, Physics Department, University of Michigan, Ann Arbor, MI, USA, May 1986.

[64] R. Richter, L.M. Sander and Z. Cheng, "Computer Simulations of Soot Aggregation," *J. Colloid Interface Sci.*, 100 (1984) 203-209.

[65] P. Meakin, Preprint provided by the author.

[66] P. Meakin and R. Jullien, "The Effects of Random Bond Breaking on Diffusion Limited Cluster-Cluster Aggregation," *J. Phys.*, 46, 1543 (1985).

[67] B.H. Kaye, "Fineparticle Characterization Aspects of Predicting the Efficiency of Microbiological Mining Techniques," *Powder Technol.*, 50 (1987) 177-191.

[68] B.H. Kaye, "The Description of the Structure of Powder Mixture Using the Concepts of Fractal Geometry." Notes for a workshop on Powder Mixing held at the Powder and Bulk Solids Conference, May 15, 1983.

[69] R.J. Cheng and A.W. Hogan, "Microscopic Study of Lead Iodide –Nucleated Ice Crystals" *Microscope*, 18, No. 4 (1970) 299-302.

[70] Dr. Cheng's picture of the fineparticle present in automotive exhaust is shown in F.P. Perera and A.K. Ahmed, "Respirable Particles," Ballinger, Cambridge, MA, 1979. The dust cover of this book shows a beautiful example of the fractal structure and several of the photographs used in the book cry out for the application of fractal geometry to the characterization of the hazards represented by some hazardous dusts encountered in our high-technology society (see reference 71).

[71] B.H. Kaye, "The Physical Significance of the Fractal Structure of Some Respirable Dusts," in prepartion.
[72] B.H. Kaye, "Erosion-Dilation Logic Strategies in the Characterization of Fumed and Precipitated Fineparticles," in preparation.
[73] J. Leblanc, "The Shape and Size Characterization of Respirable Dusts," *MSc Thesis*, Laurentian University, 1987, in preparation.
[74] A.J. Hurd and D.W. Schaefer. Diffusion Limited Aggregation in Two Dimensions. *Phys. Rev. Lett.*, 54, Number 10, March ll, 1985 pages 1043 -1046.
[75] A.J. Hurd, "The Electrostatic Interaction Between Interfacial Colloidal Particles," *J. Phys. A.*, 18 (1985) L1055-L1060.
[76] A.J. Hurd, "Two Dimensional Diffusion-Limited Aggregation of Interacting Particles," in Extended Abstracts of Fractal Aspects of Materials, 1985 Fall Meeting of the Materials Research Society, December 2-4, 1985, Boston. Book of Abstracts available from the Materials Research Society, Suite 327, 9800 McKnight Road Pittsburgh, PA 15237, USA.
[77] A.J. Hurd, "Diffusion Limited Aggregation of Silica Microspheres in Two Dimensions," in S. Safran and N. Clark (Eds.), "Proceedings of the Conference Physics of Complex and Super Molecular Fluids," June 1985, Wiley, New York 1986.
[78] Lecture notes and diagrams provided as a personal communication by A.J. Hurd, Sandia National Laboratories, Albuquerque, NM 87185, USA.
[79] B.H. Kaye, J.E. Leblanc and G.G. Clark, "A Study of the Physical Significance of Three Dimensional Signature Waveforms," *Part. Char.*, 1 (1984) 59-65.
[80] P. Meakin, "Formation of Fractal Clusters and Networks by Irreversible Diffusion-Limited Aggregation," *Phys. Rev. Lett.*, 51 (1983) 1019-1022.
[81] P. Meakin. "Structural Readjustment Effects in Cluster-Cluster Aggregation," *J. Phys.* (Paris), 46 (1985) 1543-1552.

6 Vanishing Carpets, Fractal Felts and Dendritic Capture Trees

6.1 Sierpinski Carpets and Swiss Cheese

Two "holy" pictures from Mandelbrot's book on fractal geometry are shown in Figure 6.1 [1]. Before I read the legend at the bottom of these pictures, I had two Rorschach reactions. The top picture reminded me of a picture of pesticide droplets on a glass slide that I had just finished looking at before picking up Mandelbrot's book. The lower diagram, reminded me not only of the electron micrograph of a paint film, which was shown in Figure 5.35, also but of a section through a porous metal filter being used to filter quartz dust in a uranium mine. Obviously, the mathematics behind the systems in Figure 6.1 must be applicable to coverage problems in pesticide spraying technology and to the description of filters for capturing dust fineparticles. They should also be useful for describing the structure of sections through pigment dispersions and through porous systems. When I showed a copy of Figures 6.1(a) and (b) to a class of students and asked them to tell me what they saw in the pictures, the answer given included: craters on the moon, a view of trees from an airplane, icebergs floating in the sea, machine gun bullets in a target, holes in a piece of wood, raindrops on a window-pane, bacterial colonies growing in a nutrient dish, mold on a piece of cheese, rust on a car, a cross-section through sandstone, a thin section through concrete, the appearance of a stone chip floor and currants in a Christmas cake. This range of perceived images reported by my students hints at the wide range of systems which may be describable by the fractal mathematics underlying the systems in Figure 6.1.

On glancing at the appropriate pages in Mandelbrot's book, I discovered that Mandelbrot's name for the two pictures reproduced in Figure 6.1 was Swiss cheese, and that mathematically they represented randomized Sierpinski carpets. Obviously, before I could use fractal geometry to describe deposited droplets, filters or fineparticle dispersions, I had to explore the mysterious fractal properties of Sierpinski carpets.

The **Sierpinski carpet** is named after its inventor, a Polish mathematician active in the earlier decades of the 20th century [2]. Mandelbrot chooses to regard the Sierpinski carpet as being a product of the nibbling away of pieces of a cloth by a geometrically inclined moth which he describes as a **trema** (note: Mandelbrot when coining the word trema, indicated that he took it from the same Greek root word that has given us the English word termite. He says that, "trema may be the shortest Greek word that has not yet been put to work with a significant scientific meaning"). To create a Sierpinski carpet of fractal dimension 1.87, Mandelbrot instructs the trema to nibble away at the carpet according to the attack plan illustrated in Figure 6.2. First of all, the carpet area is divided into nine squares and then the middle portion is removed. Our holy carpet

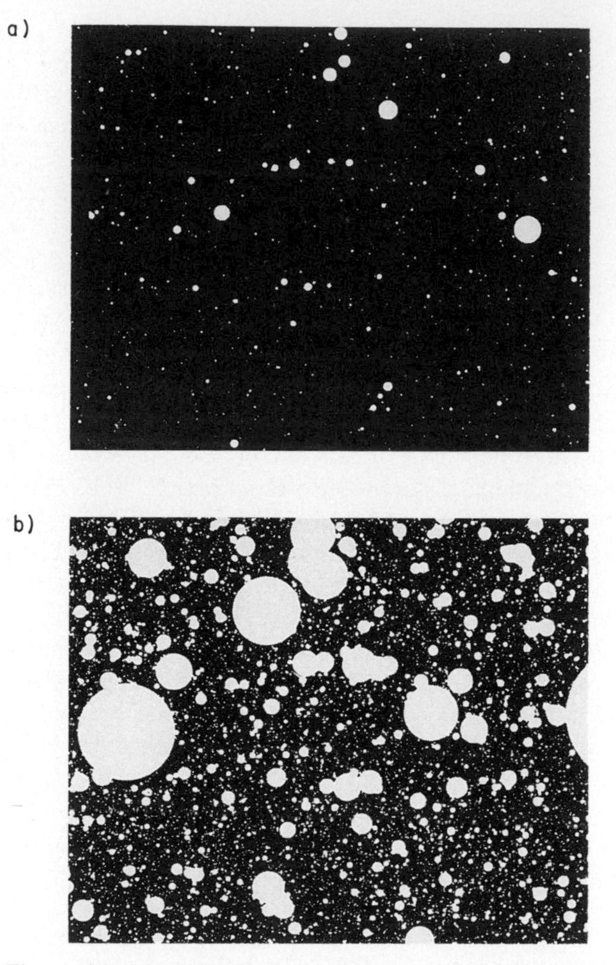

Figure 6.1. Mandelbrot's Swiss cheeses are ideal Sierpinski carpets with circular tremas (holes, patches, etc.) which can trigger "recognition reactions" from a wide range of specialists leading to applications of fractal geometry in many fields of scientific endeavour (from B.B. Mandelbrot, "The Fractal Geometry of Nature," W.H. Freeman & Co., San Francisco, 1983; reproduced by permission of B.B. Mandelbrot).

now constitutes a first-order Sierpinski carpet of fractal dimension of 1.89. The second-order carpet is constructed by removing the central ninth of each of the residual eight squares of the carpet to produce the system shown in Figure 6.2(b). The third stage in the generation of the carpet is shown as Figure 6.2(c). If one carries out this construction ad infinitum, one ends up with a system which is invisible because it has no surface area. It contains an infinite number of holes, bounded by an infinite number of threads of infinite length, in which none of the holes are connected to each other.

As my mind tried to grasp the concept of the mathematical description of an invisible carpet with an infinite number of holes, I recalled the children's story written by Hans

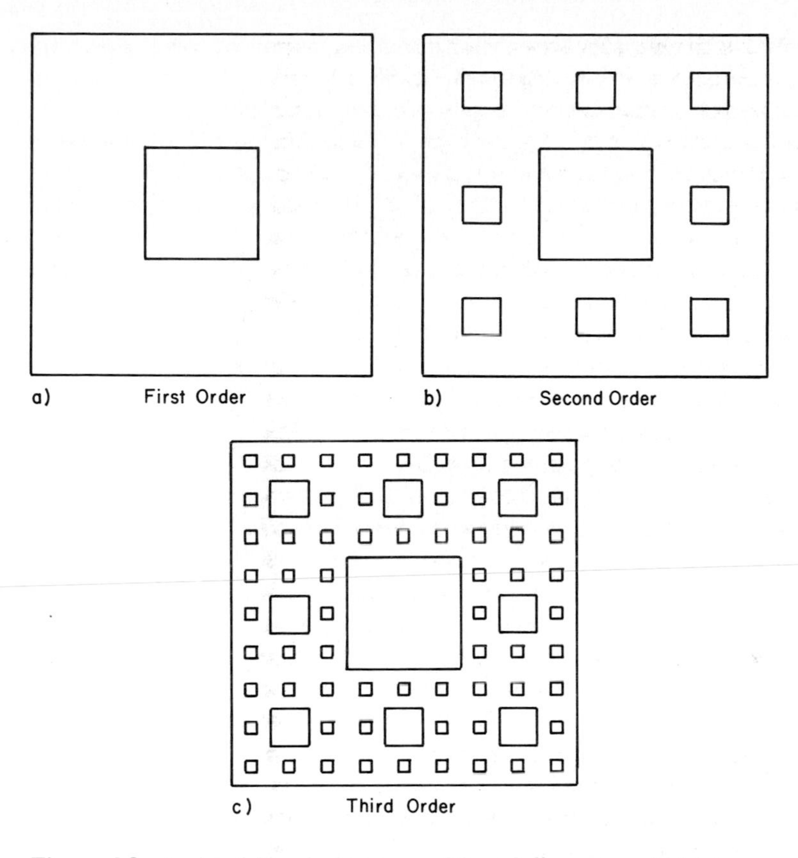

Figure 6.2. An ideal Sierpinski carpet of fractal dimension of 1.89, having no area and infinite thread perimeter, can be constructed by carrying out the construction sequence illustrated above ad infinitum.

Christian Andersen about the king's new suit of clothes. In the story, we are told about a king who paid an enormous amount of money for an invisible suit of clothes. Some unscrupulous tailors pretended to cut, handle and sew an invisible fabric to make the king this invisible suit. When the king walked around in his underwear, after pretending to put on the invisible suit, he received complimentary comments from his assembled courtiers, who pretended to be able to see the clothes. Only when a child called out that the king had no clothes on was the fraud discovered and the tailors were chased out of the kingdom. If the tailors had been able to produce a royal geometer to explain that the clothes were made out of Sierpinski silk, woven from an infinitely thin number of threads, requiring infinite labour to generate the material, the tailors might have been able to stay on in the employment of the king. On the other hand, the king's executioner may have made them into fractals using some sharp-edged Euclidean instruments!

Sierpinski carpets of different fractal dimensions can be constructed using different construction algorithms, as illustrated in Figure 6.3. Obviously, we cannot expect too many natural systems to be described by symmetrical carpets of the type shown in Figures 6.2 and 6.3. However, if one constructs a self similar randomized carpet, in which the positions of the elements of an ideal system are randomized in space, then one can arrive at a carpet system which starts to bear a resemblance to many naturally occurring systems. In Figure 6.4, the system in Figure 6.2(c) has been randomized in space by locating all of the different sized squares in Figure 6.2(c) on a square of the same size as the original carpet by selecting the x,y positions of the centres of the individual subsquares from random number tables (the randomized Sierpinski carpet in Figure 6.4 was generated by Mr. John LeBlanc of the Physics Department, Laurentian University). If we imagine that the process had been carried on ad infinitum, with smaller and smaller squares, we would have generated a system that looks like many naturally occurring systems.

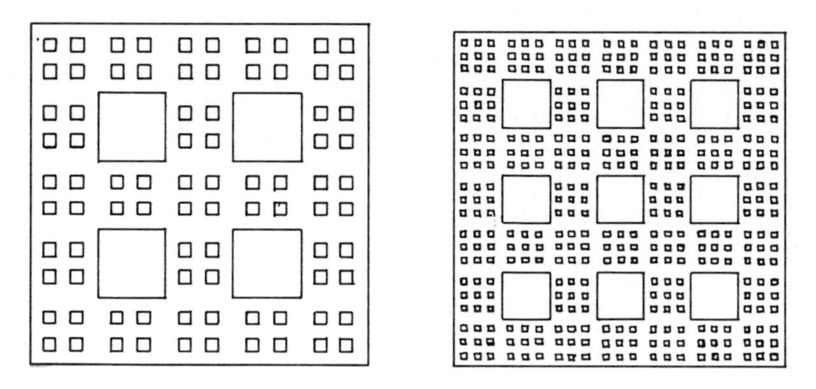

Figure 6.3. Sierpinski carpets of different fractal dimensions can be constructed by removing different fractions of the subdivided carpet as illustrated above.

The Mandelbrot figures, in Figures 6.1(a) and (b) were generated by a very similar process to that used to generate Figure 6.4. Mandelbrot took a size distribution of circles corresponding to the sizes of the successive stages of the holes in the Sierpinski carpet and randomized these circular holes in the plane of the diagram. Mandelbrot discusses the fractal dimension of the Swiss cheeses in Figure 6.1(a) and (b) as measures of the speed with which the background of the holes vanishes as one increases one's resolution of inspection to see smaller and smaller holes. It is useful, therefore, to define a quantity, which we shall describe as the **Sierpinski fractal**, as a useful parameter to describe the structure of systems similar to those in Figure 6.1.

To gain an understanding of what is meant by the Sierpinski fractal of a system, imagine that we can look at a Sierpinski carpet with a series of optical instruments of various resolving power. Imagine that our first instrument can only resolve the largest hole visible in the carpet in Figure 6.2, and that we could look through this instrument and measure what we believed to be the uneaten area β_1 of the carpet left when we could see holes of area α_1. Now let us look through the next instrument to see the next size of hole α_2 and revise our estimate β_2 of the surface of the carpet, which we assume is left intact when we can see holes of size α_2. In this way, we can investigate how fast the carpet vanishes as we look for smaller and smaller holes. From this type of data, we could plot the graph in Figure 6.5(a) and we see that we can draw a straight line through the data. The slope of this line is a measure of how fast the carpet is disappearing. In the same way that we used the rate at which a perimeter grew towards infinity as a measure of the fractal structure of the boundary, we can use the rate at which the carpet disappears, with increased resolution of inspection, as a measure of the fractal structure of the system. The slope of the dataline, in Figure 6.5(a) is 0.11. If we subtract this dataline from 2 (the classical dimension of the unnibbled carpet), we have

$$\delta = 2 - m = 1.89$$

which is the fractal dimension of this particular Sierpinski carpet, as deduced by Mandelbrot from the theories of fractal geometry. It follows that, in general, if we measure the rate at which the background of a dispersed system disappears, at a series of increasingly detailed levels of inspection, and discover a linear relationship on log-log graph paper of the type shown in Figure 6.5(a), then the structure of the system can be described by a Sierpinski fractal, the magnitude of which can be deduced from the slope of the dataline. In this book, we shall describe a data presentation of the type used in Figure 6.5(a) to evaluate the fractal of a dispersed system as a "residual area-resolved trema" Richardson plot. In Figure 6.5(b), the residual area-resolved trema plot for the data generated by inspecting the randomized, statistically self-similar Sierpinski carpet in Figure 6.4, is summarized. It can be seen that this carpet is disappearing less quickly

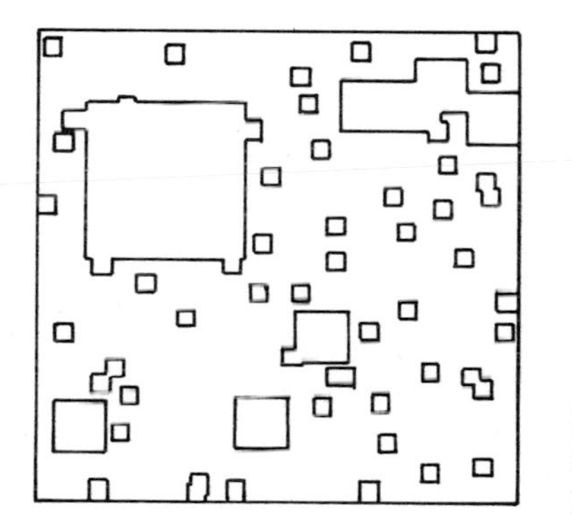

Figure 6.4. A statistically self-similar version of a Sierpinski carpet can be generated by randomizing the positions of the holes in the carpet.

than the ideal carpet because randomization permits overlapping of holes. The effect of overlapping holes can be quantified by comparing the magnitudes of the Sierpinski fractals of the two systems.

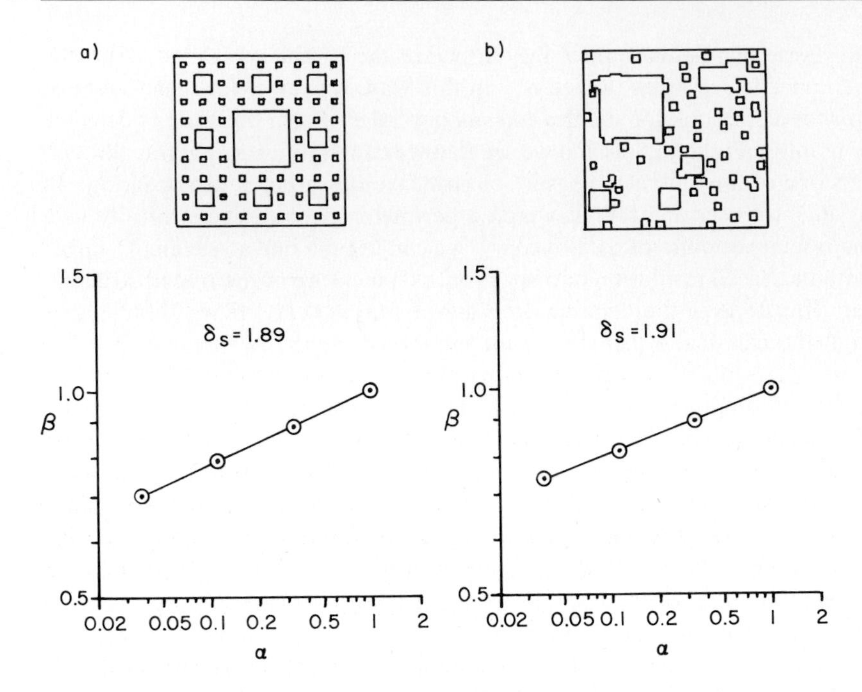

Figure 6.5. The rate of disappearance of an ideal Sierpinski carpet is related to the fractal dimension of the carpet by the relationship $\delta = 2 - m$, where δ is the slope of the dataline and m is the fractal dimension of the carpets. In a statistically self-similar Sierpinski carpet, overlapping of the holes reduces the rate of disappearance of the carpet.
(a) Disappearance rate of an ideal Sierpinski carpet.
(b) Disappearance rate of a statistically self-similar Sierpinski carpet.
α = Just resolved trema (hole).
β = Area of carpet assumed uneaten.

6.2 A Fractal Description of the Deposition Efficiency of Simulated Pesticide Spray Systems

To illustrate how we can measure the Sierpinski fractal of a dispersed system in two-dimensional space, let us assume that the systems in Figures 6.1(a) and (b) can be interpreted as simulated photographs of the deposition of pesticide droplets on a leaf. A scientist studying a deposition of pesticide droplets is interested not only in the size distribution of the droplets, which is one measure of his ability to deliver the pesticide spray, but also in the efficiency with which the droplets cover the leaf. When the size distribution of the circles in Figure 6.1(a) is measured and plotted on log-log graph paper, the size distribution data generate a straight line as shown in Figure 6.6. To look at the efficiency with which the droplets cover the leaf, we can now transform the data in Figure 6.6 into the following format. First, we list the area of the droplets, as measured under a microscope, and normalize their magnitude by dividing by the area of the field of view [we assume that the square fields of view in Figures 6.1 (a) and (b) are magnified views of areas of the leaves on which the droplets are deposited]. Let us now assume that initially when looking at a leaf the microscope is adjusted so that only the largest drop of size α_1 can be seen clearly. As mentioned earlier, this clearly visible size is described in the fractal language of Mandelbrot's book as the "just resolved trema." We now calculate the fraction of the area of the leaf not covered by visible droplets, assuming that the only droplets that can be seen are bigger than the "just resolved trema." Thus, the area of the leaf not covered by pesticide would be β_1 for the largest droplet.

We would next assume that the microscope was re-adjusted so that we could now just see the next smallest droplet of size α_2 normalized units. At this resolution of examination, the percentage of the leaf not covered by droplets is calculated from the data to give the value β_2. By increasing the resolution of the microscope, so that we could just see droplets α_3 to reveal a residual leaf area not covered by the droplets of β_3, etc., the whole of the data are transformed. Using the transformed size distribution data the graph in Figure 6.7 is plotted.

In Figure 6.7, the logarithm of the area of the just resolved trema droplet, expressed in normalized units, against the residual fractional area of the leaf not covered by droplets, at that level of resolution is plotted. It can be seen that the data generates a straight-line relationship which has a slope of 0.03. It follows from this discussion that

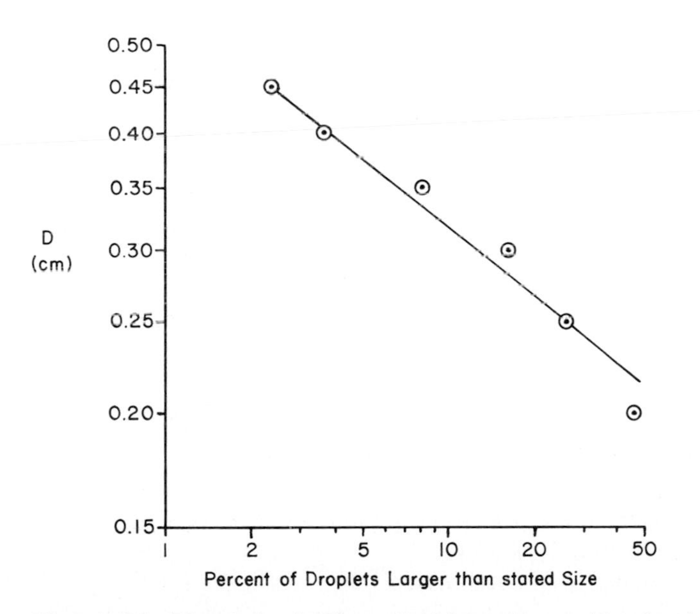

Figure 6.6. The circles in Figure 6.1(a) can be considered to be droplets of pesticide. The distribution function of the simulated droplets, measured with a Dapple image analyser, generates a straight-line relationship on log-log graph paper.

the measured fractal dimension of the system in Figure 6.1, deduced from the dataline in Figure 6.7, is, $\delta = 2 - 0.03 = 1.97$. This fractal dimension is a measure of how fast the background of the leaf disappears as we look for smaller and smaller droplets, and hence is a measure of the efficiency with which the simulated droplets are covering the leaf.

From theoretical reasoning, Mandelbrot states that the fractal dimension of the background in Figure 6.1(a) is 1.990 as compared with the experimentally determined value of 1.97. For this type of measurement, this can be considered to be a close agreement between the theoretical and measured value of the Sierpinski fractal, and indicates that the Sierpinski fractal of the background of a dispersed system can be measured accurately using the data transformation discussed in the foregoing paragraphs.

The fact that the size distribution of the dispersed elements, in Figure 6.1(a) is describable by a log-log distribution hints at the fact that whenever a set of fineparticles describable by a log-log distribution function are deposited on a surface in a random manner, the resulting system may be describable by a Sierpinski fractal.

In Figure 6.8, the measured "residual area-resolved trema" Richardson plot, of the field of view of Figure 6.1(b), is shown. The slope of the dataline in Figure 6.8 is 0.09, leading to a deduced Sierpinski fractal dimension for the background of the tremas of 1.91. This compares with the theoretical Sierpinski fractal of 1.90 reported by Mandelbrot. By comparing the data in Figures 6.7 and 6.8, it can be seen that the physical significance of a Sierpinski fractal is that the smaller the magnitude of the Sierpinski

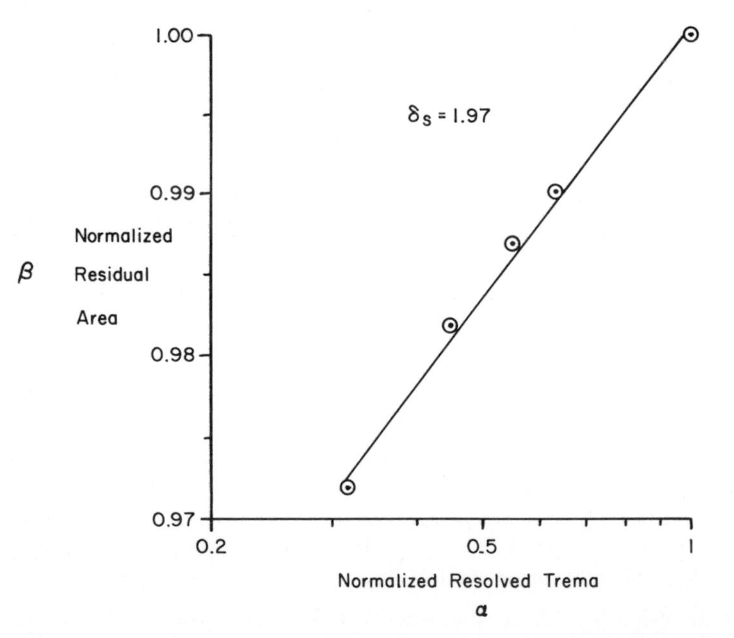

Figure 6.7. The Sierpinski fractal of the field of view shown in Figure 6.1(a) is determined experimentally to be 1.97.

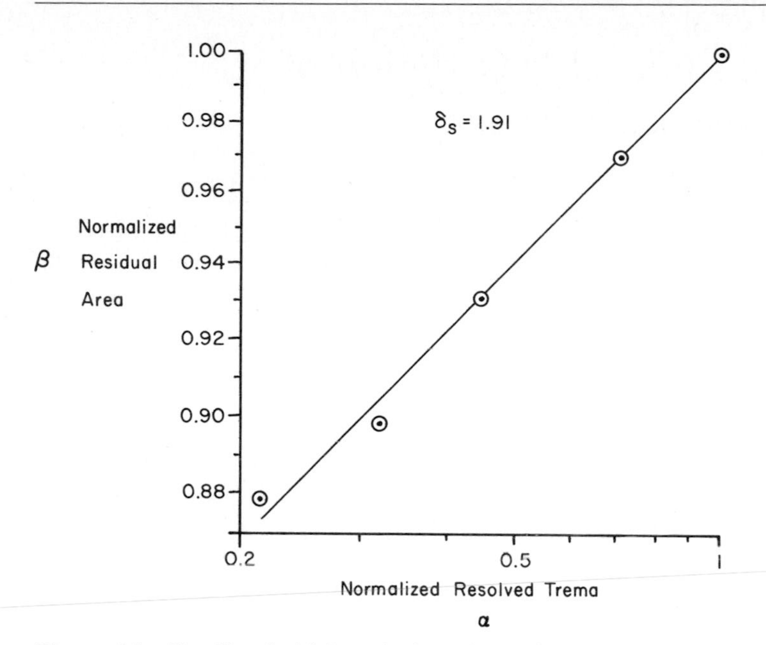

Figure 6.8. The Sierpinski fractal of the field of view shown in Figure 6.1(b) as determined experimentally is found to be 1.91.

fractal, the more quickly the background disappears as the resolution is increased. Note that in this case the measured experimental value of the Sierpinski fractal is slightly higher than the theoretical value. This difference is probably due to the fact that, in the mathematical construction of the field of view shown in Figure 6.1(b), Mandelbrot allowed some of the holes to lie on top of holes already generated. Therefore, some of the holes were not as efficient at eating away at the background as they should have been, since some of the area they were "eating" was already empty.

The difference between the theoretical and experimentally determined Sierpinski fractal of a dispersion of pesticide drops on a leaf could be a useful measure of the efficiency of changes in spray technology techniques. For example, if the spray droplet size as generated by a nozzle used to spread pesticide in an agricultural situation is known, then the theoretical fractal of the deposited droplets could be modelled on a computer. The deposition of the droplets at random on a background area could be simulated with the computer being instructed not to allow the droplets to overlap each other (one would not need actually to generate a field of view, since the known droplet size distribution function could be transformed into data for a residual area-resolved trema Richardson plot without actually measuring any deposited pattern of pesticide droplets). One could then measure the Sierpinski fractal of the deposited droplets in an actual crop or tree spraying experiment, and investigate whether electrostatic charging of the pesticide droplets increased the Sierpinski fractal of the real pictures of deposited droplets on a leaf. Any difference between the theoretical deposition Sierpinski fractal, and the experimentally determined Sierpinski fractal would be a measure of the efficiency of crop protection technology.

6.3 Sierpinski Fractal Description of Real Dispersed Systems

In Figure 6.9, an electron micrograph of a section through a single bead of dry ink used in a Xerox machine is shown (this type of dry ink is known technically as the **toner**). The magnification of the field of view is one hundred thousand. The beads of the dry ink are made by dispersing carbonblack pigment in a clear plastic. The individual fineparticles of this carbonblack dispersion manifest boundary fractal structure. The distributions of the boundary structural fractals of the individual profiles shown in this figure were discussed in Chapter 4. If we now regard the carbonblack as forming tremas in the field of view, we can calculate the Sierpinski fractal of the supportive matrix. The first step in the evaluation of the Sierpinski fractal is to measure the size distribution of the carbonblack profiles and then transform the data to construct a "residual area-resolved trema" Richardson plot of the type illustrated for the Mandelbrot Swiss cheese in Figure 6.6. The data for the carbonblack dispersion for calculating Sierpinski fractal of the support matrix are shown in Figure 6.10. It can be seen that two clear datalines are manifest on this graph. My explanation for the occurrence of these two distinct data regions is as follows.

It is probable that the original powder did not contain any grains smaller than the limit indicated by the dotted line separating the two regions. The very low Sierpinski fractal for fineparticles smaller than or equal to 0.15 represents the fact that when the carbonblack was dispersed in the plastic, it would be subjected to high shear dispersion forces. These forces could strip small fragments from the primary fineparticles to form the

Figure 6.9. The background of pigment dispersions, such as those of carbonblack in a plastic matrix, can be shown to have fractal structure. (magnification 100 000). Photograph provided by D. Alliet of the Xerox Corporation and used with permission.

second region of occupied space, represented by the Sierpinski fractal of 1.998 in the second region of the graph.

For these reasons, the two dataline regions of the graph are labeled "dispersion debris" and "prime carbonblack filler". The prime carbonblack filler region corresponds to the coarse resolution data. It can be seen that the dataline has a slope of 0.030, which corresponds to a Sierpinski fractal for the background matrix of 1.970. This Sierspinski fractal describes the way in which the original carbonblack powder is dispersed in the plastic.

Another possible source of a very fine debris in the midst of the primary carbonblack profiles is the act of microtoming the Xerox toner beads to make the section which was photographed in the electron microscope. A **microtome** is a very sharp knife, which makes a very precise cut through material to be examined microscopically. The microtome may have cut off the ends of some fineparticles protruding into the section photographed in Figure 6.9.

It is probable that the Sierpinski fractal of such systems will characterize important aspects of paints and **composite material** (see discussion of paint film structure in Section 5.4).

Probably the oldest man-made composite material is concrete. In recent years there has been a great deal of interest in improving the strength of concrete by eliminating air voids in the matrix and in using special gravel to increase the density [3]. In Figure 6.11,

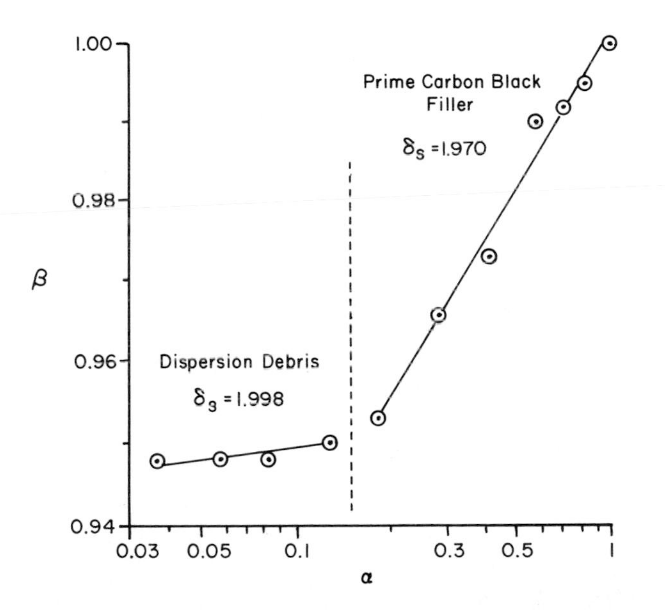

Figure 6.10. The "residual area-resolved trema" Richardson plot for the carbonblack profiles in Figure 6.9 exhibit two distinct linear data relationships.
α = Normalized residual area.
β = Normalized area of just resolved trema (in this case the trema is the carbonblack profile); α and β normalized with respect to the area of the whole field of view.
δ_s = Sierpinski fractal dimension of support matrix.

a)

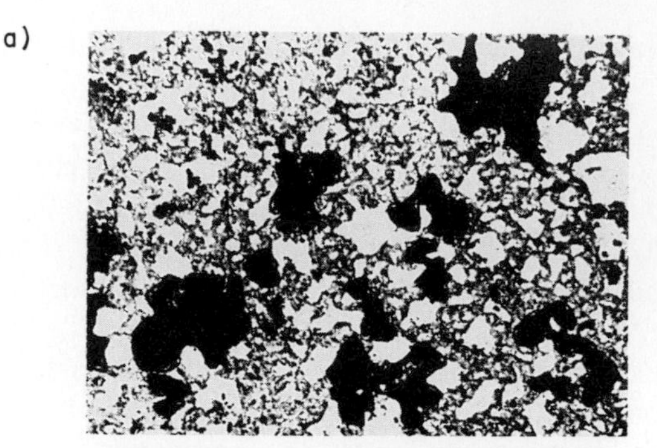

b)

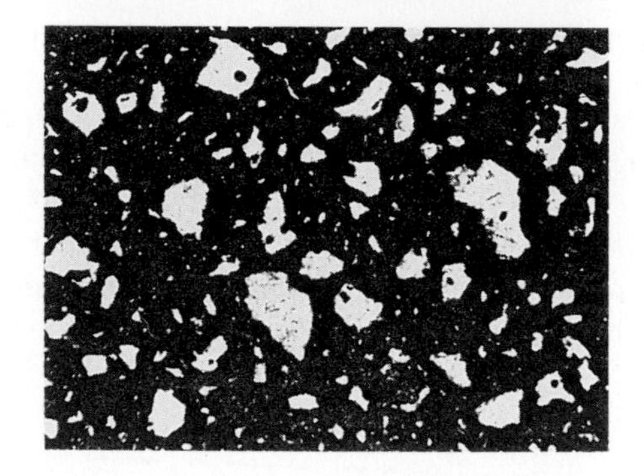

Figure 6.11. Sections through concrete systems exhibit dispersed phases which create Sierpinski fractals in the supportive matrix.
(a) Normal concrete containing air voids.
(b) High-strength concrete from which air voids have been removed, showing the sections through the gravel and sand present in the concrete matrix [3].

two sections through two different concretes are shown. The normal concrete, shown in Figure 6.11(a), has many air voids in it. Up to 30% of the volume of normal concrete can be air voids. In Figure 6.12, the air voids in Figure 6.11(a) have been drawn separately. The graph for calculating the Sierpinski fractal of the background surrounding the air voids is also shown in Figure 6.12. As in the case of the dispersed carbon-black, the individual tremas (air voids) have a fractal structure. It is probable that the fractal dimensions of the individual holes and the Sierpinski fractal of the surrounding matrix can be related to the failure strength of the concrete. The significance (if any) of the two linear data regions of the graph requires a larger field of view to ascertain the exact nature of the regions and the carrying out of related physical tests other than those available for the preliminary experiments reported here.

The concrete shown in Figure 6.11(b) is a new high-strength concrete from which the voids have been virtually eliminated. The profiles which dominate the field of view are probably sections through the gravel and sand constituents of the concrete (note: the term **aggregate** is used in the concrete industry to describe gravel used to fill out concrete. There is another meaning of the word which, throughout this book, denotes a loosely assembled conglomerate of fineparticles. To avoid confusion, we shall continue to talk of gravel or filler sand and not aggregate when we discuss concrete).

In Figure 6.13, the Richardson plot used to derive the Sierpinski fractal of the matrix surrounding the gravel and sand is shown. It is probable that Sierpinski fractals will prove useful in describing the structure of various types of concrete. Experimental

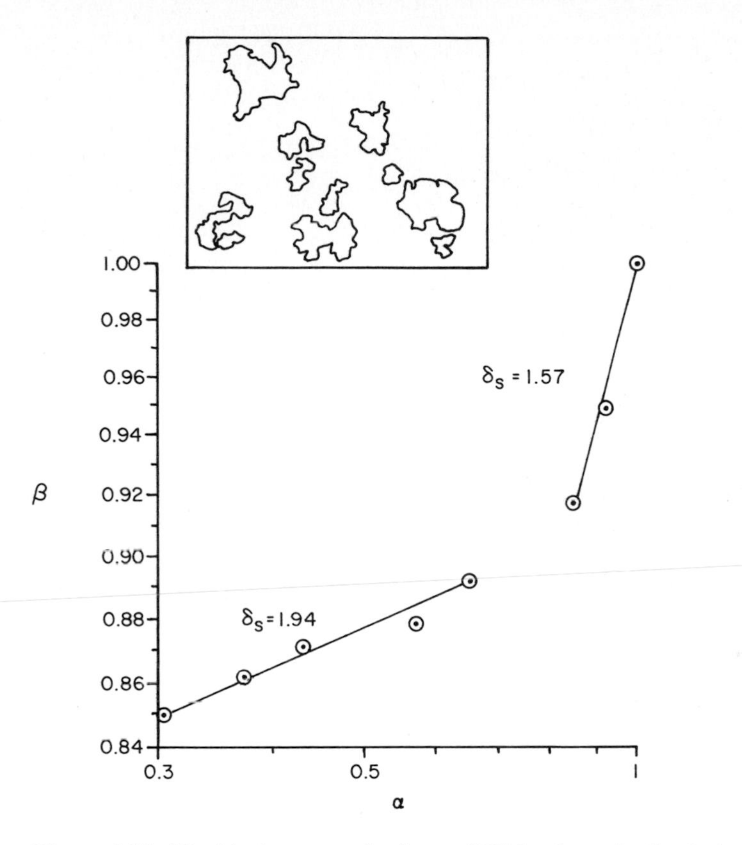

Figure 6.12. "Residual area-resolved trema" Richardson plot for the background fractal (created by the holes in the concrete). Both α and β are normalized with respect to the area of the whole field of view. δ_S = Sierpinski fractal of background matrix.

studies of the Sierpinski fractals of various types of concrete and other composite materials are in progress at Laurentian University.

Many rocks are naturally occurring composite materials. In Figure 6.14, a section through what the geologists describe as a dispersion of "feldspar in a rhyolite matrix" is shown. One of my research associates, who was developing methods for measuring the fractal structure of natural systems, picked this specimen at random from several exhibits in the geological laboratories at Laurentian University. He proceeded to measure the Sierpinski fractal of the background matrix of the rock. The data he generated are shown in Figure 6.14. At first, we were mystified by the fact that two distinct datalines existed in the "residual area-resolved trema" graph in Figure 6.14. However, when this graph was shown to Professor Beswick of the Geological Department, he immediately reacted with interest because apparently this type of rock is known to crystallize in two stages, generating two different populations of crystals of different size. Therefore, the two datalines of Figure 6.14 probably represent the two different populations of crystals in the rock. Sierpinksi fractals could prove to be a useful

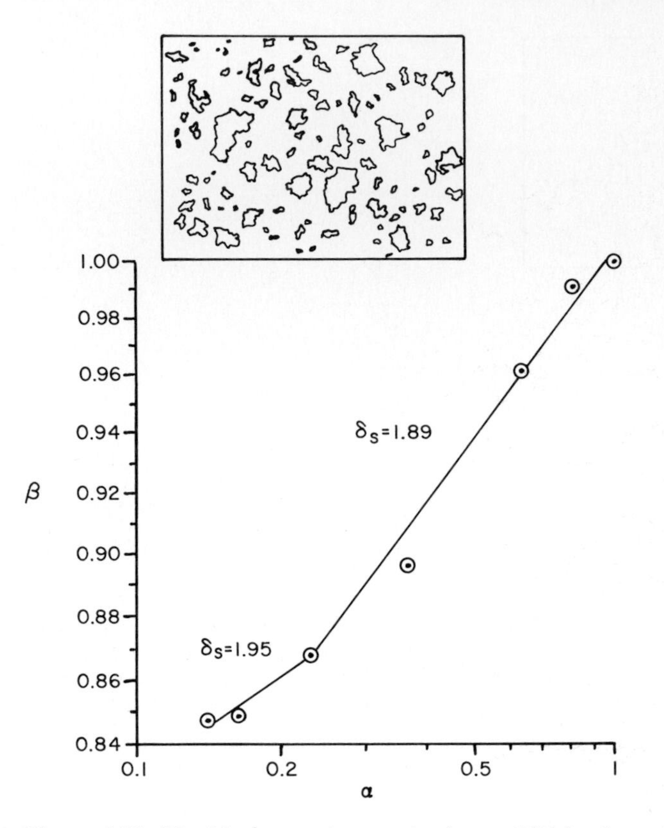

Figure 6.13. "Residual area - just resolved trema" Richardson plot for the aggregate and sand in Figure 6.11(b). The break in the dataline may represent the resolution at which one starts to inspect sand rather than gravel, the sand being much finer than the gravel in the assembled concrete.

β = The normalized residual area and the just resolved trema expressed in normalized units (the tremas in this case are sections through the sand and gravel creating silhouettes in the background matrix). α and β are normalized with respect to the whole field of view.

δ_S = Sierpinski fractals

quantitative structural factor for describing the structure of ore rocks and geological systems in general (see also the discussion of the Apollonian gasket in Signpost 4 in Chapter 10). It should be noted, however, that the Sierpinski fractal does not provide any information on the way in which the dispersed fineparticles are situated in two-dimensional space.

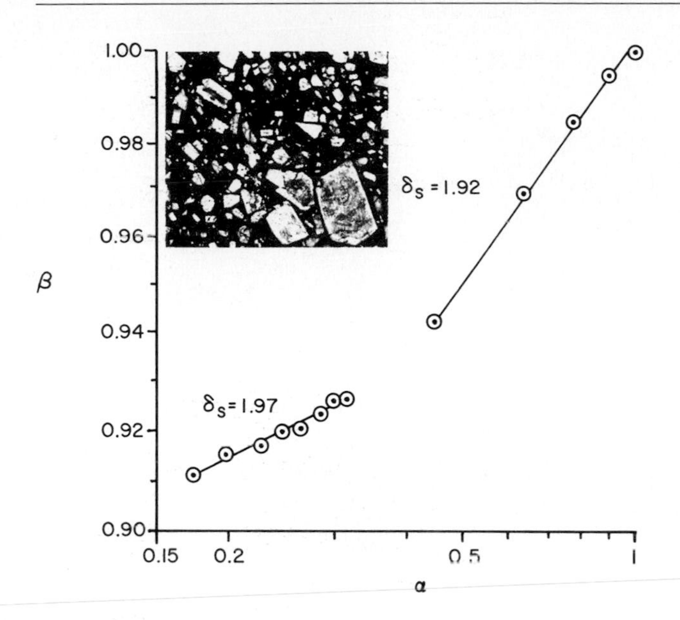

Figure 6.14. A specimen of feldspar in a rhyolite matrix exhibits two linear regions probably corresponding to the two-stage crystallization process taking place when the rock was formed.
β = The normalized residual area and a the just resolved trema, expressed in normalized units (the tremas in this case are Feldspar crystals); α and β are normalized with respect to the whole field of view.
δ_S = Sierpinski fractals.

6.4 Exploring the Fractal Structures of Filters

It only takes a small leap of the imagination to go from the fields of view that we have been looking at in Section 6.3 to match them with the systems shown in Figure 6.15, which represent various types of filters.

The filter in Figure 6.15(a) is a photograph of the surface on what is known as a **silver membrane filter** [4,5]. These filters are used extensively by specialists in occupational health and safety to study quartz dust filtered from mine air. The **nominal rating**, that is, the smallest size of fineparticles that can be collected on the filter, is 0.8 μm. When respirable hazardous quartz dust is collected on the surface of the filter, it can be inspected and quantitatively evaluated by X-ray diffraction to determine the amount of respirable quartz filtered from the mine air. The filters are made by pressing silver powder together and lightly sintering it into a continuous mat containing open continuous pores throughout its structure. The Sierpinski fractal of the pore structure will probably be related to the sizes of the powder grains used to manufacture the membrane filter, the sintering times used to fabricate the filter, the strength of the filter and the air resistance offered by the filter. The Sierpinski fractal of porous metal electrodes and

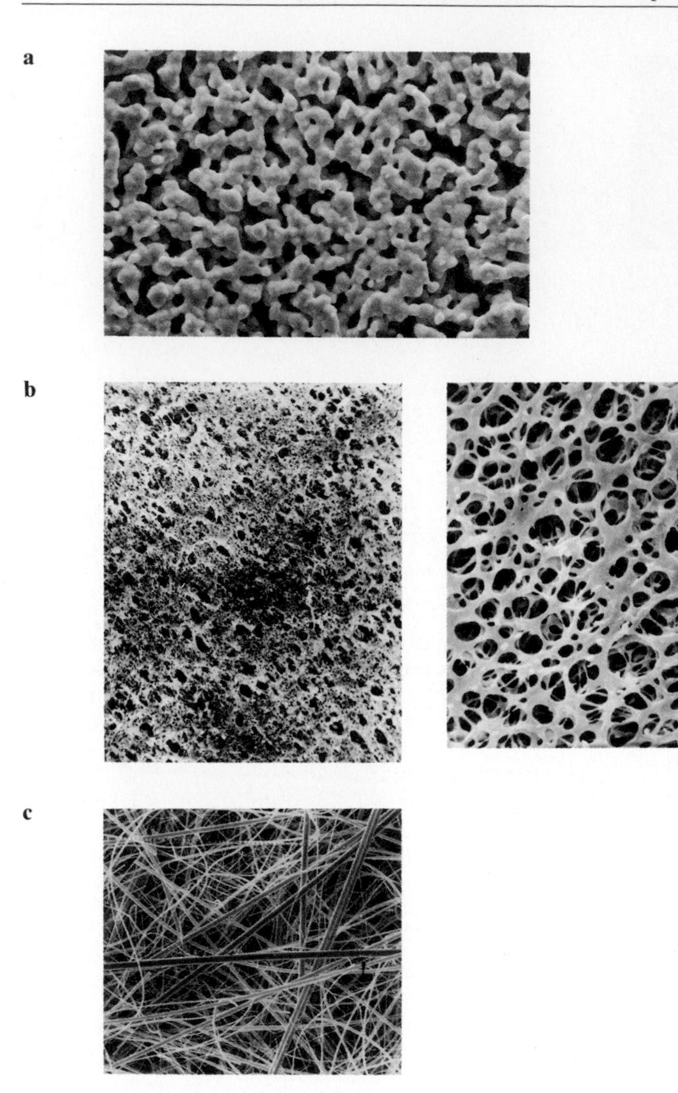

Figure 6.15. It only requires a small leap of the imagination to recognize Sierpinski carpets as being the mathematical model for filter systems.
(a) Surface photograph of a silver membrane filter used to filter quartz dust in a working environment [4, 5]. (Illustration provided by and used with the permission of The Poretics Corporation, 151 I. Lindbergh Avenue, Livermore, CA.)
(b) Plastic membrane filters manufactured by Gelman Science [6]. (Reproduced by permission of Gelman Science Inc.)
(c) Glass-fibre filter manufactured by Gelman Science Inc., [6]. (Reproduced by permission of Gelman Science Inc.)

other sintered bodies made from ceramic powders and/or metal powders may be a useful characterization of their structure.

The filters shown in part Figure 6.15(b), are **membrane filters**. This name comes from the fact that they are thin and look like membranes of the body. In discussions of filter technology, it is useful to differentiate between two main types of filters. One type of filter is known as a **surface filter**, and the other as a **depth filter**. In Figure 6.16, the two types of filter are shown. The surface filter, known as a **Nuclepore filter**, is made by bombarding a thin piece of plastic with neutrons. The passage of the neutron through the film weakens the plastic. Subsequent etching of the neutron tracks produces smooth, cylindrical pores through the plastic film. The size of the hole can be controlled by the period of etching. When filtering with a surface filter, the process is essentially a two-dimensional process with the filtered objects remaining on the surface of the filter. The effective size of a nuclepore filter is the smallest size of fineparticle that is retained by the filter, and is identical with the diameter of the hole in the filter. A depth membrane filter is sponge-like in structure and much thicker than a Nuclepore filter. The stopping power of this type of filter has to be measured using a standard aerosol or test suspension. The trapping of the filtered fineparticles, by the sponge type filter involves a randomwalk through a Menger sponge. A Menger sponge is a three-dimensional extension of a Sierpinski carpet which will be discussed briefly in the next chapter. The rated filter sizes, of both of the filters in Figure 6.16, are almost the same, although the sponge-type membrane filter obviously has much larger surface holes than the Nuclepore filter. It would be interesting to measure the Sierpinski fractal of the membrane filter to see if it could be related to the measured stopping power of the filter.

In Figure 6.15(c), a very different type of filter, made by assembling a random mat of long thin fibres, is shown. This type of structure is described technically as a **felt**.

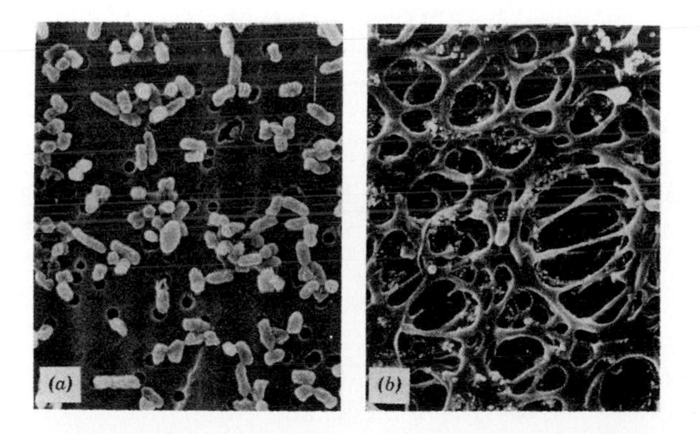

Figure 6.16. Filtration experts differentiate between surface filters and depth filters [7].
(a) Bacteria filtered on a Nuclepore surface filter.
(b) The same bacteria filtered on a sponge membrane filter. This type of filter is much thicker than Nuclepore surface filters. The two photographs were taken at the same magnification; the single holes in the Nuclepore filter are 0.4 μm in diameter. The experimentally determined stopping power of the sponge filter is 0.45 μm.
(From B.H. Kaye, "Direct Characterization of Fineparticles," reprinted by permission of John Wiley & Sons Inc., 1981. Copyright ©, 1981, John Wiley & Sons Inc., New York.)

A felt is made by filtering a suspension of fibres through a support grid, with subsequent removal of the deposited fibres once they have formed a mat. Many natural fibres, such as wool and hair, are sufficiently rough textured so that when pressed together in a felt it becomes a cohesive material. Cloth felt is made this way. The word filter is closely related to the word felt, since early man used to make a felt of animal hairs to make their clothes with, then started to use pieces of the cloth to filter dirt from their drinking water. **Paper** is a felt of cellulose fibres. Fresh cellulose fibres from newly processed wood forms a cohesive felt without the addition of an adhesive. When making paper out of recycled wood fibres, it is sometimes necessary to spray droplets of glue into the matted fibres to make them become a cohesive whole.

Over the years, one of my personal interests in research has been in the design of filtration systems to protect buildings and individuals from toxic dusts. Because many filters are made from paper, the basic structure of the filter is similar to that in Figure 6.15(c). When I first thumbed through Mandelbrot's book on fractal geometry in the autumn of 1977 and saw the diagrams that he presented on page 105, I thought I had discovered a section devoted to the fractal structure of felted fibre filters. This particular picture is reproduced in Figure 6.17(a) and its similarity to Figure 6.15(c) is obvious. An inspection of the legend under the original version of Figure 6.17, however, showed that Mandelbrot imagined that the criss-cross of lines he had drawn represented the street patterns that one meets in a European city. Mandelbrot regarded the varying widths of the lines on his diagram as wide boulevards, crossing streets, alleys, and so on, down to virtually infinitely thin pathways for ants. North Americans, used to the geometric grid patterns in their city streets, would probably not see a city plan in Figure 6.17. To construct his figure, Mandelbrot assumed that the width distribution of the "streets" was a log-log scaling function. Mandelbrot discusses the fact that, if he could have continued to draw lines which are so thin that they cannot be seen, the "white houses" between streets would be so small that they would have zero area. In other words, Figure 6.17(a) is a **Sierpinski felt** as distinct from a Sierpinski carpet.

Mandelbrot, assumes that the system in Figure 6.17 is essentially a novelty, and describes it as being mostly decorative, contrasting sharply with my feeling that this diagram was the key to describing the fractal structure of fibrous felts and hence of filters. If a real filter is a Sierpinski felt, then the Sierpinski fractal of the holes in the filter should prove to be an interesting and useful parameter for describing the structure and stopping power of the filter.

My first encounter with the street pattern in Mandelbrot's picture reminded me of two pictures that I had seen earlier in a book on air filtration, by C.N. Davies [9]. For visual comparison, these two pictures are shown in Figures 6.17(b) and (c). They first appeared in a scientific publication by Fuchs, Kirsh and Stechkaina [8].

The fact that a Sierpinski felt appears to have a more open structure than the Sierpinski carpet in Figure 6.2 is purely a matter of how far one is prepared to go in constructing the number of holes in the Sierpinski carpet. In the Sierpinski felt in Figure 6.17, the nibbling away process has proceeded to the point where the carpet is already "thread-bare."

To explore the fractal structure of felts, the simulated filter in Figure 6.18 was constructed. In this first synthesis of a felt structure, it was assumed for simplicity that

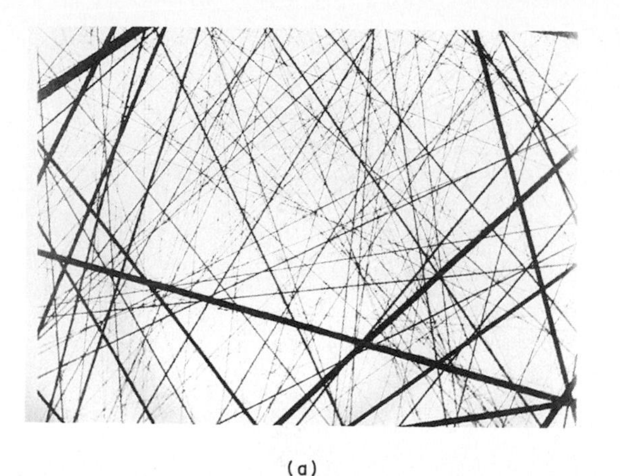

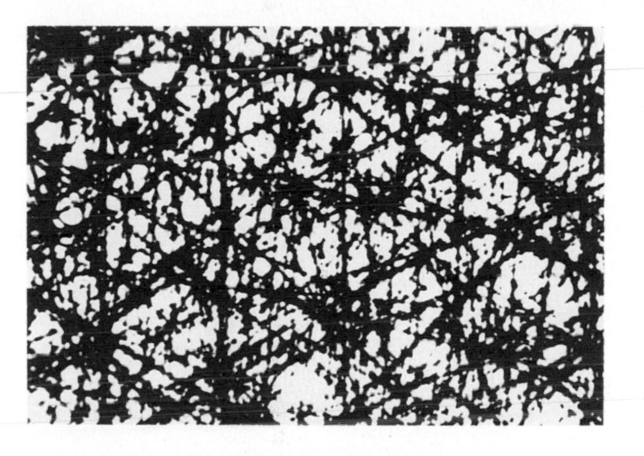

Figure 6.17. Mandelbrot's diagram of lines of decreasing thickness criss-crossing a field of view, which he called a street map, looks very similar to published diagrams of real and simulated fibrous filters.
(a) Street map generated by Mandelbrot (from B.B. Mandelbrot, "The Fractal Geometry of Nature," W.H. Freeman and Co., San Francisco, 1983; reproduced by permission of B.B. Mandelbrot).
(b) Photograph of a computer-generated model of a fibrous filter.
(c) An electron micrograph of a real fibrous filter [8]. ((b) and (c) from N.A. Fuchs, A.A. Kirsch and I.B. Stechkina, "A Contribution to the Theory of Fibrous Aerosol Filters," Faraday Division of Chemical Society, Symp. No. 7, "Fogs and Smokes," 1973. Reproduced by permission of The Royal Society of Chemistry, England.)

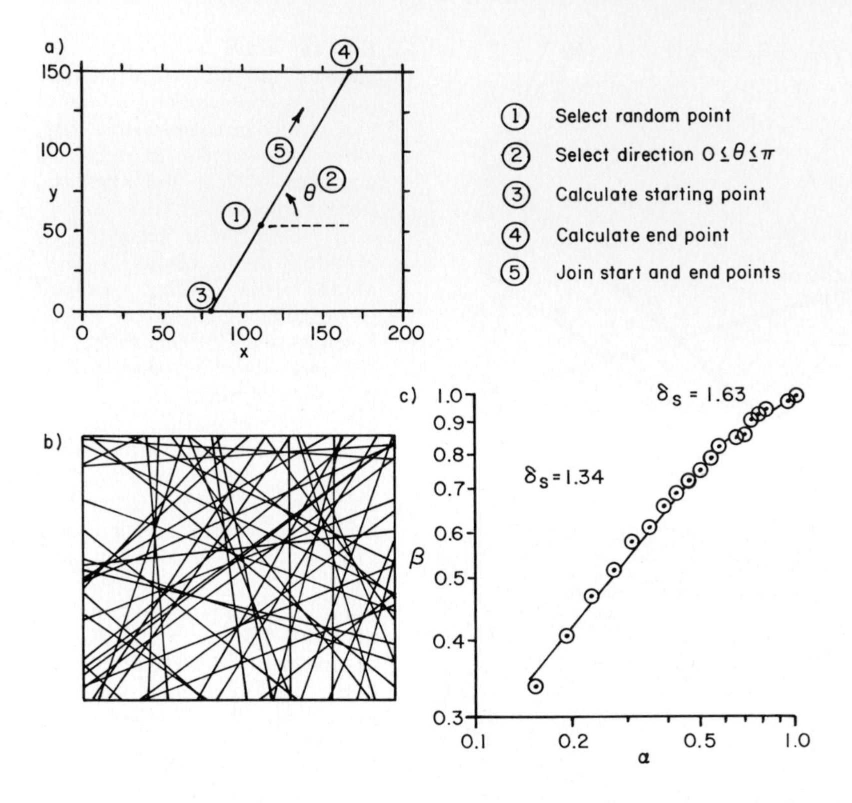

Figure 6.18. The hole structure of the simulated fibre filter can be described by two different Sierpinski fractals. The Sierpinski fractal manifest at coarse resolution appears to describe the larger holes present in the structure.
(a) The position and direction of a fibre added to the filter are selected using random number tables.
(b) Simulated fibre filter assembled according to the scheme illustrated in (a).
(c) Sierpinski fractals of simulated filter shown in (b).
α = Area of just resolved trema.
β = Remaining area at inspection resolution.
δ_S= Sierpinski fractal.

all of the fibres had the same diameter and were effectively of infinite length. The location of a fibre crossing the area of the filter was determined by the selection rules given in Figure 6.18(a). In Figure 6.18(b) a simulated filter obtained using these construction procedures is shown. The Richardson plot used to deduce the Sierpinski fractal of the holes in the simulated filter is shown in Figure 6.18(c). It can be seen that the simulated filter appears to have two different Sierpinski fractals at different levels of inspection. Until more experiments are carried out, it is not clear whether this bimodal Sierpinski fractal is an artifact of the experiment or a fundamental property of random fibre filters. Experiments are in progress at Laurentian University to test the possibility that the bimodal Sierpinski fractal of the filter is a real property corresponding to a mystifying and frustrating problem faced by scientists who attempt to build filters to trap aerosol fineparticles. If one looks at the real glass-fibre filter in Figure 6.15(c), there appear to be two or three large holes present in the fibrous mat. These few large residual holes would constitute a rapid disappearance of the Sierpinski background at coarse resolution inspection, corresponding to the high Sierpinski fractal discovered at low resolution for the simulated filter in Figure 6.18. The lower slope Sierpinski fractal, at higher resolution, could represent the high population of a smaller holes. From the point of view of filtration efficiency, these few large residual holes apparent in a filter such as those in Figure 6.15(c) constitute a problem, since most of the air passing through the filter can pass through these few holes. In industrial practice, many respirator filters have fibrous filtration pads in which there are usually larger residual holes, of the type shown in the system in Figure 6.15(c). Fortunately, dirty air often contains some larger dust fineparticles, which quickly fill the larger holes and improve the efficiency of the respirator. Hence it is often found that the efficiency of the respirator increases rapidly after the first few minutes of use in a dusty atmosphere. However, this represents a dilemma for the industrial hygienist. In his war on dust in the working environment, he tries to remove as much dust as possible, but by removing some of the coarser dust, which often constitutes the visible pollution of the working environment, he may in fact lower the efficiency of the respirator used to remove the invisible hazardous dust. Dust smaller than 30 μm is not visible to the naked eye, and some of the more dangerous dusts in the working environment are smaller than 5 μm in diameter [4].

The apparent persistence of a small number of large holes in a random mat of fibres could have a statistical explanation. Thus, if one were to be adding fibres to the mat one at a time, the chance of an incoming fibre locating across an existing large hole on the fibre mat is much smaller than its chance of lying across a smaller hole in the fibrous mat. Moreover, the chance of an incoming fibre landing on a large hole in such a position that it bisects the hole into two smaller holes is relatively small [11].

In our planning of research into the fractal structure of filters, we did not feel it useful to continue working extensively with the model in Figure 6.18(a) and we switched our attention to working with a fibre structure simulation model closer to the dynamics of paper making. In this model, short fibres are assembled at random in the paper, making space according to the model illustrated in Figure 6.19 (this type of model was first developed by Clarenburg and colleagues; see the discussion of Figure 6.20).

In the paper-making model, it is assumed that the grid used for making the paper is such that the shortest fibre arriving at the screen is going to fall broadside on to the screen, and has a length greater than the diagonal of the support grid. When making paper, a failure to meet these two requirements results in the seepage of fibres through the grid, especially in the early stages of making a web of fibres on the screen, with consequent pollution of the water leaving the paper mill. The assembly algorithm for the simulation of the paper mat from short fibres is shown in Figure 6.19(a).

There are obviously many variations that one can build into the model for assembling filters from randomly arriving fibres. One can vary the number of fibres, the length of fibres or the shape of the fibres, and one can select fibres of different width and having specified crimp and/or curvature. For the purposes of this discussion, we shall restrict ourselves to some relatively simple models. Thus, in Figure 6.19(b), the build-up of a

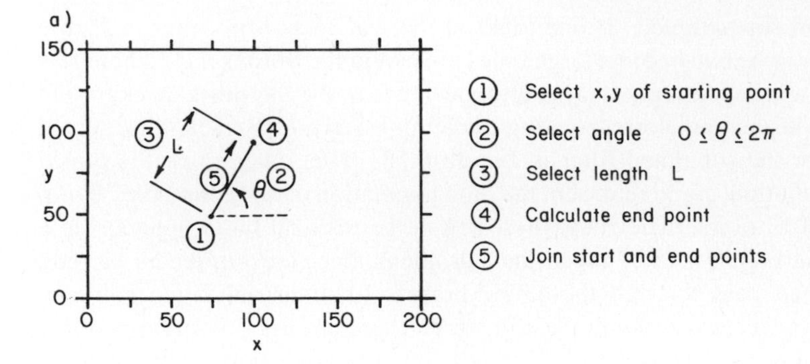

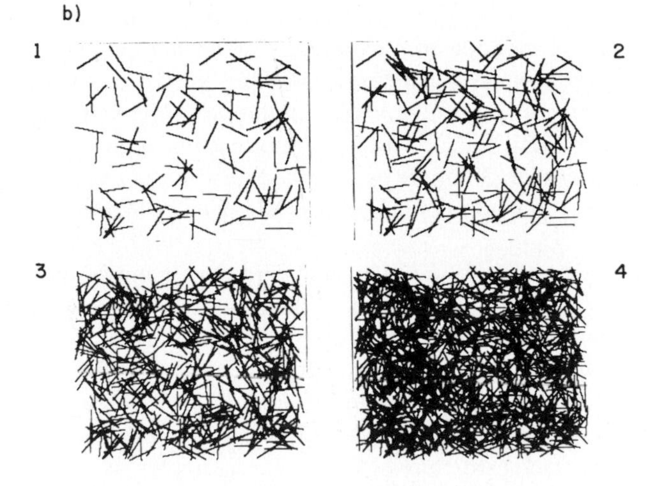

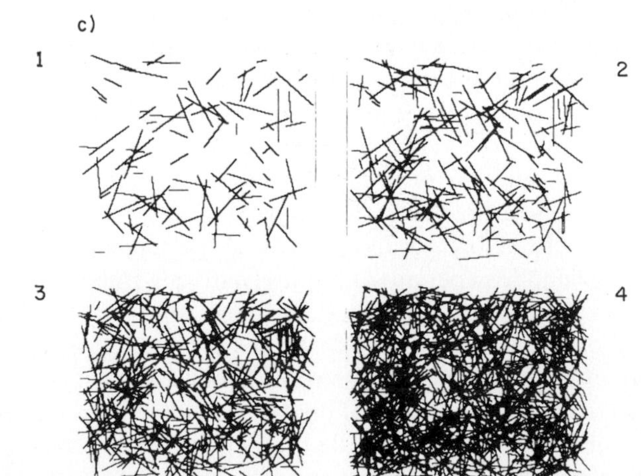

Figure 6.19. The structure of randomly assembled fibrous mats from short fibres, using a model similar to that first discussed by Clarenburg, can be simulated on the computer as illustrated above [10].

(a) Illustration of the construction algorithm followed in simulating short fibre felts.

(b) The appearance of 30-unit fibre length felts at various levels of population density.

(c) The appearance of felts at various population densities constructed from fibres of varying lengths.

filter from fibres of 30 units length when the matrix was assumed to be 200 x 250 units is shown. The only significance of this choice of units arose from the memory capacity of the small computer used to generate the diagrams [10].

In the sequence of diagrams shown as 1, 2, 3 and 4 in Figure 6.19(b), the build-up of the mat at populations up to 1000 fibres in the field of view is shown. The persistence of a few relatively large holes in the fibre mat is an obvious feature of the filter. In Figure 6.19(c), another sequence of diagrams illustrating the build-up of a felt when the fibres vary in length from 10 to 50 units is shown. In this situation, the probability of a fibre of given length is allocated a digit and then, when locating a fibre in the field of view one chooses first the fibre length, the fibre position and then the fibre direction, all from random number tables. It can be seen that the fact that the length of the fibres can vary does not alter the structure of the felt significantly, as soon as continuous paths start to exist between the fibres. Again, the persistence of a few large holes appears to be an important feature of the structure of the felt.

In recent years, specialists in occupational health and hygiene have been looking into the possibility of increasing the efficiency of respirators by using electret fibres to enhance the capture probability of an aerosol fineparticle passing by a fibre in the filter felt. An **electret** is a system which carries permanent electrostatic charge, as distinct from a transient one, created by friction or the deposition of an electric ion on a surface. The name electret was coined by comparison with the term magnet, which has a permanent magnetic field, as distinct from an electromagnetic field, which only has a field when surrounded by a coil carrying the electric current [12]. In the course of some research work at Laurentian University, it was suggested that the persistent holes in a filter might be suppressed by creating the felt from random fibres of electret material, which could be aligned by electrostatic fields [11]. The technique of aligning the fibres need not be restricted to electret fibres, since ordinary fibres could be given a charge by frictional charging prior to their arrival at the assembling screen of the felt making process. If this latter possibility were to be explored, one could not always be sure as to the charge that a fibre would be carrying, and there is need for preliminary experiments to simulate the structure of a fibrous filter in which the materials were aligned in a magnetic field, some of them aligning across the field and some of them along the field lines [13]. Small amounts of magnetic material could be incorporated into synthetic fibres and used to align the fibres in the felt-making process. In Figure 6.20, the build-up of a felt made from fibres of the same length, using random arrival with equal probability of parallel and perpendicular alignment to the field, is shown. A visual comparison of the structure in Figure 6.20, part 4, with that in Figure 6.19(c), part 4, would appear to indicate that the aligned felt has a more even structure than that of the random structured felt; however, one would need to carry out many experiments to compare the randomly fluctuating systems and to evaluate the Sierpinski fractals of aligned and randomly assembled felts to see if the visual impression was an optical illusion or a reality. Simulation experiments to compare the Sierpinski fractal of fibres aligned in different ways, with different properties, are in progress at Laurentian University [11].

As mentioned earlier, a technique for simulating a fibrous felt using randomly assembled short fibres was originally developed by Clarenburg and co-workers in a

Figure 6.20. Modern felt-making technology can align the fibres on the supported grid leading to felts of the form illustrated above.

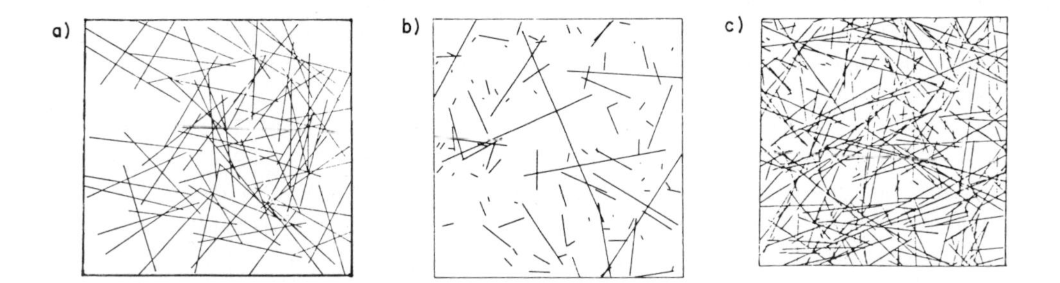

Figure 6.21. Abstract representation of technical problems from widely differing technology suggest the potential widespread use of Sierpinski felt fractals to described systems encountered in many different applied sciences.

(a) Original short fibre felt described by Piekaar and Clarenburg [17]. (Reprinted with permission from Chemical Engineering Science, Vol. 22, L.A. Clarenburg and H.W. Piekaar, "The Tortuosity Factors in Fibrous Filters," Copyright 1967, Pergamon Press plc.)

(b) Simulated fibre composite material described by Balberg and co-workers [20 – 23] (from I. Balberg and S. Bozowski, "Percolation in a Composite of Random Stick-Like Conducting Particles," *Solid State Communications*,Vol. 44, No. 4, 1982; reproduced by permission of I. Balberg).

(c) Simulated crack structure in a stressed rock described by Jaeger and et al. [24] (from Jaeger, Englman and Sprecher, J. Appl. Phys. 59, 4048 (1986). Reproduced by permission of R. Englman and the American Institute of Physics).

series of scientific publications dealing with the structures of fibrous felts [14 – 19]. A diagram from one of his papers [17] is shown in Figure 6.21(a).

The simulation of the properties of random fibre felts and aligned fibres has applications in other fields of research, as indicated by a visual comparison of Clarenburg's diagram in Figure 6.21(a) with the fields of view shown in Figures 6.21(b) and (c). Balmberg and co-workers examined the properties of a plastic material into which carbonblack has been incorporated as a filler [20 – 23]. They pointed out that, to a first degree of approximation, the carbonblack filler pigment could be regarded as stick-like and that the probability of carbonblack pigment linking up within the material was essentially a problem in bond percolation, which no longer took place on a lattice, and in which bonds of various lengths could be expected. In the manufacturing process, it is possible to align fibres by various techniques, and Balmberg and co-workers explored the possibilities of modifying current materials by varying the amount of alignment achieved in the manufacturing process. Their model is obviously applicable to any fibre modified-reinforced composite material.

Figure 6.21(c), is a computer-simulated view of cracks inside a damaged fragment of material. In their studies, Jaeger et al [24] were looking at the way in which rocks fragment and how rocks are damaged when primary fragments are released by blasting. Thus, we see that what started off as an exercise in simulating filler structure can be applied to simulating the failure properties of rocks and pounded fragments (the application of fractal geometry to the scientific study of fragmentation is explored in more detail in Chapter 9).

6.5 Dendritic Capture Trees in Filter Systems

Specialists working in the field of filtration have noticed that very small fineparticles captured by the fibres of a filter tend to build up as fern-like structures on the fibres, rather than to spread themselves uniformly over the fibre surface [25, 31]. This phenomenon is illustrated in Figures 6.22(a) – (c). These fern-like growth are described by the aerosol specialists as **dendrites**, which comes from the Greek word "dendron" meaning a tree. In this book, I shall refer to the dendrites of captured fineparticles growing on a fibre as **capture trees.**

In 1984, Dr. Ensor of the Research Triangle Institute in North Carolina drew my attention to the fact that the structure of the capture trees observed in fibrous filters could probably be described quantitatively using the concepts of fractal geometry. This fact was demonstrated later in a publication by Dr. Ensor and his colleague Dr. Mullins [25], from which Figure 6.22(a) is taken. A visual comparison between the elements of Figure 6.22 and those of the simulated deposition of metal ions on electrodes generated by Voss, shown in Figure 5.24, suggests that a major factor controlling the growth of a capture tree is the presence of electrostatic forces which attract passing aerosol fineparticles to the tip of the existing capture tree. This in turn suggests that the fractal

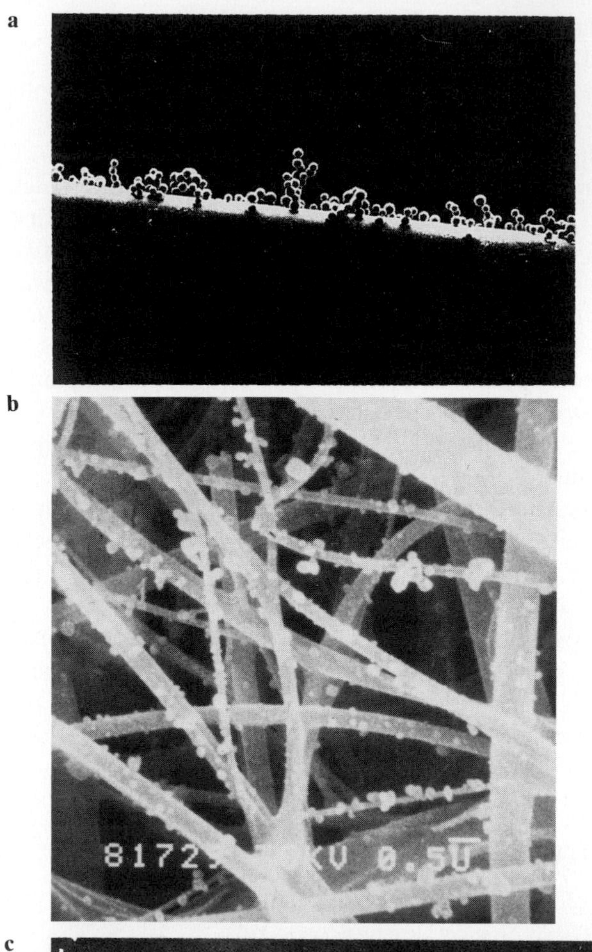

Figure 6.22. Not only can the structure of a fibrous filter be described by means of the concept of fractal geometry, but also the growing agglomerates within the fibre system appear to have fractal structure.

(a) 4.1 μm latex capture trees growing on a 25 μm diameter glass-fibre, photographed by Ensor and Mullins [25].

(b) Sodium chloride dendrites growing within a fibrous filter, photographed and used by permission of Dr. Rubow of the University of Minnesota [26]. (Photo courtesy of TSI Incorporated, St. Paul, Minnesota, USA.)

(c) Sodium chloride dendrites growing within a HEPA filter (Hepa is an acronym for high-efficiency particle filter). From the trade literature of Flanders Filter Corp. [27].

(d) In the Hansen filter powdered resin deposited on wool fibres are used to capture aerosol fineparticles by electrostatic capture [28] (used by the permission of Mr. R. Howie, Institute of Occupational Medicine, Roxburgh Place, Edinburgh, Scotland).

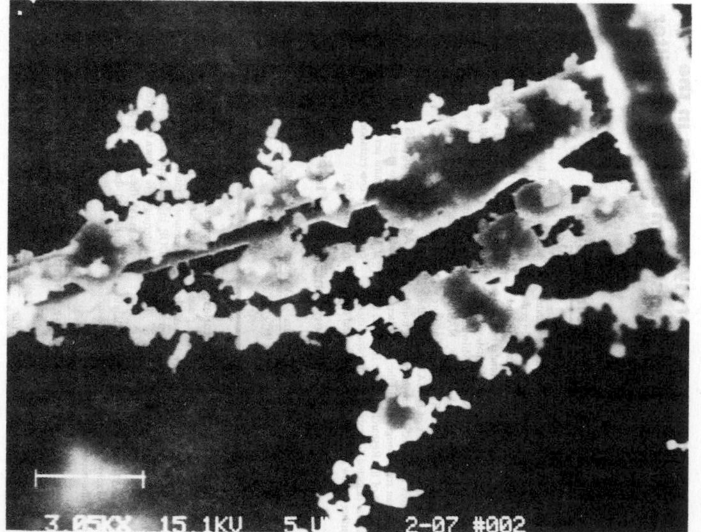

d

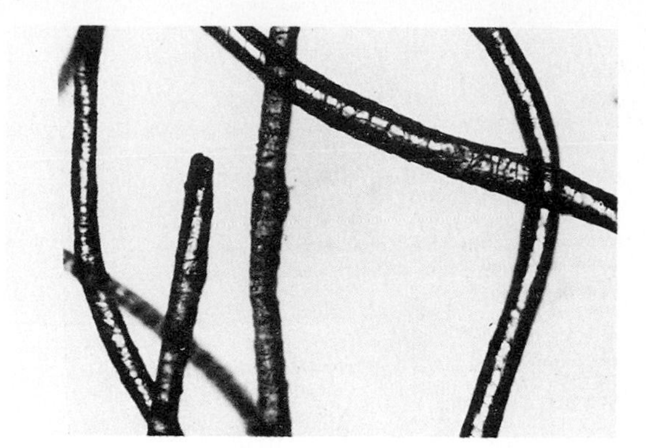

dimension of the growing tree will be a measure of the probability of bonding on collision, and also of the electrostatic forces operative in the filter system governing the efficiency of the filter for capturing specific fineparticles.

The aerosol fineparticle capture trees shown in Figures 6.22(b) and (c), are formed from sodium chloride aerosols. It can be seen from Figure 6.22(b) that the fine fibres of the filter are more effective than larger ones at capturing the passing aerosols. In Figure 6.23, the fractal dimensions of two of the larger capture trees in Figure 6.22(c) are shown.

Characterizing the fractal structure of capture trees could prove to be a very important technique for studying the efficiency of the new type of filters, based on the use of electret fibres mentioned in Section 6.4. As already mentioned, electret fibres carry permanent electrostatic charge, as a result of which they are very efficient at capturing some dangerous dusts. However, the performance of an electret filter is dust specific. This is to be expected, since if electrostatic forces are playing a major part in the capture of the fineparticles then the electrostatic properties of the dust with respect to the fibre will change from dust to dust. Changes in environmental conditions such as humidity will also alter the efficiency of an electret filter and this would become apparent in the change in structure, and hence in the fractal dimension, of the capture tree growing under different environmental conditions.

Hence it will be interesting to carry out experiments in which capture trees are grown on a single fibre, using specific dusts and controlled environmental conditions, to see if the fractal structure of observed capture trees can be related to the efficiency of capture processes operating within the filter.

It has recently been suggested that one of the major problems of filter systems, migration of fineparticles through a filter, may be related to the fractal dimension of the capture tree. In a filter system, if a dendrite grows up from a fibre such as that shown in Figure 6.24(a), in which growth takes place with very efficient capture (for example if there is a probability of 1 that an encounter between the tree and a passing fineparticle will result in a permanent capture), the tree is likely to be an open, feathered structure with a weak stem. The air flowing through the filter will exert a pressure on the fern-like tree, with possible rupture of the stem attaching it to the fibre. Thus the highly

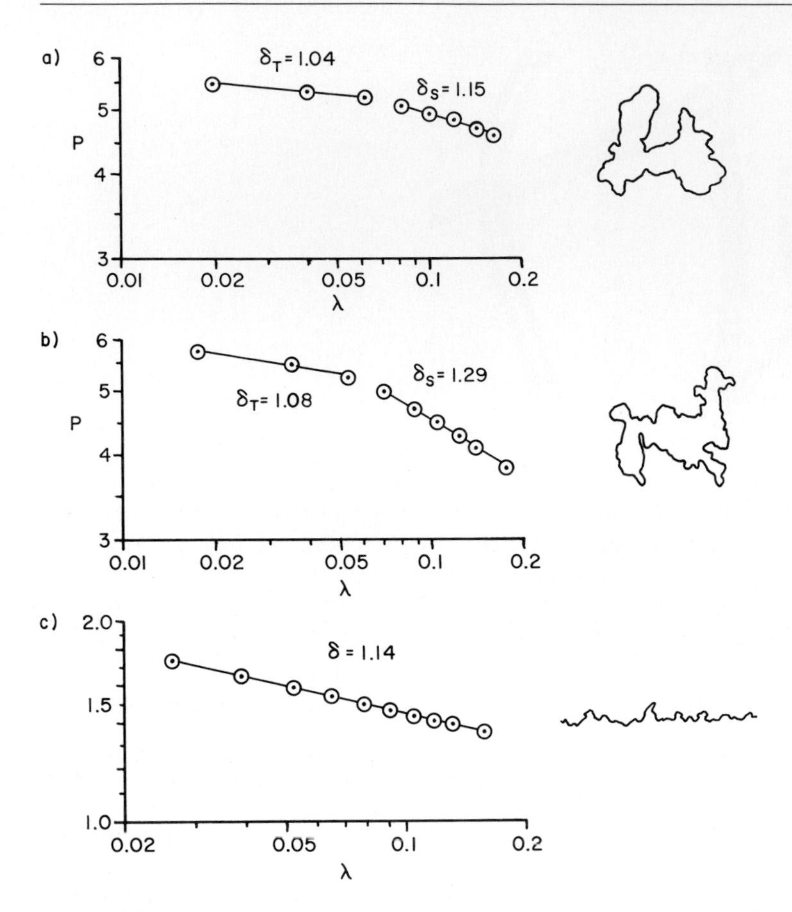

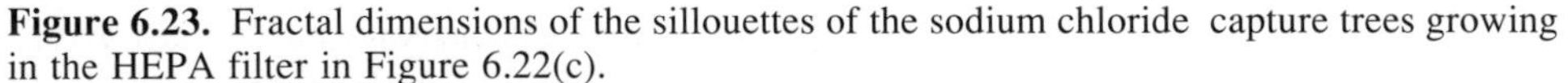

Figure 6.23. Fractal dimensions of the sillouettes of the sodium chloride capture trees growing in the HEPA filter in Figure 6.22(c).
(a) Fractal dimension of cluster A.
(b) Fractal dimension of cluster B.
(c) Profile fractal dimension of the loaded fibre in Figure 6.22 along the segment shown above.

branched fractal capture trees of the system shown in Figure 6.22(c) obviously have weak stems at the points noted by the numbers 1 – 5.

On the other hand, if the capture tree is a low-density bush of the type shown in Figure 6.24(b) (for a fineparticle on encounter, the capture probability for this type of growth is less than 1; see the discussion of Voss trees in Chapter 4), one may need more fibres to catch the same number of fineparticles, but the resulting trees with their low profile and dense structure may be less likely to snap off and migrate through the filter. By growing dendrites on single fibres and comparing them with structures modelled on the computer, in a way similar to the electrode deposition trees grown by Voss, one could assess the relative contribution of various forces to the growth of the captured tree.

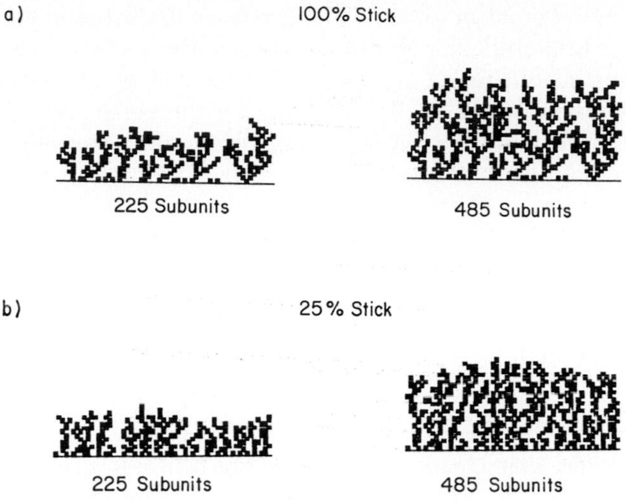

a) 100% Stick

225 Subunits 485 Subunits

b) 25% Stick

225 Subunits 485 Subunits

Figure 6.24. A design dilemma that may confront engineers using electret-based filters is that high-efficiency capture may result in low-strength, high fractal dimension capture trees, that can break off and migrate through the filter (the capture trees in this figure are models and not real, observed capture trees).
(a) High fractal structure capture trees with weak stems are generated by a high probability of capture.
(b) Low-bush capture trees generated by "low capture encounter probability" during the deposition process are less likely to snap off and migrate through the filter than the high fractal capture trees.

It should be noted that Ensor and Mullins pointed out that the overall profile of the various trees sprouting along the fibre will have a fractal dimension like that of the boundary of a fineparticle, and that this fractal dimension of the fibre and deposit profile will probably be related to the measured drag coefficients governing flow through a filter as the captured fineparticles build up in the filter.

Figure 6.22(d) illustrates a structure from another area of filtration science, where it may be useful to apply the concepts of fractal geometry when describing the structure of the filter. In 1930, N.L. Hansen developed a revolutionary filter for respirators, which proved to be much more efficient than any other respirator filter in use at that time [9,28]. In the **Hansen filter**, powdered resin is added to the surface of wool fibres. Wool fibres are very rugged and scale like in texture. The finely divided resin clings to the fibre and apparently attracts passing aerosol fineparticles by an electrostatic mechanism. The individual resin fineparticles added to the Hansen filter obviously have a fractal structure. It may be that the efficiency of different shattered resins used in this type of electrostatic enhanced filter could be related to the efficiency of the filter.

The penetration of the aerosol fineparticles through a depth filter is a randomwalk in three dimensions. This aspect of filtration theory will be discussed briefly in the next chapter.

A very different type of filtration problem to that which has been discussed in the previous sections of this chapter is the filtration of a suspension of solids where there is so much material in suspension that there is a build-up of deposited material on the filter. This **filter cake** is a porous body with a fractal surface area in three-dimensional classical space. This type of filtration problem has been studied from the perspective of fractal geometry by Houi and Lenormand [29, 30].

6.6 Cantor on the Rocks

No doubt, depending on the reader's anticipation background, the title of this section can conjure up many mental images in the mind. These images could range from a picture of a horse running over some rocky ground to a Cantor from the synagogue experiencing financial difficulties. Perhaps some readers thought that the section title is the name of a new drink concocted by fractal fanatics. In fact, the title was intended to suggest that Cantorian sets, bearing useful information, can be generated by drawing lines across rock sections that look like Sierpinski carpets. Mandelbrot, in his discussion of the "road map" in Figure 6.17(a), points out that the chord set on a line drawn across the "city map" is a Cantorian set.

The set of chords on a line intercept taken across a fibrous mat is a one-dimensional search exploring the structure of the filter. The size distribution of the chord set and the dimension of the Cantorian dust of the points of intersection on the search line are related to the Sierpinski fractal of the fibre mat. The basic problem involved in relating the dimension of the Cantorian dust of the points on the line search to the Sierpinski fractal of the filter system is one of experimental confidence. One needs to make many measurements on the fibrous filter to establish the exact value of the Sierpinski fractal, and when one switches to the use of the Cantorian dust to characterize filter structure based on a line intercept search, which is a one-dimensional search, the statistical fluctuations in the chord set is so high that one needs to make many measurements to achieve statistical confidence. The efficient design of experimental procedures to relate the Cantorian dust generated on a line search of a fibrous filter to the Sierpinski fractal of a filter system is under active study and for the sake of illustration some early data generated in this study are shown in Figure 6.25.

If a line intercept-based procedure can be developed to describe, in an adequate manner, the structure of a fibrous system by means of the appropriate dimension of a Cantorian dust, this would prove to be very useful, since the line intercept information can be generated quickly and at low cost by modern computer-aided microscope methods of inspection.

The title of this section was actually inspired by the fact that when I first looked at a Cantorian bar and a Cantorian set in Mandelbrot's book, it reminded me of a method for characterizing the structure of rocks used by geologists known as the **Rosiwal**

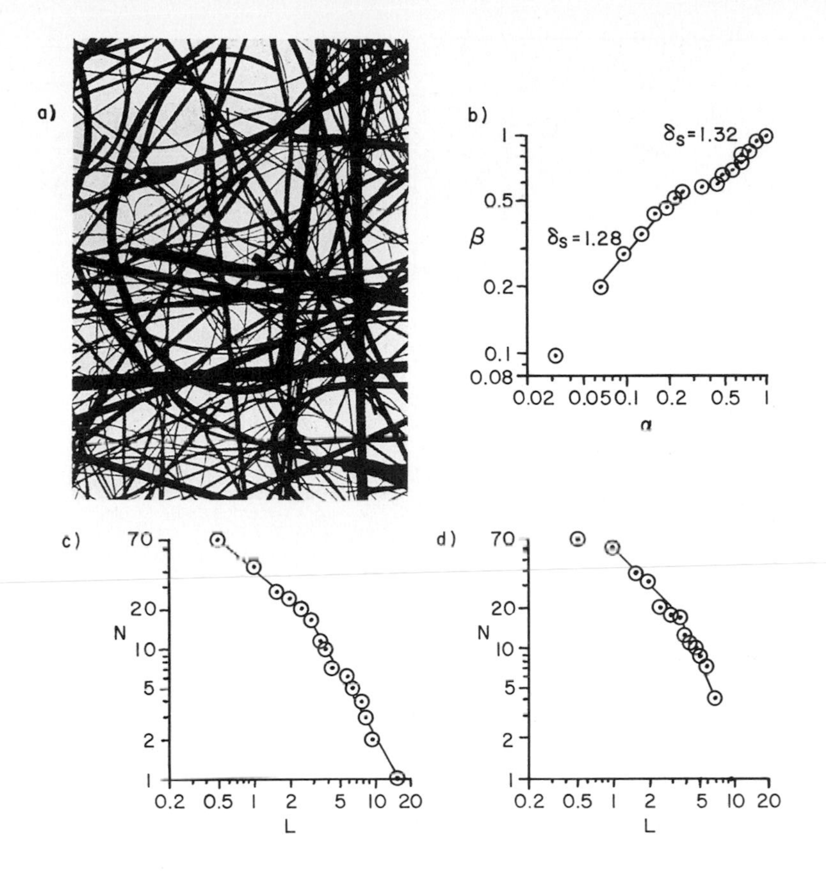

Figure 6.25. The intercept chords generated on a line search of a filter made by a random assembly of fibres is a Cantorian set of chords. The dimensionality of Cantorian dust on the line search is a measure of the structure of the filter system.
(a) Micrograph of filter made from glass-fibres [16].
(b) Sierpinski fractal of the holes in the glass-fibre filter.
(c) Chord distribution of fibre intercepts on line search of the fibrous mats in (a).
(d) Chord distribution of spaces in the filter generated by a line search of the filter structure.
(Reprinted with permission from Chemical Engineering Science, Vol. 22, L.A. Clarenburg and H.W. Piekaar, "Aerosol Filters - Pore Size Distributions in Fibrous Filters." Copyright 1967, Pergamon Press plc. Also reproduced with the permission of L.A. Clarenburg.)

intercept method [32,33]. Consider the field of view shown in Figure 6.26. One can regard this as being a section through a piece of ore which is to be processed to release the black circles which constitute a valuable mineral. Before the mineral processing engineer blasts the rock out of the ground, he wants to know the richness of the ore to decide if he is justified in mining the ore. The specialist who studies the rock specimen to assess its richness in terms of the valuable dispersed material is a **petrographer**. **Petrography** is a specialty area of geology. Long before the invention of fractal geometry, Rosiwal showed that if one draws a random set of lines across the surface of

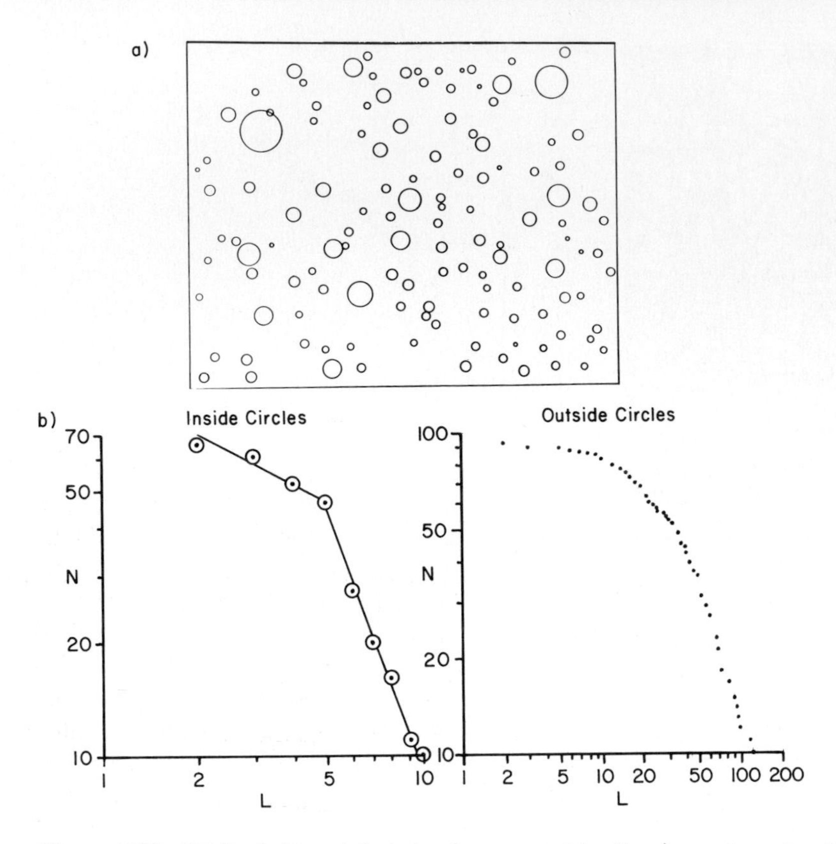

Figure 6.26. Rosiwal showed that chords generated by line inspection of a dispersed field of material (in this case ore constituents in a polished section of rock) can be used to estimate the volume concentration of the dispersed material. Thus the chord density on the set of line intercepts shown above is 0.78 and, from the way in which the field of view is constructed, it is known that the area of the field of view covered by black circles is 7.8% of the area. It can also be shown that the average chord length of the intercepts is related to the average size of the dispersed material.

a polished section of a piece of ore, that the volume percentage of the ore in the specimen is the same as the length of the chord crossing the dispersed material divided by the total length of the search line (I know from personal experience that people new to the subject of petrography find this simple relationship hard to believe. The reader can try testing this theory for himself, since the percentage of the area covered by the black profiles in Figure 6.26 is 7.8%. The reader is invited to draw random lines on the picture and to discover for himself that the total length of the chords within the profiles divided by the length of the total search line tends to this value).

If one can be sure that the profiles to be inspected in a section are randomly distributed in the section, one can inspect them with a set of regular lines, since double randomness is not required by the theory of the method. If, however, there is any possibility that the profiles are not randomly distributed, then one should use random lines to be sure there is no bias in the measured data. In many situations, however, one can assume that the profiles to be evaluated are in a random array and then the regular set of lines used to inspect the system can be the scanning lines of a TV camera, and the intercept line data can be processed by a computer linked to the TV camera.

Scientists active in the development of fineparticle characterization procedures realized that the set of chords generated by inspecting the system with a set of lines contained all the information necessary to generate the size distribution function of the dispersed material [34–37]. However, early attempts to implement size estimation procedures based on such chord intercept measurement procedures ran into difficulties because of the large number of chords required to generate the necessary information with acceptable confidence. It is interesting to that 30 years after the basic logic was worked out for line scan intercept procedures, new instruments based on this type of logic are coming on to the market in the area of fineparticle characterization [38,39]. In the new generation of instruments, scanning laser beams are being used to measure the size distribution of fineparticles in a liquid suspension by measuring the chord length generated as the laser beam tracks across the fineparticles. In this case, since the laser beam is moving, a time of flight is recorded rather than a physical magnitude. The fact that data processing equipment is much faster and cheaper than 30 years ago is making it feasible to develop this type of instrument after the original developments lapsed into obscurity because of the lack of powerful enough data processing machines at an acceptable cost level. A full discussion of the logic of line intercept techniques using a Rosiwal-type intercept information to evaluate the structure of rocks and other composite material is beyond the scope of this book. However, it can be predicted that as a general knowledge of fractal geometry spreads into the specialist fields of geology and metallography, we can expect to see widespread use of the chord set as a fractal indication of structure with appropriate use of Cantorian dust dimensions to characterize the structure of the inspected system

References

[1] The material reproduced in Figure 6.1 is taken from B.B. Mandlebrot, "The Fractal Geometry of Nature", (1983) B.B. Mandelbrot, San Francisco, 1983,pp. 306-307.

[2] Sierpinski's work is reviewed in Mandelbrot's book, [1]; details of his work can be found in the following publications: (a) W. Sierpinski, "Sur une Courbe dont Tout Point est un Point de Ramification," *C. R. Acad. Sci.* (Paris), 160, (1915), 302; (b) W. Sierpinski, "Sur une Courbe Cantorienne qui Contient une Image Biunivoque et Continue de Toute Courbe Donnée," *C. R. Acad. Sci.* (Paris), (1916) 162, 629; (c) W. Sierpinski in S. Hartman et al. (Eds.), "Oeuvres Choisies". Editions Scientifiques, Warsaw, 1974.

[3] J.D. Birchall and A. Kelly, "New Inorganic Materials," *Sci. Am.*, May (1983), 104-115.

[4] T.T. Mercer, P. Morrow and W. Stober (Eds.), "Assessment of Airborne Particles," Proceedings of the Third Rochester International Conference on Environmental Toxicity, 1972, Charles C. Thomas, Springfield, IL, 1972.

[5] Illustration in K.R. Spurny, "Aerosol Filtration by Means of Analytical Pore Filters," reference 4, Ch. 4.
[6] The filter systems in Figures 6.15(b) and (c) are taken from "The filter Book," Gelman Sciences, Ann Arbor, MI, 1986.
[7] The filters shown in Figure 6.16 are reproduced from B.H. Kaye "Direct Characterization of Fineparticles," Wiley, New York, 1981, p. 51. Nuclepore is the registered trademark of the Nuclepore Corporation, 7035 Commerce Circle, Pleasanton, CA 94566, USA. Comprehensive literature on the structure and properties of Nuclepore filters is available from the manufacturer who kindly provided the photograph reproduced in Figure 6.16.
[8] N.A. Fuchs, A.A. Kirschand I.B. Stechkaina, "A Contribution to the Theory of Fibrous Aerosol Filters," Faraday Division of the Chemical Society Symposium No. 7, "Fogs and Smokes," 1973. This diagram is also reproduced in reference 9.
[9] C.N. Davies, "Air Filtration," Academic Press, London,1973,p. 171.
[10] The data in Figure 6.18 are taken from G.G. Clarke "Fractal Geometric Description of Filters and the Filtration Process," *MSc Thesis*, Laurentian University, in preparation.
[11] This problem is explored in more detail in reference 10. See the discussion of Timbrell's work with asbestos fibres in reference 4.
[12] J. VanTurnhout, C. VanBochov and J.G. Veldhuizen, "Electret Fibres with High Efficiency Filtration of Polluted Gases," *Staub*, 36, No. 1.
[13] See the discussion of Timbrell's work with asbestos fibres in reference 4.
[14] R.M. Werner and L.A. Clarenburg, "Aerosol Filters - Pressure Drop Across Single Component Glass Fibre Filters," *Ind. Eng. Chem. Process Des. Dev.* (1965) 288-293.
[15] L.A. Clarenburg and R.M. Werner, "Aerosol Filters - Pressure Drop Across Multicomponent Glass Fibre Filters," *Ind. Eng. Chem., Process Des. Dev.* (1965) 293-299.
[16] H.W. Piekaar and L.A. Clarenburg, "Aerosol Filters - The Tortuosity Factors in Fibrous Filters," *Chem. Eng. Sci.* 22 (1967) 1817-1827.
[17] H.W. Piekaar and L.A. Clarenburg, "Aerosol Filters - Pore Size Distributions in Fibrous Filters," *Chem. Eng. Sci.*, 22, (1967) 1399-1408.
[18] L.A. Clarenburg and H.W. Piekaar, "Aerosol Filters - 1 Theory of the Pressure Drop Across Single Component Glass Fibre Filters," *Chem. Eng. Sci.* 23, (1968),765-771.
[19] L.A. Clarenburg and F.C. Schierech, "Aerosol Filters - 2 Theory of the Pressure Drop Across Multi Component Glass Fibre Filters," *Chem. Eng. Sci.*, 23, (1986), .773-781.
[20] I. Balberg and N. Binenbaum, "Computer Study of the Percolation Threshold in a Two-Dimensional Anisotropic System of Conducting Sticks," *Phys. Rev. B*, 28, (1983), 3799-3812.
[21] I. Balberg and N. Binenbaum, "Cluster Structure and Conductivity of Three-Dimensional Continuum Systems," *Phys. Rev. A*, 31, (1985), 1222-1225.
[22] I. Balberg and N. Binenbaum, "Directed Percolation in the Two-Dimensional Continuum," *Phys. Rev. B*, 32, (1985). 527-529
[23] I. Balberg and S. Bozowski, "Percolation in a Composite of Random Sticklike Conducting Particles," *Solid State Commun.*, 44, (1982), 551-554.
[24] Z. Jaeger, R. Engleman, Y. Gur and A. Sprecher, "Internal Damage in Fragments," *J. Mater. Sci. Lett.*, December (1985).
[25] D.S. Ensor and M.E. Mullins, "The Fractal Nature of Dendrites Formed By the Collection of Particles On Fibres," *Part. Character.* 2 (1985) 77-78.
[26] Micrograph courtesy of Dr. K.L. Rubow, University of Minnesota.
[27] Diagram based on a photograph in the technical literature of Flanders Filters Inc., 1985.
[28] F.J. Feltham, "The Hansen Filter; Filtration and Separation," 16 (4) July - August 1979, pp.370-372.
[29] D. Houi and R. Lenormand, "Visualization and Statistical Modeling of Particle Accumulation on a Filter Medium," Extended Abstract from the 17th Annual Meeting of the Fineparticle Society,Miami, Florida, April 1985.

[30] D. Houi and R. Lenormand, "Particle Deposition on a Filter Medium," in D.B. Landau and F. Family (Eds.), Proceedings of the International Conference on Kinetics of Aggregation and Flocculation, held in Athens, Georgia, April 2-4, 1984.
[31] C.E. Billings, *PhD Thesis*, "Effects of Particle Accumulation in Aerosal Filtration," California Institute of Technology, 1966; see also a picture from this thesis in reference [9].
[32] For a readily accessible discussion of the Rosiwal intercept method, see G. Herdan, "Small Particle Statistics," 2nd ed., Butterworths London, 1960.
[33] Many of the appropriate statistical relationships involved in the assessment of rock structure by techniques such as the Rosiwal intercept method are to be found in F. Chayes, "Petrographic Model Analysis," Wiley, New York, 1956.
[34] The science of studying two-dimensional sections using statistical search methods, with consequent description of three-dimensional properties of the system, is known as quantitative microscopy or quantitative stereology. Comprehensive texts on these two subjects are available in R.T. DeHoff and F.N. Rhines, "Quantitative Microscopy," McGraw-Hill, New York, 1968, and E.E. Underwood, "Quantitative Stereology," Addison Wesley, Reading, MA, 1970.
[35] Information on the use of chord techniques for generating information on the size distribution of dispersed material can be found in B.H. Kaye, "Some Aspects of the Efficiency of Statistical Methods of Particle Size Analysis," *Powder Technol.* 2 (1968-69) 97-110.
[36] B. Scarlett, "A Statistical Description of Particulate Systems," in M.J. Groves and J.L. Wyatt-Sargent, (Eds.), "Particle Size Analysis, 1970," Society for Analytical Chemistry, London, 1970, 101-113.
[37] G. Mason, "Random Chord Distributions From Triangles," *Powder Technol.*, 12 (1975), 277-281.
[38] A commercial instrument based on laser linescan logic was developed by Procedyne Corp. See a discussion of this instrument in B.H. Kaye "Direct Characterization of Fineparticle Systems," John Wiley, New York, 1981.
[39] A size characterization instrument based on linescan logic with the information being generated by recording the time of flight of a laser beam, scanning a three-dimensional array of the fineparticle systems is available from Brinkman Instruments.

7 An Exploration of the Physical Significance of Fractal Structures in Three-Dimensional Space

7.1 Randomwalk Theory of Powder Mixing in Three- and Four-Dimensional Space

When thumbing through my original copy of Mandelbrot's book, I was particularly interested in the illustration reproduced in Figure 7.1. I thought that this diagram, on page 136 of the first edition of Dr. Mandelbrot's book, must be a picture of the initial stages of a process in which grains of one powder were being dispersed in another. However, when I started to read the text, I found that Dr. Mandelbrot was describing what he called a **Levy dust** as a model for the dispersion and clustering of galaxies in outer space. Mandelbrot's discussion of the information summarized in Figure 7.1 involved the discussion of **Olbers' paradox.** This paradox is - "Why, if there is an infinite number of stars in space, is the night sky not uniformly bright?" As I started to read the chapter of Mandelbrot's book containing Figure 7.1, I soon found myself involved in **Levy flights** into outer space.

A Levy flight is a particular type of Brownian motion involving multilevels of random steps in which, every now and again, the randomly staggering object takes a

Figure 7.1. The fractal structure of galaxies in outer space requires the same mathematics to describe their structure as that which can be used to describe the dispersal of one powder with black grains in a large amount of another powder with white grains (from B.B. Mandelbrot, "The Fractal Geometry of Nature," W.H. Freeman & Co., San Francisco, 1983; reproduced by permission of B.B. Mandelbrot).

flying leap into another region of space. In Figure 7.2 is shown the track of a Levy flight from Mandelbrot's book, which to me looked like the time record of the dispersion of a grain of one powder in a powder mixture over a period of time. A Levy dust is a record of the positions occupied during a Levy flight without the connecting lines drawn in the diagram. Levy flights and Levy dusts are named after the French mathematician Paul Levy (1886-1971), who was one of Mandelbrot's teachers. Mandelbrot acknowledges in his book the influence that Levy had on his thinking, and gives a two-page biography of this eminent French mathematician. A Levy flight can be used to model the spread of an epidemic. Thus, when a bird can be involved in spreading a disease to a distant region, a flight to a distant location is the large leap, but then as the disease spreads slowly in a local animal population there is small random diffusion until another bird "leap frogs" the disease to a distant region once again (see Signpost 8: Butterflies, Ants and Caterpillars in the Garden of Eden).

When modelling a randomwalk involving a Levy-type flight, the probability of movement includes a small number of random number selections that will generate large steps in the randomwalk. Thus, if we had 100 random numbers representing various steps possible in a Levy flight, then numbers from 1-90 could be small unit steps with 91 being a four step, 92 being a six step and so on, with the directions of the steps also being randomized in space. Mandelbrot shows in his book how a fractal dimension can be associated with the spatial density of the Levy dust recording the progress of a Levy flight. For a randomwalk to be an exact Levy flight, according to the definition given by Mandelbrot, the probability of a set of different size steps must be determined by a given mathematical function. The simulated Levy true flight, represented by a probable path schedule listed at the beginning of this paragraph, is not a true Levy flight, but something very close to it which can be called a **Levy-type flight** or a **pseudo Levy flight.**

For many years I have been involved in studying the problem of mixing two or more powders together. The industrial problems that I have studied have ranged from such exotic problems as techniques for mixing powdered rocket fuel for space crafts, down

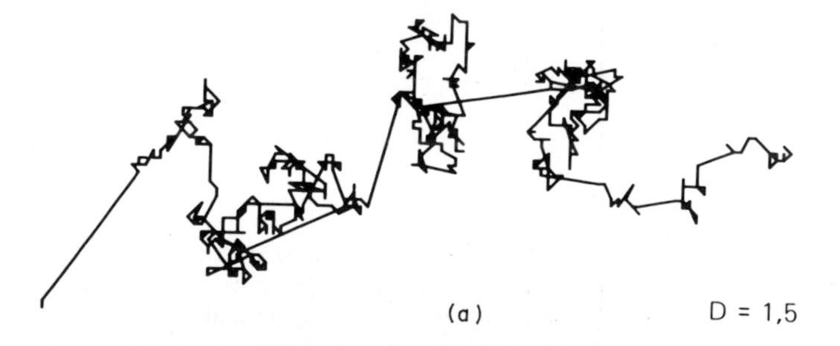

Figure 7.2. The Levy flight is a particular type of randomwalk in which there is, at any time, a small but finite probability of a large leap from the present location, so that the dispersal of the object involves a randomwalk of small steps interspersed with the occasional large leaps. The above is a Levy flight-given in Mandelbrot's book (from B.B. Mandelbrot, "The Fractal Geometry of Nature," W.H. Freeman & Co., San Francisco, 1983; reproduced by permission of B.B. Mandelbrot).

through the problems of making prepackaged cake mixes, to the problems of making headache tablets in the pharmaceutical industry. The classical engineering vocabulary for describing the progress of the dispersal of one powder in another in a mixing process uses terms taken from liquid mixing technology.

Over the years of my involvement with powder mixing technology, I had always felt that randomwalk theory was a better way of describing the dispersal of one powder in another, but I have never had the time to put together my ideas in a coherent manner. In this chapter, I shall sketch briefly how randomwalk theory can be used to describe the mixing act without using the classical terms of chemical engineering, and how the mixing process can be regarded as a randomwalk in three-dimensional space leading to the generation of fractal systems in two-, three- and four-dimensional space. However, before we can look at powder mixing from a randomwalk - fractal geometry perspective, it is necessary to review the classical vocabulary of powder mixing technology.

When describing powder mixing, mechanical engineers talk about **shearing, convective mixing** and **random diffusion**, as being involved in the randomization of the position of the individual fineparticles in a powder mixture [1]. In Figure 7.3 the ways in which these three different mechanisms can mix a set of fineparticles are illustrated. To simplify our discussion of powder mixing mechanisms, we shall confine our discussion to a two-dimensional array of fineparticles. When constructing the diagrams in Figure 7.3, it was assumed that a mixture of black and white powder grains, each of which is represented by a square in a two-dimensional matrix, was to be mixed.

When a shear force is applied to a portion of a mixture, movement of the layers of the mixture over each other breaks down agglomerates in the mixture to help disperse the material. Figure 7.3(a) represents an initially agglomerated 5% mixture of black squares in white squares. To simulate an agglomerated mixture, it was assumed that since the parent matrix had 2500 squares, the number of black squares in the mixture was 125. In the initial loading of the matrix, it was assumed that agglomerates of 4, 6, 8, 12 and 16 black squares could exist. The first step in simulating the structure of the agglomerated mixture is to allocate, for selection purposes, a random number to each agglomerate. In a simple model, it is assumed that all agglomerate sizes are equally probable, so that the random number selection table would be:

agglomerate size 4, 6, 8, 12, 16
selection digit 1 2 3 4 5

It is also assumed that all agglomerates are dense and not chain like.

Algorithm for Location of an Agglomerate in the Matrix

1. Select agglomerate size from random number table.
2. Select X and Y co-ordinates for the agglomerate from a random number table.
3. Select alignment for agglomerate if it is not square by selecting a digit from a random number table. An even number selection places the agglomerate along the X axis; an odd number places the agglomerate along the Y axis.
4. The growing total number of placed black pixels is recorded as the agglomerates are added to the matrix of the simulated mixture. When the total number of pixels required to complete the expected number of pixels in the mixture is less than sixteen, the final agglomerate of the correct size needed to complete the simulated mixture is added.

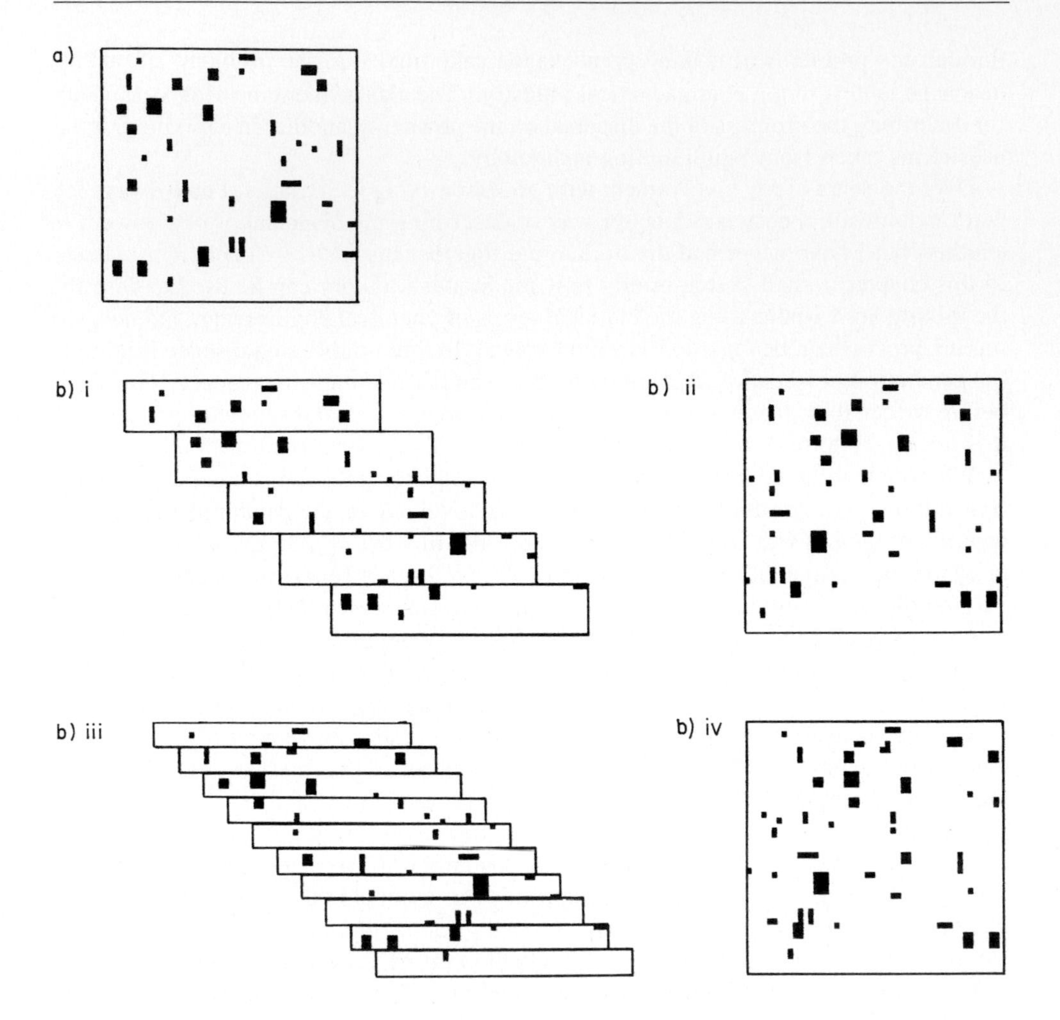

Figure 7.3. Engineers active in powder mixing technology delineate three different mechanisms as significant elements of a powder mixing process: shearing, diffusion and convection.
(a) A simulated, initially loosely agglomerated mixture used to demonstrate three different mixing mechanisms considered to be operative in a powder mixer.
(b) Shearing separates agglomerates in a powder mixture.
(i) Sheared mixture of system a. (ii) Reassembled sheared mixture showing the breakdown of the agglomerates under the shearing action. (iii) and (iv) The severity of shear dictates the efficiency with which agglomerates are broken up in a sheared powder mixture.

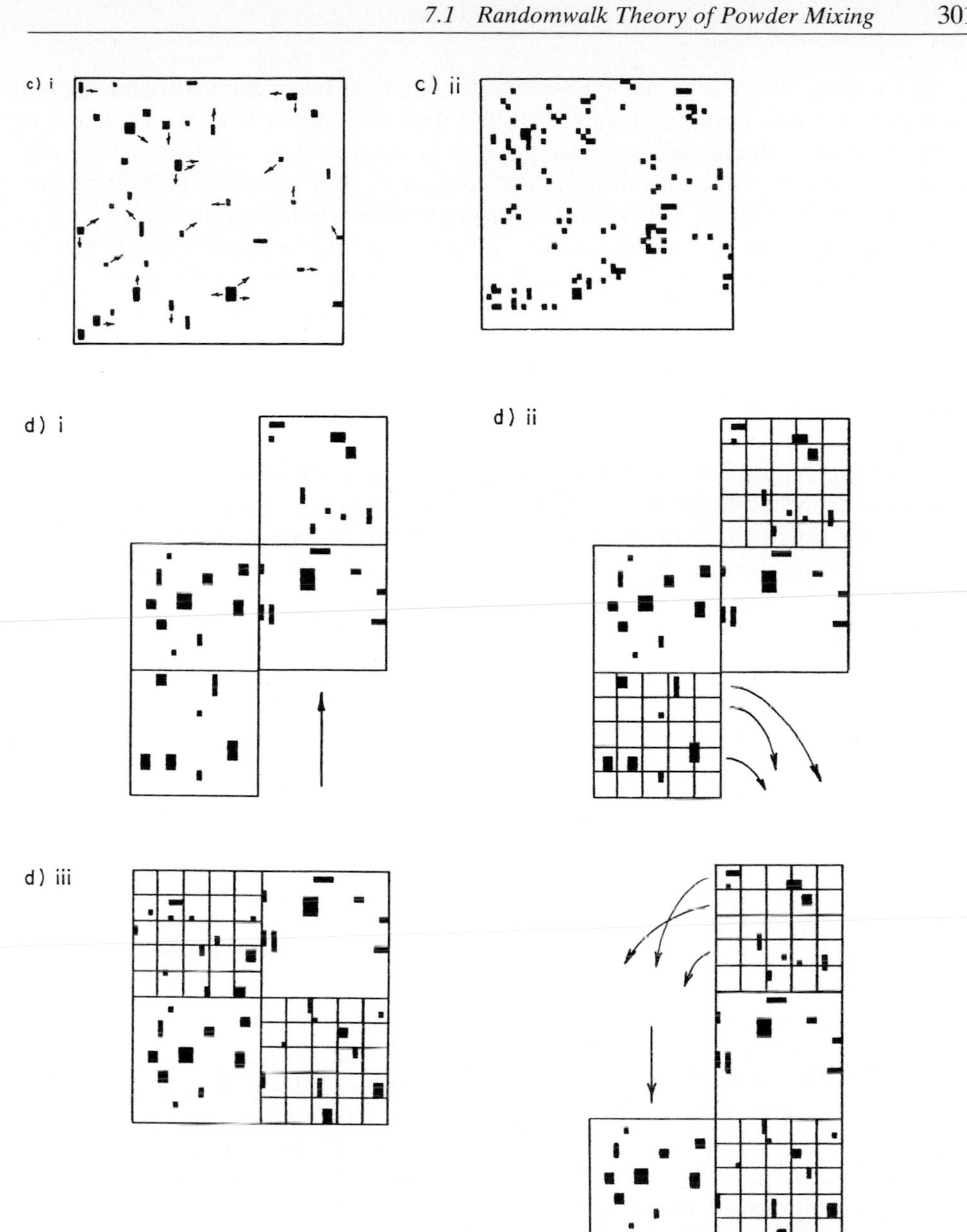

Figure 7.3 (cont). (c) Random diffusion of individual grains can take place in a dilated powder bed.
(i) Dilated mixture (dilation is achieved either by shaking or the expansion of the powder bed by air fluidization or some other technique. Turbulence creates individual randomwalks within the dilated mixture). (ii) Compact bed showing changes in structure by dilation-diffusion.
(d) In convective mixing bulk movement of the powder followed by tumbling randomization takes place.
(i) Convectively rearranged bulk motion. (ii) Tumbling randomization. (iii) New state of powder mixture.

To simulate the shear of the agglomerated mixture, strips of the matrix are moved over each other as shown in Figure 7.3(b)(i). In a real mixer, this type of action is normally carried around the circumference of a cylinder so that, in effect, the sheared material is continuous and the element marked A, B, C, in the diagram is moved to the beginning of the sheared matrix, to form the system shown in Figure 7.3(b)(ii).

The effectiveness of various shearing actions to disperse the agglomerate can be simulated by assuming that shearing material moves in strips of a given width. Thus, if each strip sheared is 10 pixels wide, this represents a milder shearing action than if we assume that every strip of 5 pixels of the matrix is sheared. Thus, in Figure 7.3(b)(iii) and (iv), the difference in the breakup of agglomerates caused by 5 pixel matrix line shearing is illustrated as compared with 10 pixel line shear mixing in Figure 7.3(b)(i) and (ii).

Powder technologists differentiate between **cohesive powders** and **free flowing powders.** In cohesive powders, inter-fineparticle forces are relatively high and it is difficult for fineparticles to move away from each other. Therefore, when mixing a cohesive powder, shearing is the only mechanism that is effective in breaking down agglomerates to disperse the material in the mixture.

In published discussions of powder mixing, random diffusion of the individual fineparticles is considered to be a major powder mixing mechanism. However, it is not always pointed out in such discussions that in a powder mixture, diffusion of the type that disperses molecules in a liquid cannot take place in the powder mixture unless the powder mixture is first opened up by shaking or tumbling or by some other mechanism. Opening up a powder mixture so that the individual grains of powder are free to move about is known technically as **dilation**. What is known as diffusion in a powder mixture is actually a two-stage process, in which the powder bed is first dilated and then the individual fineparticle grains which are now free to move about are randomized by turbulence or some other mechanism. Increasingly engineers are turning to fluidized bed mixers in which the fineparticles to be mixed are suspended in a stream of turbulent air. In this type of mixer the fluidizing air creates the dilation of the bed, and then the turbulence in the fluidized bed is the diffusing mechanism [1].

It should be noted that chemical engineers have always been a little nervous about mixing powders using fluidized bed systems, because many of the materials to be mixed, such as cake mixes, are potentially explosive. When mixing potentially explosive powder mixtures, it is necessary to use an inert gas, such as nitrogen, to fluidize the powder to be mixed. The steps involved in randomization of position by random diffusion are illustrated symbolically in Figure 7.3(c).

Very often, any tumbling action within a powder bed tends to create a dilated state in which individual fineparticle diffusion can take place. Thus, Kaye and Sparrow showed that in a barrel mixer, the effective mixing takes place in the cascading fineparticles moving over the top of the rotating mass of powder, not in the bed of powder rotating at the bottom of the cylinder [2].

The third powder mixing mechanism operating in powder mixers is defined by chemical engineers as **convection**. In a convective mixing act, a paddle or some other system physically moves one part of a mixture to another region of the mixer. This causes gross rearrangement of the constituents of the mixture, but does not normally, in itself, randomize the position of the grains. In convective mixing, the positions of the individual grains are randomized, after gross movement, by the tumbling actions of the portions of the powder moving in to fill the voids created by the movement of the paddle creating the convective change in the structure of the mixture. Thus, for a very simple situation, one can imagine that a paddle moved up through the mixture in Figure 7.3(a). Because of the movement of the paddle, one side of the mixture is elevated. The random tumbling of the convectively displaced elements of the powder mixture into the voids randomizes the positions of the individual grains of the powder as shown. (Note: in powder mixing, sometimes one has to take into account physical properties, such as electrostatic charging in which case random diffusion after convective mixing is not a normal element of the mixing process. Readers interested in this aspect of mixing theory should consult reference [3]).

As a first step in representing powder mixing as a randomwalk, it should be noted that what the chemical engineers know as "shear" can be described as a drift probability in a specified direction at a specified depth imposed upon any randomwalk. Thus, when we seek to simulate the movement of material subjected to shear, we would apply steps to the individual fineparticles in the direction of shear. We could then describe the movement of fineparticles in a mixer subjected to turbulence and shear by permitting the individual fineparticles to take random steps in many directions with an imposed higher probability of steps along the direction of shear. Convective mixing can be described from a randomwalk description of powder mixing as the possibility of a large jump in motion in a given direction, followed by subsequent randomwalking about the new position. In other words, the fineparticles of a powder mixture subjected to convective mixing, are dispersed by a Levy-flight type motion involving non-uniform step randomwalk diffusion.

Long before I had ever heard of a Levy flight, I built Levy-type flight leaps into a device for creating and delivering a powder mixture to the next stage of a manufacturing process [4]. One of the very difficult problems in powder mixing technology, is that as soon as one starts to move a mixture, there is a danger of segregation of the mixture into its initial components. For this reason, a powder mixture prepared efficiently in a mixer can sometimes be separated into its constituent parts as it is delivered to the next stage of a process. In the early 1960s I was involved in some studies of pouring ingredients for making glass into a furnace where the free-flowing constituents had to be randomized as they arrived in the reaction vessel. For our purposes, this was equivalent to taking three streams of powder and pouring them into a device to deliver them to a container. To achieve this mixing, a prototype model of a system illustrated in Figure 7.4 was developed [4]. The ingredients to be mixed were added to the top of what I call a randomizing tower. In the body of the randomizing tower were angled plates with random holes in them. A typical plate is shown in Figure 7.4(b). As the powder cascades over the sloping distributor plate, local tumbling creates a randomizing effect corresponding to randomwalk diffusion in a dilated powder system. Movement down through the random holes in the sloping distributor plate creates the equivalent of convective disturbances in the powder mixture. This leap from plate to plate I would now prefer to call a Levy-type element in the randomwalk mixing of the ingredients. The varying distances between the holes of one plate and the cascading powder in the

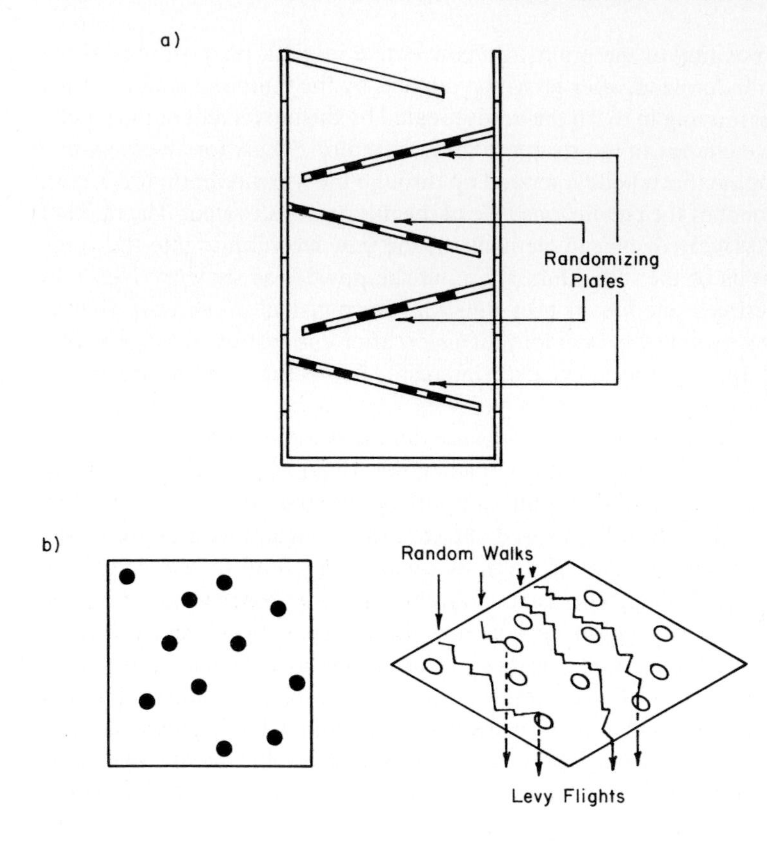

Figure 7.4. A randomwalk in three dimensions takes place in a randomizing tower used to mix powder ingredients.
(a) Randomizing tower containing randomizing plates.
(b) Typical randomizing plates.

next distributor plate constitute the variable-length Levy-type flight leaps. The secret of success in using this randomwalk in three dimensions to achieve powder mixing is that the grains of powder walking down the plates and through the holes should never move so fast that the bigger components of the mixture gain a momentum advantage to create segregation [6, 7].

In the original patent application describing the randomizing tower, it was envisaged that air jets placed at random around the wall of the tower might be a useful adjunct to create randomization within the randomizing tower [4]. From a randomwalk perspective it can be stated that the grains of the mixture moving down through the tower take part in a three-dimensional randomwalk, and that the stack of material in the container at the base of the tower constitutes a three-dimensional dispersion which can be described as a fractal system in three-dimensional space. Sectioning the powder mixture would reveal Sierpinski carpets. If one component of the mixture was rich enough for continuous paths to exist throughout the mixture from one grain to another, the paths in the mixture would form a percolating system in three-dimensional space. The tortuosity of these paths would be describable by the language of fractal geometry.

One of the first techniques used to create the sponge-type membrane filters described in Chapter 6 was to mix pulverized table salt crystals into a plastic at a concentration such that continuous paths existed between the salt grains. From our knowledge of fractal geometry, we know that this means that the salt crystals had to constitutet at least sixty percent by volume of the mixture. Only above this mixture richness could percolating paths exist in the structure of the mixture. The final stage in the making of the filter involved dissolving the salt crystals to leave paths in the filter constituting the filtration channels of the filter system. This type of system is also important in the design of controlled release drugs, in which the pharmacist wishes to have an active drug leach out of a plastic matrix to deliver the drug at a steady rate [8].

The structure of a mixture in which percolating paths exists can be described in terms of a system known as a **Menger sponge** [5]. An ideal Menger sponge is created in three-dimensional space by using the same type of construction as that used to create a Sierpinski carpet in two-dimensional space, with three-dimensional holes being removed from the three-dimensional body. The construction of the Menger sponge can be envisaged from the appearance of the final sponge as shown in Figure 7.5. In a real, natural sponge type structure, the holes of the Menger sponge would be randomized in position to produce a statistically self-similar system, the structure of which could be described using the concepts of fractal geometry.

Powder mixing engineers have developed various types of randomizing towers, some of them employing structured mixing with local diffusive randomwalk, whereas other devices achieve mixing by a free space randomwalk in three dimensions. In the language of the powder mixing specialists, randomizing towers are usually known as **passive mixers** or **stationary mixers** [1]; I prefer the first term.

It is hoped that this brief discussion of a new fractal geometry perspective on the problems of powder mixing will have convinced the reader that this is a fertile area of technology where fractal geometry will flourish and produce many interesting results.

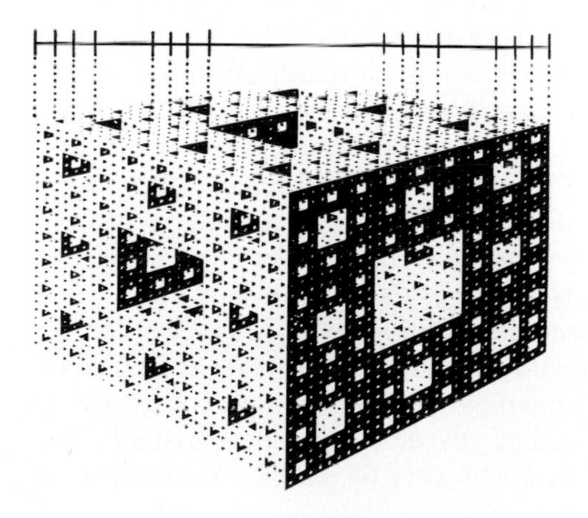

Figure 7.5. An ideal Menger sponge is a three-dimensional extension of the concepts embodied in an ideal Sierpinski carpet. Faces of the Menger sponge are Sierpinski carpets and chords drawn across the surface of the sponge form Cantorian sets. Points on this Cantorian set are Cantorian dusts (from B.B. Mandelbrot, "The Fractal Geometry of Nature," W.H. Freeman & Co., San Francisco, 1983; reproduced by permission of B.B. Mandelbrot).

7.2 Fractal Geometry and Aerosol Physics

Aerosol science is the study of the behaviour of such things as mist, smoke and airborne dust [9,10]. As discussed in earlier chapters, one of the specialist areas where scientists were quick to apply the concepts of fractal geometry was in the characterization of the structures of dust fineparticles. Dr. Otto Preining, of the Technical University of Vienna, pioneered the applications of fractal geometry in another area of aerosol science. Early in 1980, Dr. Preining sent me a copy of a scientific paper entitled "Fractals and Aerosol Characterization" [11], in which he applied the concepts of fractal geometry to the characterization of the structure of a dispersing aerosol cloud. He pointed out that under some weather conditions smoke from a chimney stack maintains a thin pencil-like structure over large distances. In such a situation, from an operational point of view, the cloud can be considered as a Euclidean system of dimension 1. Dr. Preining, in his pioneer publications, discussed how the dimensionality of a dispersing chimney plume, that is, the emitted smoke trail, varies with atmospheric conditions.

In an address to the annual meeting of the American Association of Aerosol Scientists in 1984, Dr. Preining made the following comments on the use of fractal geometry in aerosol science.

> "Suddenly, basic concepts of mathematics, dormant for nearly 50 years and considered unusable monsters, became tools for modelling and describing quantitative complexity. (Fractal concepts) can be and have to be applied to aerosols and they ought to be, and have already occasionally, been used to describe the complexities of shapes and compositions of individual particulates. Only the use of such structural parameters as fractals will facilitate the diagnosis and the applicability of mathematical models to real world aerosol systems, since all non-linear processes, like coagulation and/or certain chemical reactions, will depend strongly on the structure of the system" [12].

As Dr. Preining envisaged in these statements, aerosol science is becoming one of the fastest growing areas of fractal geometry application. The individual smoke fineparticles inside a cloud undergo a three-dimensional randomwalk which generates a fractal density structure within a cloud after a given period of time. One can model the dispersion of the various fineparticles in the dispersing cloud of smoke using randomwalk models, basically the same as those discussed in the previous section of this chapter, when looking at powder mixing as a stochastic process. One can also look at cloud dispersion structures after a period of time using the fourth dimension - time. Fractals involving space and time are described as **fractons**. A discussion of fractons is outside the scope of this introductory book, but the concept is mentioned here for the sake of completeness.

As in the case of powder mixture systems discussed in the first section of this chapter, the inspection routines for studying aerosol systems are often of reduced dimensionality, so that Sierpinski carpets and Cantorian sets appear in the experimental data. Thus, scientists are already applying "**computerized axial tomography,**" (a process known as **CAT scanning** in North America), to the study of aerosol distribution in enclosures. The word **tomography** comes from a root word meaning "a drawing of a slice through a system" from the Greek word temmein, meaning "to cut". In computerized tomography, the cut is a simulated one, generated in a computer by taking a multiple set of inspection data from many angles using penetrating radiation [13, 14]. In medical CAT scanners the penetrating inspection radiation is X-rays. In aerosol science, where the density of material in an inspected zone is much lower, it is possible to use infrared radiation to generate the multiple angle images used to generate the CAT scan image in the computer. The image generated by the use of infrared inspected – computed generated tomography is a Sierpinski carpet. This application of infra-red based tomography of aerosol clouds generating Sierpinski carpets in two-dimensional space has been pioneered by Luck, Siemund and Lorbeer [15]. (Note: these scientists do not use the vocabulary of fractal geometry in their publications, but a quick inspection of their publications immediately suggests the presence of Sierpinski carpets in their experimental data) CAT scanners based on X-rays could also be used to create Sierpinski carpet views of powder mixtures.

In a widely used technique for inspecting the structure of an aerosol cloud, the aerosol fineparticles are sucked into a narrow tube and moved past a beam of light. The size of the fineparticle is deduced from the optical energy that it scatters as it moves through the inspection zone of the instrument. Several commercially available instruments based upon this physical principle are available. This group of aerosol sizing instruments are referred to as **photozone stream counters** [16].

The currently available photozone instruments only register the total number and size of each of the fineparticles passing through the inspection zone. Many of them also work with a dilution stage to take the sample of aerosol from a cloud and dilute the concentration of the cloud to avoid multiple occupancy of the inspection zone. If one were to re-design such instruments slightly to inspect the stream of undiluted fineparticles coming out of a cloud, then a record of the gaps between the fineparticles would generate a Cantorian chord set which would be characteristic of the structure of the cloud being studied. This possibility was first pointed out by Dr. Preining, although he did not use the term Cantorian set in his 1980 publication [17]. Describing the structure of dust clouds using the concepts of fractal geometry is one of the few times in this new science where some of the terms invented by Dr. Mandelbrot become a little confusing. Thus, using Dr. Mandelbrot's terminology, we have to talk about the dimension of the Cantorian dust of the Cantorian chord set generated by the time intervals in a photozone-generated time chord record to characterize the structure of the dust cloud. In such a situation, conceptual Cantorian dusts become hopelessly intermingled with real dust. When studying the structure of dust clouds, it is probably better to talk about the dimensionality of the **Cantorian points in line space** rather than of the "Cantorian dust." In this way, one can avoid confusion between the real dust in the cloud and the mathematical dust on the inspection line.

In our discussion of randomwalk models in two-dimensional space, we discussed the modelling of the growth of agglomerates in two-dimensional space and the modelling of the coagulation of a smoke in two-dimensional space. The coagulation of smoke fineparticles is a very important area of aerosol science. The mathematical procedure used to model the growth of two-dimensional agglomerates can obviously be extended to three-dimensional space. A very simple technique for modelling the structure of agglomerates in three-dimensional space, for freely associating - dispersing fineparticles in systems such as a powder mixture, is based on using random number tables.

Thus, if we were to take six different random number tables and convert every seven into a blank pixel, then if we superimposed the six random number tables in space we could regard them as constituting Z plane slices through XYZ space. Contiguity of an aggregate exposed by the two-dimensional slices (ie. successive random number tables) in three-dimensional space could be traced and the results sketched as shown in Figure 7.6.

In Figure 7.7, this simulated agglomerate has been fed into a Macintosh computer and a commercially available program was used to generate different perspective views of the agglomerate. If the object had been an aerosol fineparticle, these different views would represent different orientations of the fineparticle falling in space.

This transient agglomeration model based on the use of several pages of a random number table could be useful to study aerosol systems which can stick to each other when formed initially, but which cool rapidly so that they no longer stick on collision. Thus, if one is looking at the coagulation of a metal aerosol formed by vaporization, it is probable that at the high temperature of initial condensation of the vapour there may be growth of agglomerates, but that as soon as they have cooled below a certain point the agglomerate will no longer stick to each other on collision.

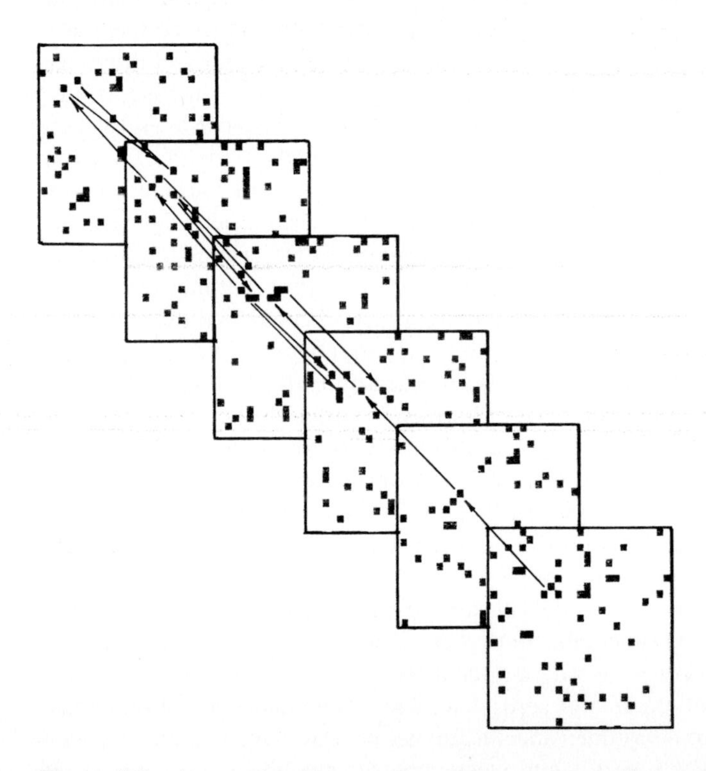

Figure 7.6. A set of random number tables can be used to model the structure of a freely associated cluster of fineparticles in three-dimensional space by regarding the individual random number table as Z plane slices exposing XY planes in three-dimensional space.

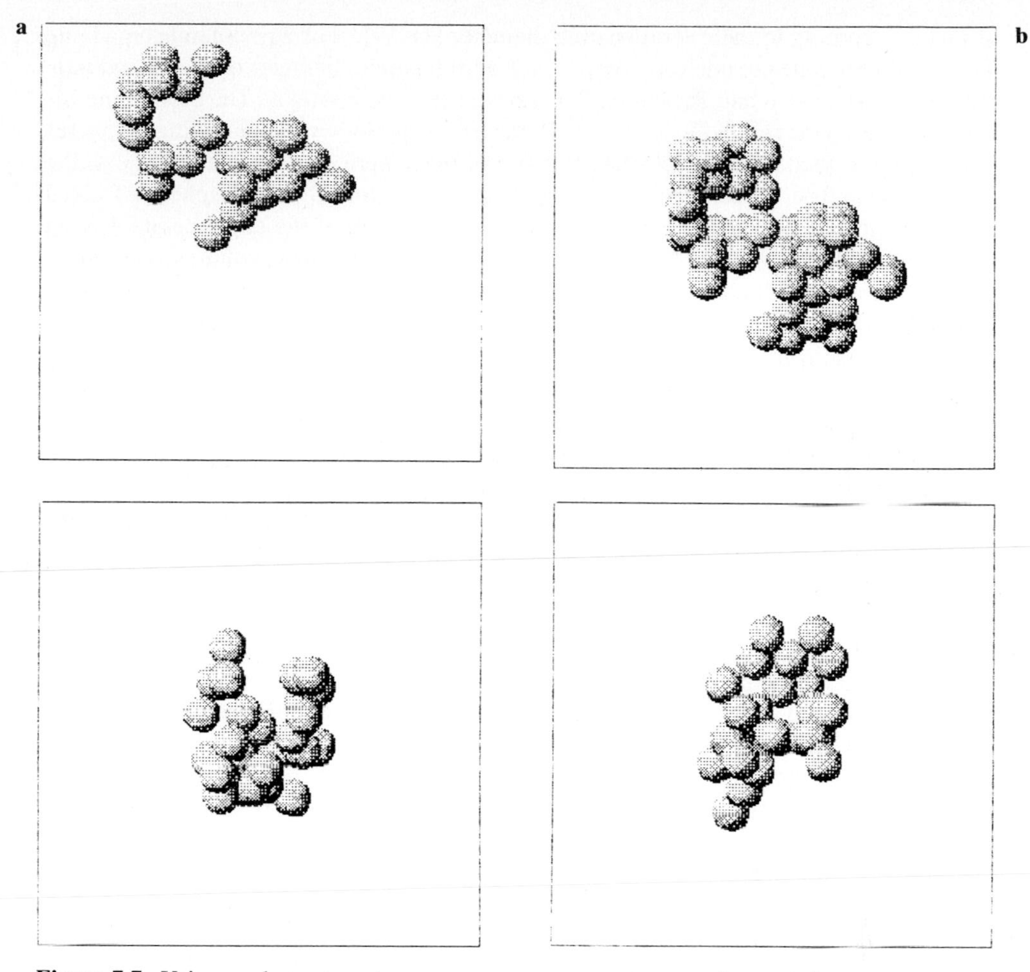

Figure 7.7. Using modern computer programs, one can rotate an agglomerate in computer space to show the appearance of a tumbling agglomerate at various orientations in space.

Scientists studying aerosol systems have developed techniques for studying agglomeration in a cloud which, from the perspective of someone interested in fractal geometry, can be used to study the formation of fractal agglomerates in three-dimensional space by randomwalk techniques. Thus, if one generates a cloud of monosized spheres which randomly collide with each other in a turbulent cloud to form agglomerates, then if at a particular time one sucks out of the cloud a stream of the agglomerates, one can inspect the agglomerates to look at their structure.

Dr. Stöber and his colleagues studied aerosol clouds of initially monosized latex spheres, and measured the aerodynamic size of the agglomerates formed in the cloud. Groups of fineparticles with the same aerodynamic diameter are known as **isoaerodynamic fineparticles**. In the device developed by Dr. Stöber and co-workers, the **spiral-disk centrifugal aerosol spectrometer,** airborne fineparticles are deposited on a strip of metal according to their aerodynamic diameter [18,19]. For agglomerates made up of several unit spheres, one can compare the aerodynamic diameter of the fineparticle with its physical size and structure. To carry out this comparison, Dr. Stöber and his colleagues used the following concepts. If one has an agglomerate containing N spheres of diameter d, then one can calculate the size of the sphere of the same volume as the agglomerate. The agglomerate will experience more drag than the sphere of equal volume as it falls through a viscous system, but the weight of the agglomerate driving it downwards is the same as the weight as the sphere of equivalent volume. Therefore, if the agglomerate falls more slowly than the sphere of equal volume, it must be due to more drag on the agglomerate which must be related to its structure. Therefore, one can use a factor κ to compare the drag on agglomerates of different structure but containing the same number of components, as defined by the equation

$$\frac{\upsilon_{A}}{\upsilon_{SE}} = \kappa$$

where υ_A = viscous drag on the agglomerate, υ_{SE} = viscous drag on a sphere of equal volume and κ = relative drag coefficient.

In Table 7.1 are some outlines of agglomerates studied by Stöber and co-workers. It can be seen that the coefficient of the increased drag factor appears to be related to the structural features of the agglomerate and that those which experience high drag are those which are beginning to develop highly branched fractal structure. One can describe the agglomerates shown in Table 7.1 as being **embryonic fractals**. The data in this table strongly suggest that if we were able to use the equipment of Stöber to study some larger agglomerates with more highly developed structure, one would probably link the viscous drag experienced by the agglomerates to the fractal structure and geometric shape factor of the agglomerates. This, in turn, would open up the possibility of developing a theory for predicting the falling speeds, and hence the dynamics, of highly structured dusts which are potentially dangerous if inhaled into the lungs.

Assessing the danger to the lungs of an airborne dust fineparticle is a complex problem which depends upon several factors associated with the structure of the dust fineparticle. Pioneer studies of the health hazards posed by inhaled dust were focused on problems associated with quartz and coal dust. These types of dusts are composed of fineparticles of relatively simple, dense structure as illustrated by a group of iso-aerodynamic coal fineparticles shown in Figure 7.8(a). The difference between the aerodynamic diameter and the physical diameter for such simple dusts is usually a relatively simple function of the density of the dusts.

It will be noted, however, that the apparently larger fragments of coal present in the iso-aerodynamic group begin to show convoluted structures that could be describable by a fractal dimension when scrutinized over a short range of inspection magnitudes. It will also be noted, that in general, the coal fineparticles are larger than their equivalent aerodynamic diameter, even though they are supposedly composed of material of density greater than 1 (1.8 to 2.0 is a representative range of specific gravity for large lumps of good quality coal). This probably is explainable by the fact that coal is a porous substance and the variation in physical size of iso-aerodynamic coal fineparticles is probably a function of shape and the varying porosity of small fragments.

Table 7.1

n	Configuration	κ
2		1.12
3		1.27
3		1.16
4		1.32
5		1.25
4		1.17
5		1.45
5		1.30
5		1.19
6		1.57
6		1.43
6		1.17
7		1.67
8		1.73
8		1.56
8	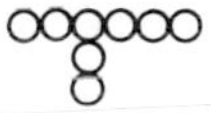	1.64

In occupational health studies of the health threat created by inhaled dust fineparticles, it is useful to consider three major structural parameters of the dust fineparticle. One of these, the aerodynamic size has already been defined. The magnitude of the aerodynamic size of a dust fineparticle governs its motion in the air during the act of breathing. Once the fineparticle has penetrated the respiratory system to the initial passages of the lung, the physical extent of the fineparticle, as well as its aerodynamic size, must be taken into account when considering the danger of **lodgability** and adhesion to the lungs. Other features of the dust structure are also important when considering the hazard posed by a dust fineparticle to the lung; thus, in the case of quartz dust, the presence of sharp edges on the dust is an important aspect of the dust structure. However, since in this book we are concerned with fractal structures, we shall concern ourselves mainly with the aspects of respirable hazards posed by dust fineparticles which can be linked to their fractal structure [21].

Shown in Figure 7.8(b) and (c), are two other groups of iso-aerodynamic dust fineparticles which have been studied by Kotrappa (these profiles were originally

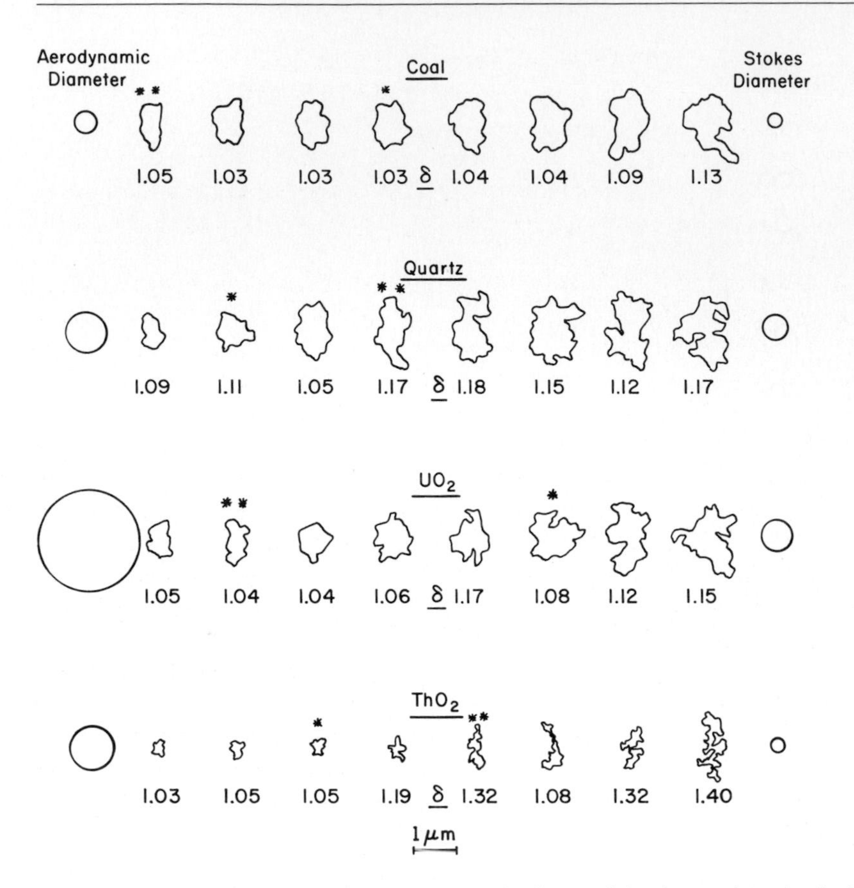

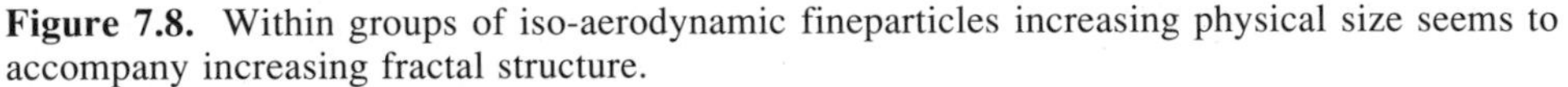

Figure 7.8. Within groups of iso-aerodynamic fineparticles increasing physical size seems to accompany increasing fractal structure.
(a) Iso-aerodynamic coal fragments [20].
(b) Iso-aerodynamic quartz fineparticles [20].
(c) Uranium dioxide iso-aerodynamic fineparticles [21].
(d) Thorium oxide iso-aerodynamic fineparticles [21].

presented and discussed briefly in Chapter 3. Both powders were produced by a precipitation process. They are not fractal agglomerates formed within a cloud, but residual fractal structures from the precipitation process by which they were made). Again, in general it can be seen that within a given iso-aerodynamic group, the larger the physical size of a fineparticle, the more pronounced is the fractal structure. When we consider the danger of an inhaled dust fragment lodging in the lung, it is obvious that the thorium dioxide dust grain indicated with an asterisk * has much less chance of becoming attached to the walls of the passages of the lung than the dust grain labeled with the double asterisk **, even though they have the same aerodynamic size.

The aerodynamic size magnitude given in Figure 7.8(c) is misleading if one is attempting to estimate the amount of thorium dioxide which would enter and be returned by the lung. If one of the physically larger iso-aerodynamic grains becomes attached to the lung the mass of thorium dioxide retained by the lung, would be grossly underestimated. It would seem that future occupational health studies should take into account the fractal structure of any deposited respirable dust of open agglomerated structure. It may be that useful parameters for helping to estimate the real hazard from deposited fractally structured dust could be the fractal dimension of the profile along with some shape factor, such as the aspect ratio describing the general shape of the agglomerate [21].

A third important property of an inhaled-deposited dust fineparticle is its surface area, since adsorbed chemicals carried into the lung by the dust grain are often the major health hazard of the dust. An estimate of the amount of adsorbed chemicals carried into the lung from the magnitude of the sphere of the same aerodynamic diameter will grossly underestimate the chemical burden of the dust if the fineparticles have fractal structure. Currently, in occupational health and hygiene, aerodynamic diameters of the inhaled dust are often the only physical parameters of the dust which are measured. This can sometimes leads to gross under-estimation of the health hazard of the dust [22]. The health hazard from fractally structured dusts becomes particularly important when looking at the health hazard from atomic reactor meltdown fumes and/or welding fume type hazards. In Chapter 3, some simulated atomic reactor meltdown fumes described by Zeller were considered. Their fractal structure was much higher than the dust fineparticles shown in Figure 7.8. Welding fumes,when viewed through the microscope, are obviously fractal structures. Thus, in Figure 7.9, a zinc oxide fume with a structure similar to that observed within welding fume fineparticles, which was photographed by Professor Bolsaitis of MIT, is shown [23]. It can be seen that assessing the health hazard from such a fineparticle, if one only measured its aerodynamic diameter, would lead to gross error.

One way in which we are attempting to characterize the hazard of such dusts associated with their surface area is to size and count the number of subsiduary

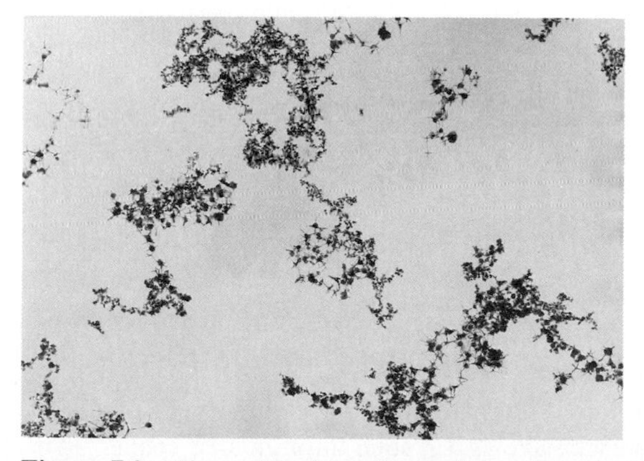

Figure 7.9. Zinc oxide fume fineparticle. The chemical – toxic health hazard from a fumed fineparticle is dependent upon the surface area within the agglomerate. One way that one can attempt to characterize the internal structure of a fumed fineparticle is to strip the agglomerate into its components, measure the area of the components and obtain a figure of surface area per unit of superficial area of agglomerate.

fineparticles in highly resolved photograph of an agglomerate. This can be achieved by eroding the contiguous units of the agglomerate system on a Dapple image analyzer. One can then count the number of spheres within the agglomerate or within a unit area of the agglomerate. One can then attempt to estimate the overall surface of the system by multiplying the superficial area of the agglomerate, as measured from a photograph, by the estimated area of the units of the agglomerate per unit of superficial area, estimated by inspecting a unit area of the profile.

Another important aspect of engineering activity in occupational health and hygiene is to study methods for removing dangerous dust from the air breathed by workers, by means of respirator filters. As already indicated earlier in this section, very fine dusts are often captured within a respirator by means of capture trees. Obviously, the probability of dust fineparticles such as those marked * in Figure 7.8 being captured is much lower than that for those marked **, because of their physical extent in space. Modelling the capture of fineparticles on a Nucleopore filter is obviously relatively easy. If, however, one were to be trying physically to filter dust such as the thorium dioxide dust in Figure 7.8(b), the aerodynamic diameter of the dust would indicate that one would have to have a filter size of 1 micron stopping power, whereas in fact the physical size of a dust fineparticle such as that marked * would require a Nucleopore filter as small as 0.5 μm in diameter. On the other hand, the fineparticle marked ** would be unable to pass through a filter with holes as large as 1.5 μm in size. Thus, by only measuring the aerodynamic diameter of fractally structured dusts, one can underestimate and/or overestimate the efficiency of the respirator that is needed to protect against a respirable dust hazard.

One of the more successful areas of occupational hygiene activity has been the success in filtering out diesel exhaust from clouds of soot produced by diesel engines. This is probably because of the fact that the size of the fractally structured diesel exhaust was underestimated by measuring the aerodynamic size of the dust, and the actual, much larger, physical size of the fineparticle meant that the diesel exhausts could be trapped by using a relatively inefficient coarse filter. One can only estimate the filtration needs when studying a potentially respirable hazard dust by fractionating the fineparticles into iso-aerodynamic groups, followed by characterization of the fractal structure and geometric shape factors of the iso-aerodynamic fineparticles by microscope examination.

Predicting the performance of a group of fineparticles, such as the thorium dioxide fineparticles in Figure 7.8(b) in a depth filter, is a very difficult problem. To study the physical capture within a filter system for such fineparticles undergoing a randomwalk through the filter, one could probably assume an equivalent profile governing lodgibility, as distinct from free space dynamics, equivalent to the Euclidean envelope of the profile.

In Figure 7.10 a simple model for creating tortuous caverns similar to those in a depth filter in a computer memory is illustrated. To create the cavern one takes several sets of random number tables and regards each page as a *Z* slice through *XYZ* space. Thus, to simulate a 60% void space, one converts 6 of the digits in the table into empty space. One can then regard the page to page contiguous areas of void space as defining the tortuous cavern through which aerosols and/or gas molecules undergo a randomwalk. In Figure 7.10(a), the several sections of a simulated throat in a depth filter are shown, and

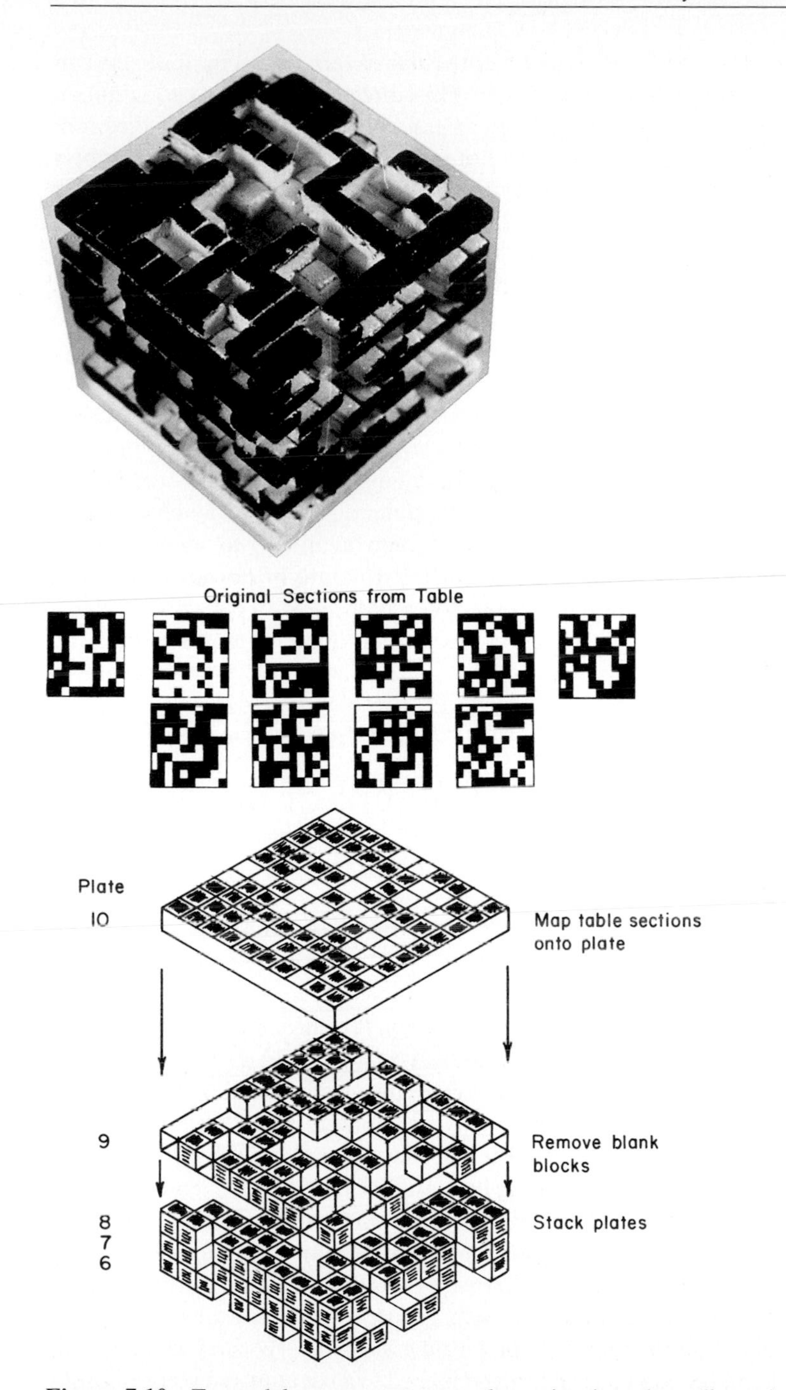

Figure 7.10. To model a tortuous cavern that exists in a three-dimensional porous body with contiguous paths through the body, one can regard random number tables as being XY slices through Z planes in three-dimensional space, after converting digits to holes according to a specified schedule.

it can be seen that the cross sectional area of a particular cavern varies in this short run model of a passageway through a depth filter. This model can be used to simulate randomwalk models of dust or gas fineparticles in fractal space. Systems as different as radon gas percolating through soil under a house and to the movement of bacteria through crushed rock in a microbiological mining process can be studied using such a three-dimensional walk in fractal space [24, 25]. All of these random walks in fractal space generate fractons. A model of the depth filter created by the simulation of a section shown in Figure 7.10(a) is shown in Figure 7.10(b).

Many problems involving reaction kinetics on solid surfaces and in solution involve the random arrival and leaving of atoms and molecules at rugged surfaces. For many of these problems, the random movement of the molecules and atoms takes place very quickly and a dominant feature of the system is the fractal structure of the surface and/or pore structure of the rugged system.

Recently, scientists have reported methods for studying the fractal structure of surfaces by computerized image analysis, and such techniques will probably grow in importance as more powerful image analyzers become available to technologists. Currently, however, the characterization of the fractal structure of porous bodies has been based upon two techniques which will be explored in the next two sections. These are surface area characterization by absorption studies and intrusion porosimetry.

7.3 Assessing the Fractal Structure of a Rough Surface from Adsorption Studies

For over 50 years, scientists have been measuring the surface area of a powder by studying the way in which molecules are adsorbed on the surface, either from a gas or from a liquid solution. Whole books have been written on this measurement technology [26]. In this brief discussion, we shall restrict ourselves to looking at a major reinterpretation of surface area measurements deduced from adsorption studies based on the concepts of fractal geometry. This new perspective on surface assessment by surface adsorption studies was pioneered by Avnir, Pfeifer and co-workers [27–42].

In a gas adsorption measurement of the surface area of the powder, the adsorption capacity of a powder at a low temperature over a series of pressures is studied. The data generated in such a study are known as the **adsorption isotherm**. For example, in 1956 I used this type of technique to measure the surface area of uranium dioxide powders by studying the adsorption-desorption of krypton gas on the surface of the uranium dioxide powder cooled to the temperature of liquid nitrogen. To carry out such an experiment, the powder to be characterized is first placed in a vacuum and all previously adsorbed gas is pumped away over a period of degassing. A controlled amount of the krypton gas is then allowed into the sample container which has been chilled to the temperature of liquid nitrogen. From the structure of the adsorption isotherm at low temperatures, the surface area of the powder can be deduced using one of several theories. In essence, the critical idea involved in the measuring of the surface area is that, as the pressure of the gas is increased, a situation is reached in which the surface of the powder is covered by a monolayer of the gas molecules. Prior to 1977, it was well known in the technology of surface area measurements by gas adsorption that the magnitude of the estimated surface area was related to the cross sectional area of the gas molecule used in the studies. Often scientists would use one of the so-called noble gases, xenon, argon, and krypton, in their adsorption studies.

The uncertainty of the measured surface area caused by the use of a particular gas in the adsorption studies was always accepted as one of the inevitable uncertainties of the method. However, soon after the publication of Mandelbrot's book on fractal geometry, Avnir and Pfeifer pointed out that what was happening was that the use of gas molecules to cover the surface of a rugged powder was rather like an extension of Minkowski's method for measuring the boundary of a rugged curve using circles. In gas adsorption studies, one was estimating the surface of a rugged system by counting the number of contiguous spheres needed to cover the surface. It is difficult to draw this system in three-dimensional space and, for illustration purposes, we can see what was happening in the gas adsorption studies by looking at the systems sketched in Figure 7.11. If one were using krypton in the adsorption studies, this would be equivalent to using a relatively large search circle to estimate the boundary of a rugged profile. Narrow fissures would be inaccessible to such relatively large molecules. If, however, the surface was explored using helium molecules, this much smaller molecule would now be able to penetrate into the narrower cracks of the surface so that a higher surface area estimate would be arrived at by multiplying the number of adsorbed molecules on the

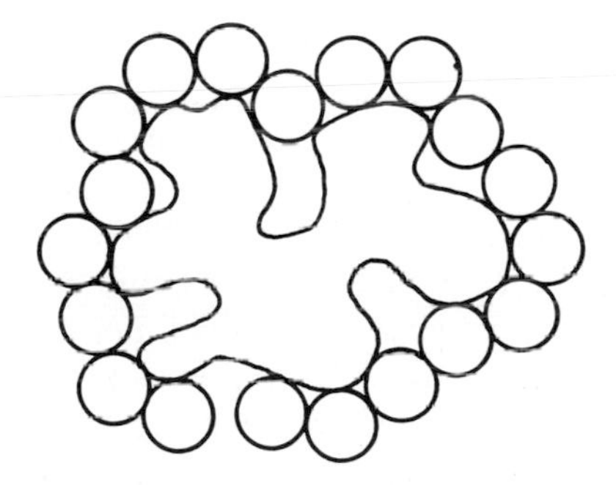

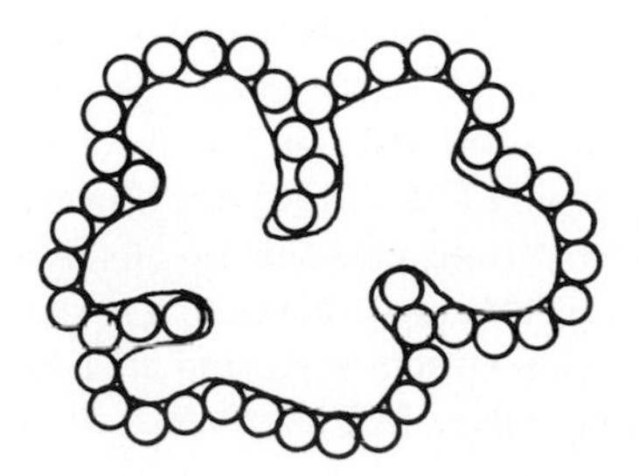

Figure 7.11. The surface area of a solid grain of powder estimated from gas adsorption studies depends upon the cross-sectional area of gas molecules used in the study. Thus, smaller gas molecules could access tiny cracks in the surface of the body, leading to higher surface area estimates as the size of the gas molecule adsorbed on the surface decreased.

surface of the powder grain. It follows that, if one were to be able to measure the surface area of a powder using a series of adsorbed molecules of different sizes, a Richardson plot of the logarithm of the estimated surface against the logarithm of the size of the gas molecule would generate a dataline related to the magnitude of the fractal dimension of the rugged surface.

Avnir, Pfeifer and co-workers have studied the published literature on gas adsorption studies of powders and pointed out that one can obtain a fractal dimension by this procedure for most powders described in the published literature on powder technology in which molecular adsorption was used to estimate surface area. They stated that non-fractal structured powders appear to be the exception rather than the rule in powder science. Suddenly, from the perspective of fractal geometry, the differences in measured surface areas due to the use of gas molecules of different sizes in adsorption studies do not represent uncertainty or error, but information on the ruggedness of the surface of a powder in terms of its fractal dimension (note: although we are discussing gas molecular adsorption in this section, the adsorption of dyes and such different substances as benzene, water and carbon tetrachloride on the surfaces of powder material is governed by the same physical processes and relationships, and the adsorption of such substances can be used to determine the fractal dimension of the surface).

In their publications, Avnir and Pfeifer have pointed out that there are two different experimental procedures which can be used to evaluate the fractal dimension of a rugged powder by adsorption studies. In one technique, one uses gases of different molecular size in a series of adsorption studies as discussed briefly above. In the other technique, the fractal dimension of a surface is deduced from the adsorptive capacity of a series of different sized fractions of the same powder. If one estimated the surface area of the powder which was composed of smooth spheres, then plotting a graph of measured surface area against the diameter of the spheres would lead to a dataline of slope –1. In general, if a sphere has a fractal surface of dimension δ, then a graph of the surface area against the size of the powder grains would lead to a slope of $\delta = 3$, where δ is the fractal dimension of the powder (for a smooth sphere $\delta = 2$, leading to the relationship that this equation would give a slope of –1 on the appropriate graph for such Euclidean bodies). A powder which has been used as a standard in many size characterization studies is Ottawa sand. This is a particularly smooth-grained, hard sand found in Ottawa, Illinois, USA (there is also an Ottawa in Canada; Ottawa is an Indian place name, and has been used to name several locations in North America). In Figure 7.12 a graph of the measured surface area of a series of different sized fractions of Ottawa sand is presented. This is compared with the measurement of the surface area of a quartz sand fractionated into several size groups using sieves. It can be seen that the surface area measurements on the Ottawa sand generate a dataline which would indicate a fractal dimension of 2.02, which is virtually a smooth Euclidean surface. On the other hand, the quartz powder obviously had a rough structure which can be characterized by a fractal dimension of 2.21. In Figure 7.13, in a graph presented by Avnir, Farin and Pfeifer [28], the relative roughness - as characterized by the fractal dimensions of various carbonate rocks as a function of place of origin- is shown. It is probably not too dramatic to claim that fractal characterization of the ruggedness of surfaces is likely to revolutionize several areas of powder technology (see the discussion of pulverizer efficiencies in Chapter 9).

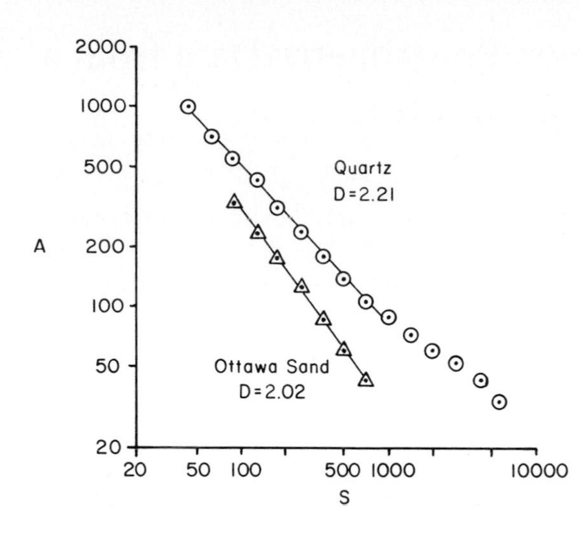

Figure 7.12. The fractal dimension for powder can be measured by fractionating the powder into different sized groups and measuring the surface area of each sized fraction.
A = the surface area in square centimeters per gram.
S = the logrithm of the diameter of the powder fractal in microns.
(Reproduced by permission of D. Avnir, The Hebrew University of Jerusalem, Israel.)

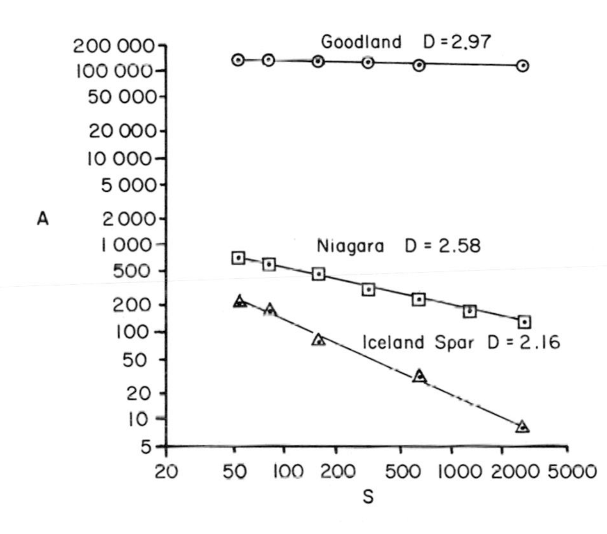

Figure 7.13. The relative ruggedness of rocks of different origin can now be described by means of their fractal dimension (reproduced by permission of D. Avnir, The Hebrew University of Jerusalem, Israel).

7.4 Interpreting Intrusion Porosimetry Data from a Fractal Geometry Perspective

One of the techniques that powder technologists use to study the structure of the pores in a body is known as **intrusion porosimetry** [26, 43]. The physical bases of the methodology can be appreciated from the simple systems shown in Figure 7.14. If a closed evacuated capillary tube is inverted in a reservoir of mercury to which external pressure can be applied, then one can show that if one increases the pressure above the surface of the mercury, the mercury is unable to enter the tube until one reaches the situation where

$$P = \frac{2\gamma\cos\theta}{R}$$

where P is the applied pressure, R is the radius of the tube, γ is the surface tension of the liquid and θ is the contact angle made by the mercury with the substance of the tube. This equation is known as the **Washburn equation** after the scientist who first deduced the equation from physical theory [26]. Many liquids can be used in intrusion porosimetry. Mercury is widely used as an intrusion liquid and this discussion of intrusion porosimetry will be limited to examples based on the use of mercury as the intrusive liquid.

It can be appreciated from Figure 7.14 that when a pressure is sufficient to enter a tube of diameter R_1, this tube will fill with mercury and the level of mercury in the

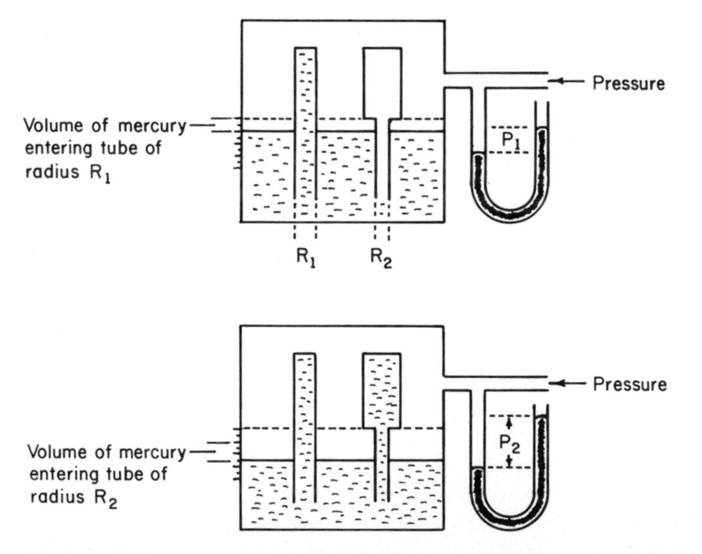

Figure 7.14. Mercury intrusion data have usually been interpreted generally as data on the size distribution of pores within a porous body. The concepts of fractal geometry can be used to reinterpret the data to generate a fractal dimension description of important aspects of the pore structure of a powder system.

reservoir will fall. This fall in the mercury level can be measured and used to calculate the volume of the tube of radius R_1. As the pressure is increased by a further amount, a pressure is reached at which the narrower tube, of diameter R_2, fitted with the bulbous reservoir, becomes flooded with mercury. Again, the changes in the mercury level, as the mercury is forced into narrower and narrower tubes, can be used to measure the interval volumes of a set of tubes. To investigate the pore structure of a porous body, a piece of the material, after its internal spaces have been evacuated of gas using a vacuum pump, is placed in a pool of mercury in a sealed evacuated container. Next, the pressure is increased above the surface of the mercury, forcing the mercury into narrower and narrower channels of the porous body. The volume of mercury entering the porous system after each increase in applied pressure is recorded. In the simplest model used to interpret this type of data, it is assumed that the pores in the porous body are circular channels of uniform radius and that, as the intrusion pressure is increased, the mercury volume measures the volume of holes of that size. However, as can be seen from the example of the tube of size R_2 in Figure 7.14, penetration through a narrow throat may lead to an extensive space behind the throat of much larger pore diameter. From this simple illustration it can be seen that this throat-volume interpretation of the data is obviously a very simplistic model for characterizing the pore structure of a porous body. It would be better to say that the technique measured access throat diameter rather than pore size. However, in the absence of better interpreting models, extensive studies have been made of various types of porous bodies and the data interpreted using this simple model are usually reported as 'pore size distributions.' In Figure 7.15(a), a mercury porosimetry data set for a relatively coarse powder with individual grains that were porous. The data were taken from a study carried out by Clyde Orr, Jr.[43].

It should be noticed in Figure 7.15(a) that the pressure applied to the mercury is given on the top abcissa and that this has been interpreted as an access throat pore diameter in microns, assuming appropriate values of θ and γ, on the bottom abcissa. If one replots the data of Figure 7.15(a) on a log-log scale the graph in Figure 7.15(b) is obtained. It has been suggested that this demonstrates that the invaded volume - access throat data for mercury porosimetry can be interpreted from the perspective of fractal geometry, since there are obviously two linear regions of the graph which are scaling functions [44, 45]. Thus, when presenting these data in a scientific publication, I suggested that one could now use two fractal dimensions, δ_1 to represent the property "invaded volume - access throat structure of the porous body from the perspective of the between grain voids" and that the within-grain voids were represented by the fractal dimension δ_2. By using the term fractal dimension of "invaded volume - access throat distribution" one avoids having to be specific about the physical significance of the data with respect to the pore size distribution of the beds. The deduced fractal dimension is characteristic of the pore structure - ruggedness of the overall body but is not descriptive of the pore structure itself. Friesen and Mikula have suggested that one can interpret the fractal dimensions for data such as those of Figure 7.15(b) by using an interpretive model based on Menger's sponges [46]. Recently it has been suggested that the Appolonian gasket (discussed in Chapter 10) could be a better model for arriving at an interpretive hypothesis for explaining the physical significance of the fractal dimension for porous body derived from intrusion porosimetry data [47]. Whatever the ultimate interpretive hypothesis attached to the scaling functions of mercury porosimetry data, such as those in Figure 7.15(b), it will involve the concepts of fractal geometry. In the next chapter we discuss an obvious area where the fractal dimension of a porous body may be linked to an important industrial technology - the recovery of oil from porous rocks.

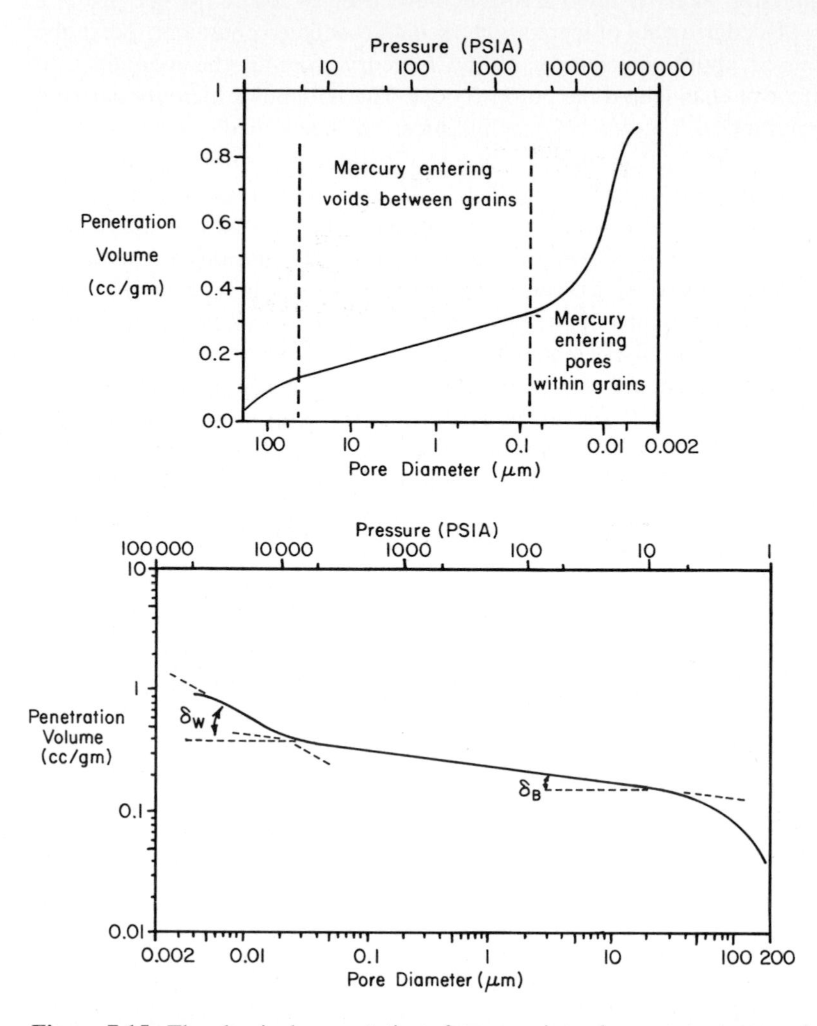

Figure 7.15. The classical presentation of mercury intrusion porous structure investigation hides the fact that the intruded volume versus the access throat magnitude can be a scaling function over a range of access throats. It has been suggested by Kaye that the slope of the scaling function is probably a fractal dimension characteristic of the porous structure of the porous system. δ_B is the fractal dimension of the between grain void space; δ_W is the fractal dimension of the "within grain porous structure."

References

[1] A good introduction to the classical chemical engineering approach to the mixing of powders is given by S.S. Weidenbaum, in M.E. Fayed and L. Otten (Eds.), "A Handbook of Powder Science and Technology," Van Nostrand Reinhold, New York, 1984, Ch. 8.

[2] B.H. Kaye and D.B. Sparrow, "Role of Surface Diffusion as a Mixing Mechanism in a Barrel Mixer, Part 1," *Ind. Chem.* 14 (1964) 200-205; "Part 2," *Ind. Chem.* 14 (1964) 246-250.

[3] G.G. Enstad, "Electrostatic Mixing of Powders," in "Mixing of Particulate Solids, Second European Symposium 1981," European Federation of Chemical Engineer-ing Publication Series, No. 15, Institute of Chemical Engineers, Rugby, 1981.

[4] B.H. Kaye, "Improvements in/or Relating to a Method and Apparatus for Handling Particles," Br. Pat., 38871-63, October 3, 1963.

[5] It should be noted that in the first edition of Mandelbrot's book (1977), Mandelbrot referred to the Menger sponge as a Sierpinski sponge. In the second edition of the book, he points out that it was brought to his attention that the system in Figure 7.5 was first described by Karl Menger (see L.M. Blumenthl and K. Menger, "Studies in Geometry," Freeman, San Francisco, 1970) and not Sierpinski.

[6] B.H. Kaye, G. Clarke and, A. LaRocque, "Simulating Powder Mixing Processes Using Randomwalk Models," paper presented at the Rosemont Conference on Powder Technology, Chicago, Illinois, 1988.

[7] B.H. Kaye and G. Clarke," Characterizing Powder Mixing Progress Using Tracer Agglomerates and Poisson Graph Paper," Rosemont Conference on Powder Technology, Chicago, Illinois, 1988, Powder Mixing Workshop Notes.

[8] A. Bunde, S. Havlin, R. Nossal, H.E. Stanley and G.H. Weiss, "Controlled Diffusion - Limited Drug Release From a Leaky Matrix," *J. Chem. Phys.*, 83 (1985) 5909-5913.

[9] Several books have been written on aerosol science. A good introductory text is P.C. Reist "Introduction to Aerosol Science," Macmillan, New York, 1984.

[10] A classical textbook on aerosol science, which is now out of print, but which contains a wealth of information, is H.L. Green and W.R. Lane, "Particulate Clouds, Dust Smokes and Mists," Van Nostrand, New York, 1957.

[11] The preprint of the scientific paper sent to me in 1980 was eventually published under a slightly different title: O. Preining, "On the Use of Optical Single Optical Counters to Acquire Spatial Inhomogeneities of Particulate Clouds," Atmospheric Sciences Research Centre, State University of New York at Albany, Publication No. 813. See also O. Preining and G. Reischl, "Aerosol Description and Descriptives," *J. Hung. MET Service*, 86/04 (1982).

[12] O. Preining, "Aerososl Characterization - A System Analytical Approach; Aerosols of the Atmosphere," by B.Y.H. Liu, D.Y.H. Pui and H.J. Fissan. (Eds.), in "Aerosols," Elsevier, Amsterdam, 1984, pp.253-256.

[13] For an introduction to the concepts and mathematical operations involved in computerized tomography, see R. Gordon, G.T. Herman and S.A. Johnson "Image Reconstruction From Projections," *Sci. Am.* October (1975) 56-68.

[14] H. Henderson, "A Sideways Look at Scanners," *New Sci.*, December 6 (1979) 782-785.

[15] H.O. Luck, B. Siemund and G. Lorbeer, "The Measurement of Spatial Aerosol Distributions in Enclosures by Means of Computed Tomography," *Part. Charact.*, 2 (1985) 137-142.

[16] For a discussion of the technical details of this group of instruments and the various commercial sources of the instrument see a discussion of this type of instrument in B.H. Kaye "Direct Characterization of Fineparticles," Wiley, New York, Chichester, 1981.

[17] B.H. Kaye, "Characterizing the Fractal Structure of Aerosol Clouds," in preparation.

[18] W. Stöber and H. Flachsbart, *Environ. Sci. Technol.*, 3 (1969) 1280.

[19] The Stöber Spiral Duct Aerosol Centrifuge is available commercially from Heraeus Christ GmbH P.O. Box 1220, D3360 Osterode Harz FRG.

[20] P. Kotrappa, "Shape Factors for Aerosols of Coal, Uranium Dioxide in the Respirable Size Range." in T. Mercer. E. Morrow and w. Stöber (Eds.) "Assessment of Airborne Particles," Charles C. Thomas, Springfield, Illinois, 1973, Ch. 16.

[21] B.H. Kaye, "The Physical Significance of the Fractal Structure of Some Respirable Dusts," in preparation.

[22] This aspect of the health hazard posed by fractally structured dusts is discussed in the essay B.H. Kaye, "What is the Size of an Inhaled Fineparticle," in "Contemporary Issues in Occupational Health and Safety" lecture notes for a short course, Laurentian University, Spring 1988 (a copy of the essay is available from the author).

[23] P.P. Bolsaitis, J.F. McCarthy, G. Mohiuddin and J.F. Elliott, "Formation of Metal Oxide Aerosols for Conditions of High Supersaturation," *Aerosol Sci. Technol.*, 6. (1987) 225-246.

[24] B.H. Kaye, "Fineparticle Characterization Aspects of Predictions Affecting the Efficiency of Microbiological Mining Techniques," *Powder Technol.*, 50 (1987) 177-191.

[25] B.H. Kaye, R. Trottier and T. P. Lim, "A Randomwalk Model for Radon Decay and Diffusion in Fractal Space Leading to a Fracton Measure of Diffusion," presented at Fineparticle Society Annual Meeting, Boston, August 4-7, 1987.

[26] A recent exposition of the theory of gas adsorption and mercury porisimetry for studying the surface of a pore structure of rough solids is the book Y.S. Lowel and J.E. Shields, "Powder Surface Area and Porosity," Chapman and Hall, London, New York, 2nd ed., 1984.

[27] P. Pfeifer, D. Farin and D. Avnir, "Chemistry in Noninterger Dimensions Between Two and Three. Part 1. Fractal Theory of Heterogeneous Surfaces," *J. Chem. Phys.* 79 (1983) 3558-3565.

[28] D. Avnir, D. Farin and P. Pfeifer. "Chemistry in Noninteger Dimensions Between Two and Three. Part 2. Fractal Surfaces of Adsorbance," *J. Chem. Phys,* 79 (1983) 3566-3571.

[29] D. Avnir and P. Pfeifer, "Fractal Dimension in Chemistry an Intensive Charac-teristic of Surface Irregularity," *Nouv. J. Chim.*, 7 (1983) 71-72.

[30] D. Avnir, D. Farin and P. Pfeifer, "New Developments in the Application of Fractal Theory to Surface Geometric Irregularity," in "Symposium on Surface Science," P. Braun G., Betz, W. Husinsky, E. Sollner, H. Stori and P. Varga (Eds.), Technical University of Vienna, Vienna, (1983) 233-236.

[31] P. Pfeifer, D. Avnir and D. Farin, "Ideally Irregular Surfaces of Dimension Greater Than Two, in Theory and Practice," *Surf. Sci.* 126 (1983) 569-572.

[32] D. Avnir, D. Farin and P. Pfeifer. "Surface Geometric Irregularity of Particulate Materials. A Fractal Approach," *Journal Colloid Interface Sci.*, 103 (1985) 1112-1123.

[33] S. Peleg, J. Naor, R. Hartley and D. Avnir, "Multiple Resolution Texture Analysis and Classification," *IEEE Trans. Pattern Anal. Machine Intelli.,* PAMI-6, No. 4, (1984) 518-523.

[34] D. Avnir, D. Farin and P. Pfeifer. " Molecular Fractal Surfaces," *Nature* (London), (1984) 261-263.

[35] P. Pfeifer, D. Avnir and D. Farin. "Scaling Behaviour of Surface Irregularity in the Molecular Domain: from Adsorption Studies to Fractal Catalysts," *J. Stat. Phys.* 36 (1984) 699-716.

[36] D. Farin, S. Peleg, D. Yavin and D. Avnir, "Applications and Limitations of Boundary Line Fractal Analysis of Irregular Surfaces: Proteins, Aggregates, and Porous Materials," *Langmuir*, 1 (4), (1985) 399-407.

[37] A.Y. Meyer, D. Farin and D. Avnir, "Cross Sectional Areas of Alkanoic Acids: a Comparative Study Applying Fractal Theory of Adsorption and Considerations of Molecular Shape," *J. Amer. Chem. Soc.*, 108 (1986) 7897-7905.

[38] D. Rojanski, D. Huppert, H. Bale, X. Dacai, P.W. Schmidt, D. Farin, A. Sori-Levy and D. Avnir, "Integrated Fractal Analysis of Silica: Adsorption, Electronic Energy Transfer, and Small Angle X-ray Scattering, *Phys. Rev. Lett.*, 56 (1986) 2505-2508.

[39] D. Avnir, D. Farin and P. Pfeifer, "Fractal Dimensions of Surfaces. The Use of Adsorption Data for the Quantitative Evaluation of Geometric Irregularity." Preprint provided by D. Avnir, Institute of Chemistry, Hebrew University of Jerusalum, Jerusalum, Israel.

[40] D. Avnir, "Fractal Aspects of Surface Science - An Interim Report," in C.J. Brinker (Ed.), "Better Ceramics Through Chemistry," Materials Research Society, 1986.

[41] P.Pfeifer, "Fractal Dimensions as Working Tool for Surface Roughness Problems," *Appl. Surf. Sci.* 1984.

[42] M. Silverberg, D. Farin, A. Ben-Shaul and D. Avnir, "Chemically Active Fractals. Part 1. The Dissolution of Fractal Objects," in R. Engleman and Z. Jaeger (Eds.), "Fragmentation Form and Flow in Fractured Media," Proceedings of conference held at Neve Ilan, Israel, 6-9th January, 1986, *Ann. Is. Phys. Soc.*, 8 (1986) 451-457.

[43] C. Orr, "Application of Mercury Penetration in Material Analysis," *Powder Technol.*, 3 (1969-70) 117-123.

[44] B.H. Kaye, "Fractal Geometry and the Characterization of Rock Fragments," in R. Engleman and Z. Jaeger (Eds.), "Fragmentation Form and Flow in Fractured Media," Proceedings of conference held at Neve Ilan, Israel, 6-9th January, 1986, *Ann. Is. Phys. Soc.*, 8 (1986) 490-516.

[45] S.H. Ng, C. Fairbridge and B.H. Kaye "Fractal Description of the Surface Structure of Coke Particles," *Langmuir*, 3 (3), May-June (1987) 340-345.

[46] W. Freisen and R.J. Mikula, Canmet Divisional Report, ERP-CRL 86-128. Avail-able from Energy Mines and Resources Canada, Canmet Technology Information Division, Technical Enquiries, Ottawa, Ontario K1A 0G1, Canada.

[47] B. H. Kaye, "The Appolonian Gasket as a Model for Evaluating Mercury Intrusion Porosimetry Data," paper presented at the Rosemont Powder Technology Con-ference, Chicago, Illinois, 1988.

8 Fractal Fingers and Floods

8.1 Fractal Fingers

If one were to take a look at the three profiles in Figure 8.1(a), one could be forgiven for thinking that the first one was a modern impressionist painting of a cockerel floating on an insubstantial fence. To some readers it could represent a ghoulish spirit rising from the morass of a bog. In fact, it is nothing so exotic. It is the pattern made by chocolate syrup draining in a Hele-Shaw cell. The diagram was generated in an experiment carried out by Walker [1]. The "bloody fingers" of Figure 8.1(a), draining from the interface of the syrup and air are advance streamers of fluid and the Picasso type shape is draining air being engulfed by the racing syrup.

The Hele-Shaw cell was invented by an English engineer in 1898 to facilitate the study of the physical properties of flowing liquids. In order to be able to examine them in a relatively simple flow system, he made a flow cell composed of two parallel plates separated by a narrow gap. He used it to study the flow around objects after making the motion visible by injecting dye to produce coloured streamlines in the moving fluid [2].

Figure 8.1. Fantastic fractal fingers can be created in a Hele-Shaw cell by liquid boundaries (diagrams based on photographs published by Walker [1]).
(a) Chocolate syrup draining from a Hele-Shaw cell.
(b) Air displacing corn syrup.
(c) Dyed water in cellulose.
(Reproduced by permission of Dr. J. Walker.)

For over 50 years after its invention, the Hele-Shaw cell constituted a convenient, but little used device for observing fluid flow situations. However, it became the centre renewed interest when a British physicist, Sir Geoffrey Taylor, started to use the Hele-Shaw cell to study an important problem of interest to the oil industry [3]. Deposits of oil are found underground in porous rocks such as limestones and sandstones. Initially, when an oil well is first drilled, there is usually pressurized natural gas above the oil which forces the oil up through the well [4]. When the oil stops flowing from the well there is usually a great deal of oil left in the porous rock, which can constitute between 30 and 60% of the total oil in the oil-field. In the early years of the oil industry, engineers were able to find a sufficient number of new oil wells each year from which oil flowed readily that they had little interest in recovering the oil left in the porous rock in an exhausted (no more natural pressure) oilwell. However, in recent years the oil industry has become increasingly interested in what are known as secondary methods for recovering oil. In one method of secondary oil recovery, gas or water is injected into the old oil well and the oil is driven out of the rock and over to another drill hole with pressure being applied to the fluid being injected down the driving well. Carbon dioxide under pressure has been used in this type of technology and water flooding has also been used.

Sir Geoffrey Taylor recognized that one could study the way in which a gas or another liquid would drive out a liquid in a narrow space by using the Hele-Shaw cell. In Figure 8.2 some of the interesting results obtained in pioneer experiments into the problems of

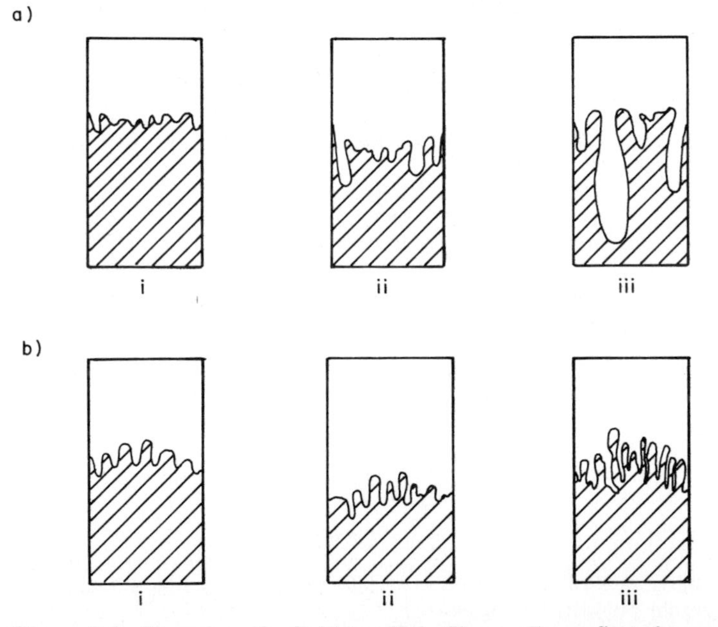

Figure 8.2. Fingering of a fluid in a Hele-Shaw cell was first observed by Saffman and Taylor [3].

(a) Air displacing glycerol (flow rate 0.1 cm/s) at various times after flow initiation.

(b) Water-glycerol solution (shaded) displacing oil [Chuoke et. al, see Ref. 5].

secondary oil recovery using the Hele-Shaw cell are summarized [3, 5]. In the sequence of diagrams shown in Figure 8.2(a), the Hele Shaw cell has been placed in a vertical position. Air is displacing glycerol from the top downwards at a flow rate of 0.1 cm/s. It can be seen that in this type of flow, long fingers of fluid develop. This is bad news for the oil engineer, since it means that the fingers groping into the rock ahead of the moving fluid face being used to drive the oil out will tend to intertwine, leaving pools of oil trapped forever in the rock as the driving fluid sweeps on towards the recovery well [6, 7]. The physical theory underlying the production of liquid fingers in such a moving interface has been studied extensively. It has been shown that the fingering depends upon the ratio of the viscosities of the driven and driving fluids [8] and also on the relative surface tension of the interface between the two fluids. Oil engineers are obviously interested in suppressing the fingers by modifying the driving fluid used in the secondary recovery of oil [5, 6, 7].

In the mid-1980s, it was discovered that if one modified the Hele-Shaw cell to have a central injection point with the fluid moving out from the injection point, fractally structured interfaces were created. Thus, the draining chocolate syrup in Figure 8.1(a) was photographed by Walker in a home-made Hele-Shaw cell (the complete descriptions of the construction of both linear flow and radial ejection versions of the Hele-Shaw cell are given in Walker's paper [1]). The system in Figure 8.1(b) is the embryonic fractal pattern created by fingering generated when air is injected into dark corn syrup in a radial Hele-Shaw cell. The pattern in Figure 8.1(c) was obtained by injecting water, dyed with methyl violet, into a solution of cellulose.

The publication of patterns similar to those generated in Figure 8.1(b) and (c) obtained using radial Hele-Shaw cells has created great interest in generating this type of interface from the points of view both of improving technology for secondary oil recovery and of studying applications of fractal geometry in the natural world [8, 9]. Thus, in Figure 8.3, three patterns generated by Daccord, Nittman and Stanley are shown [8].

The fractal pattern shown in Figure 8.3(a) was created by injecting water into a cell filled with a solution of scleroglucan (a polysaccharide) at a flow rate of 20 mm/min.

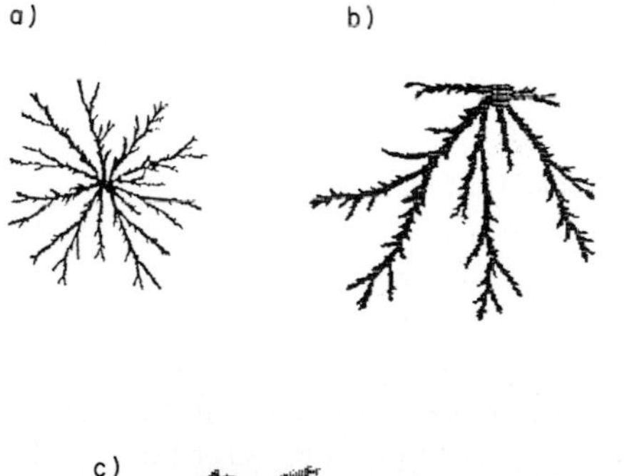

Figure 8.3. Fractal fingering observed in a Hele-Shaw cell by Daccord et-al [8].
(a) Water injected into scleroglucan.
(b) Same two fluids as for (a) with the water injected at the top edge of the Hele-Shaw cell.
(c) Water driven into a solution of guar gum.
(from H.E. Stanley and N. Ostrowsky, (Editors), "On Growth and Form: Fractal and Non-Fractal Patterns In Physics." Martinus Nijhoff Publishers, Dordrecht, Holland, (a) pg. 205, (b) p. 206, (c) pg. 208. "Copyright © 1985 Martinus Nijhoff Publishers, Dordrecht, Holland". Reproduced by permission of Martinus Nijhoff Publishers and H.E. Stanley.)

It was determined experimentally that the fractal dimension of this water spider invading the scleroglucan solution was 1.7. The fractal pattern in Figure 8.3(b) was observed for the same two fluids under different flow conditions when the liquid was injected at a point on the edge of the Hele-Shaw cell. This fractal pattern of invasion had a fractal dimension of 1.5. The pattern in Figure 8.3(c) was generated when water was driven into a solution of guar gum. The interesting aspect of this pattern is that the fingers have closed in on themselves to trap pools of fluid between the fractal fingers. This type of fractal fingering of one fluid penetrating another is receiving intensive study, and is likely to generate a great deal of fractal literature since the fractal dimension of the pattern produced by invading fluid seems to be related in a significant manner to the physical properties of the two liquids constituting the driven and driving fluid.

8.2 Fractal Floods and Fronts in Porous Media

My first encounter with the fact that fractal geometry was important to scientists studying the secondary oil recovery problem occurred one day when I was looking through a copy of the scientific journal *Powder Technology*. As I turned the pages I chanced upon a paper by Lincoln Paterson, in which fractals seemed to be growing on every page [10, 11]. I started to read the paper expecting to find a discussion of fractal geometry, only to find that the patterns in the pictures, two of which are shown in Figure 8.4(a), were described as “viscous fingers.” There was no mention of fractal geometry. (however, since that early paper, Dr. Paterson has since published several papers on the fractal structure of viscous fingers [12]). Since Dr. Paterson was interested in the secondary oil recovery program, he had filled his Hele-Shaw cell with plastic cylinders and had watched the movement of a fluid boundary created as he injected one fluid to drive out the other. The front of the moving fluid was a fractal boundary and we measured the fractal dimension of the moving front in the pairs of pictures presented by Dr. Paterson. I was pleased to see that as the front between any two liquids moved, the fractal dimension of the front remained independent of time [13, 14]. Two of my experimental measurements are shown in Figure 8.4(b). At the time that I published my results I was not aware of the fact that fractal fingering of a fluid penetrating into another, of the type discussed in Section 8.1, could occur. However, because of my work on the measurement of the surface area of powders using the permeability method and the problems of interpreting mercury porosity intrusion data (see the discussion of this latter method in Section 7.4) I was aware of the problems of driving a fluid through a porous body [15–17]. Therefore, as I looked at the pattern occurring in the system studied by Dr. Paterson, I had a visual image of liquid attempting to enter the cavities in the porous assembly created by the packed cylinders. I interpreted the physical significance of the fractal fingers as being a record of the most probable paths favouring the fluids movement through the interconnecting random pores. I suggested that it could

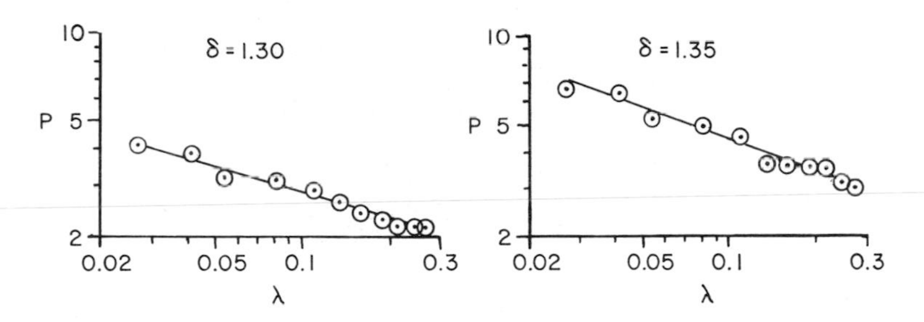

Figure 8.4. Fractal fluid fronts photographed using water injected into a Hele-Shaw cell filled with plastic cylinder and paraffin oil (photographs provided by, and used with the permission, of L. Paterson [10]).
(a) Photograph of fronts after a time interval of 1000 seconds.
(b) Richardson plots for the two fractal fronts showing that the fractal dimension of the front is essentially unchanged after 1000 s of flow [13, 14].
(From L. Paterson, "Dispersion and Fingering in Miscible and Immiscible Fluids Within a Porous Medium," Powder Technol., 36, (1983), 71-78.)

be regarded as a randomwalk involving jets of fluid finding most probable paths. The appearance of the fingers of fluid in the porous media is so similar to that of the fingers produced by the fractal intrusion of one fluid into another reported by Daccord and co-workers and other investigators that it is easy to confuse the two phenomena. They look alike but are not identical phenomena. Feder and co-workers have brought out the difference between the two similar phenomena; in a recent publication [5]. They point out that one has to be very careful with the term "permeability." In the original Hele-Shaw cell the term permeability refers to the resistance to flow of the whole cell, which is rather like a rectangular pipe. When one is looking at the flow of a liquid through packed beads or other porous systems, the permeability is now the resistance to flow of local openings in the packed porous media, that is, the pores of the porous media. In their paper, Feder and co-workers make the statement that

"the validity of using the Hele-Shaw cell for a model of flow in porous media is questionable".

One of the problems encountered when packing the Hele-Shaw cell with granular material is the fact that it is very difficult to achieve the random packing of fineparticles, whether they be spheres or irregularly shaped material, in a two dimensional plane without encountering difficult wall conditions. Thus, if one attempts to pack spheres into a cylinder the presence of the rigid wall tends to impose a structure on the random packing of the spheres. This usually creates a channel down the side of the containing vessel which has a lower fluid resistance path for the invading fluid than those available for flow through the bulk structure of the porous media. The presence of this wall channel can be seen clearly in the two photographs in Figure 8.4(a) of the Hele-Shaw cell experiment described by Paterson. The liquid is moving relatively fast along the side walls with fern like growth into the porous media from this "side-stepping" fluid flow.

The other problem one encounters when trying to create random packing in two dimensions is that the same type of problem occurs with the space between the top and bottom surface of the cell and the fineparticles packed into the space. This appears to have been minimized in the work reported by Paterson.

If one studies the flow of fluid through packed beads in a Hele-Shaw cell, it should be realized that the preferred paths through the body may have nothing to do with the type of fractal fingering that occurs in the Hele-Shaw cell when one liquid is injected into another as described in Section 1 of this chapter. The fingering may be caused by

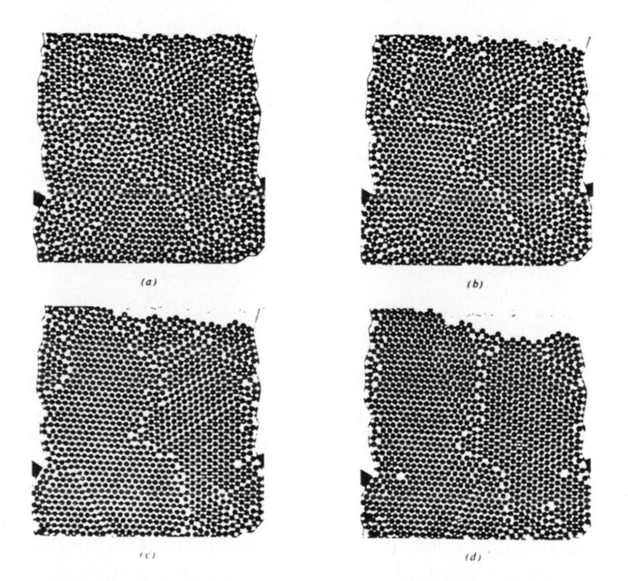

Figure 8.5. Consolidaton of a powder bed by vibration or applied pressure can result in a lower voidage (porosity), but with increased flow because of channels created between adjacent regions of tightly packed fineparticles. Wall effects can also create preferred wall flow channels. These two effects are demonstrated by the vibrated two-dimensional arrays for monosized spheres studied by Nowick and Mader [21].
(a) As assembled array.
(b) 5 s of vibration.
(c) 15 s of vibration.
(d) 30 s of vibration.
(Copyright 1965 by International Business Machines Corporation reprinted with permission.)

dislocations in the packing of the material which will create preferred paths along which the liquid will flow. Thus, if we look at the two-dimensional packings created by Nowick and Mader, as discussed in Chapter 5 (see Figure 5.21), the preferred paths can be seen clearly and they change if the array is subject to vibration (simulated annealing). In Figure 8.5 one of the Nowick and Mader two-dimensional arrays is shown, and the effect of the regular packing near the confining walls can be clearly seen in that there are runs of aligned spheres interspersed with gaps as illustrated on the left-hand side. It should be recalled that the Nowick and Mader assembly table had irregular walls to try to prevent the imposition of regular packings on the two-dimensional array by the side of the container but they obviously were not able to eliminate the effect entirely. The fact that dislocations between regular packings in a general assembly of disks may contain fractal paths has recently been demonstrated by Onoda and Toner [18]. In Figure 8.6 some fractal patterns obtained by injecting liquid into a two dimensional porous medium record by Maløy, Feder and Jossang [5] is shown. The growing fingered fractal pattern represents the paths being sought out by liquid epoxy in a two-dimensional porous medium consisting of 1.6 mm glass spheres in a monolayer between two glass plates 1 mm in diameter. These fractal fingers had a fractal dimension of 1.62. Maløy and co-workers also demonstrated that if the packing array of the monosized spheres is very regular, the fractal fingering occurring when the liquid is injected into the array has the symmetry of a snowflake rather than the random patterns in Figure 8.6 [5].

It is likely that the patterns in Figure 8.6 can be influenced by any slight deviation from monosize in the set of spheres and in the assembly technique used to create the array. Vibration of the system would also be important. We are currently planning experiments at Laurentian University to create a Hele-Shaw cell based upon the random assembly table of Nowick and Mader. The side walls of their assembly table will be

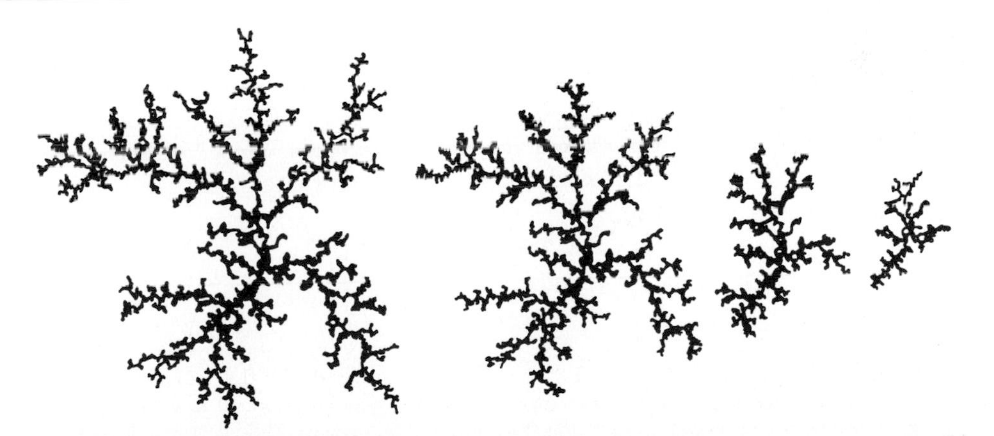

Figure 8.6. Epoxy fingers penetrating a two dimensional array of glass beads in a Hele-Shaw cell obtained by Maløy et-al. [5] (from J. Feder, T. Jossang, K.J. Maloy and U. Oxaal, "Models of Viscous Fingering" in R. Englman and Z. Jaeger (Eds). Proceedings of Conference "Fragmentation, Form and Flow," held Neve Ilan, Israel, January 6-9, 1986; *Ann. Isr. Phys. Soc.*, 8 (1986)).

replaced by soft rubber walls, which can be hydrostatically driven in sideways to create a tight wall around the edge of the array, and also to subject the packing in the new type of two-dimensional Hele-Shaw cell to different amounts of compression and vibration.

Kaye and co-workers studied fractal fronts in sandstone [19]. In their experiments, a thin section of sandstone was prepared and a central well drilled in the top surface of the sandstone. The system is shown in Figure 8.7(a). The dye solution was then fed into the centre of the sandstone and the moving dye solution created a fractal front which is probably related to the pore structure of the sandstone. The fractal dimension of the front created by the moving dye in the sandstone is shown in Figure 8.7(b).

When discussing the fractal fingering patterns obtained by injecting epoxy into a bed of monosized spheres, Feder pointed out that in the fingering studied in normal Hele-Shaw cell the fingering is a function of the spacing between the plates of the cell, whereas in the porous medium the length scale of the fingering is always set by the pore size. He went on to state that the randomness in the pore structure is a requirement for fractal fingering and that the action of displacing a fluid from a given pore by the driving fluid at the moving interface is made not on the basis of an absolute value of the pressure difference between the air and the fluid alone, but rather is due to the value of the pressure relative to the capillary pressure associated with the pore neck leading to that pore.

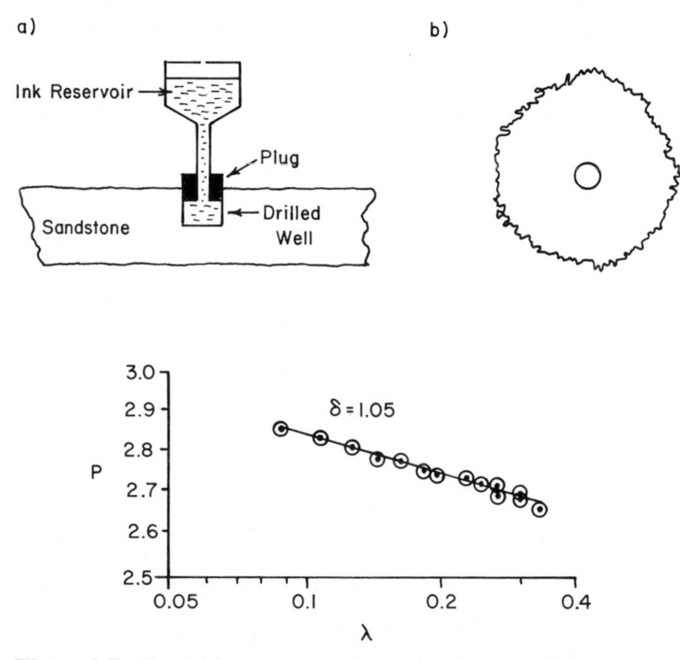

Figure 8.7. Fractal fronts created in sandstone by elution of an invasion fluid carries information on the pore structure of the rock and hence on the dynamics of secondary oil recovery by liquid flooding [19].

(a) Experimental system for eluting fluid in a sandstone specimen.

(b) Fractal front created by eluted fluid.

(c) Richardson plot for the evaluation of the fractal front in (b).

The movement of liquid through a packed porous bed has been described by Wilkinson and Wilemson as invasion percolation [20]. Feder pointed out that it is important to stress that in invasion percolation (the type of fractal fingering illustrated in Figure 8.6) the decision to enter any given pore is a local decision at the pore level, and the viscosity of the fluids plays no role in the invasion probability. In other words, the penetration of the moving fluid front into a given pore is a function of the intrusion pressure needed to pass through the neck and is related to that "pressure of penetration" discussed when looking at mercury intrusion at the close of Chapter 7. It therefore seems that the pore size distribution as determined by mercury intrusion may be directly related to the fractal dimension of fractal fingers created by an invading or driving fluid moving through a porous medium. Although the parallelism may not be immediately obvious to those not expert in the study of porous media, Feder's description of the movement of a liquid front as a series of local decisions to enter individual pores is essentially the same as the model that I suggested in my first discussion of Paterson's work in which I envisaged the front to be moving by a series of randomwalk jets by fluid finding passages accessible to the moving fluid under encounter conditions involving surface tension, radius of access and applied pressure. One should therefore anticipate that the fractal dimension of the reinterpreted mercury intrusion data will be related both to the fractal dimension of the Sierpinski carpet revealed in a section through the porous body and to the fractal fingering achieved when a fluid is injected into the porous media to drive out another fluid. Invasion percolation and the study of viscous fractal fingering represent one of the more rapidly growing areas of applied fractal geometry (21-31).

References

[1] J. Walker. "Fluid interfaces, Including Fractal Flows, can be Studied in a Hele-Shaw Cell," Amateur Scientist section, *Sci. Am.*, November, (1987) 134-138.

[2] H.S. Hele-Shaw, *Nature* (London), 58, (1898) 34.

[3] P.J. Saffman and G.I. Taylor, *Proc R. Soc.* London 245,(1958) 312.

[4] For an introduction to the basic processes which are believed to have caused the presence of oil in porous rock, see C.E. Zobell, "The Origin of Oil," in R. Colburn (Ed.), "Modern Science and Technology," Van Nostrand, New York, (1965) 595-602.

[5] J. Feder, T. Jossang, K.J. Maløy and U. Oxaal. "Models of Viscous Fingering," in R. Engleman and Z. Jaeger (Eds)., "Fragmentation, Form and Flow in Fractured Media," Proceedings of conference held Neve Ilan, Israel, 6-9th, January, 1986, *Ann. Isr. Phys. Soc.*, 8 (1986)

[6] See the diagrams and brief discussions dealing with secondary oil recovery in E.E. David, Jr., "The Federal Support of Mathematics," *Sci. Am.* 252 (1985).

[7] K.J. Maløy, J. Feder and T. Jossang, *Phys. Rev. Lett.* 55,(1985) 2688.

[8] G. Daccord, J. Nittman and H.E. Stanley, "Fractal Viscous Fingers-Experimental Results" in H.E. Stanley and N. Ostrowsky (Eds.), "On Growth and Form," Martinus Nijhoff, Boston, 1986.

[9] A.L. Robinson, "Fractal Fingers in Viscous Fluids," *Science*, 228 (1985) 1077-2080.

[10] L. Paterson, "Dispersion and Fingering in Miscible and Immiscible Fluids Within a Porous Medium," *Powder Technol.*, 36 (1983) 71-78.

[11] L. Paterson, V. Hornoff and G. Neale, "A Consolidated Porous Medium for the Visualization of Unstable Displacements," *Powder Technol.*, 33 (1982) 265-268.

[12] L. Paterson, "Fingering with Miscible Fluids in a Hele-Shaw Cell," *Phys. Fluids,* 28 (1985) 26-30.

[13] B.H. Kaye, "Application of Recent Advances in Fineparticle Characterization to Mineral Processing," *Part. Charact.* 2(1985) 91-97.

[14] B.H. Kaye, "Fractal Description of Fineparticle Systems in N-dimensional space," Preprints, 3rd European Symposium on Particle Characterization, Nuremberg, May 9-11, pp. 131-148.

[15] B.H. Kaye, "Permeability Techniques for Characterizing Fine Powders," *Powder Technol.*, 1(1967) 11-22.

[16] B.H. Kaye and P.E. Legault, "Real Time Permeametry for the Monitoring of Fineparticle Systems," *Powder Technol.*, 23 (1979) 179-186.

[17] B.H. Kaye, "Fractal Geometry and the Characterization of Rock Fragments," in R. Engleman and Z. Jaeger (Eds.), "Fragmentation, Form and Flow in Fractured Media," Proceedings of conference held at Neve Ilan, Israel, 6-9th January, 1986, *Ann. Isr. Phys. Soc.*, 8 (1986) 490-516.

[18] J.Y. Onada and J. Toner, "Deterministic Fractal Defect Structures in Close Backings of Hard Disks".

[19] B.H. Kaye, G.G. Clarke, A.E. Beswick and R.A. Trottier, "Examination of Eluted Fractal Fronts in Porous Bodies as a Means of Characterizing the Pore Structure and Permeability of the System," in "Proceedings of the Conference on Fractal Geometry in Material Science held in Boston, December 4, 1986," Materials Research Society, Pittsburgh, 1987.

[20] D. Wilkinson and J.F. Willemson, *J. Phys. A, Math. Gen.*, 16 (1983) 33-65.

[21] A.S. Nowick and S.R. Mader, "The Hard Sphere Model To Simulate Alloy Thin Films," *IBM J.*, September, November (1965), 358-374.

[22] R. Lenormand and C. Zarcone, "Role of Roughness and Edges During Imbibition in Square Capillaries,"Society of Petroleum Engineers of AIME, Society of Petroleum Engineers Reprint 13264, paper presented at the Annual Technical Confer-ence and Exhibition of the Society of Petroleum Engineers of AIME, Houston, Texas, September 16-19, 1984. This paper contains some pictures of fractal fronts and a simulated section of growing clusters which looks exactly like a Sierpinski carpet.

[23] R. Lenormand and C. Zarcone, "Invasion Percolation in Etched Network Meas-urement of a Fractal Dimension," Reprint supplied by the authors.

[24] R. Lenormand, C. Zarcone and A. Sarr, "Mechanisms of the Displacement of One Fluid by Another in a Network of Capillary Ducts," *J. Fluid Mech.*, 135 (1983) 337-353.

[25] R. Lenormand and C. Zarcone, "Growth of Clusters During Imbibition in a Network of Capillaries," in D.P. Landau and F. Family (Eds.), "Proceedings of the International Conference on Kinetics of Aggregation and Agglomeration 1984,".

[26] P.G. deGennes, "Imperfect Hele-Shaw Cells," preprint provided by the author, College de France, 75231 Paris Cedex 05, France.

[27] P.G. deGennes, "Partial Filling of a Fractal Structure by a Wetting Fluid," in D. Adler, H. Fritzche and S.R. Ovschinsky (Eds.),"Physics of Disordered Materials," Plenum Press, New York, (1985), p. 227-241.

[28] M. Murat and A. Aharony, "Viscous Fingers and Diffusion Limited Aggregates Near Percolation," *Phys. Rev. Lett.*, 57, (1986), 1875-1877.

[29] E. Guion, J.P. Hulin and R. Lenormand, "Application de la Percolation à la Physique de Milieux," *Ann. des Mines*, Mai-Juin (1984) 17-40.

[30] A.J. Katz, and A.H. Thompson, "Fractal Sandstone Pores. Implications for Conductivity and Pore Formation," *Phys. Rev. Lett.*, 54 (1985) 1325-1328.

[31] A discussion of the problems of the movement of fluid through porous systems from a fractal perspective is available in B.H. Kaye, "Fineparticle Characterization Aspect of Prediction Affecting the Efficiency of Microbiological Mining Techniques," *Powder Technol.*, 50 (1987) 177-191.

9 Fracture, Fragments and Fractals

9.1 The Fractal Structure of Fractured Surfaces

Mandelbrot tells us in his book that he coined the word "fractal" to describe his new geometry of rugged systems, from the Latin adjective *fractus*. *Fractus* comes from the same rootword as fraction and fragment, and means irregular or fragmented. *Fractus* itself comes from *frangere*, which means to break. In the coining of this word, Mandelbrot directed attention to the fact that surfaces created by breaking an object are rugged, indeterminate and describable by fractal dimensions [1]. The study of the structural strength of materials and the ability of different alloys and composite materials to resist fracture represent a special branch of material science. Specialists in this area use the term **fractography** to describe the quantitative study of the structure of fractured surfaces [2, 3]. One of the first groups of applied scientists to use fractal geometry in their scientific studies were metallurgists concerned with advancing the art and science of fractography [2, 4 and 5].

One of the major obstacles faced by the scientist studying fracture surfaces is the need to reduce the dimensionality of the characterization process. Chermant and Coster chose to reduce the dimensionality of the problem of characterizing a fractal surface by studying a vertical section through the fracture surface [4]. This transforms the characterization problem into that of looking at the fractal dimension of a boundary in two-dimensional space. If the difference between the peaks and troughs of a cross section through the surface is not too extreme then, as discussed when studying sandblasted surfaces in Chapter 3, one can use a profilometer trace to study the roughness of the surface.

In Figure 9.1, some data generated in a study of a photograph of a fractured surface by Erwin Underwood is shown. The structured walk technique was used to measure the fractal dimension of this fracture boundary. The Richardson plot for the investigation is shown in Figure 9.1(b). The fractal dimension of the rugged surface of a fracture exists in three-dimensional space and the fractal dimension of the fracture surface is a quantity between 2 and 3. It can be shown that if δ_B is the fractal dimension of a rugged profile generated by making a section through a cracked surface, then

$$\delta_A = 1 + \delta_B$$

where δ_A is the fractal dimension of the fracture surface.

By studying a boundary in two-dimensional space to generate information on a fractal dimension in two- or three- dimensional space, one incurs the penalty that if the rugged surface varies erratically in several directions, one has to take many sections through the surface to generate enough data to obtain an adequate description of the ruggedness of the surface in three-dimensional space. Preparing the several sections through a fracture

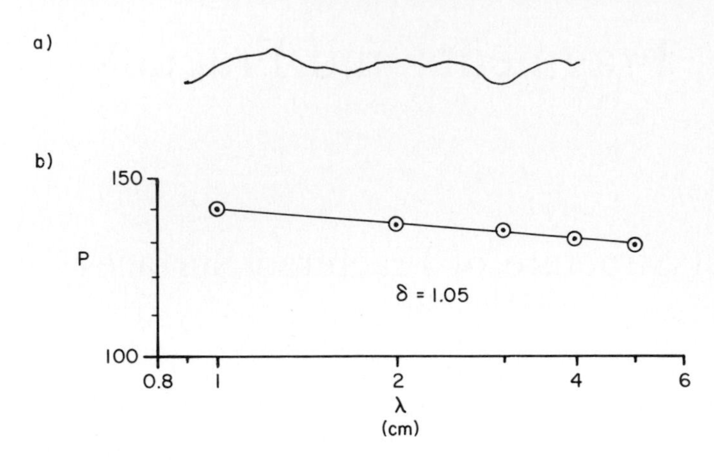

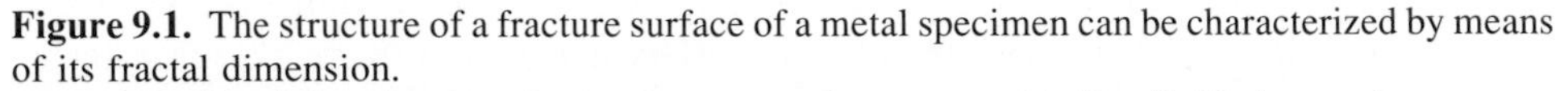

Figure 9.1. The structure of a fracture surface of a metal specimen can be characterized by means of its fractal dimension.
(a) Profile of the photograph of a fracture surface as provided by E. Underwood.
(b) Richardson plot for the exploration of the structure of the profile shown in (a).

in a relatively hard material such as a rock or a piece of metal is the major cost in such a characterization study. Underwood also determined the fractal dimension of a metal fracture by making vertical sections through the fractured surface [2].

Passoja and co-workers adopted a different strategy from that of workers such as Chermant and Coster and Underwood in their study of fracture surfaces. They chose to embed the fracture surface in a resin matrix. The embedded material was then sectioned parallel to the general direction of the fracture surface using standard metalographic techniques [5 – 10]. As one proceeds down through the resin fracture surface by making a series of cuts or by using polishing equipment, one reaches the first section in which the fracture surface pokes up through the resin. The peak of the fracture surface looks like an island surrounded by a sea of resin. Subsequent removal of another layer of the composite resin-fractured material reveals a series of growing islands. These islands may also contain lakes within their structure. The way in which Passoja and co-workers deduced the fractal dimension of the rugged fracture surface from the structural features of the islands appearing in successive sections taken through the surface is illustrated in Figure 9.2.

To simplify the explanation of the method, let us assume first of all that we were studying a set of islands which had Euclidean profiles, such as the circles drawn in the simulated sections shown in Figure 9.2(a). As the sectioning of the surface proceeded downwards through the embedded fracture surface, the islands would grow and new islands would appear as illustrated. If we measured the area of the islands and the length of coastlines for each section then, if we were to plot a graph of the logarithm of the length of the coastline of the islands against the logarithm of the total area of the islands, we would obtain a dataline of the type illustrated in Figure 9.2(c). If we had used squares or ellipses to portray the growing islands, we would have obtained the same slope dataline shifted over as illustrated in Figure 9.2(c). Consider now what happens

a) Sections through a theoretical body containing spheres.

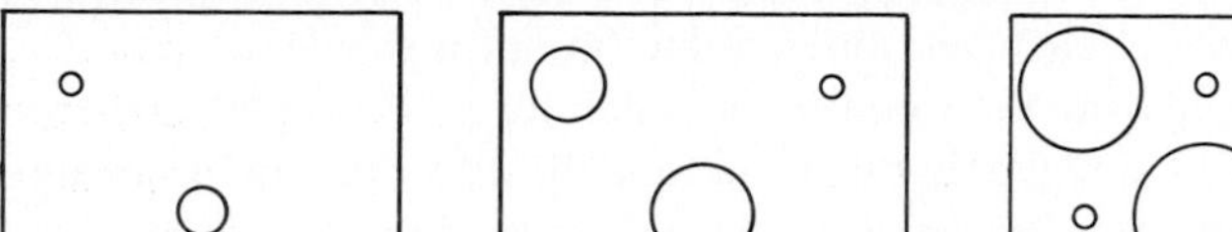

b) Sections through a body containing irregular fracture material.

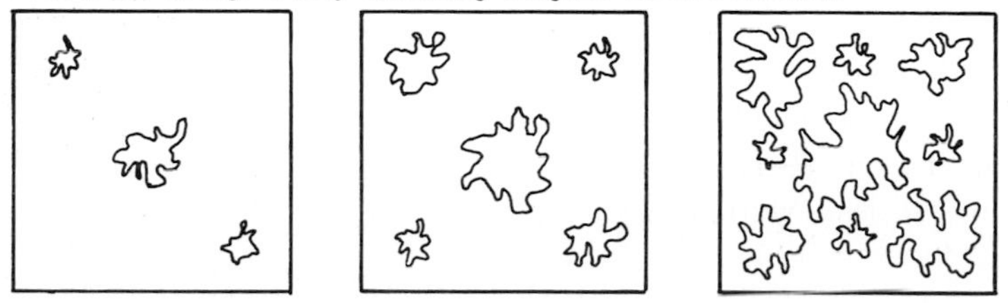

c) Total perimeter versus total area for successive sections through 3 bodies.

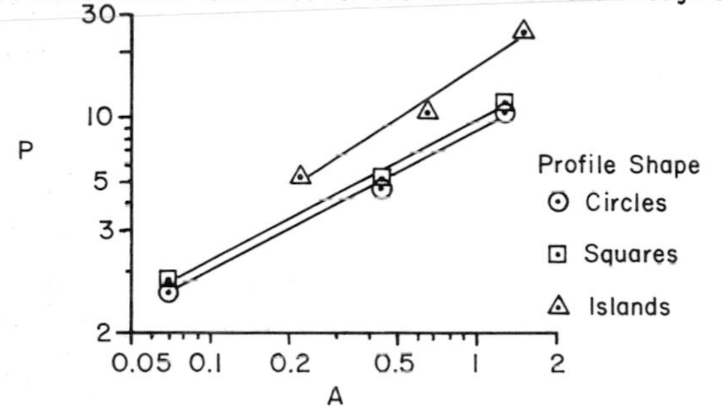

Figure 9.2. Passoja and co-workers derived the fractal dimension of a fracture surface by examining the islands which appeared in successive sections of a fracture surface embedded in resin.
(a) Appearance of growing islands in successive theoretical sections through a fracture surface in which Euclidean circular islands appeared during the polishing stages.
(b) Typical appearance of growing islands manifest in sections through a real fracture surface embedded in a resin matrix.
(c) For the growing islands in a polished section, a graph of the total area of the islands against the total perimeter of the islands for any one section results in a dataline the slope of which can be related to the fractal dimension of the boundaries of the islands.

when the islands appearing in a real section through a fracture surface have fractal boundaries. This would mean that the measured perimeter of the island, at a given fixed level of resolution, would be greater than the circle of equal area by an amount related to the fractal dimension of the rugged island boundary and the scale of scrutiny used to inspect and characterize the areas and boundaries appearing in the section. Therefore, when we measure at a fixed level of image resolution, the increasing area and perimeter of the set of sequentially revealed sections, for the series of simulated images in sequence (b) in Figure 9.2, then the dataline of the logarithm of the measured perimeter against the logarithm of the area in each section is a dataline of a steeper slope. The fractal dimension can be deduced from the slope of the dataline, as illustrated in Figure 9.2(c). It should be clearly understood that in this technique the experimental procedure for generating data does not involve measuring the fractal dimension of a given island by inspecting the boundary of the island area in a range of magnifications. It is essential either to keep the resolution magnification the same for a sequence of polished section, or to make an adjustment factor if one has to use a different level of magnification and resolution to inspect the boundaries manifest in series of sections.

Passoja and co-workers have given the name **slit-island technique** to their method for evaluating the fractal dimension of a fracture surface. A typical set of data generated by Passoja and co-workers is presented in Figure 9.3(a). The way in which the fractal dimensions of the fracture surface can be used to investigate the properties of a substance can be seen by the data reported by Passoja and co-workers for a set of heat-treated metals shown in Figure 9.3(b).

Passoja and co-workers have shown in some of their studies that the fractal dimension of the exposed islands did not depend on the actual section that was studied. In such a situation, when using the slit-island technique, it is a sufficient experimental procedure to measure the fractal dimension of any island boundary, or part of a boundary manifest in a polished section. In this situation one can use dilation logic to measure the fractal dimension of the boundary. In Figure 9.4, a section through a resin-filled fracture surface of a rock core revealing rugged islands is shown. The data were generated during a study of a fracture behaviour of a test cylinder of a rock subjected to compressive stress. The fractal boundary of the islands, as measured by the dilation logic on the Dapple image analyzer, is shown in Figure 9.4(b).

Passoja and co-workers have also used a technique which they call "**fracture profile analysis**" (FPA) to quantify the fractal structure of a fracture surface [5]. In this technique, they embed a fracture surface in resin in the same way as when preparing to use the slit-island technique. Next, a section is made perpendicular to the gross plain of fracture. The revealed profile of the fracture exposed by this technique looks like the profilometry traces described in Chapter 3. Passoja and co-workers then subject this "noisy waveform" profile to Fourier analysis. It has been shown that using the spectrum of the components in the Fourier analysis of the "noisy waveform profile," the fractal structure of the profile can be deduced from the power spectrum of the components in the waveform. The discussion of the theory of the FPA method of fractal dimension characterization is beyond the scope of this introductory discussion of applications of fractal geometry and fracture; it has been mentioned here for the sake of completeness of presentation. Readers interested in the details of the technique are referred to the publications of Passoja and co-workers [5 – 10].

Another group of specialists interested in the structure of fractured surfaces are mineral processing engineers, who must crush rock to a fine powder to liberate the valuable minerals trapped in an ore [11, 12]. When one wishes to characterize the surface structure of a relatively large fragment produced in a crushing operation, one can embed the fragment in resin and make a section through the composite material and inspect the boundaries of the fragments using dilation image analysis logic. Another

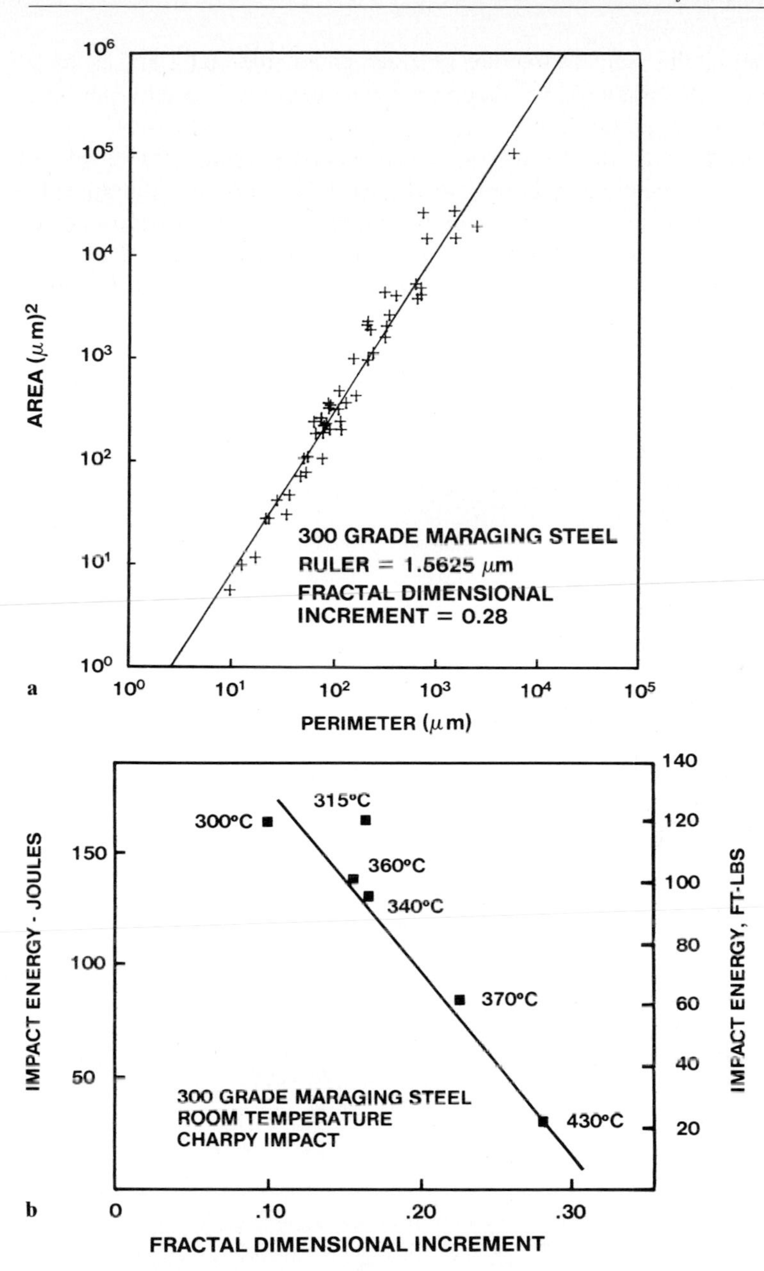

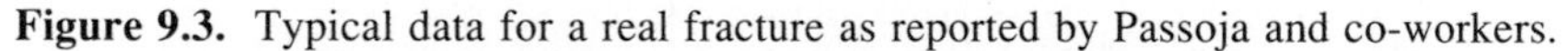

Figure 9.3. Typical data for a real fracture as reported by Passoja and co-workers.
(a) Area perimeter plots for successive sections through a fracture surface.
(b) Experimentally determined relationship between the fractal dimension of a fractured specimen of heat-treated steel and the temperature of the heat treatment.
(From B.B. Mandelbrot, D.E. Passoja and A.J. Paullay, "Fractal Character of Fracture Surfaces of Metals," in B.B. Mandelbrot and D.E. Passoja, (Eds.), in "Extended Abstracts of the Meeting on Fractal Aspects of Materials: Metal and Catalyst Surfaces, Powders and Aggregates," Materials Research Society, Pittsburgh, (1984), pp. 7-9. Reproduced by permission of D.E. Passoja.)

technique for measuring the fractal structure of fragmented material such as rock powder is to use the gas adsorption or dye adsorption techniques developed by Pfeiffer, Avnir and co-workers (see Chapter 7).

The various technical experts who are interested in using fractal geometry to quantify crack structures range from geologists through civil engineers to food technologists. Thus, geologists who are attempting to recover thermal energy from hot rocks drill down into the body of hot rocks beneath the surface of the earth. An explosive charge placed at the bottom of the well is detonated to create cracks all around the bottom of the bore hole. They then either pump down cold water and take up the hot water from another well drilled close to the first, or they take out hot water already in the porous rock which moves through the cracks created by the explosion in the bore hole [13].

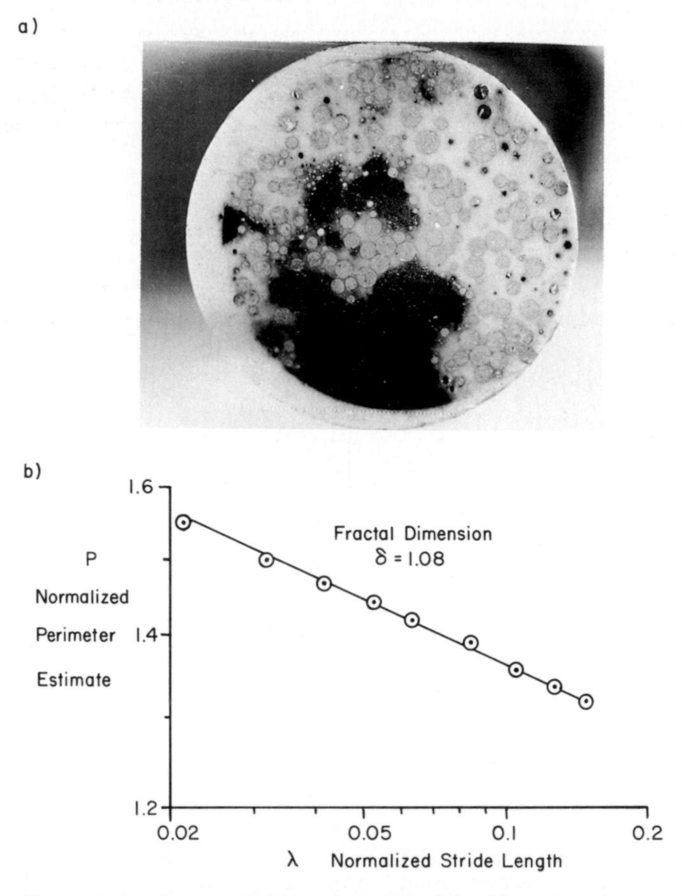

Figure 9.4. The fractal dimension of the island boundary in a section through a fracture surface can also be measured using Dapple image analysis logic. (Data in this diagram provided by Miss Kim Parker, Laurentian University student.)
(a) Polished section through a resin embedded fracture surface of a Norite rock core.
(b) Richardson plot for the boundary of the island in (a) as measured using the Dapple image analyzer.

In the same way, engineers wishing to improve the flow of natural gas trapped within sandstone sometime crack the sandstone to create more cracks in order to increase the flow rate of the gas. Obviously, both groups of geologists would like to be able to characterize the cracks obtained by the use of explosives, or other cracks created by other techniques, in the rocks with which they are working. They could use a disk-like sample of a rock from a core removed from a bore hole and study how it cracks when a missile is fired at the rock specimen.

Civil engineers are beginning to use fractal dimensions to characterize cracks in concrete [11, 15, 16]. Ceramics engineers are very interested in characterizing crack structure and crack progress dynamics [17, 18].

An application of fractal dimensions in food science with which I was associated, was the problem of characterizing the breakage characteristics of cookies (biscuits) for customer acceptance. Customers who are used to a certain type of cookie like it to break in a consistent manner, and in a way that is similar to the fracture pattern they have always encountered with a particular brand. The breaking pattern of a cookie can be quantified using the concept of fractal dimension and image analysis of a picture of a broken cookie. Perhaps in the future the popular phrase "that's the way the cookie crumbles" will be accompanied by a fractal dimension to quantify the exact way in which it crumbles!

9.2 Describing Progress of a Fracture Process From a Fractal Perspective

The fragmentation of a body occurs when a system of cracks propagating through the body creates a set of interconnecting crack surfaces which, as they meet, causes the body to fall apart [19]. Long before a body fails completely, cracks can often be seen on the surface of a body. We often think of a crack as a set of lines, whereas in fact a crack is a rugged surface which we see as a set of lines in a two-dimensional inspection surface. Thus, the cracks that appear on the outside of a cup are only the surface indication of failure surfaces building up in the body of the wall of the cup. Incidentally, it is worth noting that the fractal caverns, growing from the surface cracks down into the body of the cup, are honeymoon havens for hostile bacteria. In these caverns, the bacteria breed and subsequently emerge to infect the fluid in the cup. In another situation when the glaze on the cup is cracked, it is the surface area of the glaze exposed to chemical attack by acid drinks, such as orange or lemon juice, which often result in the leaching out of lead compounds, present in the body of the cup and in the glaze layer, to create the hazard of lead poisoning created by cracked domestic pottery. (See the discussion of the randomwalk of bacteria through fractal space as an example of a fractal geometry-based parameter affecting the efficiency of microbiological mining, given in Chapter 6.)

It is obvious that any study of the fragmentation process must begin with a study of the way in which a crack spreads through a body, and that a major characterization problem facing the scientists in this area is an adequate method of characterizing the structure of cracks. For example, the ceramicist is often interested in the ability of his products to resist fracture when hit by an object. One of the tests that the manufacturers of ceramic tiles, for decorating bathrooms and kitchens use to investigate the cracking behaviour of their products involves the dropping of a steel ball on to a tile from a specified height. If one of the engineers who was used to carrying out this test was to encounter the Whitten-Sander fractal as an unlabeled diagram, he would not think that he was looking at a soot fineparticle; he would think that it was an abstract drawing of the cracks that he generates in his everyday work of testing tiles. He would quickly realize that fractal geometry could be used to describe his cracks. The "tile testing" engineer could adapt the mathematical model of the growth of the soot fineparticle to simulate crack growth in his tiles. In his modified model the growth of the crack structure would be caused by the random branching growth at the tip of an existing crack as it fingered out into uncracked space.

Before the ceramicist's model could be an adequate description of real crack growth, he would have to specify how the crack was able to extend from its present location out into the uncracked material. The apparently simple question, "how does a crack grow?," is one of the more complex questions of materials science. Only a simple discussion of various crack growth mechanisms can be presented in this book.

For the purpose of this elementary discussion of fracture physics, it is useful to distinguish between two types of crack propagation through a stressed material. To describe the difference between these two types of structural failure, consider the simple sketches in Figure 9.5. If we place a piece of rock in the jaws of a vice and apply pressure to the rock, then when a certain compressive stress is achieved, the body starts to crack. Further application of pressure will cause the cracks to spread until the piece of rock breaks into many pieces. Breaking a rock in this manner is described as applying **compressive stress**. It is the type of breakage that occurs when a support pillar of a building, or a rock support pillar in an underground mine, causes the structural system to collapse. It is also the fracture mechanism operating when a piece of rock is crushed in a piece of equipment used in the mining industry known as a jawcrusher. The discussion of the exact mechanism by which a crack starts to grow in a body subjected to compressive strength, is beyond the scope of this book. Interested readers with no background in material science will find an informative and entertaining introduction to the importance of crack growth in materials science in the book "The New Science Of Strong Materials, Or Why You Don't Fall Through The Floor," by J. E. Gordon [20].

Engineers working on the production of powdered ore found experimentally that it was difficult to break rocks by crushing action alone, once the fragments had reached a certain lower size limit. Below this limiting size, they found it more economic to switch to what is widely known as **impact grinding** or **impact crushing**. The terminology used by engineers crushing rocks and making other industrially important powders varies from one industry to another. In this book, I shall use the term **ballistic fracture** to describe the type of breakage achieved in impact grinding. Ballistic fracture is illustrated in Figure 9.6.

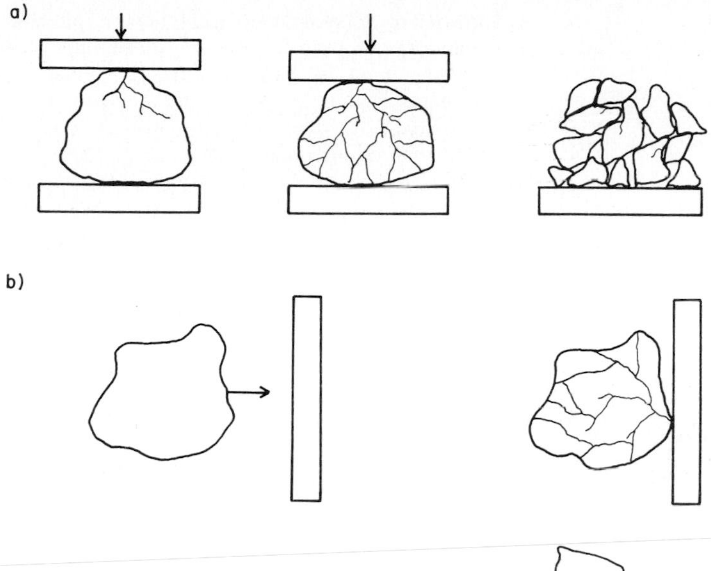

Figure 9.5. When studying the fracture of rock specimens one can make a useful distinction between compressive and ballistic fracture.
(a) Schematic representation of failure by dynamic crack propogation due to the application of compressive stress.
(b) In ballistic impact body fracture, multiple reflections of collision shock waves cause body fragmentation.

If one hits a piece of brittle rock material with a hammer, it appears to shatter into many pieces as soon as the hammer makes contact with the rock. In the absence of any other information, most people would surmise that the type of breakage occurring in the piece of rock hit by a hammer was essentially the same as the failure mechanism operating in a piece of rock being crushed in the jawcrusher, except that the process takes place much more quickly. In fact, however, fracture from a sharp blow with a hammer is an example of fracture by ballistic impact. Ballistic fragmentation involves different failure mechanisms to those operating in a jawcrusher.

To explore the physical basis of the mechanisms operative in ballistic fracture we can explore a curious fact discovered in ceramic science. Dr. Gordon tells us that in the test used by the manufacturer of square ceramic tiles, the tile is struck at the centre of the square face with a sharp blow. In describing this, Dr. Gordon reports that:

"in many cases, the tile does not break in the middle with cracks spreading out from the point of impact to the edges of the tile. When hit in the centre, in many cases four corners of the tile will drop off."

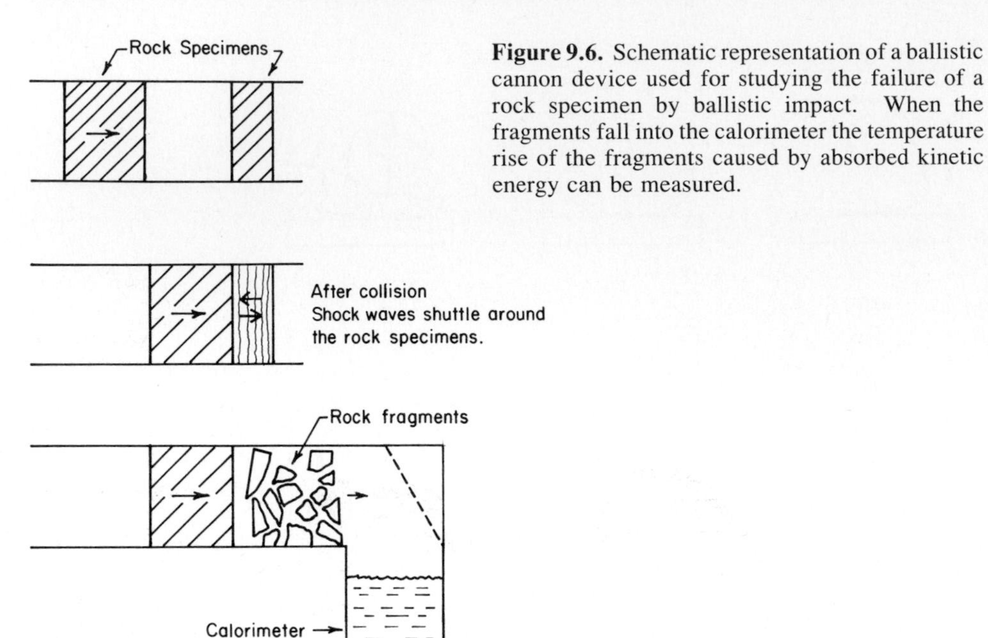

Figure 9.6. Schematic representation of a ballistic cannon device used for studying the failure of a rock specimen by ballistic impact. When the fragments fall into the calorimeter the temperature rise of the fragments caused by absorbed kinetic energy can be measured.

No doubt the first ceramic engineer to observe this strange result would be have been mystified as he picked up the pieces of his tile. To gain an understanding of the physical processes that went on in the tile after the blow had been applied to the centre of the tile, consider a sequence of events that takes place in the **ballistic cannon** (a device for studying ballistic fracture of rock specimens) equipment shown in Figure 9.6. This equipment is being used at Laurentian University to study the type of forces which cause disintegration in ballistic fracture situations. This equipment is a modification of the test equipment used to study the strength of ceramic armour used to protect tank crews from the effect of missiles hitting the tank.

To carry out an experiment with the ballistic cannon two cylindrical pieces are cut from a rock core, one piece being twice the width of the other. They are both mounted in a smooth bore cylinder and the air is evacuated from the space between the two pieces of rock. The more massive piece of the rock core is now driven by a small explosion in such a way that it travels down the tube and hits the smaller piece of rock cylinder. On impact, a shock wave starts to spread out through the second piece of rock, long before it starts to move a significant distance from the fired piece of rock. As Dr. Gordon tells us when discussing such shock waves:

"the highest speed at which a stress can be transmitted through any substance is usually the speed of sound in that substance. Indeed, sound is best thought of as a wave or a series of waves of stress passing through a substance at its natural speed. It can be shown that the speed of sound in steel, aluminium and glass is of the order of 5000 m/s in each instance. This is much faster than the speed of any hammer blow and considerably faster than the speed of most bullets."

In our ballistic cannon, the shock wave caused by the collision of the two pieces of rock crosses to the other side of the piece of the rock cylinder, and starts to bounce back and forth across the cylinder. In some positions, the crossing shockwaves create a region of tension within the rock.

Most materials are much weaker in tension than in compression, since compression tends to close any crack beginning in the structure whereas tension stress amplifies the presence of a crack.

The rock specimen being studied in our ballistic cannon breaks down because of **tension** after impact, not **compressive stress**. When we hit a piece of rock with a hammer, we in fact create a series of shockwaves in the struck piece of rock. In the words of Dr. Gordon:

"when we strike a solid with a hammer, a whole series of stress waves radiate from the point of impact and move off into the body of the material. They reach the further boundary of the solid in a time which is probably between a 10 000th and a 100 000th of a second. They are reflected back as a kind of echo with very little loss of energy. What happens next depends upon a great many things, depending on the shape of the solid, exactly where the blow was struck, and so what may happen is that the returning reflected stresswaves repeatedly meet the outgoing ones at some critical point and thus the stress may pile up at this point until fracture occurs".

Dr. Gordon explains that in the case of the ceramic tile which loses its corners, what happens is that the shockwaves spreading out from the point of impact are reflected and crowded into the corners causing the corners to break away.

In Figure 9.7, two photographs, taken by Dr. Lee A. Cross, illustrate dramatically the stages involved in the shattering of glass plates by ballistic fracture [21, 22]. In the experiments carried out by Dr. Cross, a steel sphere 4.5 mm in diameter was fired at a glass plate 5 cm square and 6.35 mm thick. At the moment of impact the ball was traveling at 200 m/s. Photograph 9.7(a) was taken 20 μs after impact (a **microsecond** is one millionth of a second). The area immediately near the colliding sphere is black because of a multitude of cracks in the glass. After 20 μs, the shockwave in the glass has already traveled across the plate and has been reflected. The interactions of the direct and reflected shockwaves have started to create cracks on the top and rear surfaces of the glass plate. Dr. Gordon tells us that a crack in glass travels and grows at about thirty percent of the speed of sound in the material. Therefore, the growing cracks shown in Figure 9.7 are moving at approximately 1700 m/s. It should be realized that energy from the tip of a growing crack will itself generate stresswaves which, in the words of Dr. Gordon,

"are probably racing about in the material in all directions at the speed of sound, being reflected on both old and new surfaces so that we are likely to end up with not one crack but many. In other words, the material shatters."

The photograph shown in Figure 9.7(b) was taken 70 μs after impact. By this time the cracks have spread throughout the glass plates. In a paper published by Dr. Cross describing his high-speed photography technique, another photograph taken 7 μs later than the photograph in Figure 9.7 is shown. In that photograph, the fragment labeled A, in Figure 9.7 is the first fragment to begin to separate from the parent glass plate. It would be very interesting to take a whole series of fracture images similar to those in Figure 9.7 using various missiles and, various specimens of different shapes and to characterize the fractal dimension of the crack growth.

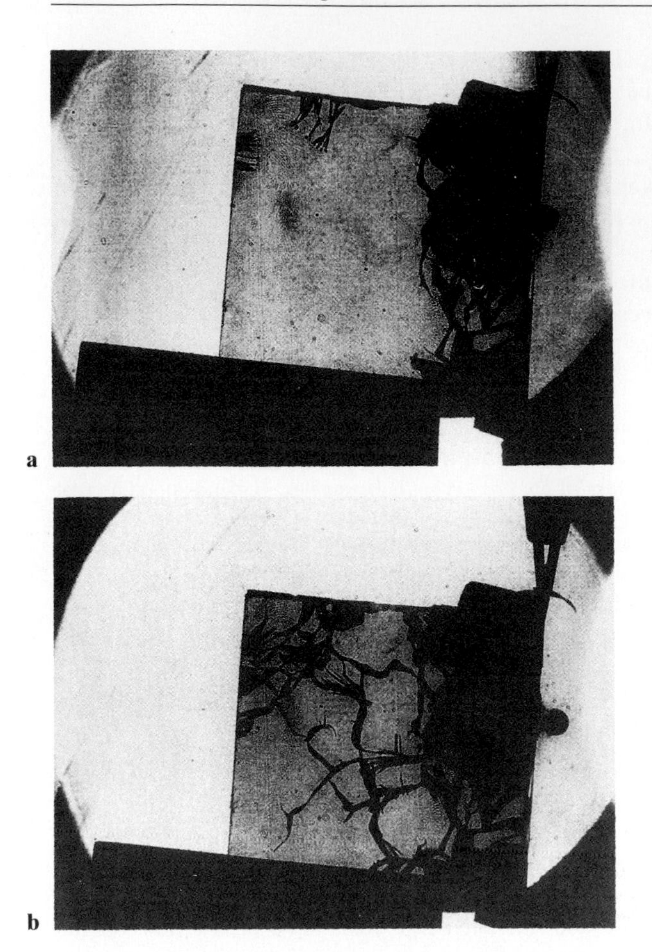

Figure 9.7. High-speed photographs of the failure of a glass plate hit by a rapidly moving steel ball, illustrating the stages involved in the failure of a body by ballistic fracture [21].
(a) Cracks present in the glass plate 20 μs after impact.
(b) Cracks present in the plate 70 μs after impact.
(Photographs furnished by Special Illumination Systems, Inc., Dayton, Ohio.)

The growing crack in the top left-hand corner of Figure 9.7(a) looks like the branching fractals to be found in Dr. Mandelbrot's book and reproduced in Figure 9.8. My Rorschach reaction on seeing Dr. Cross's pictures, and remembering Dr. Mandelbrot's "trees" in Figure 9.8, was that one should be able to model crack growth using fractal tree mathematics. Studies aimed at characterizing fractal growth in this way are under way at Laurentian University.

The apparently random branching of the cracks and the sudden changes in direction manifest by the growing cracks in Figure 9.7 illustrate how difficult it is to make a realistic model of fragmentation created by ballistic impact. We can anticipate that mineral processing engineers will start to carry out ballistic collisions of pieces of rock with different sized slices of ore material at different speeds of collision, with possible high speed photographic studies of the disintegration process. Characterization of the fractal dimension of fractured surfaces and the structure of the surface of fragments by their fractal dimension would give the engineer a much better understanding of the way in which his material fractures in his pulverization process (see the discussion of energy conservation in a pulverization process given in Section 9.4. **Pulverize**, from the Latin

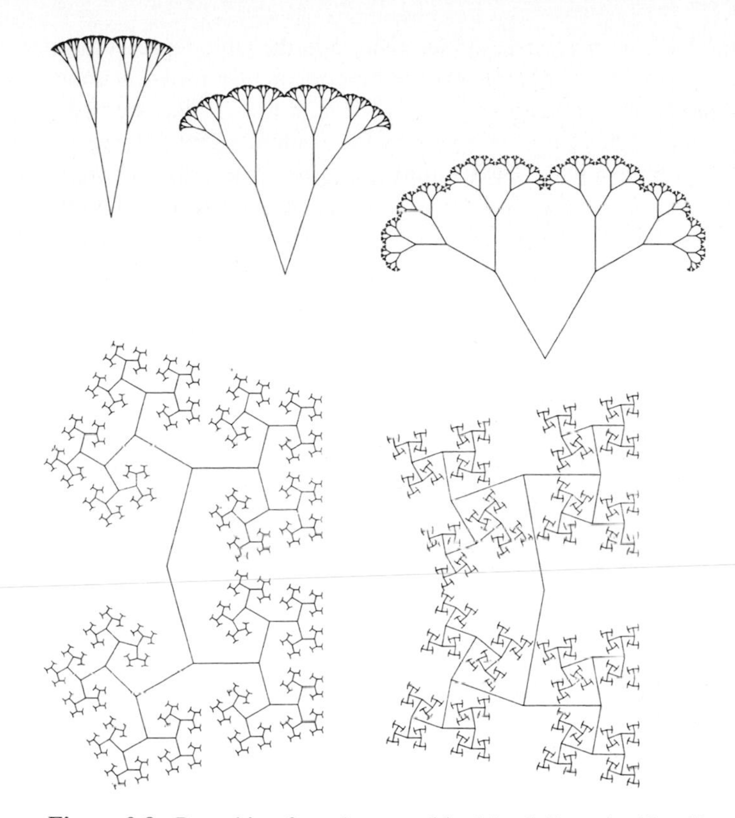

Figure 9.8. Branching fractals created by Mandelbrot, looking like different types of spreading cracks (from B.B. Mandelbrot, "The Fractal Geometry of Nature," W.H. Freeman & Co., San Francisco, 1983, reproduced by permission of Dr. Mandelbrot).

word *pulver*, means to turn into powder. The term pulverization is used in this book to denote any size reduction process such as crushing or grinding).

In a ball mill, as the ball falls on to a piece of the material to be crushed, most of the energy transferred between the ball and the piece of rock to create fragments probably takes place by ballistic impact rather than compressive failure. As ball milling proceeds, it is well known that the efficiency of fracture falls off because the build up of fine material in the mill creates a powder cushion which slows down a fracturing ball and reduces the effectiveness of collision impact between the larger fragments in the ball mill.

The way in which cracks grow in a stressed body can be modelled using percolation theory of the type developed in Chapters 5 and 6 [23]. For the sake of simplicity, the discussion of fracture modelling using percolation theory given here will be limited to a brief discussion of fragmentation modelling in two dimensions. In one model that can be considered to simulate the growth of cracks occurring within a stressed body, we can assume that the cracks grow at random from flaws situated at random in the body of a piece of material. Thus, we can imagine that if we considered our random number table to be a two-dimensional model of a rock, we could let every 5 in the table represent the location of a flaw. We could then let every crack grow at random on this initiation point in the two-dimensional plane of our mathematical model. Using this model, the body will crack into pieces when continuous paths are made by the randomly growing cracks which grow outwards from each postulated flaw-initiating centre. A full discussion of this type of modelling is beyond the scope of this book but, for the sake of illustration, consider the system shown in Figure 9.9(a). It is assumed that the percolation model has proceeded until the body is crossed by connected paths at a concentration of 60% pixel occupation. In this model of crack propagation, the pixel occupation levels in the model no longer represent a volume effect; they represent a pixel through which a crack can pass. For this type of model, corner to corner (diagonal) touching of pixels probably constitutes a viable path. It should be noted that the model gives no information on the location of the crack within the pixel.

In Figure 9.9(c), some of the fragments which would be created by the completed crack paths are shown separated from each other, to give some idea of the fragments which would be created by such a fracture pattern (the fragments with a straight edge are not shown in the second part of the figure, since we do not know their true size as the size the "model rock" limits our knowledge of what happens outside of the model boundary). Various models for simulating the growth of cracks under different physical conditions using percolation theory incorporating various parameters governing the

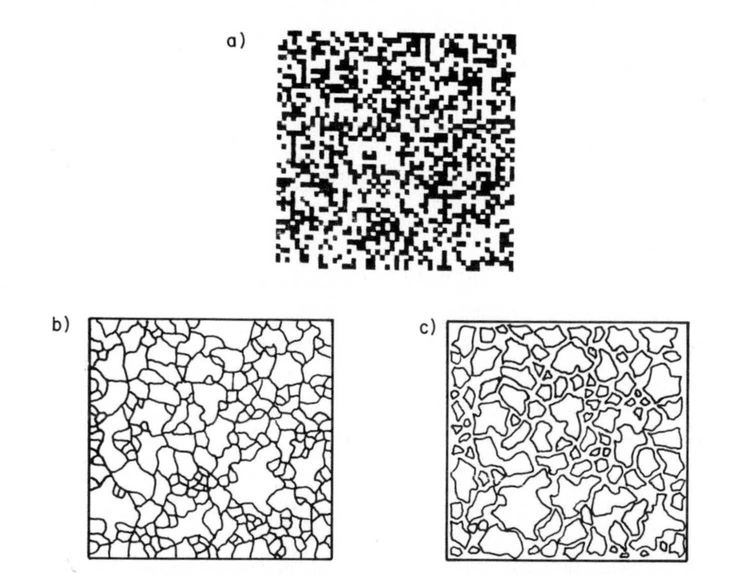

Figure 9.9. In a simple percolation model of fracture, the fractures are considered to spread out from randomly distributed flaws present in the original unstressed body.
(a) Random distribution of initiation flaws for crack growth (created by letting every 5 in a random number table represent the centre of a flaw).
(b) Completed crack pattern in a crack growth model in which cracks were allowed to grow outwards from each initiation center.
(c) Some of the fragments produced by the fragmentation model in (b).

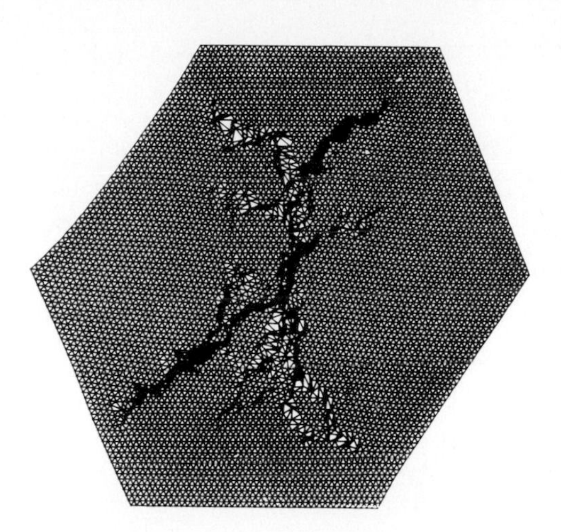

Figure 9.10. Sophisticated computer modelling of fragmentation dynamics space upon considerations fractal geometry can produce some very realistic models of a fragmentation. Shown is a fragmentation by Louis et al. [27].
(Picture 1 from E. Louis, F. Guinea and F. Flores, "The Fractal Nature of Fracture," in L. Petronero and E. Posati (Eds.), "Fractals in Physics," Elsevier Science Publishers B.V., Amsterdam, 1986. Reproduced by permission of Elsevier and E. Louis. Picture 2 from E. Louis and F. Guinea, Europhysics Lett., 3, 871 (1987).)

strength of materials have been suggested by several workers. The interested reader can find more information in references 23 – 28. Various levels of sophistication can be built into a fractal fragmentation model. For example, in Figure 9.10 we show the results of some more advanced modelling of fracture patterns and strain patterns, as created in the computer by Louis and colleagues [27]. Dr. Mandelbrot has suggested that another fractal system, which he calls a "squig" system, could be used as a model for predicting dynamic fragmentation. In Figure 9.11 (see Plate 7 at the beginning of the book) the structure of a typical squig system is shown. Its resemblance to a cracked body is obvious. The interested reader is invited to squiggle around fracture space to explore the potential of squig models of fracture dynamics [29, 30].

9.3 The Fragmentation Fractal. A New Fractal Dimension for Characterizing a Fragmented System

In Figure 9.12, another high-speed photograph taken by Dr. Cross, of a different type of fragmentation process, is shown. In this case, a steel ball was fired through a small piece of phenolic plastic board. The ball was traveling at 200 m/s on impact. By the time the picture was taken, the ball had moved a measurable distance from the piece of board through which it had passed. The ball was 4.5 mm in diameter. It can be seen clearly to the left of the board. In a fragmentation process, of the type shown in either Figure 9.7 or 9.12, the size reduction engineer is obviously interested in the size distribution of the fragments. An unlabeled display of the fragment profiles of Figure 9.12 could be mistaken for a map of a group of islands. This suggests that fractal

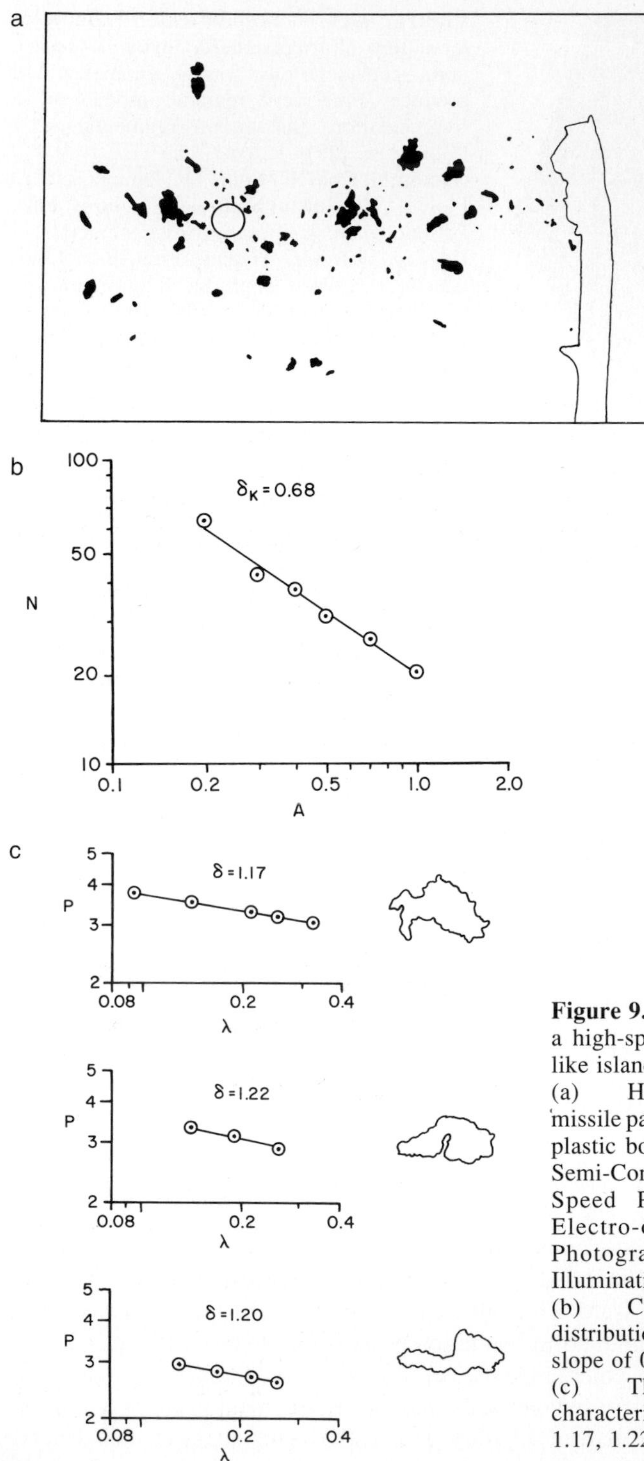

Figure 9.12. The fragments produced by a high-speed fragmentation process look like islands in a sea.
(a) High-speed photograph of a missile passing through a piece of phenolic plastic board (from L. A. Cross, "Pulsed Semi-Conductor Lasers Find Use in High Speed Photography," Laser Focus-Electro-optics, August (1985). Photographs furnished by Special Illumination Systems, Inc., Dayton, Ohio).
(b) Cumulative undersize size distribution of the fragments having a slope of 0.68.
(c) Three fragments, big enough to characterize, show fractal dimensions of 1.17, 1.22 and 1.20

geometry techniques used to describe the structure of island groups, and **archipelagoes**, may be useful for describing the state of fragmentation of the disintegrated piece of board.

Geographers have often looked at the structure of an island archipelago and attempted to describe the size distribution of the islands by means of numerical relationships. In Figure 9.13(a), a randomly assembled array of the group of islands that gave the term "archipelago" to the English language – the Aegean islands – is shown (it should be noted that the word archipelago actually means "the most important sea" and referred originally to the water around the islands, not the cluster of islands). We have already discovered that the coastline of an island is infinite, but from our study of Koch Islands we know that the area of an island is finite. Mandelbrot points out that the surface area of the island is itself a fractal structure and indeterminate, so that it is only what he calls the **map area** or the **projected area of an island** which is finite. In 1938, Korcak claimed that all the islands of the world could be described by one numerical relationship [31]. He discovered experimentally that if one plotted the logarithm of the number of islands greater than or equal to a stated size, A, against the logarithm of A, one obtained a linear relationship. In Figure 9.13(b) this relationship for the Aegean Islands

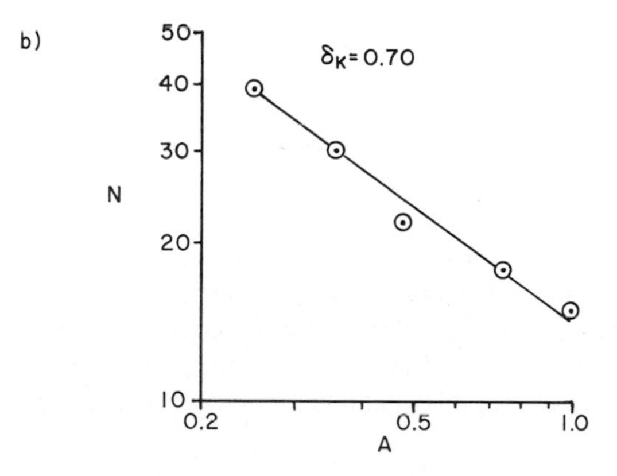

Figure 9.13. The size distribution of the Aegean Island illustrates the generation of a dataline similar to that suggested for island systems by Korcak.
(a) A randomized array of the islands of the Aegean.
(b) Log-log plots of the size distribution of the islands.
N = the number of islands greater than or equal to A; A = the map area of the islands (the projected area of the island).

is shown. Furthermore, Korcak stated that the slope of the line was 0.5 for all islands. Written numerically, this relationship is

$$N_R\ (A \geq a) = F'a^{-B}$$

where A = area of the island of a given size, a = resolved area of an island, N_R = number of islands and F' and B are constants.

Mandelbrot tells us that when he first read Korcak's scientific publication, he found the claim that "the slope of the dataline for all archipelagos was the same" to be incredible. In his book, Mandelbrot criticizes the original work of Korcak and states that the claim regarding the dataline slope is unfounded. Mandelbrot states that, in fact, the value of B in the equation above varies from one group of islands to another, and is always greater than 0.5. Mandelbrot then showed mathematically that one can define a new type of fractal dimension which is a measure of the fragmentation of a system. For reasons that will become clear as this discussion proceeds, we shall call it the "**fragmentation fractal dimension**" (a measure of the size distribution of a set of fragments).

To understand the mathematical theory underlying Mandelbrot's development of the fragmentation fractal of a dispersed system, consider the sketch in Figure 9.14, which shows, algorithms developed by Mandelbrot for constructing sets of islands. To construct a simple island system, we start with a square and divide the top side of the square into eight intervals, as shown. We then construct a small square off-shore island, which has a coastline equal to that of the original side of the square. This pattern is called the "island generator algorithm." If we carry out this construction all round the original mother island, we obtain a square with four satellite islands as shown in the sequence of sketches in Figure 9.14(a). We can now repeat the construction and construct the system in which the main island has four large islands and 48 smaller islands. This process can be carried out ad infinitum to produce a set of islands which have an infinite number of small islands. For such a system, the number of islands of magnitude equal to or greater than a stated area plotted against the stated area according to **Korcak's relationship** on log-log graph paper, is illustrated in Figure 9.14(b). An alternative method of defining the size of an island is to consider the effective diameter of the island to be some constant factor of the square root of the area. Mandelbrot uses the symbol λ to represent this island length or diameter.

It should be noted that we can take the set of islands constructed in Figure 9.14 and randomize them in space to make them statistically self-similar. The average distribution function would still obey the same data relationship, since the Korcak-type plot tells us nothing about the individual position of any given island on the map within an archipelago. Furthermore, we could take all the islands and change their shape to any other Euclidian shape, or any irregular structured but not fractally structured shape, and the same Korcak relationship would still hold. From the information in Figure 9.14, we can see that the Korcak-type relationship tells us only about the fragmentation and not the location and structure of individual islands. The slope of the dataline of the plot will depend on what mathematical generator is used to transform the original parent figure into the set of islands. In the second edition of Mandelbrot's book, several different island systems which can be generated by changing this island generation algorithm are shown.

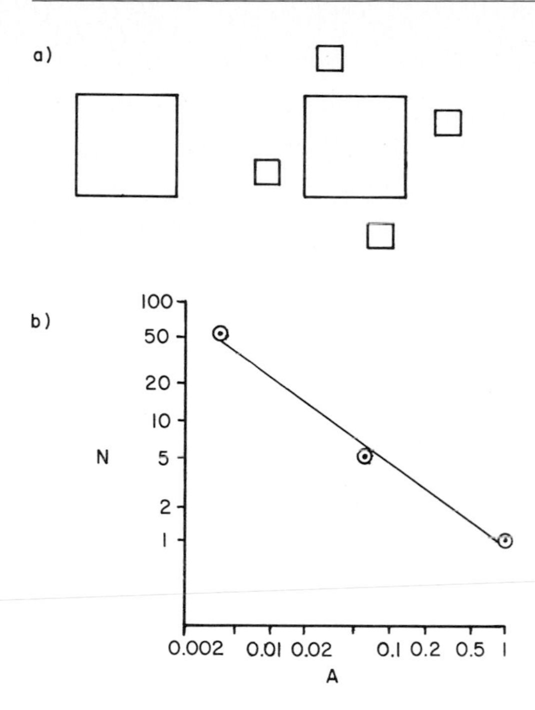

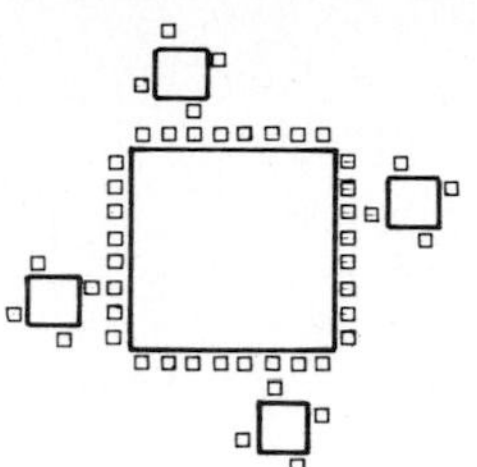

Figure 9.14. Mandelbrot's algorithm for constructing an island system having a fragmentation fractal dimension of 4 divided by 3.
(a) First-, second- and third-order island systems.
(b) Size distribution function of the island system.
N = Number of islands greater than or equal to A; A = The area of an island (from B.B. Mandelbrot, "The Fractal Geometry of Nature," W.H. Freeman & Co., San Francisco, 1983. Reproduced by permission of B.B. Mandelbrot).

Mandelbrot has shown that we can link the relationship describing the numerical distribution of the areas of a group of islands to a relationship describing the number size distribution of the effective diameter of the island, by the following simple equation:

$$N_R\ (\Lambda \geq \lambda) = F\lambda^{-\delta}$$

where δ is the fractal dimension of the coastline of islands and $N_R\ (\Lambda \geq \lambda)$ is the number of islands of size greater than or equal to λ. Furthermore, he showed that the constants $-B$ and δ in this equation and that given earlier for the area size distribution are linked by the relationship $B = \delta/2$. Mandelbrot tells us that from his measurements the average fractal dimensions of the coastlines of the islands of the earth are of the order of 1.2 . He also states that if we plot the size distributions of the islands of the whole earth, a value of B of the order of 0.6 is obtained, so that for the islands of the earth the relationship $B = \lambda/2$ holds. Mandelbrot also discusses the use of coastline – island generators in which the islands themselves acquire fractal structure. Thus, in Figure 9.15, one of the fragmentation-fractal generators discussed by Mandelbrot is illustrated together with a construction algorithm for drawing a system of islands including lakes. It should be stressed that the fractal dimension of the coastline is a property of an individual island, whereas the fragmentation fractal deduced from the Korcak plot is a property of the group of islands.

It follows from the above discussion that experimental studies of the distribution of the map area of a group of islands will yield information on their characteristic structure and possibly on the mechanisms that formed the island.

We can now make an interesting leap from one technology to another. It seems reasonable to assume that in general, the same reasoning with regard to the number

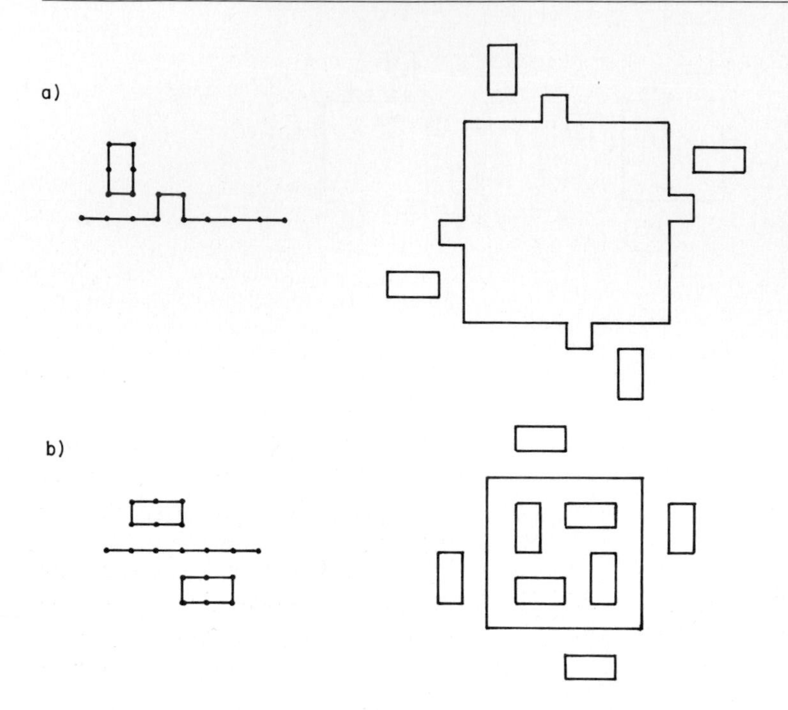

Figure 9.15. Mandelbrot describes algorithms for constructing island systems with fractally structured boundaries and island systems which include lakes.
(a) Construction algorithm for an island system with a fracturally structured coastline and the first-order island system created using this algorithm.
(b) A typical system for creating islands with lakes included the island system described by Mandelbrot. Basic construction algorithm and the appearance of the first-order island-lake system.
(From B.B. Mandelbrot, "The Fractal Geometry of Nature," W.H. Freeman & Co., San Francisco, 1983. Reproduced by permission of B.B. Mandelbrot.)

distribution of islands will apply to the number distribution of fragments of a pulverized rock. This leap is suggested intuitively by the fields of view shown in Figures 9.12 and 9.13. Thus, the slope of a Korcak plot of the appropriate size parameter of a set of fragments can be used to describe the way in which a rock has shattered or fragmented. The fractal dimension deduced from the size distribution of the fragments could be called a Korcak fractal or a fragmentation fractal. When describing geographic systems, I use the term Korcak fractal, but when looking at fragments I shall use the term fragmentation fractal. In Figure 9.12(b), the size distribution of the fragments produced by a steel ball passing through a phenolic resin board is shown. It can be seen that the size distribution function is a linear dataline of slope –0.68. There really is not enough data to let us explore the possibility that the slope is $\delta/2$ when δ is the fractal dimension of the rugged boundaries of the fragments, but it is interesting to note that the three largest fragments of the picture which exhibit fractal structure have "coastlines" of 1.22, 1.20, and 1.17. If the reader will forgive the pun, this "fragmentary evidence" hints that it may be worth carrying out a more comprehensive investigation into the relationship between the ruggedness of the fragments and the fragmentation fractal of a set of fragments, where the data plot of fragment size can be shown to be describable by a scaling function.

In a personal communication Dr. Russ of North Carolina State University has pointed out that the projected, that is the silhouette, fractal dimension of the profile manifest if the fragment is embodied in resin and sectioned. This is because some of the features of the structure of the profile lie in the shadow of the other features when creating the silhouette of the fragment. It is the sectioned profile fractal dimension that would be operative in any link between the Korcak fractal of the fragment distribution and the fractal dimension of the fineparticle fragments. Based on these considerations the values of 1.22 for the fractal dimensions of the three fragments studied etc. are probably lower than the boundary fractal dimension which would be manifest on inspection of a sectioned fragment. In turn this suggests even stronger possibilities that the sparse evidence of Figure 9.12 supports the hypothesis of a link between the fractal dimension of the profile and the size distribution of the fragments.

The mineral processing industry has expended great effort in developing numerical relationships describing the size distribution of fragments generated in a pulverization process [32]. Some of these relationships have involved log-log type plotting of accumulative size distribution data, in a similar way to the data display given in Figure 9.12 and 9.13.

Sometimes, distinguishing between different distribution functions of this kind is very difficult because of the lack of accuracy in the data and/or the insensitivity of the data scales used to plot the distribution function of variations in experimental data. It may be that some of the experimentally reported distribution functions produced by a pulverization process could be reinterpreted, and a Korcak-type plot used to obtain a numerical estimate of a fragmentation fractal dimension which will be descriptive of the fragmentation process occurring in any particular pulverization device. If the suggested relationship between the fractal dimension of a fragment boundary and the size distribution of the fragments holds for a fragmentation process, then we would have the interesting possibility that the fractal dimension of the projected boundary of a fragment produced by a pulverization process would be directly related to the fragmentation from a Korcak-type plot of the size distribution of the fragments. This in itself would open up possibilities for predicting the fragmentation behaviour of an ore body from an investigation of the fractal dimension of the projected boundary of a fragment. Alternatively, it would enable the engineer to estimate the fractal structure of the surface of the fragments from the measured distribution function. Obviously, to establish the validity of such a revolutionary approach to interpreting the size distribution data obtained from fragmentation studies will require a major investigation of the fracture behaviour of single lumps of ore. It would also be interesting to explore the size distribution functions obtained from a study of crushing stress failure and ballistic impact failure of similar rock specimens. From the fact that a crack which spreads through a disintegrating body generates the boundary of a fragment, and considering that the overall disintegration process involves the interaction of a multiplicity of cracks having the same dynamic structure as the single crack forming a boundary of fragments, it may not be unreasonable to anticipate that the type of fragmentation produced in the disintegration of the ore body will be related to the structure of the fractal boundary of an individual fragment. Stated in another way, the fragment topography and the size distribution of the fragments are obviously both consequences of the dynamics of crack propagation in a failing structure and should be interrelated.

The Sierpinski fractal and the fragmentation fractal are complementary descriptors of a fineparticle system dispersed in a matrix. The Sierpinski fractal focuses attention on the way the fineparticles are packed into an available space, whereas the fragmentation or Korcak fractal focuses attention on the structure of the elements of the dispersed system.

9.4 Fractal Geometry and New Perspectives in the Mineral Processing Industry

9.4.1 Dust Explosions

An important problem encountered in many mines is the danger of explosion from the dust and or gases in the work areas. Dust explosions are a working hazard in many different industries. When I was working in Chicago, two explosive hazard problems with which I was involved were the investigations of why a sugar factory and a pizza factory had been the sites of disastrous explosions. My colleagues and I discovered that the sugar factory had blown up because a worker illegally used a fork-lift truck to unload powdered sugar. A spark from the fork lift machinery triggered an explosion in a cloud of powdered sugar. In the case of the pizza factory, engineers were blowing powdered cheese from one point in the factory where it was produced to another where it was to be sprinkled on the pizzas. An electrostatic charge built up in the conveyer device triggered a spark which blew the factory up. Again, the largest industrial explosion which occurred in my hometown of Hull, England, occurred when a cocoa factory blew up and completely demolished one wall of the factory. I believe that the fractal structure of the powdered cheese and cocoa fragments was a significant aspect of the explosive hazards formed by these powders. Although this discussion concentrates on the relevance of fractal geometry perspectives for the understanding of the physical aspects of dust structure that are pertinent to the dangers of explosions in the mining industry,most of the information presented here is also relevant to gaining a better understanding of explosion hazards in granaries, the food industry and many other industries.

To gain an understanding of the mechanics involved in a dust explosion, consider an array of fineparticles as shown Figure 9.16. These fineparticles could be coal dust, flour or powdered cheese. The basic requirement for an explosion to propagate through a cloud of dust is that radiative heat transfer from the first source of heat in the cloud must

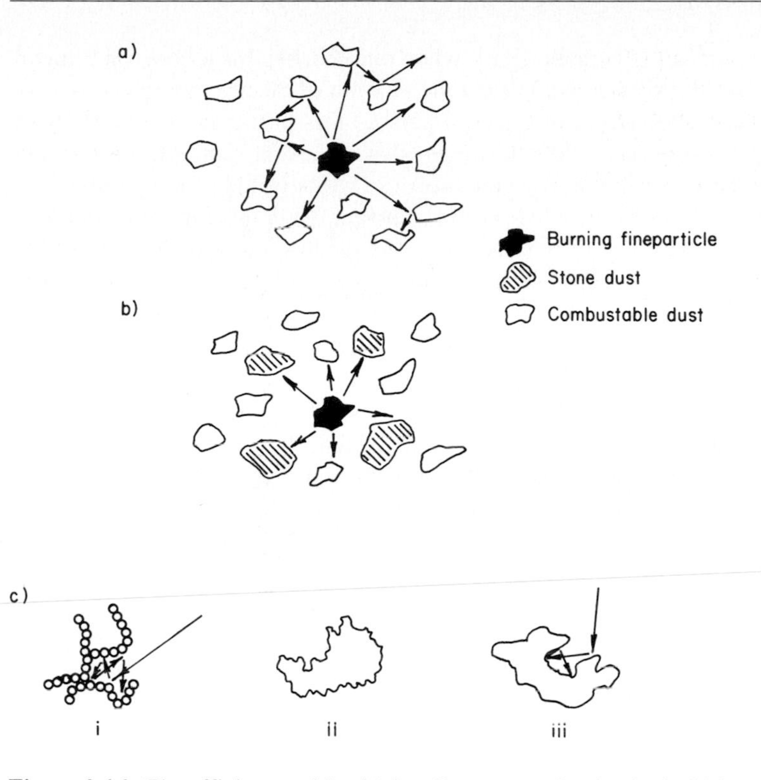

Figure 9.16. The efficiency with which radiant energy is adsorbed within a cloud of combustible material is an important factor to be considered when estimating the danger of dust explosions.
(a) A chain reaction setting off a dust explosion occurs when radiative transfer from a burning dust fineparticle in a cloud heats adjacent fineparticles to the ignition temperature.
(b) One mechanism for reducing the danger of dust explosions involves the scattering of stone dust fineparticles amongst potentially explosive dust. The stone dust acts as a radiative energy sink to supress the overall temperature rise within a potentially dangerous cloud of dust.
(c) Fractally structured dust of combustible material is potentially more likely to trigger a dust explosion, since it is a more efficient absorber of radiant energy. Thus, a diesel soot fineparticle has low mass and high surface area, with radiation entrapment pockets within its structure, causing a rapid rise in temperature when subjected to heat radiation.

heat up adjacent fineparticles to their **ignition temperature.** Many people fail to realize that a body has to be relatively hot before it will burn and that, for example, the temperature in the flame of a match which ignites other material is very high. The match causes combustion because the heat of the flame raises the temperature of a small area of the object to be ignited up to the temperature known as the ignition temperature of the object, at which it begins to burn. Because one can snuff out a match with wet fingers, the popular idea is that the match flame is not at a high temperature. Most people confuse the amount of heat given out by an object with its temperature.

The same situation occurs in assessing electrical hazards. The voltage that a person generates by moving around a building on a cold dry day can be of the order of 30 000 volts, which is why one can generate a spark when one reaches for a book on a metal bookshelf. In this case the voltage is high but the amount of electric charge is low, so that although the spark generated can startle a person, this particular source of high voltage is not usually dangerous. Note, however, that electrical sparks from a highly charged surgeon through a scalpel have been known to cause trouble for a patient. On the other hand, one can be killed by a low-voltage current when the current discharged from the source is high, because of the energy in the electric current flowing at low voltage. Condensing steam can cause severe scald damage not because of the temperature of the steam, but because of the enormous amounts of energy released as steam changes to water. The heat energy of the burning match is low even though the temperature is high.

Returning to our dust cloud, let us assume that the central object of the cloud is a burning piece of coal dust which is radiating energy to the surrounding dust fineparticles. If the dust fineparticles are smooth, the amount of energy that they absorb is limited by reflection. The radiant energy bounces around the dust cloud and is spread out in the dust cloud in the way indicated symbolically in Figure 9.16(a). One of the ways in which engineers lower the danger of explosion from coal dust is to sprinkle stone dust along the roadway of the mine. The dust is fine enough to become suspended in the air currents created by moving machinery but not small enough to constitute an inhalation hazard for the miner. When the stone dust is suspended within the coal dust cloud as shown in Figure 9.16(b), some of the radiant energy is absorbed by the relatively large, unburnable, stone dust fineparticles. This robs the cloud of coal fineparticles of energy, so that none of the adjacent coal dust material is able to reach its ignition temperature.

If the potentially explosive dust suspended in a cloud has a fractal structure, then the radiative absorbance of the individual grains is much higher than for a simpler smooth-textured dust fineparticle. Thus, if inadvertently some soot fineparticles were to become entrained in a cloud of ordinary coal dust, the flow of energy from an ignited fineparticle would be efficiently absorbed by the soot fineparticles, which themselves could burn quickly to radiate energy and thus trigger a chain of events leading to a disastrous explosion. The fractal dimension of the rough surface of a dust and or the structural fractal of the overall dust fineparticle will control the radiant energy absorption efficiency. Any study of the potential explosive hazard of a combustible dust, the grains of which have a rugged structure, should include a characterization of the fractal dimension of the structure and or the texture of the fineparticle. In the past, studies of explosive hazard factors due to the structure and texture of the dust have been hard to quantify. The availability of fractal dimension parameters for linking physical structure to the efficiency of radiant energy adsorption offers the prospect of enhanced quantitative characterization of hazardous dust materials.

9.4.2 Energy Efficiency in a Pulverization Process

Engineers responsible for the production of powders, especially in the mining industry, are obviously interested in the efficiency with which the energy they put into a pulverizing device is used to cause the breakup of the piece of material to be pulverized. One basic approach to the measurement of the energy efficiency of pulverizing devices has been to look at the energy represented by the creation of and stored in the new surfaces formed when a large lump of ore breaks down into several fragments. Over the years, several theories relating energy utilization to the efficiency of pulverization by measuring the surface area of the fragments produced have been put forward by different investigators. In Figure 9.17, a useful graphical summary, based on data presented by Kelly and Spottiswood, of the various numerical relationships which have been suggested for relating energy input to surface produced is presented [19]. The various relationships summarized in Figure 9.17 have always been controversial. For example, Austin, an expert in crushing and grinding, stated in a recent review of pulverization practice:

"Rittinger's law, that the energy of size reduction is proportional to the new surface produced, has no correct theoretical basis" [32].

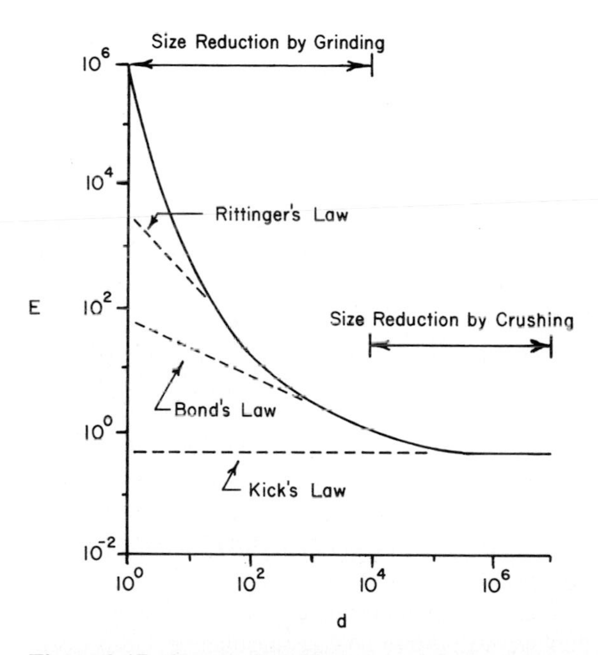

Figure 9.17. Over the last 50 years engineers, studying pulverization processes have put forward several possible relationships linking the energy invested in a pulverization process with the average size of the fragments produced [19] (from E.G. Kelly and D.J. Spottiswood, "Introduction to Mineral Processing," John Wiley & Sons, Inc., New York, 1982. Reproduced by permission of John Wiley & Sons, Inc).

When scientists unfamiliar with powder production in the mining industry are introduced to the problems of energy efficiency in that industry they are often surprised to find that pulverizing processes, in general, appear to be very inefficient. Thus, again quoting from Austin's review:

"experiments on mills show that the fraction of the electric power input to the mill, which is used directly to break bonding forces, is very small (smaller than one percent) and is usually less than the error involved in the measurement of the energy balance".

Mineral processing engineers have always recognized that one of the basic problems to be faced in attempting to relate the energy invested in a pulverizing device to the energy of the surface created is the accurate measuring of the fineness of the powder produced by the pulverization process. Another problem in this approach to energy conversion efficiency evaluation is that not all of the energy transferred to the powdered material appears as new surface. Very often, the fragments can have internal cracks and distorted crystal lattices which themselves have absorbed energy, which is released later as the powder ages (freshly produced powder is often much more chemically reactive than aged powder because of the strain energy trapped in the former).

However, probably much of the controversy concerning the various relationships linking surface energy created with energy input to the pulverizing device can be traced to uncertainty and/or systematic bias in the procedures used to characterize the surface area of the fragments produced by the pulverizer. Most methods for characterizing the surface area of the fragments do not generate absolutely accurate data. Therefore, most of the studies carried out with a given pulverizer – rock system and a specified surface area measuring device can only generate data of relative value.

Although uncertainties in the instrumental methods for characterizing surface areas, and in the definition of exactly what constitutes a new surface, are serious problems in any attempt to quantify the efficiency of pulverizing devices, the discussion of the fragmentation process, from a fractal geometry perspective, presented in this chapter suggests that a far more serious criticism of any attempts to quantify the efficiency of a pulverizing device, in terms of the surface created in the fragmentation process, is the indeterminate nature of the surface of a rough fragment. Since the surface area of a fractally structured fragment is essentially infinite, then one reaches the paradox that all pulverizers will be infinitely efficient, since they all generate infinite surface area during the pulverizing process for a finite imput of energy. As a statement of physical reality this statement is obviously nonsense. It is similar to the crisis that occurred in physics when they reached the strange conclusion, from their trusted and exact theories, when studying blackbody radiators, that all radiant heat sources should instantaneously radiate infinite energy (a **blackbody radiator** is the most efficient absorber and emitter of radiant energy that exists; see the discussion of quantum geometry in Chapter 10). The physicists at that time were very frustrated by this paradoxical impasse. At first, they were reluctant to realize that the problem arose from their failure to understand the basic structure of the energy universe that they were studying. As a consequence of this crisis of their inability to predict the energy emission from a blackbody radiator, physicists had to redefine the basic properties of energy. Reluctantly, they came to the conclusion that energy was not infinitely subdivisible but existed in "lumps" which they called "quanta." As a consequence of this re-definition of their ideas concerning the structure and nature of energy, they were able to open up new vistas in physics in an area of science which is now know as "**quantum theory**." Their philosophical impasse proved to be a gateway to new opportunities. In the same way, mineral processing engineers will probably be frustrated to discover that their attempts to measure the efficiency of their pulverizing equipment in terms of the energy of surfaces created is a false trail leading nowhere because of an overlooked philosophical problem. From the perspective of fractal geometry, mineral processing engineers have been chasing fractal ghosts because the concept "surface area" turns out to be indeterminate. However, perhaps in the long run, a realization that the surface area of a powder is an indeterminate quantity will refocus their attention into paths of investigation which in the next decade will prove as fruitful to the mineral processing engineer as quantum physics has been to the physicist.

Perhaps we can indicate the direction in which pulverization research could move in the future by considering Mandelbrot's discussion of the Korcak relationship describing the structure of an archipelago of islands. He casually mentions under the heading "an innocuous diversion" (page 119 of the second edition of Mandelbrot's book), when discussing the number of islands in an archipelago, that from a practical point of view there are always an infinite number of islands in any group. After I had read this particular statement I re-examined the islands of Lake Ramsey on which Laurentian University is built. From my office window, I could see several large islands in the bay. When I went down to the shoreline I could see that there were several more, smaller islands visible from the shoreline. Down on the shoreline itself, it was very difficult to decide whether a boulder sticking up through the surface of the lake was an island or not. I counted a hundred of these indeterminate islands near my part of the lakeshore and was driven to the conclusion that indeed the number of islands even in one lake was infinite. At the same time, this study of Lake Ramsey forced me to recognize that the problem of "infinite island population" revolves around the definition of an island. The question "How many islands are there in Lake Ramsey?" appears to be a precise question but it has no answer until one establishes what is the operational definition of what constitutes an island. Mandelbrot goes on to point out that it follows from the fact that there is always an infinite number of islands in an archipelago that the coastline of all island groups is infinite. Therefore, the coastline of the largest island is negligible compared with the coastline of the infinite number of smaller islands. We can in effect rewrite Mandelbrot's statement and make it relevant to the mining engineer by deleting the word "island" and replacing with "fragment" in Mandelbrot's original text. Thus, from a fractal geometry perspective, the projected perimeter of a rock fragment and hence the generated surface area are infinite. Therefore, the surface area of all of the fragments is infinite, with the measured surface of area of the few large fragments being negligible compared with that of the overall set of fragments.

At Laurentian University, we recently carried out a study of the compressive failure of a rock core. As I read Mandelbrot's comment on the indeterminancy of the island coastlines, I recalled that in our experiment, as the rock core failed under the applied compressive stress, there was something like a puff of smoke that came from the crushed rock core. Although we had carefully collected the fragments of the shattered rock core to make a study of the fragmentation process, that puff of very fine dust escaping from our experiment probably carried away more surface area than all of that which remained in the collected fragments. Our failure to collect the puff of dust is typical of the studies involving the efficiency of pulverizing equipment in terms of surface area created. The surface area of the fragments that the engineers have collected for study may have been a negligible fraction of all the surface area that escaped from their experiments.

The relationships developed by Mandelbrot in connection with the description of an archipelago hints that the mineral processing engineer should look at a Korcak-type relationship, i.e. a fragmentation fractal, to characterize the fragmentation achieved in a given process. From the perspective of fractal geometry, it is apparent that if the engineer wishes to estimate the energy invested in the fragmentation process, then he must make a direct measurement of the energy which is not used to achieve fragmentation during ballistic impact. Consider, for instance, the ballistic cannon experiment discussed briefly in Figure 9.6. If one measures the velocity of the prime missile section just before impact (this can be done readily by letting the missile intercept two laser beams just before impact), then one can calculate the kinetic energy of the ballistic missile producing fragmentation. We can then let all of the fragments produced by the disintegration process move along the column and hit a screen, which causes them to drop into a liquid bath. One can then determine the temperature rise in the fragments in comparison with their temperature before impact. From the calculated difference one can measure how much energy went into the fragmentation process as distinct from that which ended up as heat in the fragments. The heat rise in this type of equipment could be considerable. For example, a golf ball hit with a standard golf club will experience a 1 °F rise in temperature due to the energy from the blow which does not become transformed into kinetic energy of the moving golf ball.

During the disintegration of the rock, there will be some sound energy given out ("crack" is probably an imitative word based on the sound made when a body breaks) and there may even be some light energy created during the collision, but overall most of the energy in the original missile will either become stored in the surface energy of the fragments or will be transformed into heat during the collision. Increasingly, we can expect pulverization research to focus on arranging a better energy transfer between the object bearing the energy intended to achieve fragmentation and the object to be shattered.

The fragmentation fractal could possibly turn out to be a useful measure of what is achieved during fragmentation since the engineer often is not particularly interested in creating surface energy in his fragmented powder, but rather in achieving a useful set of fragments. For example, if crushing a rock is a preliminary step before microbiological leaching, the engineer would like to be able to fragment the rock to produce a relatively monosized set of fragments which themselves are cracked to provide access into the interior of the fragments for the bacteria which will eventually result in the retrieval of valuable mineral from the fragmented rocks. The engineer interested in microbiological mining definitely does not want to produce several large fragments along with a puff of dust since the large fragments will have a lot of material inside them which is not accessible to the bacteria and the fine dust will be gobbled up quickly by the bacteria with little return for the energy invested in the fragmentation of the rock. Engelman et al. have discussed the type of fragmentation that is the aim of engineers working on the oil shale rocks of southern Israel [26]. They pointed out that the engineers do not like large rocks since they are difficult to handle, nor do they want a lot of fine material since this clogs the oil recovery equipment used to extract the oil from the shattered shale. Engelman et al. described how a group of workers in Israel are aiming to design blasting of the rock in the quarry in such a way that fragments of desirable characteristics are created right in the quarry, to facilitate subsequent fragmentation processes. This integrated approach to the production of rock fragments would appear to be an area of important research potential.

9.5 Brainstorming About Fractal Geometry and the Fracture Resistance of Composite Materials

Originally, powder was added to plastics to lower the cost of the moulded article, because powder, such as crushed calcium carbonate, was cheaper than the plastic. However, it was soon discovered that the calcium carbonate added strength to the plastic. We now know that the strengthening came from the fact that as a crack attempted to travel through the plastic its energy was dissipated when it encountered a powder fineparticle because the latter tended to divert the crack. The energy of the crack was dissipated as it ruptured the interface between the limestone fineparticle and the plastic. This process is illustrated schematically in Figure 9.18. In his book on the new science of strong materials, Dr. Gordon reviews the history of the technology of adding fibres to weak materials to increase their crack resistance and hence their effective strength [20]. He starts his review by considering why the ancient Egyptians used to put straw inside their sun-dried mud bricks. He also reviews the history of the first successful commercial plastic material, **Bakelite** material, discovered in 1906 by Dr. Baekeland. Dr. Gordon tells us that the turning point in the commercial exploitation of Bakelite occurred when Baekeland noticed that the addition of wood fibres to the resin raw material before it was hardened transformed its strength and hardness. Dr. Gordon does not tell us why Dr. Baekeland was exploring the effect of adding wood flour (ground-up sawdust) to his plastic, but it is very probable that he was attempting to cheapen the cost of moulded products by using this filler. Dr. Baekeland had the astuteness to notice that the addition of the wood flour improved the mechanical properties of the substance at the same time as it lowered costs. Dr. Gordon tells us that initially there was consumer resistance to "wood fibre filled" Bakelite. He recounts how people used to say when discussing Bakelite, "they say they put sawdust into the stuff to make it cheaper." Dr. Gordon then tells us how he had to explain to such grumbling consumers that without the sawdust the plastic would have had poorer mechanical properties.

Dr. Gordon in his discussion of how fibrous fillers, such as fibre-glass, improve the strength of composite materials tells us:

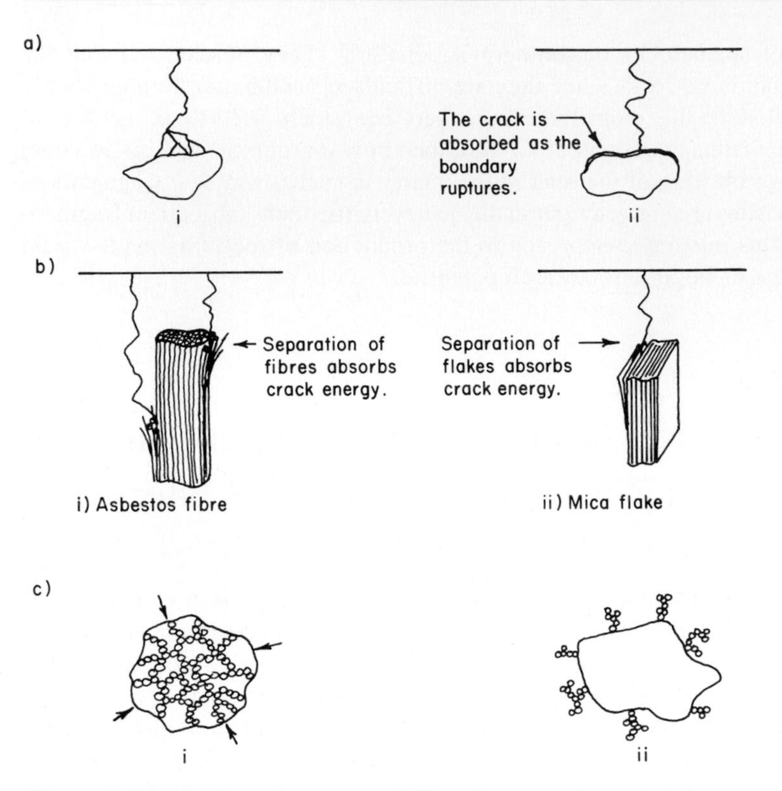

Figure 9.18. The fractal structure of fillers in composite materials, may play an important role in the way in which a filler absorbs crack energy when cracks attempt to move through a composite material.
(a) When a crack meets the filler fineparticle in a composite material the crack can be stopped if the boundary between the matrix and the filler fineparticle ruptures to absorb the energy.
(b) Fillers such as asbestos fibres and mica flakes in composite materials probably are very efficient crack stoppers because the sub-structures of the filler fineparticles rupture as the crack hits the filler fineparticle.
(c) A fractally structured pigment may retain some air internally when incorporated into a plastic. This would give it some compressive resilience with enhanced ability to absorb crack structure, both from its springiness and its very large surface area.
The performance of even simple crushed limestone-type fillers may be enhanced by coating the fillers with silica flow agents which have fractal structures so that they can weaken the bond between the plastic and the filler and also adsorbs considerable crack energy by their own fractally high surface area.

"if we suppose a crack to be proceeding through a resin, it will very soon encounter a fibre. If the material has been properly made, that is to say there is not too much and not too little adhesion between the fibre and the matrix then the fibre will not be broken at that point but the material will crack at the interface between the glass and the resin, the crack will spread along the fibre so that the fibre becomes detached from the matrix often for a considerable distance.

The crack in the resin is very apt to fork wherever it meets a fibre so that the number of cracks in the resin is greatly multiplied. We can generally see this process quite easily whenever a fibre-glass article has suffered from a blow because the material in that region though not broken usually turns white. This whiteness is due to the reflection of light from the surfaces of the many internal cracks. Material in this condition is not much weaker than it was before, although it has already absorbed a good deal of energy simply in providing all those internal surfaces".

If the reader bends a piece of plastic from a cheap plastic product, which probably contains pigment powder and filler powder such as calcium carbonate, he will notice that as he bends the piece of plastic back and forth, it will whiten at the bend. This is because the absorption of the crack energy ruptures the bonds between the filler fineparticles and the plastic.

Some authorities recommend that a crash helmet made from fibreglass resin composite, or other powder-filled material, should not be used a second time after a severe impact has been absorbed by the helmet, because the composite material will have used up some of its ability to absorb energy in absorbing the energy of the crack. However, the structural design and physical properties of composite materials are changing all the time and anyone wishing to investigate this aspect of the use of fibre-glass helmets should consult the currently available information from the manufacturer.

From the foregoing paragraph, it is obvious that studies of the fractal structure of a crack near an interface between plastic and fineparticle filler will be of interest in predicting the energy absorptive capacity of the interface with respect to its ability to dissipate the energy of the crack and hence prevent its further propagation through the material.

Dr. Gordon tells us that one of the reasons why asbestos fibres and mica (a flaky mineral) are effective as fillers in plastic material is that as the crack hits the filler fineparticle, not only does the interface between the plastic and the fibre dissipate energy, but, because of the low energy required to separate an asbestos fibre into its constituent sub fibres the ability to absorb energy from a vigorous crack is enhanced because the various sub-fibres, will rupture from each other. In the same way, the mica flakes will cleave from each other inside the plastic to absorb the energy of the crack.

After reading Dr. Gordon's comments on how fibrous fillers work, it occurred to me that each asbestos fibre is potentially a fractal fibre. Thus, one can keep on dividing an asbestos fibre into sub-fibres ad infinitum, leading to the paradox that any individual fibre has potentially an infinite surface area and constitutes a self-contained fractal system. Indeed, the health hazards from asbestos becomes worse as the asbestos is worked over, because the fibres break down into finer and finer sub-fibres. The finer the sub-fibre generated by working the original fibre, the greater is the health hazard to the worker. This explains why insulation workers are more at risk from asbestos than the original asbestos miner, and why shipwreck salvages taking asbestos fibres out of old ships, and textile workers, are even more at risk than workers preparing new insulation material. Mica flakes are also individual fractal systems since one can keep on splitting each sub-flake split off until one generates a powder of self-similar flakes which, in theory, is composed of an infinite number of infinitely thin flakes having an infinite surface area. The superb energy adsorption capacity of fractal fillers, such as asbestos and mica, is illustrated in Figure 9.18(b). At this point in my consideration of what constitutes a desirable filler fineparticle for a filled plastic, my imagination went into the brainstorming mode to consider some ideas as to why it may be advantageous deliberately to design filler powders having a fractal structure for the plastic industry.

Brainstorming is an idea-generating procedure which I first encountered on moving to the United States in 1963. At that time, when the government put out a request for contract submissions in a new area, the Director of Research of an appropriate division would call together a set of people who might be able to contribute to the preparation of a contract bid. They would all sit around a table, with ample supplies of coffee and donuts, and "brainstorm." In such a discussion, anything that occurred to a member around the table could be put forward and discussed, even if it seemed in retrospect to be a brainless idea. Focusing on a means of getting rid of a silly idea was often as productive as waiting for an inspiration to generate a brilliant idea. Sometimes in these brainstorming sessions, the discussions would start slowly but then as the exchange of ideas heated up, they became an exciting way of exploring new ideas. Sometimes, for the sake of discussion, someone would put forward an idea which seemed to be so way out that it hardly deserved serious attention. These ideas were called blue sky ideas. They often turned out very productive ideas. Here, for what they are worth, are a few **blue sky ideas** that arose as I indulged in some brainstorming concerning the utility of "fractal fillers" for composite materials.

I seem to recall when writing this chapter that back in the days when I worked with calcium carbonate, it was found that precipitated calcium carbonate powder was superior in performance to pulverized limestone powder as a re-enforcement filler. I also knew from a lecture that I had heard in Toronto that ordinary limestone powder, when placed in a typical plastic, will be under considerable stress when the plastic is cooled, because the expansion coefficient of the plastic is much higher than that of the powder. The plastic "shrink-fits" itself around the grains of filler powder as it cools to become rigid in the same way that carpenters used to shrink-fit hot metal bands around an assembled wooden wheel to put it into a compressive state. A limestone filler fineparticle in a piece of plastic can actually bc under stress equivalent to 22 times the pressure exerted by the atmosphere around us. Because of this compressive strength, the boundary between the plastic and the limestone is tenacious and the fineparticle is as likely to crack as the boundary is to rupture. Perhaps calcium carbonate in precipitated form had a fractal structure similar to that of carbon black. This might result in two physical phenomena. First of all, it is highly unlikely that all the internal air will be eliminated from a conglomerate filler fineparticle when a fractal structured fineparticle is blended into a plastic substance. This would lead to the individual fineparticle inside the final composite being able to act somewhat like a spring to absorb the compressive strength generated by the cooling plastic [see Figure 9.18(c)]. This would also mean that there could be a relatively weak boundary between the conglomerate and the plastic because of the air inside the conglomerate (remember that Dr. Gordon tells us that the bond between the fibre and the resin in a fibre-glass composite should not be too good to make good fibre-glass composites). This blue sky idea in turn generated the thought that one might be able to enhance the reinforcing properties of simple limestone powder by adding small quantities of silica flow agent powder to the powder bulk to be fed into the moulding machine. The coating of silica flow agents around a filler fineparticle may act to generate a compressive interface around a filler fineparticle and strengthen the composite material by weakening the surface bond. This would be cheaper than filling the entire plastic with the fractally structured silica fineparticles [see Figure 9.18(c)].

Even if these ideas turned out to be incorrect, it is almost certainly true that the extensive surface area of a filler powder such as carbon black or the nickel powders described briefly in Chapter 1 would be much more efficient at stopping a crack than a simple dense sphere, because of their much higher surface area.

Perhaps composite material engineers will discover that fractal geometry is a powerful tool as they attempt to tailor-make the composite materials of the future and discover new fractalicious fillers!

References

[1] B.B. Mandelbrot, "The Fractal Geometry of Nature," Freeman, San Francisco, 1983.

[2] E.E. Underwood and K. Banerji, "Fractals in Fractography," *Mater. Sci. Engin.*, 80 (1986) 1-14.

[3] J.J. Mecholsky and S. R. Powell, "Fractography of Ceramic and Metal Failures," ASTM Special Technical Publication, No. STP 827, American Society for Testing and Materials, Philadelphia, 1984.

[4] J.L. Chermant and M. Coster, "Fractal Objects in Image Analysis," in "Proceedings of the International Symposium on Quantitative Metallography, Florence, 1978," pp. 125-137.

[5] B.B. Mandelbrot, D.E. Passoja, and A.J. Paullay, "Fractal Character of Fracture Surfaces of Metals," Nature (London), 308 (1984) 721-722.

[6] D.E. Passoja and J.A. Pasioja, "Fourier Transform Techniques in Fracture and Fatigue," in (Eds.) L.N. Gilbertson and R.D. Zipp "Fractography and Material Science," ASTM Special Technical Publication, No. STP 733, American Society for Testing and Materials, Philadelphia,1981, pp. 355 -386.

[7] B.B. Mandelbrot, D.E. Passoja and A.J. Paullay, "Fractal Character of Fracture Surfaces of Metals," in B.B. Mandlebrot and D.E. Passoja (Eds.), "Extended Abstracts of the Meeting on Fractal Aspects of Materials: Metal and Catalyst Surfaces, Powders and Aggregates," Materials Research Society, Pittsburgh, (1984) pp. 7-9.

[8] K.S. Feinberg, "Establishment of Fractal Dimensions for Brittle Fracture Sur-faces," Senior Thesis, Pennsylvania State University, 1985.

[9] J.J. Mecholsky, T.J. Macken and D.E. Passoja, "Self-Similar Crack Propagation in Brittle Materials in Fractography of Glasses and Ceramics," American Ceramic Society, 1987, in press.

[10] J.J. Mecholsky, D.E. Passoja and K.S. Feinberg, "Quantitative Analysis of Brittle Fracture Surfaces Using Fractal Geometry," *J. Am. Ceram. Soc.* (1987) in press.

[11] An international conference on "Fragmentation, Form and Flow in Fractured Media" was held at Neve Ilan, Israel, 6-9th January, 1986. The proceedings of this conference, edited by R. Engleman and Z. Jaeger, have been published as Volume 8 of the *Annals of the Israel Physical Society,* 1986, available from the Israel Physical Society, P.O. 16105, Jerusalem 91160, Israel. There are several papers in this publication relevant to the modelling of fracture and the study of the fragmentation.

[12] B.H. Kaye, "Fractal Geometry and the Characterization of Rock Fragments," in R. Engleman and Z. Jaeger (Eds.), "Fragmentation, Form and Flow in Fractured Media," Proceddings of conference held at Neve Ilan, Israel, 6-9th January, 1986, *Ann. Isr. Phys. Soc.*, 8 (1986) 490-516.

[13] See, for example, news item, "Britain's Geothermal Energy Resources," *New Sci.,* 1st May (1966) 30; this is a review of the book by R.A. Downing and D.A. Gray, Geothermal Energy, the potential in the United Kingdom," published by H.M. Stationery Office, England.

[14] J.W. Dougill, "Structural and Continuum Aspects of Fracture in Brittle Matrix Composites," paper presented at the symposium "Structure and Crack Propagation in Brittle Matrix Composite Materials," held in Warsaw, November 12-15, 1985.
[15] See also D.C. Spooner, C.D. Pomeroy and J.A. Dougill, "Damage and Energy Dissipation in Cement Paste in Compression," *Conc. Res.*, 28, No. 94 (1976) 21-29.
[16] G.C. Sih and A.D. Tommasi, (Eds.), "Fractal Mechanics of Concrete Structural Application and Numerical Calculations," Martinus Nijhoff, Boston, 1985, 276 pp.
[17] See, for example, the review article by B. Johnstone, "Ceramics Faces Its Big Test," *New Sci.,* 12th September (1985) 41-43.
[18] B.H. Kaye, R. Trottier and T.A. Wheat, "Quantitative Characterization of Ceramics by Fractal Geometry," paper presented at the Canadian Ceramic Society Meeting Toronto, February 23-25, 1986, *J. Can. Ceram. Soc.* 55 (1986) 57-66.
[19] A clear and comprehensive introduction to the science of crushing and grinding equipment used in the mineral processing industry is to be found in the book by E.G. Kelly and D.J. Spottswood, "Introduction to Mineral Processing," Wiley, New York, 1982.
[20] J.E. Gordon, "The New Science of Strong Materials, or Why You Don't Fall Through the Floor," 2nd ed., Penguin Books, 1976. In the United States this book (1984) is available as a Princeton University Paperback, Princeton University Press, 41 William Street, Princeton, NJ, 08540.
[21] The photographs used in this section were provided by Dr. Lee A. Cross, Vice President, Special Illumination Systems Inc., P.O. Box 501, Dayton, OH 45409-0501, and are used with his permission.
[22] L.A. Cross, "Pulsed Semi-Conductor Lasers Find Use in High Speed Photography," *Laser Focus Electro-optics,* August (1985).
[23] A. Ahrony, A. Levi, R. Englman and Z. Jaeger, "Percolation Model Calculation of Fragment Properties," preprint provided by Dr. A. Ahrony, School of Physics and Astronomy, Tel Aviv University, Tel Aviv, Israel; R. Englman, Z. Jaeger and A. Levi, "Percolation Theoretical Treatment of Two Dimensional Fragmentation in Solids," *Philos. Mag.* 50 (1984) 2307-2315.
[24] Y. Gur, Z. Jaeger and R. Englman, "Fragmentation of Rock by Geometrical Simulation of Crack Motion, 1," *Engi. Fracture Mech.*, 20, (1984) 783-800.
[25] Z. Jaeger, R. Englman and Y. Gur, "Internal Damage of Fragments," *J. Mater. Sci.*, (1985).
[26] R. Englman, Y. Gur, and Z. Jaeger, "Fluid Flow Through a Crack Network in Rocks," *J. Appl. Mech.*, 5 (1983) 707-711.
[27] E. Louis, F. Guinea and F. Flores, "The Fractal Nature of Fracture," in L. Petronero and E. Posati (Eds.), "Fractals in Physics," Elsevier, Amsterdam, 1986.
[28] Y. Termonia and P. Meakin, "The Formation of Fractal Cracks in a Kinetic Fracture Model," preprint provided by Dr. P. Meakin.
[29] B.B. Mandelbrot, "Squig Sheets and Other Squig Fractal Considerations," *J. Stat. Phys.*, 36 (1984) 519-539.
[30] The suggestion that Squig Fractal systems might be useful in fracture dynamics was made by Dr. Mandelbrot during a lecture on Fractal Systems given at the Boston meeting on Fractals in Material Science, 1985.
[31] J. Korcak, "Deux Types, Fondamentaux, de Distribution Stastitique," *Bull. Inst. Int. Stat.*, 3 (1938) 294-299.
[32] L.T. Austin, "Size Reduction of Solids, Crushing and Grinding Equipment," in M.E. Fayed and L. Otten (Eds.), "Handbook of Powder Science and Technology," Van Nostrand Reinhold, New York, (1984), 562-606.

10 Signposts to More Rambling Explorations of Fractal Space

Signpost 1 General Ramblings

On the morning I finished Chapter 9 of this book, I realized that I had already gone past the tentative page limit suggested by the publisher. As I contemplated all the pathways through fractal space that we could explore, if we had the time and the energy, I felt like the man shown in Figure 10.1.1. I realized that I would have to bring this randomwalk through fractal dimensions to a conclusion, but I was frustrated by the fact that my planning notebook contained draft outlines of several more chapters. Unwilling to abandon these notes entirely, it occurred to me that I could sneak in a final chapter

Figure 10.1.1. The possibilities for future rambles through fractal dimensions seem infinite.

outlining some ideas on where we could ramble in a second book which could be called "More Rambling Explorations of Fractal Space," (my dictionary defines "ramble" as "to wander at will for pleasure or to be incoherent." I had the first definition in mind when selecting the title of this chapter).

The signpost shown in Figure 10.1.1 is like those to be found in England, where signposts wait for one at the end of pleasant country lanes. As you rest for a moment their wooden fingers invite you to explore more potentially fascinating pathways. Because the signpost in the figure invites the reader to a fractal experience, you will notice that the pathways leaving the signpost start to branch in such a way that, whichever path you choose, you would soon lose yourself in infinite possibilities.

The sections of this chapter labeled Signpost 2, etc., are clearly more than simple signposts. They are more like the brochures displayed by a travel agent to tempt you to take various tours of exotic lands. The short discussions of possible tours through fractal land given in each signposted section of this chapter are intended to achieve two objectives. In the original planning of my book I collected a great deal of material from various scientists intending to use the material in the chapters that I would have written. I feel a little embarrassed that I am not able to use all of this material which was graciously provided to me by the many scientists active in fractal land. By listing their papers in the references for each signposted section, I can at least acknowledge their help and direct the reader with specialist interests into the scientifically advanced areas of applied fractal science. Secondly, as I wrote the book I found myself in a situation where new scientific publications were coming out faster than I could incorporate up-to-date details of scientific work for a given area discussed in chapters I had already written. Initially, as I received new material relevant to chapters of the book already written, I went back and rewrote sections of the book to incorporate the new material. It soon became apparent, however, that this was not going to be a possible strategy if a book was to be finished in finite space and time. Therefore, I decided that the best thing to do would be to discuss briefly new work relevant to all the existing chapters in some of the signpost sections, and to keep adding references to each signposted reference section up to the very last minute permitted by the publisher's schedule. Since the intention of this chapter is to give guidance to areas other than those discussed in this book in addition to giving an entry into specialist literature for those who want to delve further into applied fractals, the references are organized by signpost heading and not as a continuous list. Where appropriate, notes and extra references not quoted in the text of the chapter are added to the appropriate section to give extra information to the reader. Each signpost section is in reality a minichapter for tomorrow's rambles.

I hope that in these signposted sections I can indicate some interesting material on fractals of general interest which the readers might discover easily for themselves. Some fractal articles appear in journals not readily available to the public. In this situation, I have also given the address of the writer where it was available. For example, an article by Professor J.J. Mecholsky was published in a magazine published by the College of Earth and Mineral Sciences, Pennsylvania State University [1]. In this article, Professor Mecholsky shows some interesting visual comparisons of similar fractal systems from widely different applied sciences. Thus, in one of his diagrams he shows side by side pictures of the surface of **pyrolytic graphite** (a special type of carbon) next to that of the surface of a **cauliflower**. It is impossible to decide which fractal system is which, unless one looks at the key to the diagram. Again, he shows side by side a photograph of a crack in a metal surface and a picture of the **Red Sea** (a "crack" in the earth's crust) taken from a satellite. The structure of the two are amazingly similar. To enable the reader to track down such fascinating visual comparisons of twin fractal systems, Professor Mecholsky's address is quoted in addition to the appropriate reference.

The way in which fractal geometry concepts have traveled from one subject to another is itself an interesting story. A personal experience involving surprising transfer of fractal concepts from one field of science to another was triggered by a lecture that I gave on Fractal Geometry at a staff development seminar at the research headquarters of the Goodrich Rubber Company in Ohio. The lecture was concerned with the fractal structure of agglomerates of latex spheres. One of the scientists attending the seminar was Dr. John A. Davidson, who is an expert in microphotography and who is also involved in the production of educational films. He immediately saw that fractal geometry could be applied to the problem of characterizing the sharpness of an image in a photographic film. A good quality photograph gives the observer the impression that the outlines of the various items on the photographs are sharp boundaries. However, if one looks at the boundary of a figure in a photograph through a microscope, it can be seen that the image is made up of clouds of silver crystals. Determining the exact limit of the boundary of a photographic image is similar to deciding where a cloud begins and clear air ends. Subsequently in a scientific publication, Dr. Davidson and his colleague Denis Keller discussed the problems of determining the fractal dimension characteristic of the sharpness of the boundaries of a photographic image. Their discussion of boundary sharpness ranged from a quantitative discussion of the sharpness of a saw blade through to the boundary of a photographic film. The exact location of the edge of a boundary in a photographic film is an important problem when attempting to analyse satellite pictures using colour vision cameras and computer data processing systems, so that Dr. Davidson's work, which started with the quality of educational film images has applications in space science and in determining the location of boundaries in medical X-rays. Beginning fractal enthusiasts might like to know that the paper by Davidson and Keller contains an appendix giving an introduction to the mathematics of the Koch Triadic Island [2].

Late in 1985, Dr. John Davidson telephoned from Ohio to tell me about his discovery of fractal imagery in a science fiction story. The book in which he had discovered this fractal imagery was "Sentenced to Prism," by Allan D. Foster [3]. (Mr. Foster also wrote the book, "The Black Hole," which was made into a movie of the same title.) In "Sentenced to Prism," which is about a visit of Evan Orgel (I suspect this name came from the phrase "evolved structured organic jelly" - a biochemical description of human flesh?) to a planet teeming with silicon-based lifeforms. Fractal systems form part of the scenery of the planet Prism. On page 5 of the novel we find this portion of text:

"Something was coming toward him (Evan). Whatever was approaching wasn't very big but then it wouldn't have to be to do severe damage given his helpless semi-comatose state. He couldn't see it clearly because the special discriminatory visor on his suit helmet wasn't functioning properly. The visor was necessary because many of Prism's lifeforms were organized according to fractal not normal geometry. They tended to blur if you stared at them for very long, as the human eye sought patterns and organizations where none existed. Fractals existed somewhere between the first and second dimension, or the second and the third. No one, not even the mathematicians, were sure."

Of course, the readers of this book will know that they exist between all of the dimensions not just between any given pair. The episode goes on:

"It didn't matter so long as you looked through the Hausdorf lenses. They were built into the visor of his suit's helmet, which was broken. As a result the fractally organized figures didn't look quite right when viewed through unadjusted transparencies."

If you read this science fiction book you will find that fractal concepts rear their head every now and again during the evolution of the story. When Dr. Mandelbrot visited Laurentian University in the fall of 1986, the Physics students presented him with a copy of the book. He was amused to find that his theories were affecting even the artistic styles and concepts of science fiction.

There is one change I would have recommended to the author of, "Sentenced to Prism" if I had read his book in draft form. He should have given the space traveller "Mandelbrot lenses," not a Hausdorf visor, to be able to see fractal reality. **Hausdorf** was a mathematician who did explore the theory of mathematical dimensions theory, but it was through Mandelbrot's vision that the world has been able to see the beauty and utility of fractal geometry. Sometimes in the scientific literature on applications of fractal geometry in science, a fractal dimension is described as a Hausdorf or a **Besicovitch dimension**. I think it is better to describe the allocation of numbers to describe ruggedness space as a fractal dimension rather than to invoke the name of Hausdorff and Besicovitch, who explored dimensional space, but who did not invent fractal geometry.

References

Specific References

[1] J.J. Mecholsky, "Fractals Fact or Fiction in Earth and Mineral Sciences," *News Magazine of the College of Earth and Mineral Sciences*, University Park, Pennsylvania, 55 (3), (1986) 29-33.

[2] J.A. Davidson and D.J. Keller, *Soc. Motion Picture Television Eng.*, August (1985) 802 - 809.

[3] A.S. Foster, "Sentenced to Prism," Ballantine Books, New York, 1985.

References of General Interest

[1] D.J. Albers and G.L. Alexander (Eds.), "Mathematical People; Profiles and Interviews," Birkhauser, Boston, Basle, Stuttgart, p. 207 - 225; this is a record of a conversation between Benoit Mandelbrot and Anthony Barcellos.

[2] P. Bak, "Doing Physics with Micro Computers," *Phys. Today*, December (1983) 25-28. This article contains several pictures of computer-generated Monte Carlo simulations of Ising models of interest to the solid-state physicist. There pictures of solid-state systems look exactly like simulated powder mixtures to fineparticle specialists. The computer programs discussed in the article can be used to simulate powder mixtures structures.

[3] P. Bak, "The Devil's Staircase," *Phys. Today*, December (1986) 38-45. This article discusses how the range of massive frequencies of a driven oscillator can be related to the infinite possibility of steps on a devil's staircase. In this article the use of Cantorian sets and fractal dimensions is applied to the range of frequencies manifested by various systems of interest to physicists.
[4] M. Batty, "Fractals Between Dimensions," *New Sci.*, April 4 (1985) 31-35.
[5] P.W. Carlson, "IBM Fractal Graphics," *Compute*, March (1986) 78-80.
[6] C. Davis and D. Knuth, "Number Representation and Dragon Curves," *J. Recreational Math.*, 3 (1985), 66-81 and 33-149.
[7] A.K. Dewdney, "Computer Recreations - Exploring the Mandelbrot Set," Sci. Am., August (1985) 16-24; "Computer Recreations, Beauty and Profundity," *Sci. Am.*, November (1987) 140-145.
[8] R. Dixon, "Geometry Comes Up To Date," *New Sci.*, May 5 (1983) 302-305.
[9] E. Edelson, "The Ubiquity of Nonlinearity," *Mosaic*, 17, No. 3 (1986) 10-17.
[10] A colleague, Dr. Derek Wilkinson, browsing through a bookstore in Toronto, discovered a delightful, but small book which, although it does not mention fractals, has a great deal of information on fractal systems disguised under the title "Physics and Geometry of Disorder Percolation Theory," by A.L. Efros, translated from Russian by V.L Kisin, Mir, Moscow, 1986. (Mir Publishers, 2 Pervy Rizhsky Pereulk 1-110, GSP, Moscow 129820, USSR).
[11] K.J. Falconer, "The Geometry of Fractal Sets," Cambridge University Press, Cambridge, 1985, 162; this is a book on the mathematics of fractal sets for specialist scientific readers.
[12] M. Gardener, "Mathematical Games in which Monster Curves Force Redefinition of the Word Curve," *Sci. Am.*, December (1976) 124-133.
[13] Fractal generating algorithms are outlined in J. Holbrook and J. Weiner, "A Fractal Workbook," Mathematics and Statistics Department University of Guelph, Ontario, Canada, N1G 2W1, 1983.
[14] R. Jullien and R. Botet, "Aggregation and Fractal Aggregates," World Scientific Publishing Co., Farrer Rd., P.O. 128, Singapore 9128, 1987, 144 p. In North America this book can be ordered from Princeton University, Princeton, NJ.
[15] M. Kac, "More on Randomness," *Am. Sci.*, 72(3) (1984) 282-283.
[16] The reader without a mathematical background may find a copy of an essay by B.H. Kaye, "From Euclid to Mandelbrot, a History of the Evolving Vocabulary of Fractal Geometry," to be a useful semantic survival kit to keep by their elbow as they venture into fractal geometry. Copies are available from the author at Laurentian University, Sudbury, Ontario, at a price of $15 Canadian to cover printing and handling costs.
[17] D. Malmberg "Fractals and Other Diabolical Designs," *Commodore Power Play*, June-July (1986) 88-92.
[18] B.B. Mandelbrot, "Getting Snowflakes Into Shape," *New Sci.*, June 22 (1978) 808-810.
[19] Interview with Benoit B. Mandelbrot, *Omni*, September (1984).
[20] J. McDermott, "Geometrical Forms Known as Fractals Find Sense in Chaos" *Smithsonian* (a journal published by the Smithsonian Institute, Washington D. C.), 12, No. 9, December (1983).
[21] K. McKean, "The Orderly Pursuit of Pure Disorder," *Discover*, January (1987) 72-81. This article discusses the problem of generating random numbers and of applying random numbers theory to the generation of secret codes for protecting bank records and secret messages sent over telephones. The article is written at an introductory level and has some excellent graphics.
[22] A.K. Mon, "Self-Similarity and Fractal Dimension of a Roughening Interface" by Monte Carlo Simulations," *Phys. Rev. Lett.*, 57, (1986) 866-868. The author points that fractal dimensions are involved in interfaces in such different areas of applied physics as adsorption, diffusion, surface crystal growth, wetting and pore structures of rocks. The references listed enable the working scientist to begin an in-depth study of the fractals in various areas of applied physics.
[23] A. Norton, "Generation and Display of 3-D," *Computer Graphics*, 16(3) (1982) 61-66.

[24] M.M. Novack, "Advanced Graphics with the Commodore 128," MacMillan, New York, 1986.
[25] M.M. Novack and J. Weber, "Fractal Sets," *P.C.W.*, December (1986) 196-199.
[26] L.N. Sander, "Fractal Growths," *Sci. Am.*, January (1987) 94-100.
[27] B. Schechter, "A New Geometry of Nature," *Discover*, June (1982) 66-68.
[28] Sorensen, "Fractals," *Byte*, September (1984) 157-172 (This contains instructions on how to generate fractal geometric shapes.
[29] L.A. Steen, "Fractals, A World of Non-Integral Dimensions," *Sci. News*, August 20 (1977).
[30] K. Stein, "The Fractal Cosmo," *Omni*, February (1983) 62-71, 115.
[31] To find algorithms for generating fractal geometric patterns on an Apple computer, see Thornberg, "Discovering Apple Logo, an Invitation to the Art and Pattern of Nature," Addison-Wesley, New York, 1983, Ch. 11.
[32] S.M. Thorpe, "Mandelbrot Graphics for the Commodore," *Compute*, July (1986) 98-101.
[33] J.B. Tucker, "Computer Graphics Achieve New Realms," *High Technol.*, June (1984) 40-53.
[34] R. Williams, "The Electronic Chalkboard Fractals," *Intell. Instrum. Comput.*, 13, February-March (1985) 2.
[35] T.A. Witten and M.E. Cates, "Tenuous Structures from Disorderly Growth Processes," *Science*, 232, (1986) 1607-1602.
[36] D. York, "Rough Edge of Math Leads to Scenery by Computer," *Appl. Math Notes,* June (1983) 35-38.
[37] P.W. Carlson "IBM Fractal Graphics," *Compute*, March (1986) 78-80.

Signpost 2 Fractal Scenery and Artistic Vision

In Figure 10.2.1 (see Plate 8 at the beginning of the book), one of the whirling, intricate graphic displays created by using the theories of fractal geometry is shown [1]. It is the visual beauty of this type of fantastic fractal that has caught the imagination of many people, both in the original diagrams of Mandelbrot's book, and other graphic creations published in many articles and books [2]. In a lecture given at Laurentian University in 1986, Dr. Mandelbrot indicated to the audience his personal pleasure at watching the evolution of a fractal pattern from an algorithm given to the computer. Apparently the mathematician can only anticipate in general terms the pattern which he is going to generate using the theories of fractal geometry and sometimes, when something totally unexpected goes wrong, non-reproducible patterns appear on the computer screen. Thus, Mandelbrot included in his lecture on fractal geometry the picture shown in Figure 10.2.2. His comments on this picture are as follows [3]:

"This picture can be credited in part to faulty computer programming. The **bug** (a term used by computer scientists to indicate an error in their computer program) was promptly identified and corrected but only after its output had been recorded. The change that had been brought by a single tiny computer bug in a critical place was well beyond anything we had expected. It is clear that a very strict order is designed into this picture; here this order is hidden and no other order is apparent. The fact that at least at first blush the picture could pass for high art cannot be an accident. My thoughts on this account are sketched elsewhere and are to be presented fully in the near future."

Figure 10.2.2. Mandelbrot tells us that this picture is not a product of the artist's skill. Its structure is due to a "bug-disturbed" fractal algorithm (from B.B. Mandelbrot, "The Fractal Geometry of Nature," W.H. Freeman & Co., San Francisco, 1983. Reproduced by permission of B.B. Mandelbrot).

There is no doubt that there will be many studies in the future of the implications of visual patterns generated by experts in fractal graphics for the theories of art, and in the analysis of the esthetic appeal of various designs.

When I heard Dr. Mandelbrot lecture on the role of error in producing the creative style of "high art" shown in Figure 10.2.2, I recalled reading somewhere that when computer scientists attempted to design automated systems for the production of Persian-type carpets using computer-controlled looms, the computer-controlled carpet pattern was geometrically perfect but lacked "brilliance." The computer experts were then able to give the equivalence of "brilliance" to their patterns by introducing small errors into the programming of the positioning of the individual tufts used to create the carpet. Apparently the presence of the small flaws in the pattern distracts the eye in its scanning inspection pattern, and gives the equivalence of brilliance to the geometric structure of the pattern. I also remember reading somewhere that the carpet makers of the East deliberately build small errors into their geometric patterns while making carpets. They do this on supposedly religious grounds based on the belief that man should not make a perfect thing. They believe that only God can or should make a perfect object. This superstition may correspond to an empirical discovery by ancient weavers that the brilliance of the carpet to the human eye depends upon small random deviations in the structure of the pattern.

There have been some discussions of the philosophical implications of fractal geometry and the structure of the universe. For example, Dr. Mandelbrot discovered a picture in an ancient manuscript, "Bible Moralisee," which was written between 1220 and 1250 in the Eastern Champagne dialect of French, in which God is shown building the universe using both Euclidian geometry and fractal shapes [4]. Perhaps fractal geometry is a far more ancient art than modern man suspects.

"WE DID THE WHOLE ROOM OVER IN FRACTALS."

Figure 10.2.4. The visual impact of fractal design is such that we can anticipate the commercial availability of fractally decorated fabric and wall paper. However perhaps a fractally decorated room by the cartoonist S. Harris might be too overwhelming! (Used by permission of S. Harris.)

The intricacy and beauty of fractal graphics has caught the imagination of the scientific world. This is illustrated by the fact that several major scientific journals have used fractal graphics to decorate their front pages. Thus, fractal graphics have appeared on the front covers of such journals as *American Scientist, Science, Nature, Physics Today, Research and Development* and *Scientific American* [5]. Scientists are also using fractal graphics to demonstrate the resolution of computer graphics and image display systems. For example, the graphic shown in Figure 10.2.3 (see Plate 9 at the beginning of the book) is taken from the commercial literature produced by a manufacturer of image analysis systems [6]. When I first saw this picture I thought it was an actual photograph of a nerve cell. Some scientists think that fractal geometry may help them to understand the structural order observed in the patterns of normal living cells, and in the structure of diseased cells constituting cancer tissue. Thus, by modelling the growth of fractal systems that look like healthy or diseased tissue they may be able to understand the growth guiding information built into the genetic code, and how cancers develop when such structure-generating information is disrupted.

I remember that the first time I saw a "fractalscape" such as that shown in Figure 10.2.1, I thought about the commercial possibilities of selling various items bearing fractal designs. One of the products I thought of was wallpaper covered with fractal designs. I think, however, that to use such designs in domestic decoration to the extent illustrated in Figure 10.2.4 would be a little too overwhelming! By the end of 1986, fractal postcards and fractal posters were beginning to hit the market and the annual fractal calendar cannot be too far behind. Thus, in Figure 10.2.5(a) and (b) (see Plate 10 at the beginning of the book), both sides of a fractal New Year's greeting card, received by a geologist colleague, wishing him appropriately enough "infinite happiness in the New Year," are shown. Another **fractalicious** (a new word meaning a delicious fractal) artistic creation appears in Figure 10.2.5(c), which shows a fractal sea-shell on a postcard sent to me by a **fractophilic** friend (fractophilic – a new word meaning "lover of fractals").

The use of fractal logic to create landscapes that look out of this world has already been exploited by Lucas films in designing scenarios for their movie "The Return of the Jedi." In the colour section placed in the middle portion of Mandelbrot's book, some of the beautiful simulated landscapes created using fractal geometry algorithms are shown. Figure 10.2.6 (see Plate 11 at the beginning of the book) shows some mountains generated by computer simulation which where used by IBM in some widely publicized advertisements. These graphics were generated by Richard Voss and perhaps a future article on "Rambling Through the Voss Mountains" would prove interesting.

References

[1] Alan Flook tells me that he generated the artistic pattern in Figure 10.2.1 when testing the quality of various colour printers he was considering purchasing.

[2] For a very comprehensive book full of beautiful computer-generated fractal graphics, see H.O. Peitgen and P.H. Richter, "The Beauty of Fractals," Springer-Verlag, Berlin., 1986.

[3] Dr. Mandelbrot has given some of his views on the relationship between fractal geometry and art in B.B. Mandelbrot, "Scalebound or Scaling Shapes. A Useful Distinction in the Visual Arts and in the Natural Sciences," *Leonardo*, 14 (1981) 45-47.

[4] See B.B. Mandelbrot, "Fractal Geometry of Nature," Freeman, San Francisco, 1983, p. 276.

[5] See the front covers of *Scientific American*, October 1985; *Physics Today*, Febru-ary 1986; *American Scientist*, January-February 1986 and the cover of Science, February 1986.

[6] Reproduced and taken from the commercial literature (by permission) of Tracor Northern, 2251 West Belt Line, Highway, Middleton Wisconsin, 53562-2697, USA.

[7] A.K. Dewdeney, "Fractal Mountains, Graftal Plants and Other Computer Graphics," *Sci. Am.*, December (1986) 14-20.

[8] Pictures from the greeting card used by permission of ACDS Graphics Systems Inc., 100 Edmonton St., Hull (Quebec), Canada, J84 6N2.

[9] Fractal design decorated postcards and other fractal graphics are available from Art Matrix, Post Office Box 880, Ithaca, New York 14851, USA. Picture used with permission.

[10] H.O. Petigen (Ed.), "The Science of Fractals, A Computer Graphical Introduction," Springer-Verlag, Berlin (1987).

[11] B. Schechter, "Fractal Fairy Tales," *Omni*, October (1987) 86-92.

Signpost 3 "Fractal Gamblers"

If one looks at the record of stock market fluctuations, as shown in the financial column of a newspaper, its jagged structure invites fractal analysis. Mathematicians who would like to make their fortune in the stock market have frequently tried to apply signal processing procedures to the stock market record to see if one could break it down into component structures, into basic rhythmic components that could be useful in predicting the future behaviour of the stock market. Thus, Fourier analysis has been applied to the stock market records on the assumption that they contain certain cyclic phenomena. Even without Fourier analysis it is obvious that there is a four year cycle built into the United States stock market which can be linked to the four year Presidential election pattern. Note, however, that the four year economic cycle of the stock market is about one year out of phase with the actual date of the election. This phase lag occurs because of the fact that all the gifts given out by government seeking to be re-elected stimulates the economy well into the year after the election, at which time it becomes safe to break election promises.

It is interesting to break the stock market record into several sequential sections, and to compare the fractal dimension of the fluctuations in one period with that over another period. However, if we recall that the basic idea which is being developed throughout this book is that phenomena that show fractal boundary structure are probably generated by the random unpredictable interaction of many causes, the obvious fractal nature of the stock market record seems to emphasize that perhaps the basic underlying force in any stock market is the interaction of many random causes. Therefore, the person who makes a fortune on the stock market is not necessarily a genius, but in all probability is a child of random chance. The financial expert may well be the winner thrown up by random chance. (remember, a traitor is a hero who lost.) When listening to the advice of the stockbroker, one should realize that his advice is usually no more reliable than that of the witch doctor looking at a pattern of chicken bones thrown on to the ground. If the stockbroker was really good at predicting stock movement he would be so rich that he wouldn't have to work for a living advising you what to do with your money. However, for those who would like to explore the fractal dimensions of the stock market and other economic systems, there is a chapter in Dr. Mandelbrot's book dealing with the fractal structure of the economic system.

Signpost 4 Lakes, Islands and other Geofractals

As we discovered in Chapter 1, one of the first fractal problems studied by Mandelbrot involved the study of the infinite nature of the coastline of Great Britain. If the reader browses through any geography and or geology book, picture after picture will be discovered that can be characterized by fractal dimensions. I had intended to write a whole chapter on the applications of fractal geometry to geography and geology, but all we can do in this signpost section is point the reader in several interesting directions.

The reader interested in fractal geometry and geography will find a wealth of information in the publications of Dr. Goodchild and co-workers [1, 2].

The speed at which the geographers are applying fractal geometry to their systems can be appreciated from the fact that as early as 1984, Dr. Mark of the State University of New York had already compiled a three page bibliography on the use of fractal geometry in geography and cartography [3] (**cartography** is the subject dealing with the theory and technology of constructing maps). In a course on fractal geometry held at Laurentian University students have measured many fractal coastlines, quantifying by fractal dimensions, for example, the difference between the Atlantic side and the western side of Newfoundland, the fractal dimension of Norway's highly indented coastline and the structure of Manitoulin Island. This island, shown in profile in Figure 10.4.1, is the largest island in a fresh water lake in the world. It can be seen that the

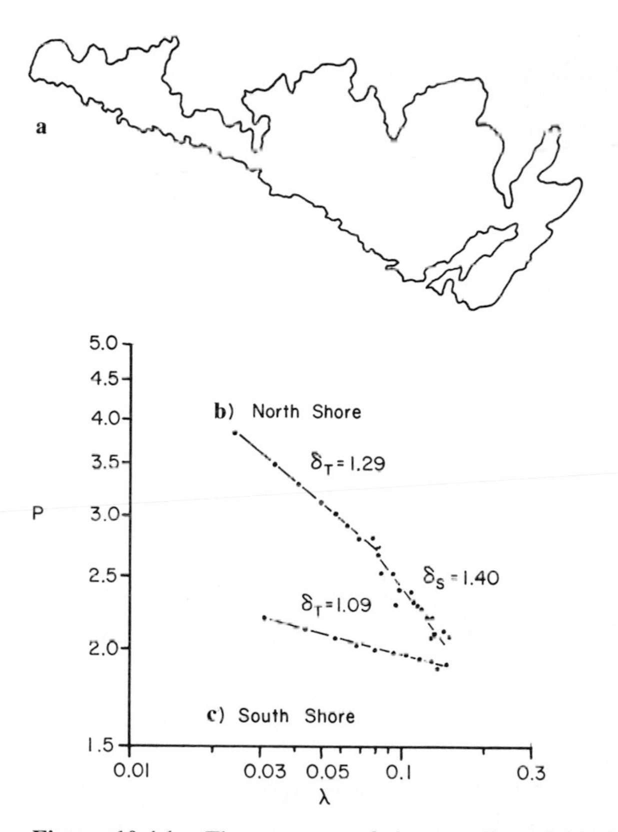

Figure 10.4.1. The structure of the coastline of Manitoulin Island shows that the fractal dimension of two coastlines exposed to very different erosive forces manifest very different fractal boundaries. It is probable that the ruggedness of the coastline can be related to the erosive forces.

(a) Profile of the island.

(b) Richardson plot for the north shore of the island.

(c) Richardson plot for the southern shore of the island showing that at coarse resolution the boundary appears to be virtually Euclidean.

erosive forces forming the two sides of the island have produced shorelines of very different ruggedness. In Figure 10.4.1(b), a set of data generated by a student studying the fractal structure of the coastlines of the island is shown.

Several geologists have started to apply fractal geometry to the study of the structure of rocks. Sandstone in particular appears to be an excellent model of a porous body with fractal structure.

In Figure 10.4.2(a), a famous mathematical figure known as an **Apollonian gasket** is shown (Apollonius was a Greek mathematician who worked on geometric problems in the city of Alexandria in the year 200 B.C.). Mandelbrot discusses the fractal structure of an Apollonian gasket system in his book [4]. Mandelbrot also talks about the fact that the circles in the gasket are an example of **Apollonian packing**. The first time I browsed through Dr. Mandelbrot's book, I remembered seeing the same diagram on the front of an article reviewing research into the crushing and grinding of powder.

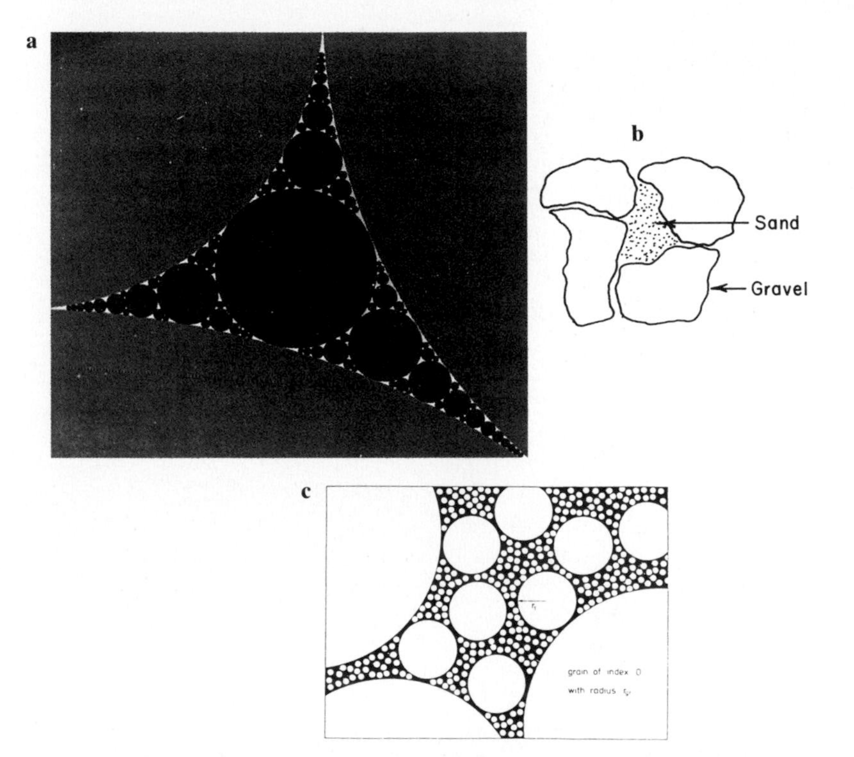

Figure 10.4.2. Several apparently dissimilar systems can be described by the fractal system known as an Apollonian gasket.
(a) Mathematically perfect Apollonian gasket. (From B.B. Mandelbrot, "The Fractal Geometry of Nature," W.H. Freeman & Co., San Francisco, 1983. Reproduced by permission of B.B. Mandelbrot.)
(b) Diagram from a technical paper discussing the structure of concrete [6].
(c) Figure taken from a research publication on the fractal structure of sandstone [7]. (From H. Pape, L. Riepe and J.R. Schopper, "The Role of Fractal Quantity as Specific Surfaces and Tortuosities for Physical Properties of Porous Media," Part. Character., 1, 66-73 (1984); reproduced by permission of H. Pape.)

Obviously, to chemical engineers the Apollonian gasket represents a crushing process in which the large grey balls are squeezing the smaller ones to create smaller and smaller fragments. However, the diagram also recalls a problem I studied when working on the design of nuclear reactors. Engineers building nuclear reactors were anxious to create concrete shielding walls having the maximum density per unit volume. Such dense concrete has a maximum stopping power when surrounding an atomic system giving off radioactive fragments from the atoms generated by nuclear fission. It has been shown that the best way to create really dense concrete is to assemble a primary packing of large lumps of rock into which smaller rocks, which can fit into the gaps between the bigger rocks, can be poured. This second stage is then followed by infiltration into the structure of yet smaller sand fineparticles which fill the gaps between the first and second sets of rocks [5]. To aid the infiltration process, the dried packed spheres are vibrated as each component is fed into the structure. The final additive to this structure is a thin paste of cement in water to which detergent has been added to aid the wetting process. This final thin paste (called a slurry) then drains down into the packing and sets the whole system to a solid mass.

During the time that I was writing this signpost, I was also involved in buying a new house which the builder built upon a rock foundation. First of all he cleared the clay and packed about 20 feet of blasted rock pieces of average size about 20 cm to fill up the ravine on which he was to build my house. He then infiltrated crushed slag of average size 2–3 cm into the gaps between the large rock fragments. He subsequently infiltrated sand into the crushed slag with water to help the finer sand grains to infiltrate down into the smaller voids. Thus, I am able to tell my students that my new house is built upon an Appollonian gasket foundation representing a many stage-fractal system! I am sure that the workers who built the foundation of my house are unaware of the Appollonian gasket, but as in many other areas of applied science they had by trial and error found out the best way to build a solid rock foundation using a cascade of smaller and smaller infiltrated fineparticles.

Even in ordinary concrete structures, the size ratio between the crushed gravel, fine sand and cement constituents of the concrete mix are chosen so that when the whole assembly is poured into a mould, the sand fills up the gaps between the crushed gravel and the cement grains fill up the gaps between the sand grains. A sketch from a scientific paper discussing the structure of concrete is shown Figure 10.4.2(b) [6]. This appears to be a statistically self-similar Apollonian gasket.

If one focuses attention on the theoretical Apollonian gasket figure, one sees that in any one pocket formed by the bigger disks there is an infinite number of smaller and smaller disks. One cannot tell the magnification with which one would look at an overall structure of an Apollonian gasket because the whole system is self-similar at any scale of scrutiny. Thus, the size distribution of the different sized disks packing the gasket is a fractal system. It follows from the similarity of the elements (a) and (b) in Figure 10.4.2 that the required size distribution of the various components of concrete should be a fractal system similar to that of an Apollonian gasket system, even though some departure from the ideal size distribution function may be necessary to ensure accessibility of the various components to the various internal voids, as one successively infiltrates into the structures smaller and smaller components.

In practice, it would be necessary to use a series of fractions rather than a continuous distribution of components, and there is obviously a limit in real concrete technology to the size of the smallest grains that one can make and infiltrate into the overall structure. Whole books have been written on the packing of powder systems. It now appears that many of the empirically determined desirable size distributions, for different concretes having different properties, are fractal systems. Scientists interested in creating dense ceramics or metal alloys with different properties use the same type of sequential assembly of various size grain components infiltrated one after the other into the packed container. Therefore, the ceramic and powder metallurgist specialist can use the concepts of fractal geometry to study the powder mixtures they use to make materials with desired properties.

In the late 1950s, scientists studying the fabrication of dense concrete took out patents for processes for use in the small-scale construction of such items as concrete garden paths using the concept of dry assembly of the various components. Thus, one would lay down the crushed rock forming the coarsest part of a concrete mixture in the excavated hollow to contain the path. One would then add sand to fill up the holes between the crushed rock. The final cement slurry, plus detergent, is then poured into the rock-sand mixture using a watering can. Apparently the process failed to catch on, since workers who were used to hauling barrow loads of concrete around could not be persuaded of the effectiveness of the method for making concrete – perhaps the process did not seem consistent with the traditional image of the strong-armed construction worker man-handling the raw wet concrete en route to its final destination.

Figure 10.4.2 (c) comes from yet another different aspect of applied technology. It is a model of sandstone structure developed by Dr. Pape, Professor Schopper and co-workers [7], who are geophysicts and mathematicians interested in the structure of oil-bearing sandstone. They developed their model in an attempt to be able to describe the tortuosity of the paths within the sandstone, through which the oil has to move, and the internal surface area of the sandstone, since this determines the holding capacity and the resistance to oil flow as the oil moves through the sandstone. In their paper they state:

"The theory of fractal dimensions in three-dimensional space, originally limited to topologicial curves, surfaces and volumes, are extended into physical properties like specific surfaces, tortuosity, porosity and formation factor."

When discussing the evolution of their model of sandstone they use the term "pigeon hole model." This **pigeon hole model** is virtually identical with the one I have used to discuss the Apollonian structure of dense concrete. In nature, sandstone will often be composed of one-sized grains deposited by moving water with the later infiltration of smaller sized grains with the subsequent arrival of even smaller grains, and so on. Thus, we have the deposition of smaller and smaller sizes of grain within the holes left by each successive grain size. Thus, mother nature solved the problem of how to create dense concrete structures long before nuclear engineers attacked the problem (see also the discussion of the fractal structure of rocks in references 8 and 9).

The sandstone model developed by Pape, and Schopper and co-workers is obviously relevant to the type of system discussed in Chapter 8 when we were considering the application of fractal geometry to the problems encountered when discussing techniques for the secondary recovery of oil. The discussion of their model for sandstone was delayed until the discussion of the structure of an Apollonian gasket, since it seemed helpful to bring together in one diagram various related models found in different areas of engineering technology.

Agricultural soil is a mixture of different sized fragments of different sized materials. It is the structure of the holes and channels within the soil which governs such important things as water drainage and the aeration of crop roots. Thus, drainage models in soil systems developed by soil scientists will be similar to the models developed by Dr. Pape and co-workers.

The reader might think that the graphic display shown in Figure 10.4.3 (see Plate 12 at the beginning of the book) was produced in a manner similar to that of the fractal pattern in Figure 10.2.1. In fact, it is taken from an article on prospecting for valuable minerals using "false colour" computer processing of pictures taken from a satellite [10]. It may be that the fractal structures of the several sub-areas of the systems shown in various colours in this satellite picture may not have any significance, but on the other hand it may be a way of identifying significant structural features in this type of photograph.

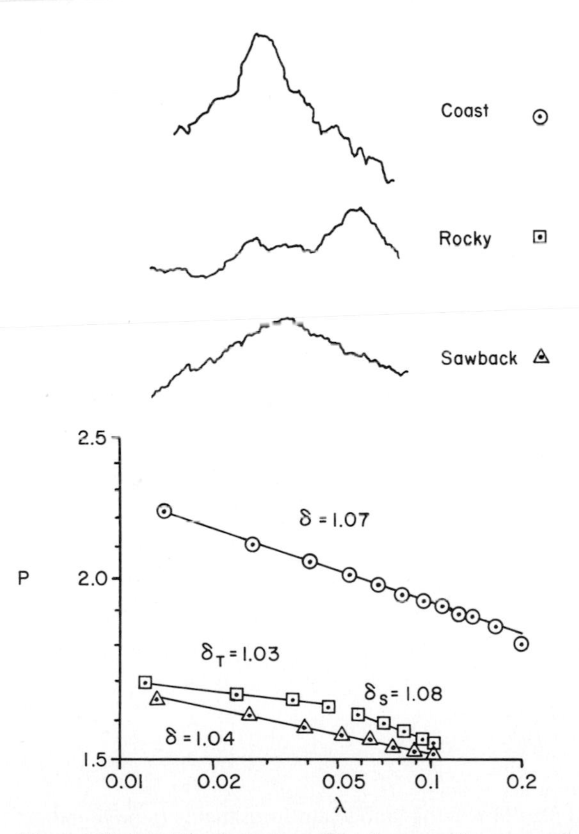

Figure 10.4.4. The fractal profile of mountains can probably be related to their formation kinetics and the erosive forces to which they have been subjected since their formation.

A photograph showing the outline of mountain ranges obviously manifests fractal structures. If one compares a photograph of the sharp-edged peaks of the Himalayas, which from a geological point of view are young mountains, with the rounded mountains of South Africa, which are geologically old, then one can see that from the perspective of fractal geometry, erosion reduces the fractal dimension of the mountain range. In Figure 10.4.4, the profiles of several different mountains are shown, together with the Richardson plots of the Dapple dilation data for evaluating the fractal structure of the mountain profiles [11].

Not only are the elevation profiles of mountains fractals, but the ground plan (iso-height contour lines) are also fractals. Thus, Figure 10.4.5 shows the outline of a Bornhardt, a type of worn-down old mountain to be found in South Africa, and the Richardson plot, from which the fractal dimension of the growth plan is determined. One feels intuitively that the fractal dimensions of such geological shapes have to be related to the geological forces that formed the Bornhardt [12]. Perhaps fractal dimension summaries of a series of contour profiles climbing the topography of a mountain could be a convenient summary of the structure of the mountain.

River erosion cutting through the rocks can create new, very high dimension fractal structures. The profiles of lake bottoms generated by ultrasound scanning, and the profile of landscapes estimated by using laser equipment mounted on aircraft can all be characterized by fractal dimensions [13, 14]. Mandelbrot discusses the fractal structure of river systems in his book.

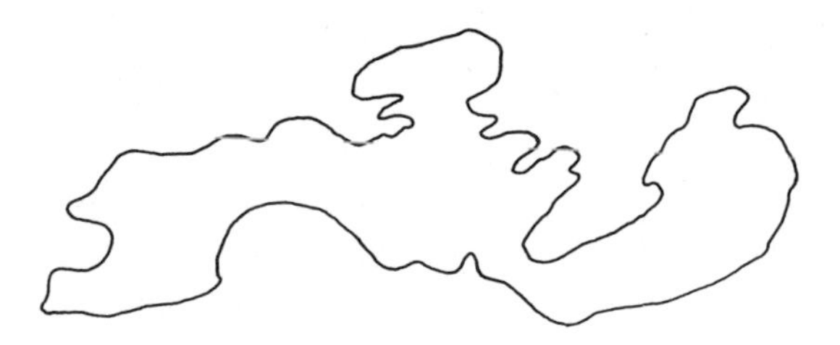

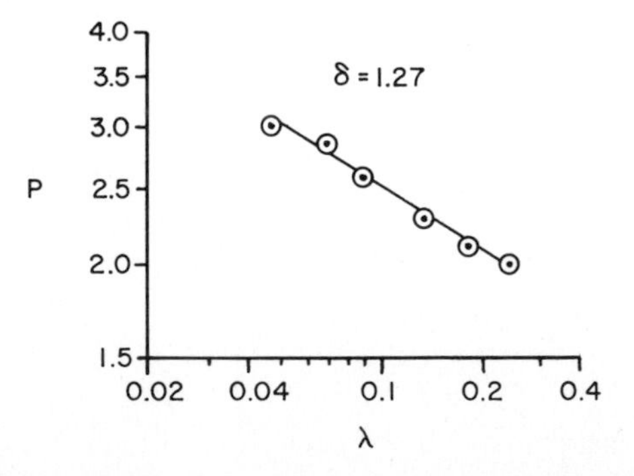

Figure 10.4.5. The outline of the ground plan of a unique mountain form that is to be found in places such as South Africa and parts of Australia. A Bornhardt, named after its discoverer, has fractal dimensions that may be related to the geological forces which formed the mountain [12].

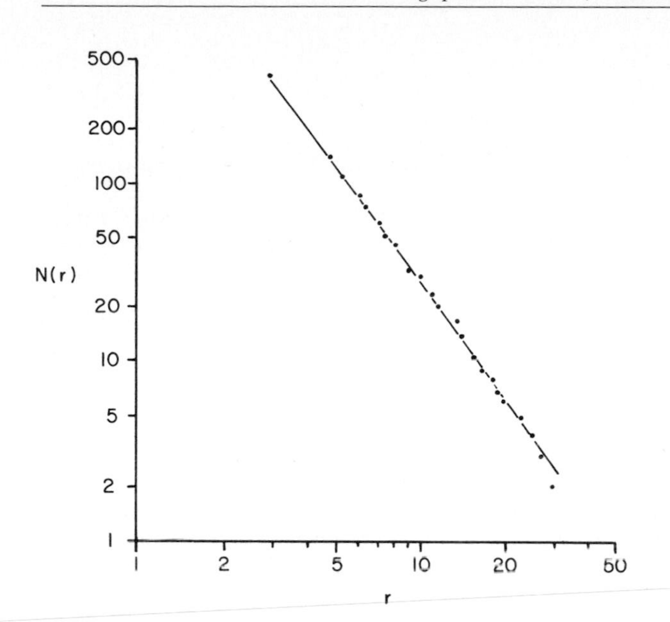

Figure 10.4.7. The size distribution of drift ice as measured from a satellite photograph obeys a Korcak relationship. (From M. Matsushita, "Fractal Viewpoint of Fracture and Accretion," *J. Phys. Soc.* Jpn., 54, 857-860 (1985); reproduced by permission of M. Matsushita.)

Clouds and lightening flashes are also fractal systems of interest to geographers, geologists and meteorologists. Lovejoy and co-workers have made extensive studies of the fractal structure of clouds. The study of the movement of clouds and weather systems is closely related to the study of turbulence in fluids, and for this reason the references to the publications of Lovejoy and colleagues on cloud physics are given at the end of Signpost 9. Systems similar to lightening flashes have been studied by Niemeyer and co-workers [15].

Sometimes the crystallization process producing a rock results in fractally structured crystals For example, in Figure 10.4.6 (see Plate 13 at the beginning of the book) a fern-like fractal crystal studied by Professor Fowler of Ottawa University is shown. There is no doubt that the fractal dimension of the fingered crystal will be related to the crystallization dynamics [16].

In Chapter 9, the Korcak fractal of an island system was discussed and in Figure 10.4.7 another "island-type" system which appears to have a fractal size distribution is shown. It is the size distribution of pieces of drift ice viewed from a LANDSAT satellite as reported by Matsushita [17]. If the reader will forgive the pun, this is a good example with which to bring to a close this brief discussion of geological fractals, which has only been able to look at the "tip of the iceberg" of fractal applications in geology, geography and meteorology.

References

[1] M.F. Goodchild "Fractals and the Accuracy of Geographical Measures," *Math. Geol.*, 10, No. 2 (1980) 85-98.
[2] M.F. Goodchild and D.M. Mark, "The Fractal Nature of Geographic Phenomenan," in press. Dr. M.F. Goodchild, Department of Geography, University of Western Ontario, London, Ontario N6A 5C2, Canada. Dr. D.M. Marks, Department of Geography, State University of New York, Buffalo, New York, USA.
[3] Dr. D.M. Marks, Department of Geography, State University of New York, Buffalo, New York 14260, USA.
[4] B.B. Mandelbrot, "Fractal Geometry of Nature," Freeman, San Francisco, 1983, p. 170.
[5] For a classic discussion of the technology involved achieving maximum density in a packed powder system, see C.C. Furnas, "Grading Aggregates Number 1, Mathematical Relations For Beds of Broken Solids of Maximum Density," *Ind. Eng. Chem.*, 23 (1931) 1052-1058. (Note that in this paper aggregate refers to crushed gravel, not the type of aggregate grown by Meakin and other workers).
[6] D. Buslik, "The Validity and Usefulness of Buslik's Equation," in "Proceedings of the Second European Symposium on the Mixing of Particulate Solids," Symposium Volume No., 65, Institute of Chemical Engineers, Rugby, 1981.
[7] H. Pape, L. Riepe and J.R. Schopper, "The Role of Fractal Quantity as Specific Surfaces and Tortuosities for Physical Properties of Porous Media," *Part. Character*. 1, 66-73 (1984).
[8] A.J. Katz and A.H. Thompson, "Fractal Sandstone Pores: Implications for Conductivity and Pore Formation," *Phys. Rev. Lett.*, 54 (1985) 1325-1328.
[9] I. Balberg and N. Binenbaum, "Direct Determination of the Conductivity Exponent in Directed Percolation." *Phys. Rev. B.*, 33, (1986) 2017-2019. This article contains a discussion of the electrical conductivity of porous rocks.
[10] G. Rochen, M. Rheault and P. St. Jullien, "La Prospection Miniere par Satellite," *Geos*, No. 3 (1986) 1–4. Geos is published by Energy Mines and Resources, Booth St., Ottawa, Ontario, K1A E04, Canada.
[11] Profile data generated by Miss Christine Hrytsak a guest student at Laurentian University in the spring of 1986.
[12] C.R. Twidale, "The Evolution of Bornhardts," *Am. Sci.*, 70, 268 – 276 (1982).
[13] See, for example the article "Laser Takes the Legwork Out of Mapping," *New Sci.,* May 4 (1978) 293.
[14] N.P.A. Burrough "Fractal Dimensions of Landscapes and of Other Environmental Data," *Nature* (London) 294 (1981) 240-242.
[15] L. Niemeyer, L. Peitronero and H.J. Weismann, "Fractal Dimension of Dielectric Breakdown," *Phys. Rev. Lett.*, 52, (1984) 1033-1036.
[16] Picture provided by, and used by kind permission of Dr. T. Fowler, of Ottawa University.
[17] M. Matsushita, "Fractal Viewpoint of Fracture and Accretion," J. Phys. Soc. Jpn., 54 No. 3 (1985) 857-860.

Other References of Interest to Geographers and Geologists

[18] A.C. Armstrong, "On the Fractal Dimensions of Some Transient Soil Properties," *J. Soil Sci.*, 37 (1986) 641-651.
[19] P.A. Burrough, "Fractal Dimensions of Landscapes and Other Environmental Data," *Nature* (London), 294 (1981) 240-242.
[20] P.A. Burrough, "Multi-Scale Sources of Spatial Variation in Soil: I. The Application of Fractal Concepts to Nested Levels of Soil Variation," J. Soil Sci., 34 (1983) 577-598.
[21] P.A. Burrough, "Multi-Scale Sources of Spatial Variation in Soil: II. A Non-Brownian Fractal Model and its Application to Soil Surveys," *J. Soil Sci.*, 34 (1983) 599-620.

[22] P.A. Burrough, "Fakes, Facsimiles and Facts: Fractal Models of Geophysical Phenomena," in S. Nash (Ed.), "Science and Uncertainty," Science Reviews, Middlesex, England, 1985.
[23] K.C. Clarke, "Computation of the Fractal Dimension of Topographic Surfaces Using the Triangular Prism Surface Area Method," *Comput. Geosci.*, 12 (1986) 713-722.
[24] K.C. Clarke, "Scale-Based Simulation of Topography," in "Proceedings, AUTOCARTO-8, Baltimore, Maryland, March 29-April 3, 1987," pp. 680-688.
[25] N.A. Dodd, "Texture Generation Using Fractal Concepts," in "Second International Conference on Image Processing and its Applications, London, England, June 24-26, 1986," pp. 251-257.
[26] G.J. Edwards, "Fractal Based Terrain Modelling", in "Conference on Computer Animation and Digital Effects, London, England, October 1984,"pp. 49-56.
[27] P. Fredericksen, O. Jacobi, and K. Kubik, "A Review of Current Trends in Terrain Modelling," ITC J., 2 (1985) 101-106.
[28] M.F. Goodchild, "The Fractional Brownian Process as a Terrain Simulation Model," "Proceedings of the Thirteenth Annual Pittsburgh Conference on Modelling and Simulation," (1982) 13 (3) pp. 1133-1137.
[29] M.F. Goodchild, "Lakes on Fractal Surfaces: a Null Hypothesis for Lake-Rich Landscapes," paper presented to the Annual Meeting of the Canadian Association of Geographers, Hamilton, Ontario, May, 1987.
[30] H.M. Hastings, R. Pekelney, R. Monteciolo, D. Vun Kannon and D. Del Monte, "Time Scales, Persistence and Patchiness," *Biosystems*, 15 (1982) 281-289.
[31] D.M. Aronson and P.B. Aronson, "Scale-Dependent Fractal Dimensions of Topographic Surfaces: an Empirical Investigation with Application in Geomorphology and Computer Mapping," *Math. Geol.*, 16 (1984) 671-683.
[32] A.P. Pentland, "Fractal-Based Description of Natural Scenes," *IEEE Trans. Pattern Anal. Machine Intelli.*, PAMI-6 (1984) 661-674.
[33] A.P. Pentland, "On Describing Complex Surfaces," *Image and Vision Comput.*, 3 (1985) 153-162.
[34] A.P. Pentland, A.P. 1986. "Perceptual Organization and the Representation of Natural Form," in "Artificial Intelligence, Austin, Texas, August 6-10, 1986," 1986, pp. 269-273.
[35] A.G. Roy, G. Gravel and C. Guathier, 1987. "Measuring the Dimension of Surfaces: a Review and Appraisal of Different Methods," "Proceedings, AUTOCARTO-8, Baltimore, Maryland, March 29-April 3, 1987," 1987, 68-77.
[36] M.C. Shelber, H. Moellering and N. Lam, "Measuring the Fractal Dimensions of Surfaces," "Proceedings, AUTOCARTO-6," 1983, pp. 319-328.
[37] J.C. Simon, and J. Quinqueton, "On the Use of a Peano Scanning in Image Processing," in R.M. Haralick and J.C. Simon (Eds.), "Issues in Digital Image Processing," Sijthoff and Noordhoff, Germantown, Maryland, 1980, pp. 357-366.
[38] C.C. Taylor and P.A. Burrough, "Multiscale Sources of Variation in Soil: III. Improved Methods for Fitting the Nested Model to One-Dimensional Semivariograms," *Math. Geol.*, 18 (1986) 811-821.
[39] M. Vandepanne, "3-D Fractals," *Creative Comput.*, ll (1985) 78-82.
[40] R. Voss, "Random Fractal Fongeries: From Mountains to Music," in S. Nash (Ed.), "Science and Uncertainty," Science Reviews, Middlesex, 1985, pp. 69-85.
[41] C.A. Aviles, C.H. Scholz and J. Boatwright, "Fractal Analysis Applied to Characteristic Segments of the San Andreas Fault," *J. Geophys. Res.*, 92 (1987) 331-344.
[42] R.H. Bradbury and R.E. Reichelt, "Fractal Dimension of a Coral Reef at Ecological Scales," *Mar. Ecol. Prog., Ser.*, 10 (1983) 169-171.
[43] R. Curl, "Fractal Dimensions and Geometries of Caves," *Math. Geom.*, 18 (1986) 765-783.
[44] S.L. Demko, L. Hodges and B. Naylor, "Construction of Fractal Objects with Iterated Function Systems," *Comput. Graphics,* 19 (1985) 271-278.
[45] G.H. Dutton, "Fractal Enhancement of Cartographic Line Detail," *Am. Cartogra.*, 8 (1981) 23-40.

[46] F.S. Hill Jr., and S.E. Walker, Jr., "On the Use of Fractals for Efficient Map Generation," "Proceedings, Graphics Interface '82, May 17-21, 1982, Toronto, Ontario," 1982, pp. 283-289.
[47] S.K. Kennedy and W.H. Lin, "FRACT - a Fortran Subroutine to Calculate the Variables Necessary to Determine the Fractal Dimension of Closed Forms," *Comput. Geosci.*, 12 (1986) 705-112.
[48] E. Jakeman, "Scattering by Fractals," in L. Pietonero and E. Tosatti (Eds.), "Fractals in Physics," North-Holland, Amsterdam, New York, 1986, pp. 55-60.
[49] J.P. Muller, "Fractal Dimension and Inconsistencies in Cartographic Line Representations," *Cartogr. J* , 23 (1986) 123-130.
[50] P.G. Okubo and K. Aki, "Fractal Geometry in the San Andreas Fault System," *J. Geophys. Res.*, 92 (1987) 345-355.
[51] W.L. Power, T.E. Tullis, S.R. Brown, G.N. Boinott and C.H. Scholz, "Roughness of Natural Fault Surfaces," *Geophys. Res. Lett.*, 14 (1987) 29-32.
[52] F.S. Rys and A Waldvogel, "Analysis of the Fractal Shape of Severe Convective Clouds," in L. Pietronero and E. Tosatti (Eds.), "Fractals in Physics," North-Holland, Amsterdam, New York, 1986, pp. 461-464.
[53] P.Z. Wong, "Surface Roughening and the Fractal Nature of Rocks," *Phys. Rev. Lett.*, 57 (1986) 637-640.
[54] T. Jeffery, "Mimicking Mountains," *Byte*, December (1987) 337-344.

Signpost 5 Trees, Crabs, Cauliflowers and Camouflage

The heading of this signpost was again a draft chapter heading in my planning notebook. It was partly inspired by the fact that several of the early publications on fractal geometry showed pictures of cauliflower and broccoli. These pictures were used to illustrate the fact that biological forms such as the structure of ferns and trees could be quantitatively described using the concepts of fractal geometry. Thus, in discussing branching systems looking like deciduous trees in the winter, Mandelbrot coined such phrases as "fractal canopies" to describe the fractal surface of the support system for the leafy structure of a tree in the spring. An abstract drawing of a branching system can be, for example, the skeleton of a flower system, an outline of the structure of the lung or the drainage system of a river. Mandelbrot has applied fractal geometry to all of these systems. The interested reader is directed to a discussion of biological form in Mandelbrot's book and in articles by Nicklas [1] and West and Goldberger [2]. The crowns of many different types of trees as viewed from the air obviously have a fractal structure. This fact could be used by a computer employed to process aerial photographs to take a decision as to which trees were present in the photographs without the need for a skilled human to interpret the visible profiles [3]. At the other end of the tree, and plants in general, another fractal system is to be found - the branching root system. Some diseases suffered by plants distort the root system. It may be that the fractal dimension of the root system in healthy and diseased plants might be a new quantitative way of characterizing the progress and treatment of a plant disease.

In Chapter 3, it was mentioned briefly that the structure of some cancer cells can be described by a fractal dimension. Browsing through any medical textbook will soon show many biological cell systems which can be characterized using the concepts of fractal characterization. Recently we carried out some characterization studies on the fractal structure of white blood cells which have been activated by the presence of a foreign body in the blood flow. This work is being carried out in co-operation with Dr. D. McIver of the University of Western Ontario [4].

In the early 1960s, I was concerned with the problem of characterizing the patterns of camouflage designs. Early in my studies on fractal geometry it seemed to me that one could characterize the structure of the camouflage blobs, which were part of an overall design using fractal geometry. It seems to me that the fractal dimension of the blob may be linked to its ability to distract the eye in any given situation. Thus, the patterns vary with regard to the background against which they are viewed, but overall their patterns have a structure relevant to the ability to distract the eye. I was reminded of this possible use of fractal geometry when I received a letter from Dr. Von H. Pape, who tells me that he is able to use the concepts of fractal geometry to characterize the coloured patches of hair on mixed breed dogs [5]! Coloured patches on animals represent the evolution of natural camouflage long before man started to use fractal patches to disguise his presence in jungle warfare.

References

[1] K.J. Nicklas, "Computer Simulated Plant Evolution," *Sci. Am.*, 54 (1986) 78-86.

[2] B.J. West and A.L. Goldberger, "Physiology in Fractal Dimensions," *Am. Sci.*, 75 (1987) 354-364.

[3] See, for example, the photographs on pp. 25 and 26 in V.G. Zsilinszky, "Photographic Interpretation of Tree Species in Ontario," Department of Lands and Forest, Ontario, 1966.

[4] D.J.L. McIver, B.J. Rogers, R. Trottier, B. MacFarlane and B.H, Kaye, "A Multifractal Analysis of Blood Cell Activation, paper presented at the Boston Conference, "Fractal Aspects of Disorder Materials," December, 1987, extended abstracts published as a booklet by the Materials Research Society, Pittsburgh.

[5] Personal communication from Dr. H. von Pape, Institute for Geophysics, Technical University of Clausthal, D-3392, Clausthal Zellerfeld, FRG.

Signpost 6 Fractal Geometry and the Structure of Catalysts

One of the mysteries encountered by a student in a first-year course on chemistry is the role of a catalyst in achieving chemical reactions. A chemical dictionary defines **catalysis** as:

"the acceleration or promotion of chemical action by a reagent, called the catalyst, which itself remains unchanged at the end of the action."

The term catalyst was coined in 1835 by the Swedish chemist Jacob Berzelius (1779–1848). As the student continues his study of catalysis, he discovers that the catalyst molecules or atoms very often link up temporarily with reacting atoms. During the overall reaction there is often the formation of an intermediate compound which may involve some transient changes in the structure of the catalyst compound. Subsequently, secondary reactions return the catalyst material to its original state and the new, desired, compound emerges from the system. Usually the catalyst is a finely divided material dispersed on a support system of high surface area. I have a very vivid memory of carrying out a catalysed organic reaction during my high-school days using platinized asbestos. The asbestos was the support system for the platinum, which was present as a very finely dispersed powder. I remember the experiment well, since it blew up! If I had known about the dangers of asbestos at the time, my concern for the hazard created by the flying fragments of glass would have been compounded by a dread of the asbestos fibres. Back in those days, the asbestos was used as a catalyst support because of its very high surface area, and also possibly because its surface structure activated some of the molecules participating in the chemical process. This "excitation" property of asbestos fibres is important in occupational health studies, since it has been shown that a combination of cigarette smoke and asbestos fibres is a particularly dangerous inhalation hazard because the adsorption of the potentially carcinogenic (cancer-causing) chemicals on to the surface of the asbestos fibres occurs in such a way that the activity of the dangerous chemical molecule is increased. Thus, the inhaled coated asbestos fibre serves not only to carry the smoke chemicals into the lung, but also to catalyse their destructive work. These properties of asbestos fibres are typical of the desirable properties of catalysts in general, even though asbestos as such has turned out to have very undesirable health problems.

When designing a catalyst system, one seeks to have a very large surface area so that chemical reaction can occur at many places in the catalyst system. The catalyst must be deposited on the surface of the catalyst support in an activated form and the support system must be such that the chemical reactants in the gaseous form passing into and out of the catalytic system will flow through in a relatively easy manner. Because of these three requirements:

(1) high surface area;
(2) finely divided material of the actual catalyst; and
(3) the permeability of the catalyst support system to the reactant gases,

the catalyst scientist is wallowing in fractal systems, since all three physical properties can now be quantified using the concepts of fractal geometry.

In the early days of my involvement in the applications of fractal geometry in applied science, I gave a lecture in South Africa to a group of scientists which contained many specialists tackling the problems of the catalytic conversion of coal gas into synthetic gasoline. As I outlined the potential applications of fractal geometry to catalytic systems, such as the measurement of the available surface area by fractal interpretation of gas adsportion studies using the techniques developed by Pfeifer and Avnir, and the possibility of describing the pore structure that controls the permeability of the catalyst by re-interpreting the mercury intrusion data on the pore structure from a fractal geometry perspective, the catalyst specialists became very excited. I gained the impression that as soon as the lecture was over, they streamed off to their laboratories to apply fractal geometry to their systems.

In present-day studies aimed at improving the knowledge of the structure of catalyst systems, specialists are realizing that an important question to be answered is, "When does a cluster of atoms change from being a cluster to a minute crystal?" The answer seems to be at about a cluster size containing 700 atoms [1 – 5]. There is also some indication that clusters of atoms below this size are arranged in space in such a way that their structure can be described by fractal geometry. One can build models of the atom clusters which mimic active catalytic systems which in appearance are identical with the models of diesel exhaust soot discussed in Chapters 5 and 6. Such fractally structured models of atom clusters would help to gain an understanding of the role of structure in catalytic mechanisms. Scientists are investigating the possibility that some fractal clusters of atoms may be more catalytically active than others containing the same number of atoms but having a different fractal dimension.

To achieve finely divided catalyst material of the right atomic cluster size, one has to arrange the vapour condensation dynamics on to the catalyst support in such a way that the chances of a given cluster size arising by turbulent collision of the atoms in the vapour phase will generate clusters of the desired size and fractal structure. It may be that a modelling of the collision probabilities to achieve different types of clusters can be carried out with a monosized latex spheres in aerosol form. (see the discussion of cluster formation in aerosols in Chapter 6). The interested reader will find more general information on the fractal aspects of catalytic science and the characterization of rough surfaces in references 6 – 11.

References

[1] T.H. Maugh, "When is a Metal Not a Metal?," *Science*, 219 (1983) 1413-1415.

[2] T.H. Maugh, "Clusters Provide Unusual Chemistry," *Science*, 220 (1983) 592-595.

[3] W.D. Knight, "Development of Extended Solids from Micro Clusters," Paper H-1-1, American Physical Society, Detroit, 1984. Dr. W.D. Knight, Department of Physics, University of California-Berkeley, Berkeley, California 94720, USA.

[4] T.H. Maugh, "Catalysts, no Longer a Black Art," *Science*, 219 (1983) 474-477.

[5] T.H. Maugh, "A New Picture of Catalysts Begins to Emerge," *Science*, 219, (1983) 944-947.

[6] S.H. Ng, C. Fairbridge and B. H. Kaye, "Fractal Description of the Surface Structure of Coke Particles," *Langmuir*, May/June (1987) 340-345.

[7] D.L. Jordon, R.C. Hollins and E. Jakeman, "Measurement and Characterization of Multiscale Surfaces," *Wear*, 109 (1986) 1-4.

[8] R.L.C. Flammer and N.N. Clark, "Computer Based Algorithms for Fractal Analysis for Surfaces," Rosemont International Conference on Powder Technology, Rosemont, May, 1987. Preprints provided by authors. Particle Analysis Centre, White Hall, Morgantown, WV 26506, USA.

[9] J.J. Gagnepan and C. Roques-Carmes, "Fractal Approach to Two-Dimensional and Three-Dimensional Surface Roughness," *Wear*, 109 (1986) 119-126.

[10] D. Romeu, A. Gomes, J.G. Perespamires, R. Silver, O.L. Perez, A.E. Gonzales and M. Jose-Yacamen, "Surface Fractal Dimension of Small Metallic Particles," *Phys. Rev. Lett.*, 57 (1986) 2552-2555.

[11] J.C. Russ and John C. Russ, "SEM Interpretation of Fractal Surfaces," Preprint provided by the author, Material Science Department, North Carolina State University, Raleigh, NC, USA.

Signpost 7 Solid-State Physics

As mentioned several times throughout this book, workers in solid-state physics and statistical physics welcomed fractal geometry with open arms because it described their randomly structured systems in an efficient manner. The literature on fractal applications in solid-state and statistical physics continues to increase at a tremendous rate. The purpose of this signpost is to point the reader to a recently made movie of fractal activity in solid-state systems and to list some recent references for the specialized reader.

In Chapter 4, the pioneer work carried out by Sapoval and co-workers on the diffusion of atoms across the interface between two metals was mentioned. After this chapter had been completed, I received some very interesting papers from Dr. Sapoval. Using their diffusion model to generate graphic demonstrations of atomic diffusion, Sapoval and co-workers have made a colour movie film. In their words:

"The movie on the time dynamics gives the direct observation of a new surprising character of diffusion in solids: large-scale erratic fluctuations."

I am told that this twelve minute movie (also available as a video cassette) is as interesting as the still shots reproduced in our Figure 10.7.1 (see Plate 14 at the beginning of the book) would suggest [1].

References

[1] B. Sapoval, M. Rosso, G.F. Gouyet and J.F. Colonna, "Fractal Structure of a Diffusion Front Duration," movie and video cassette versions available from Imagiciel, 91128 Palaiseau Cedex, France.

[2] B. Sapoval, M. Rosso, J.F. Gouyet and J.F. Colonna, "Dynamics of the Creation of Fractal Object by Diffusion and 1/f Noise," *Solid State Ionics*, 18 and 19 (1986) 21-30.

[3] J.F. Gouyet, M. Rosso and B. Sapoval, "Percolation in a Concentration Gradient," L. Pietronero, E. Tosatti (Eds.), in "Fractals in Physics," Elsevier, North-Holland, Amsterdam, 1986, pp. 137-140.

[4] M. Rosso, J.F. Gouyet and B. Sapoval, "Determination of Percolation Probability from the Use of Concentration Gradients," *Phys. Rev. B.*, 32 (1985) 6053-6054.

[5] B. Sapoval, M. Rosso and J.F. Gouyet, "Simulation of Fractal Objects obtained by Intercalation in Layered Compounds," *Solid State Ionics*, 18 and 19 (1986) 232-235.

[6] B. Sapoval, M. Rosso and J.F. Gouyet, "The Fractal Nature of a Diffusion Front and the Relation to Percolation," *J. Physi. (Paris) Lett.*, 46 (1985) L-149-146.

[7] J.Y. Onoda and J. Toner, "Deterministic, Fractal Defects Structures in Close Packings of Hard Disks," *Phys. Rev. Lett.*, 57 (1986) 1340-1343.

[8] This paper discusses the fractal dimensions of defect structures in crystalline structures. I found this paper particularly interesting because it referred back to the old classical work of Nowick and Mader which we discussed at length in Chapters 4 and 8.

[9] See book of Extended Abstracts for the meeting "Fractal Aspects of Materials: Disordered Systems," Boston, December 1-4, 1984, Editors A. J. Hurd, D. A. Weitz and B.B. Mandelbrot, Materials Research Society, Suite 327, 98000 McKnight Road, Pittsburgh Pennsylvania 15237, USA.

Signpost 8 Butterflies, Ants and Caterpillars in the Garden of Eden

The first part of the title of this signpost is taken directly from the title of an interesting scientific paper entitled "Growth Perimeters Generated By a Kinetic Walk: Butterflies, Ants and Caterpillars," written by Dr. Alla Margolina [1]. I could not complete my book on fractal geometry without quoting such an interesting title. For scientists who prefer a more technically informative title, I could have labeled this signpost "Fractal Geometry in Epidemiology and Pathological Invasion Studies." One must agree that the latter title does not stimulate the imagination quite as effectively as one conjuring up fractal bugs and butterflies. Moreover, such an ominously learned title invoking medical specialities would probably be a direct invitation for the average reader to walk straight past the signpost. In Chapter 5, when we began to fill up two-dimensional space with randomly arriving pixels, we referred to the growing clusters as "fractal animals" and "fractal insects." The term "fractal animals" became firmly established in the vocabulary of fractal geometry as scientists applied fractal geometry to the study of the spread of cancer in living tissue.

The first scientist to study this problem using modern computer simulation techniques was Dr. Eden (for a readable discussion see [2, 3]). In the first lecture I heard on the fractal model of cancer spread, the lecturer mentioned the growth of fractal animals according to the **Eden model.** Since I was new to the subject, and also probably because I am a theologian by training as well as a physicist, my imagination conjured up pictures of fractal animals exploring the garden of Eden with Adam and Eve. Perhaps even the snake in the Bible story was a fractal having an infinite number of ways of tempting man. I soon found out, however, that the Eden model had nothing to do with Adam and Eve and was named after the scientist who developed the growth model. Although the Eden model was originally developed to study the spread of cancer tissue, it is a mathematical model that can also be used to study the spread of epidemics amongst humans, or of a disease in a system such as a forest.

I discovered that the title "Ants, Caterpillars and Butterflies" heading the paper by Dr. Margolina refered to the different possibilities built into the model of the spread of an epidemic or a cancer growth using computer simulation technolgy. An ant can only go around a mathematical maze and go from one adjacent site to another. A caterpillar is envisaged as having the ability to crawl over obstacles in its path. A butterfly model can study the spread of a disease to distant points by carriers fluttering from one location to another. A butterfly model can also consider intermediate stages where a butterfly has caterpillar offspring which crawl around until they become butterflies, and then initiate further spread by a fluttering flight to distant, uninfected regions. In Chapter 6, we mentioned briefly the mathematics of pseudo Levy flight in which variations in probability of movement always contain two levels of jump. One element was a highly probable local movement and the other was a less probable finite jump in space away from the local movement. In Chapter 6 we pointed out that the movement of a powder grain being dispersed in another powder could involve both local random diffusion movement and distant convective jumps corresponding to the elements of a pseudo Levy flight. Using the language of this chapter, diffusion in a powder mixture is an ant and caterpillar mechanism, whereas convection is a butterfly leap.

By chance, one of the first epidemic patterns generated by allowing finite distant jumps in addition to local diffusion generated the outline of Figure 10.8.1, which looks like a butterfly. However, in general the structure of the spreading cancer or epidemic using various probability models of spreads can have generally fractal structures as shown in the other elements of Figure 10.8.1 and it is becoming apparent that important problems as different as the spread of AIDS and of cancer in the body can be studied using fractal geometry.

The patterns in Figure 10.8.l are reproduced from a paper entitled "Universality Classes for Spreading Phenomenon: a New Model for Fixed Static but Continuously Tunable Kinetic Exponents" [4]. This title illustrates one of the major problems of modern science. The average medical research scientist studying cancer, or the forestry expert concerned with the spread of spruce bud worm infection on his trees, would never dream of turning to a scientific paper with such a forbidding title. However, the first line of the abstract of this paper is:

"A new and quite tractable model for spreading phenomenon is proposed which contains as a special case the Eden Model and a model for epidemics."

The first line of the paper states:

"How does a disease (fluid etc.) spread through a randomly heterogeneous material?"

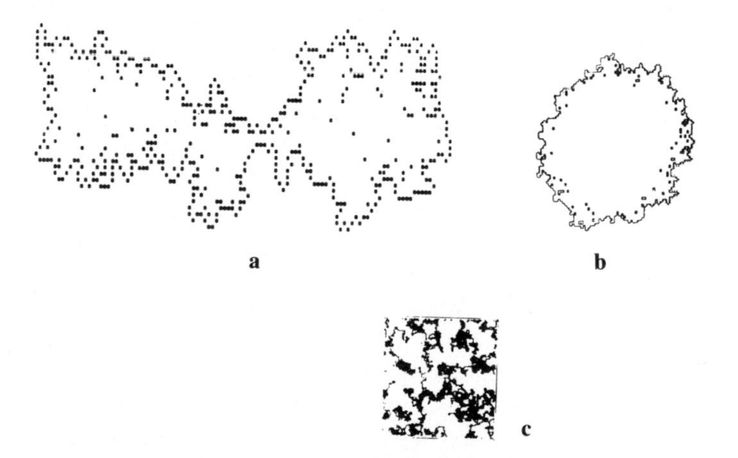

Figure 10.8.1. Animal imagery has been invoked to describe some of the complex patterns of fractal systems [4].

(a) Butterfly pattern of a spreading epidemic (1).

(b) Eden model of a growing cancerous tumour (2).

(c) Mole's labyrinth of a percolation model (3).

(From A. Bundle, H.J. Herrmann, A. Margolina and H.E. Stanley, "Universality Classes for Spreading Phenomenon; A New Model for Fixed Static But Continuously Tunable Kinetic Exponents," *Phys. Rev. Lett.*, Vol. 55, No. 7 (1985); reproduced by permission of American Physical Society and H.E. Stanley.)

This simple question is of relevance to a wide range of disciplines ranging from epidemeology, oncology (cancer) and cardiovascular physiology on the one hand, to signal propagation and network mechanics on the other. If some of these words had been used in the title, it might have attracted the wide readership it deserved.

As I have discussed at length elsewhere, I believe that the scientific literature must start to list keywords from the abstract of the paper to facilitate computer-organized information search systems. I have also shown that long words carry more information than short ones and that if one listed long words that occurred in the first paragraph of a scientific paper one would have a more efficient set of keywords than can be scavenged from the strict scientific title, which is written from the perspective of the specialist doing the work.

References

[1] A.E. Margolina, "Growth Perimeters Generated by a Kinetic Walk: Butterflies, Ants and Caterpillars" in H.E. Stanley and N. Ostrowsky (Eds.), "On Growth and Form," Martinus Nijhoff, Boston, (1986) 284-287.

[2] H.J. Herrmann, "Growth, An Introduction," in H.E. Stanley and N. Ostrowsky (Eds.), "On Growth and Form," Martinus Nijhoff, Boston, (1986), 3-20.

[3] H.J. Herrmann, "The Moles Labyrinth: a Growth Model," *J. Phys. A.*, 16 (1983) L611-L616.

[4] A. Bunde, H.J. Herrmann, A.E. Margolina and H.E. Stanely, "Universality Classes for Spreading Phenomena: A New Model for Fixed Static but Continuously Tunable Kinetic Exponents," *Phys. Rev. Lett.*, 55 (1985) 653-656.

Signpost 9 Turbulence and Chaos

We can make the movement of the fluid past an object visible by injecting streamers of dye, as illustrated in Figure 10.9.1. The thin line of dye in such a system represents what the mathematicians call **streamlines in the fluid.** When the liquid flow past the sphere is slow, we say that the motion is laminar. In **laminar flow**, the movement of the fluid follows well defined, smoothly changing streamlines, as indicated in Figure 10.9.1(a). As we increase the rate of flow of the fluid past the object, the complexity of the streamlines flowing around the sphere increases. At one stage, there is a shedding of a steady set of eddies, such as those shown in Figure 10.9.1(b). It is this systematic shedding of turbulent eddies on each side of the object which creates the fluttering of a flag when the wind blows past a flagpole at high speeds. This type of motion is called "**turbulent**." One cannot always clearly distinguish between the beginning of turbulent flow and the end of laminar flow conditions. A stream of eddies, of the type shown in Figure 10.9.1(b), is known as a "**Karman street**" after the famous scientist Von

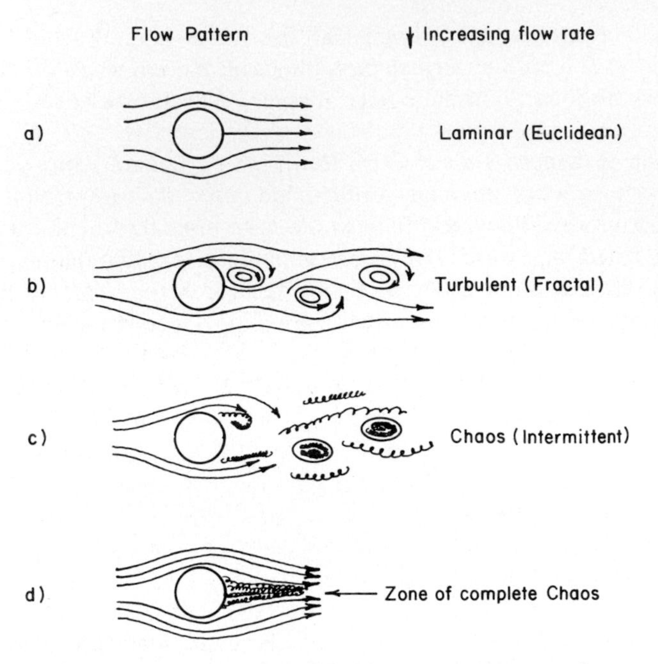

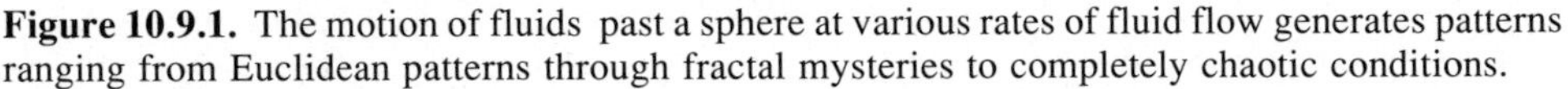

Figure 10.9.1. The motion of fluids past a sphere at various rates of fluid flow generates patterns ranging from Euclidean patterns through fractal mysteries to completely chaotic conditions.

Karman, who first studied fluid flow of this kind. As the speed of the fluid moving past the object increases beyond the speed at which a Karman street of eddies is formed, or if we consider movement past a rough or geometrically complex figure, the turbulence becomes much more chaotic. Mandelbrot tells us in his book that his own fascination with turbulence was one of the contributory factors leading to his development of fractal geometry. Thus, at the start of Chapter 10 of his book, he tells us that:

"The study of turbulence is one of the oldest, hardest and most frustrating chapters of physics. Turbulence is necessarily foreign to the spirit of the old physics (by this Mandelbrot means pre-fractal physics) that focused upon phenomena having well defined scales – many geometric shapes involved in turbulence are easily seen or made visible and cry out for a proper description. I immediately surmised that turbulence involves many fractal facets – this chapter (of Mandelbrot's book) begins with pleas for a more geometric approach to turbulence and for the use of fractals. These pleas are numerous but each are brief because they involve suggestions with few hard results as yet."

As Mandelbrot anticipated, there have been many studies attempting to link the physics of turbulence with the concepts of fractal geometry. Many of the scientific publications in this area, however, are very complex. I must confess that, as I have listened to some of the lectures on the fractal facets of turbulence given at the various conferences that I have attended, some of the words swirled around my head like a turbulent eddy. As I listened, comprehension was apparently always just beyond my reach. However, there is no doubt that if one looks at review papers on the problem of turbulence in fluids, one does see many pictures that cry out for fractal interpretation.

In my planned chapter on turbulence, my intention would have been to demonstrate the applicability of fractal geometry to turbulent phenomenon to the interested reader, leaving all mathematical interpretations to others who are better equipped to explore what appears to be, to the casual reader venturing into a mathematical paper on turbulence in fluids, a dense symbolic jungle. In my planning notes for the intended chapter on turbulence, I had glued a copy of the system shown in Figure 10.9.1. On this diagram, I drew the series of random lines through the vortex as shown in Figure 10.9.2. With these lines, one can illustrate the idea expounded, but not illustrated by Mandelbrot, that a line drawn through a vortex is a Cantorian set. The various Cantorian dusts that can be generated in this way are shown in Figure 10.9.2. The position at which one draws search lines through the vortex changes the structure of the cantorian set. This can be appreciated by comparing the patterns on the various lines as illustrated in Figure 10.9.2. These Cantorian sets are one-dimensional explorations of the structure of a turbulent eddies.

The picture in Figure 10.9.2 is based on a photograph in a review article by B. Freymuth and colleagues [1]. The interested reader will find many interesting patterns in this article, and in another by Lugt [2]. The pattern in Figure 10.9.2 was generated in a wind tunnel study of the movement of air along the upper surface of an air foil model of an experimental airplane wing. The pattern was made visible using a smoke-generating liquid. As the air begins to flow over the surface of the model of the airplane wing, smoke from the liquid is trapped in the swirling air to make the vortex visible.

In Lugt's review article, there is a satellite photograph of two typhoons spinning around each other in the North Pacific, and again one can generate various Cantorian sets by drawing different search lines through the swirling patterns.

Anyone who flies on an aircraft through thunder clouds in their early stage of formation can see the obvious curly turbulent patterns at the boundary between clear air and the condensed droplets of moisture constituting the cloud being formed by a rising column of moist air. Cloud patterns observed from the ground illustrate their fractal

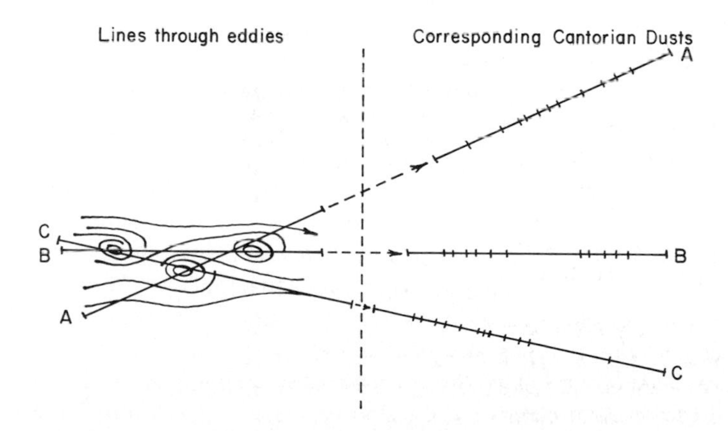

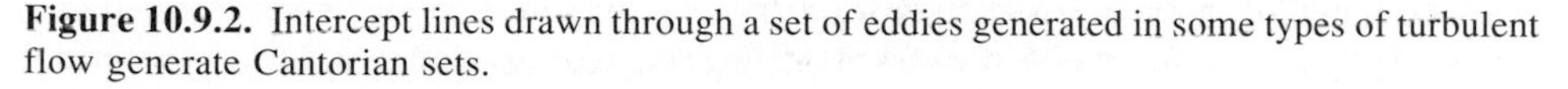
Figure 10.9.2. Intercept lines drawn through a set of eddies generated in some types of turbulent flow generate Cantorian sets.

structure, but the involvement of turbulence in their formation is not so obvious to the earth-bound observer. Scientists who have studied the structure of clouds have found an amazing similarity between the structure of clouds as small as a few metres across to those hundreds of kilometres in diameter. Turbulent in fluids is a very important aspect of weather forecasting and a study of weather systems, where fractal geometry is finding application [4 – 11].

It only requires a minor leap of the imagination from the vortex pattern in Figure 10.9.2 to the structure of a spiral galaxies in outer space (Mandelbrot has a chapter in his book in which he looks at the fractal structure of the clustering of galaxies in outer space). One student said to me in class when looking at a photograph of a galaxy, "It is almost as if God put his celestial finger into the cosmic soup and swirled it to give us the galaxies." That is an imaginative piece of poetry that may not appeal to all readers, but it does serve to stimulate the imagination to see that the pattern created in a stirred coffee cup is similar to the observed structures of the outer galaxies, and that both can be described in a fresh way using the concepts of fractal geometry. Having brought the reader this far in these turbulent aspects of physics, I feel rather like a guide who has taken an intending explorer through open pasture land to the boundary of a dangerous jungle. At the edge of the foothills the guide shows the explorer the pathway that leads into the difficult and probably dangerous jungle. The guide tells the explorer that there is great treasure to be found in the jungle but that the explorer must proceed from this point on his own. Lack of space makes it necessary for me to leave the reader to explore the fractal structure of the galaxies on his own. Bon voyage through the turbulent symbolic jungle of swirling Navier-Stokes equations and Laplace transforms [12, 13]!

A technical term for the subject dealing with the structure of the universe is cosmology. In Greek mythology, it was assumed that before the Gods came along, the universe was completely unorganized. The Greeks used the word chaos to describe the disorganized universe. The Greek word for the organized universe as created by the Gods is **cosmos.** From this word we have cosmology as the term for the study of the organization of the universe. It is interesting to note that **cosmetics**, the creative art of creating a new face, comes from the same Greek word. Thus cosmetics deal with creating order out of chaotic starting conditions. One of the fastest growing areas of modern physics theory is known as **chaos theory** [15], which studies the surprising organization which occurs within apparently completely chaotic systems.

To illustrate the various organized – disorganized levels of behaviour that one can encounter in a study of physical systems, consider the various types of flow that can occur when studying the flow of fluid past a sphere. We have already discussed the flow patterns that develop at very low rates of flow [see Figure 10.9.1(a)]. In such laminar-type flow, any element of the fluid moves along strictly Euclidean pathways. We have already discussed how, as we increase the fluid velocity, one reaches the stage where "Karman streets" of eddies develop and how the structure of the eddies shed into the fluid behind the sphere can be described using the concepts of fractal geometry. If the velocity is increased still further, one passes through the two stages shown in Figure 10.9.1(c) and (d). For such situations, one moves from a system describable by a fractal geometry to a chaotic state. Thus, from one perspective, fractal geometry is the bridge

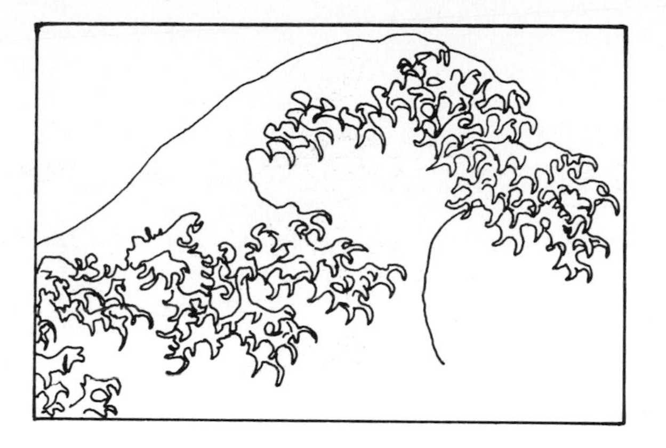

Figure 10.9.3. A Japanese artist in a famous painting anticipates fractal fingering as a breaking water wave becomes a foaming surge [9].

between Euclidian geometric concepts of streamline flow and chaos theory needed to describe completely turbulent systems in the physical world. It has been mentioned several times in this book that a fractal description of a system is complementary to a rhythmic description of the system combined with a Fourier analysis of the component structural rhythms underlying an observed pattern of complete wave type behaviour. A description of the type of flow past a sphere that occurs all the way from laminar to chaotic flow conditions in terms of Fourier analysis of the observed motion has been presented in an interesting article entitled "Roads to Chaos" by Kadanoff [3].

Both Kadanoff and Mandelbrot include in their discussion of turbulence a famous painting by a Japanese artist showing the foaming cap of a tumbling wave. An outline of the top of the dominant wave taken from this famous picture is shown in Figure 10.9.3. The stylized abstraction shown here illustrates the artist's anticipation that the initial breakdown of the steady wave cap into foam takes place by fractal fingering. This may only be artistic license but it does show how fractal geometry describes the intermediate stage between the smooth wave approaching the shore, which can be described using the theory of Euclidian geometry, and the complete chaos of the foam cap of the wave rushing to the shore. There is no doubt that many scientific publications using the concepts of fractal geometry to understand patterns of turbulence and even the structure of chaotic systems will flood the scientific literature of tomorrow [16 – 18].

References

[1] P. Freymuth, W. Bank and M. Palmer, "Vortices around Air Foils," *Am. Sci.*, 72 (1984) 342-248.
[2] H.J. Lugt, "Vortices and Vorticity in Fluid Dynamics," *Am. Sci.*, 73 (1985) 162-166.
[3] L.T. Kadanoff, "Roads to Chaos," *Phys. Today,* December (1983) 46-53.
[4] S. Lovejoy, "Area Perimeter Relation for Rain and Cloud Areas," *Science*, 216 (1982), 185-187.
[5] L.J.S. Bradbury, F. Durst, B.E. Launder, F.W. Schmidt, J.H. Whitelaw Eds.), "Turbulent Shear Flows 4," Selected papers from the Fourth International Symposium on Turbulent Shear Flows, September 12-14, 1983, Springer-Verlag, Berlin, 1985.
[6] D. Schertzer and S. Lovejoy, "The Dimension of Atmospheric Motions," in T. Tatsumi, (ed.) "Turbulence and Chaotic Phenomena in Fluids," Elsevier North-Holland, Amsterdam, 1984 pp. 505-511.
[7] S. Lovejoy and B.B. Mandelbrot, "Fractal Properties of Rain, A Fractal Model," *Tellus*, 37a (1985) 209-232.
[8] S. Lovejoy, D. Schertzer and P. Ladoy, "Outlook Brighter on Weather Forecasts," *Nature* (London), 320 (1986) 401.
[9] S. Lovejoy and D. Schertzer, "Generalized Scale Invariance in the Atmosphere and Fractal Models of Rain," *Water Resour.*, 21 (1985), 1233-1250.
[10] B.B. Mandelbrot, "Fractal Geometry of Nature," Freeman, San Francisco, 1983, p. 16.
[11] S. Lovejoy and D. Schertzer, "Scale Invariance in Climatological Temperatures and the Local Spectral Plateaus," *Ann. Geophys.* B4, (1986) 401-410.
[12] S. Lovejoy and D. Schertzer, "Scale Invariance Symmetries Fractals and Stochastics Simulations of Atmospheric Phenomena," Bulletin of the *Am. Meteorol. Soc.*, 67 (1986) 21-32. This paper contains some very beautiful simulated cloud structures.
[13] H.G. Hentschel and I. Procaccia, "Intermittencey Exponent in Fractally Homogeneous Turbulence," Phys. Rev. Lett., 49 (1982) 1158-1161.
[14] H.G. Hentschel and I. Procaccia, "Fractal Nature of Turbulence as Manifested in Turbulent Diffusion," *Phys. Rev. A* (Rapid Commun.), (1983).
[15] J.P. Crutchfield, J.D. Farmer, N.H. Packerd and R.S. Shore,"Chaos," *Sci. Am.,* December (1986) 46-57.
[16] "The Mathematics of Mayhem," *Economist*, September 8 (1984) 84-89.
[17] For some interesting chaotic patterns, see D. Avnir and M. Kagan, "Spatial Structures generated by Chemical Reactors at Interfaces," *Nature* (London) 307 (1984) 717-720. Students can generate the chaotic patterns created by Avnir and Kagan in their own laboratories using relatively simple equipment.
[18] D. Lees, "A Science called Chaos," *Maclean's* (a weekly Canadian news magazine) October 1 (1984), 76-79.
[19] B.F. Madore and W.L. Fredman, "Self Organizing Structures," *Am, Sci*, May-June (1987), 252-259.

Signpost 10 The Philosophical Impact of Fractal Geometry

In my lectures on fractal geometry to various audiences, I have found that many of the younger students are excited by the fact that fractal geometry challenges them to wrestle with the concepts of infinity. Older scientists sometimes have the opposite problem. In their training and career, they have mastered what they consider to be the smoothly changing orderly world of differential calculus. Some of them find it difficult to start living in a world where curves which do not have differential functions can dominate the properties of the system. Fractals seem to pose infinite challenges in more ways than one. Older scientists are disturbed by the fact that many of the things that they had come to accept as having definite values, such as the length of the Mississippi River or the coastline of Great Britain, suddenly become indeterminate. Their intellectual resistance to fractal geometry is similar to the intellectual problems that challenged scientists in the late 1920s when Heisenberg's uncertainty principle seemed to destroy the direct link between cause and effect, which seemed to be the necessary foundation of the physical universe. Many scientists in the 1920s rejected the philosophical implications of Heisenberg's uncertainty principles. For example, Einstein, when rejecting the idea that the universe was one large stochastic bowl of soup, said, "God does not play dice with the universe". I had hoped to add a chapter to this book entitled "Black Bodies, Infinite Radiation and Quantum Geometry." Because of the lack of space, this embryonic chapter of the book has mutated into a lecture essay given to an audience of physics teachers. In this essay, the crisis that occurred in theoretical physics at the turn of the century, when classical physics predicted that bodies should be radiating infinite energy (a problem referred to as the violet or ultraviolet catastrophe in physics textbooks) is reviewed. At the turn of the century, it was shown that this "infinite radiation" paradox was found to hinge upon the fact that nobody had ever stopped to question the "divisibility of energy." Everybody had always assumed that one could always keep on halving a quantity of energy until one was left with an infinitely small packet of energy. Planck was able to point out that if one made a new assumption, that energy is not infinitely divisible but that a certain minimum amount of energy, which he called a **quantum** (from the Latin word meaning a lump), is the smallest packet of energy that could exist, then if one worked out the property of blackbody radiators using this new idea, one arrived at a formula that fitted the measured radiation spectrum of a black body.

We are told that Planck did not immediately accept the philosophical implications of his own mathematical innovation which led to modern "quantum physics." For a long time Planck regarded his "quantum" idea as a temporary innovation in the mathematics of the system, rather than a fundamental reassessment of the structure of the Universe. Eventually, however, physicists had to accept the fact that if they want to describe the real world that they experienced, then they had to accept the fact that energy is not infinitely subdivisible.

Many students who begin to study the paradoxes of geometric set theory and fractal geometry, such as the fact that all lines have an infinite number of points and that all areas contain an infinite number of points, emotionally reject these ideas as being contrary to their experience of the real world. In fact, one can show that if one decides that the size of a point cannot be infinitely small and that there is a physical limit to the size of a geometric point, then one can start to construct a **quantum geometry** which does not contain the paradoxes of geometric set theory and which limits the infinities of fractal boundaries in the same way as the resolution of the computer graphic printout system or the dots used by a printing press limit our ability to draw or print ideal fractals. The paradoxes of geometric set theory basically come from the fact that one assumes implicitly that one can have infinitely small points. A "quantum geometric world view" based on finite-sized points can explain the fact that only certain orbits around an atom are possible in quantized space. One wonders why the mathematicians did not react to their geometric paradoxes to create quantum geometry in the way that the physicists reacted to the miss-match of theory and reality of blackbody radiators to arrive at quantum physics. Probably, many mathematicians will reject the ideas of a quantum geometry in the same way as many physicists continued to reject some aspects of quantum physics for many years after the original work of Planck. It should be noted however, that in the real world, scientists working with "pixels" in automatic picture processing of satellite pictures and the transmission and generation of television pictures have already created their own extensive quantum geometry. In this quantized geometry, their pixel size plays the same role that Planck's constant plays in a quantified energy universe. All through this book, when modelling diffusion-limited agglomeration, etc., we have been using quantum geometry in which the pixel size was the undefined smallest point possible in the geometric universe being studied. Perhaps a future book will contain a chapter on "pixel structures in the world of quantum geometry."

Signpost 11 Fun with Fractal Logic

Mathematicians who do not take themselves too seriously have always realized that some of the strict logic of their profession leads one into strange contradictions. Thus, "Alice in Wonderland" was written by a mathematician. Much of the humour of that book revolves around illogical use of logical arguments [1]. After reading Mandelbrot's book, I felt that there ought to be a new chapter written for Alice in Wonderland which would be described as "Topological Alice" [2]. One could imagine poor Alice having infinite adventures when lost in a Menger sponge. Imagine the difficulties she would encounter when trying to remove Cantorian dust from a Sierpinski carpet or running up a devil's staircase taking an infinite number of steps to arrive at her destination.

Mandelbrot has always enjoyed some of the fun of fractals as well as its logic. He was intrigued, when during a lecture I gave in the South of France to an audience of fractal experts, I was able to point out to him that he had overlooked the fractal nursery rhyme enjoyed by English children. Apparently he had not heard of:

> "There was a crooked man who walked a crooked mile who found a crooked sixpence upon a crooked style,
> he bought a crooked cat that caught a crooked mouse,
> and they all lived together in a crooked little house."

I pointed out in my lectures that, now that the world had fractal geometry, one could rewrite that rhyme with fractal replacing crooked. The mathematical plan for the fractal house bought by the fractal man would obviously be a Menger sponge with an infinite number of rooms. The only problem is that it would have taken the fractal man an infinite time to walk the fractal mile to his fractal house. Also, since a style is a kind of staircase over a fence, then a "fractal style" would probably have been built by the

Devil and would have an infinite number of steps. Therefore, again it would have required infinite time for the fractal man to pass over the fractal style. However, the rhyme is left indeterminate since the fractal man would have infinite time in which to complete his journey and arrange his domestic affairs.

If the analogy of the fractal man having a house with an infinite number of rooms appears to be far-fetched, it should be realized that expounders of set theories have used the idea that in outer space one could accommodate a newly arrived guest at a hotel which is full but has infinite rooms by simply moving everyone up one room in the list of rooms [3].

In "Alice in Wonderland," you will remember that Alice gets herself into trouble by eating things which make her grow and shrink. Perhaps in "Topological Alice" she would eat almond bread (the literal translation of Mandelbrot's name into English is almond bread) and she would then find herself exploring different levels of fractal space. Perhaps, in her new adventures, she might meet Dr. Mandelbrot, who would be able to point out to her that fractally she did not exist. In one of his discussions of the implications of fractal systems, Mandelbrot points out that the human body is actually the intertwining of two fractally branched systems, the veins and the arteries. Therefore, he concludes that the human body is a vanishingly small amount of flesh between the two infinitely branched vascular networks. What a fractalicious idea!

In the second book of Alice's adventures, called "Alice Through the Looking Glass," we are told of Alice's meeting with Humpty Dumpty. Humpty Dumpty is a character English children learn about in their nursery rhymes:

"Humpty Dumpty sat on a wall,
Humpty Dumpty had a great fall,
All the King's horses and all the King's men,
Couldn't put Humpty Dumpty together again."

Although the rhyme doesn't say so, Humpty Dumpty is always shown in picture books as a large egg dressed up in human clothes. In "Alice Through the Looking Glass," he is still in one piece. He is sat on the wall and talking like an obscure (and therefore learned) philosopher. In "Topological Alice," he could become a fractal by falling off the wall. All the King's horses and all the King's men, as they attempted to put him back together again, would discover that the size distribution of the eggshell fragments followed a scaling function describable by means of a fragmentation fractal dimension [4].

Another classic mathematical fairytale is the book "Flatland." The revised edition of this book was written in 1884. The author, E.A. Abbott, was a schoolmaster whose primary interests were literature and theology [5].

The basic plot of "Flatland" is that a square, who lives in two-dimensional space, is visited by a sphere. The three-dimensional sphere, who to the two-dimensional square looks like a mysteriously appearing and vanishing circle as he moves through the squares plane of existence, teaches the square how to imagine dimensions greater than his own. All the inhabitants of Flatland are Euclidean. The simplest figures in Flatland are needle-type objects who are women (the reader who ventures from this book into an exploration of Flatland is warned that its author seems to have a pre-women's liberation view of ladies). Isosceles triangles are policemen and soldiers. Squares are respectable

Figure 10.11.1. In "Topological Alice," after suffering dimensional collapse, perhaps Alice would learn how to deal with fractal dragons using Minkowski sausages.

tradesmen. The more sides that a personality in Flatland has, the nearer they are to perfection. Thus, in Flatland, a circle is considered to be a perfect shape since it is a polygon with an infinite number of infinitely small sides. Because of their infinite number of sides, circles are the philosophers and priests of Flatland.

In a dream, the sphere takes the square to Pointland and Lineland. The sphere teaches the square how to extrapolate the dream sequence of various dimensional experiences into three dimensions. As soon as the square understands three dimensions, he preaches the reality of higher dimensions to his fellow squares. Because of his preachings, he is considered a lunatic by his flat relatives and is locked up in an asylum.

In an added "fractal experience" chapter for "Topological Alice," it would be interesting to let Alice suffer dimensional collapse to take her to Flatland [6]. There, she could befriend the square. Together they could take a trip to a Koch island and spend an infinite time exploring the infinite beaches around the island. They could have infinite joy exploring the fractal river systems. Perhaps they would meet a fractal dragon such as that shown in Figure 10.11.1. They could discover that the **fractal dragon** was really a pathological curve in disguise and that it could be tamed by throwing a Minkowski sausage at it! (Mandelbrot tells us in his book how an early generation of mathematicians considered curves such as the Koch triadic island as pathological or "sick." Minkowski's sausage was an early attempt to use quantum geometry to domesticate pathological curves such as the Koch Triadic Island). The dragon in Figure 10.11.1 is one of the better known fractal curves which have been reproduced in many articles on fractals. A coloured version of a fractal dragon shown in Mandelbrot's book was used in the late 1980's by Guelph University (Canada) on a

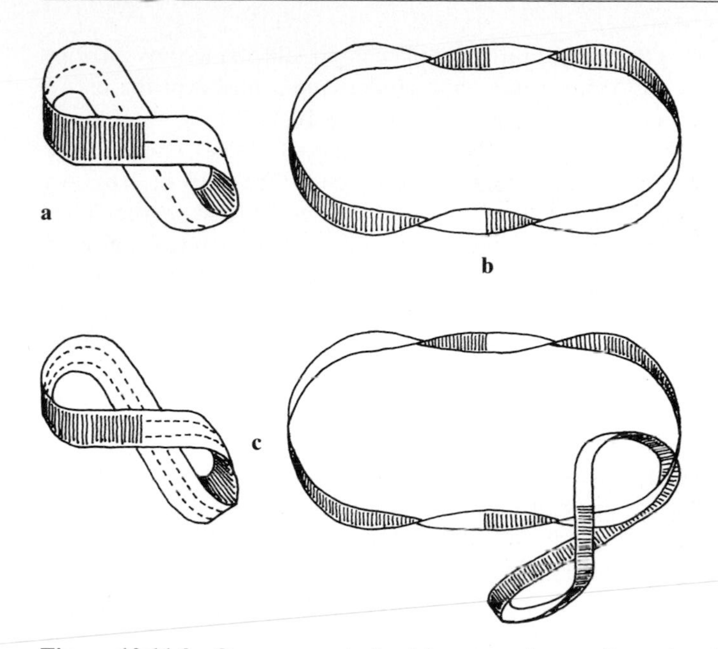

Figure 10.11.2. One can create fractal systems in one-dimensional space by sequential cutting a Mobius ring.
(a) A Mobius ring is a one-dimensional space with one edge, as shown by the fact that the single line drawn from one point meets itself as it covers both sides of the ring.
(b) Cutting the Mobius ring down the middle results in one large circle, not two.
(c) Cutting one-third-in around the Mobius ring results in linked circular ribbons, the smaller one of which is itself a Mobius ring.

student recruitment poster. Copies of this poster placed on display in the corridors of our Physics Department had the mysterious power of vanishing overnight to reappear at random in student rooms all over the campus [7].

Together, Alice and the square could revisit Lineland in a dream and view the lifestyle of a Cantorian set shunting back and forth in their infinite (randomized?) but limited space.

One day, while thinking about the possible adventures that Alice could have in Flatland, it occurred to me that one of the adventures that she could have would be to discover that the flat space inhabited by the square was actually a large Mobius ring. A Mobius ring is a mathematical shape which is of interest to mathematicians. Mobius was a German mathematician who lived from 1790 to 1868, considered to be one of the founders of the branch of mathematics that we call **topology**. You can make your own Mobius ring by taking a long strip of paper and giving it a single twist before joining the ends together as shown in Figure 10.11.2. Although the twisted circular ribbon looks to you and me as if it has two surfaces and two edges, to a mathematician it only has one surface and one edge. I have seen professional conjurers exploit the unexpected properties of Mobius rings in their magic shows. One trick that magicians do with a Mobius ring is to ask a child in the audience to take a crayon and draw a line all the way around the topside of the ring. Of course it turns out to be impossible, since any attempt to draw a continuous line on the Mobius ring ends up drawing a line over the entire surface of the Mobius ring. This fact is illustrated in Figure 10.11.2(a).

If one makes a continuous cut along the line shown in Figure 10.11.2(a), one does not end up with two rings but with one complete ring, as illustrated in Figure 10.11.2(b). An even more fascinating result occurs if one starts to cut the Mobius ring at a position one-third in from the edge, continuing the cutting operation until one has been twice around the Mobius ring. This cutting procedure is illustrated by the cutting pattern shown in Figure 10.11.2(c). When this happens, the final result is that one creates a large ring interlinked with a small ring. The small ring is the same length as the original Mobius ring but thinner. It is also a Mobius ring.

I have seen both cutting attacks on a Mobius ring outlined in the foregoing paragraphs carried out on a large Mobius ring used in magic shows. The audience reaction indicates that the result is totally unexpected and contrary to common sense (Einstein once said that "common sense" is made up of the mental prejudices we acquire before the age of 16). One day, when amusing myself with a Mobius ring and a pair of scissors, I discovered that if one cut the secondary Mobius ring, produced by the one-third in cut operation on the original Mobius ring one can create in theory an interlocked system of rings with infinite perimeter. At any stage in the creation of the rings, except for the initial Mobius ring, the system is self-similar since, by inspection, you cannot tell which stage of cutting you are inspecting. This multiple cut system obviously has at least some attributes of fractal space. If one has an afternoon with nothing in particular to do, one can take scissors and tape and demonstrate with a Mobius ring that different segmentation routines lead to various fractal systems all tending to infinite perimeter.

My afternoon discovery with scissors and tape may be trivial from the mathematician's point of view, but certainly seems to intrigue students when one demonstrates its existence in the class. I have used this multi ring space based on the Mobius demonstration as a useful way to start students on the road to thinking topologically and fractally.

In Signpost 2, we discussed the possible use of fractal wallpaper in the future. In early 1987, a short note published in the *New Scientist* discussed random knitting which generated a fractal pattern sweater, which had an attractive pattern on its surface. Perhaps future knitting patterns will include random number tables along with instructions "knit one, purl one and take a fractal chance."

I would like to end this book with the two cartoons, shown in Figure 10.11.3. The problem facing the gentleman attempting the crossword puzzle is symbolic of the fact that fractal logic and fractal geometry terms have already invaded the English language, and they are likely to persist for a long time. However, many classically trained scientists are still a little suspicious of fractal patterns of thinking. They may well be still looking at individuals like myself from the sidewalks of classical research, and wondering whether or not they should plunge into fractal pattern analysis. There is no doubt that in the first rush to apply fractal geometry to natural systems there have been intellectual excesses and attempts to force everything into a fractal logic framework. Sometimes, people have switched to a fractal description when there was no inherent advantage in changing the classical description of the system. Even in such cases,

Figure 10.11.3. These cartoons are from the front and end pages of the book on Growth and Form edited by Eugene Stanley and Nicole Ostrowski published by Martinus Nijhoff Publishers. (From H.E. Stanley and N. Ostrowsky (Eds.), "On Growth and Form; Fractal and Non Fractal Patterns In Physics," "Martinus Nijhoff Publishers, Dordrecht, Holland. Copyright (c) 1985 Martinus Nijhoff Publishers, Dordrecht, Holland." Reproduced by permission of Martinus Nijhoff Publishers and H.E. Stanley.)

however, the intellectual stimulus provided by exploring an alternative description of a system will probably be a useful exercise even if ultimately fractal geometry proves to be of little help in that particular situation. Apart from anything else, it is one of the themes of this book that looking at things from the fractal perspective can be fun, even if the fractal systems are no more useful than the multi-segmented Mobius ring.

Certainly, during the period from 1977 – 1987, as I have explored fractal concepts in applied physics with my students, I have had many stimulating discussions about the various models of fractals to be discovered in fineparticle systems.

I hope that the material presented in this book will help others to add fractal thinking to the array of techniques they use as they explore the physical world around them. I hope the reader will enjoy his own rambles through fractal space as much as I have enjoyed writing this record of my early randomwalks through fractal dimensions.

References

[1] There are many editions of "Alice In Wonderland" and its companion "Alice Through the Looking Glass" by Lewis Carroll [the penname of Ludwig Dodgson, an Anglican (Episcopalian) Minister and Professor of Mathematics at the University of Oxford. The well worn copy of these two books that I have on my desk is the combined issue, Companion Library Edition, published by Grosset and Dunlop, New York, 1965.

[2] B.H. Kaye, "Topological Alice – a New Dimension in Mathematical Nonsense," in preparation; part of a planned series of essays on "Fractalicious Thinking and Fun").

[3] For a discussion of this extraordinary hotel located in galactic space, and the development of the ideas of set theory, see N. Ya, Vilenkin, "Stories About Sets," Academic Press, New York, 1968, p. 40.

[4] Physicists might like to note that when Humpty Dumpty becomes a fractal, not only is he now describable by a Korcak fractal, but also his entropy has increased. Putting Humpty Dumpty back together is an exercise in lowering the entropy of the fragments, akin to the lowering of entropy when water freezes to form ice.

[5] E.A. Abbott, "Flatland," a romance of many dimensions with illustrations by the author – a "Square;" 6th ed. revised with Introduction by B. Hoffman, Dover, New York, 1952.

[6] I am exploring Alice's possible adventures in Flatland in a planned chapter for "Topological Alice" which would be called, "Flat Alice Goes Exploring With A Square" (see reference 2).

[7] The beautifully coloured fractal dragon used on the poster is shown in B.B. Mandelbrot "Fractal Geometry of Nature," Freeman, San Francisco, 1981, p. C4.

Bibliography

The following bibliography of fractal references is not intended to be comprehensive. The subject of fractal geometry is growing so quickly that it is impossible to keep such a list completely up to date.

It should be noted that the word "fractal" does not necessarily appear in the titles of the papers. The first reference, for example, is to a book on particle size which has a brief discussion of the application of fractal geometry to the characterization of fineparticles.

I would be pleased to hear from anyone who would like to have their paper quoted in future editions of this bibliography.

Allen, T., *Particle Size Analysis,* 4th ed., Chapman and Hall, London (1991).

Avnir, D. (Ed.), *Fractal Approach to Heterogeneous Chemistry,* Wiley, Chichester, UK (1989).

Bak, P., Chen, K., "Self-Organized Criticality", *Sci. Am.* Jan. (1991) 46–53.

Bak, P., Chen, K., "The Physics of Fractals", *Physica D* 38 (1989) 5–12.

Bak, P., Chen, K., Tang, C. "A Forest-Fire Model and Some Thoughts on Turbulence", *Phys. Lett. A* 147 (1990) 297–300.

Bak, P., Tang, C., Weisenfeld, K., "Self-Organized Criticality", *Phys. Rev. A* 38 (1988) 364–374.

Barnsley, M. F., Sloan, A. D., "A Better Way to Compress Images", *Byte* Jan. (1988) 215–233.

Bassingthwaighte, J. B., "Physiological Heterogeneity: Fractals Link Determinism and Randomness in Structures and Functions". *NIPS* 3 Feb. (1988) 5–10.

Bean, C. J., McCloskey, J., "Power-law Random Behaviour of Seismic Reflectivity in Boreholes and its Relationship to Crustal Deformation Models", *Earth Planet. Sci. Lett.* in press.

Berry, M. V., "Falling Fractal Flakes", *Physica D* 38 (1989) 29–31.

Berry, M. V., Percival, I. C., "Optics of Fractal Clusters such as Smoke", *Opt. Acta* 33 (1986) 577–591.

Blenkinsop, T., "The Fractal Distribution of Gold Deposits", in *Int. Symp. on Fractals and Dynamic Systems in Geosciences,* J. H. Kruhl (Ed.), Johann Wolfgang Goethe University, Frankfurt am Main, FRG (1993).

Blenkinsop, T. G., "Cataclasis and Processes of Particle Size Reduction", *Pure Appl. Geophys.* 136 (1991) 59–86.

Bonczyk, P. A., Hall, R. J., "Measurement of the Fractal Dimension of Soot Using UV Laser Radiation", *Langmuir* 8 (1992) 1666–1670.

Briggs, J., *Fractals: The Patterns of Chaos.* Harper and Row, New York (1993).

Brown, C. A., Savary, G., "Describing Ground Surface Texture Using Contact Profilometry and Fractal Analysis", *Weir* 141 (1991) 211–226.

Brown, G. J., Miles, N. J., Hall, S. T., "Fractal Characterization of Pulverized Materials", *Part. Part. Syst. Charact.* 10 (1993) 1–6.

Chen, H. Y., Iskander, M. F., Penner, J. E., "Empirical Formula for Optical Absorption by Fractal Aerosol Agglomerates", *Appl. Opt.* 30 (1990) 1547–1551.

Chen, H. Y., Iskander, M. F., Penner, J. E., "Light Scattering by Fractal Agglomerates of Smoke Aerosols", *J. Mod. Opt.* 37 (1990) 171–181.

Chen, J., Wilkinson, D., "Pore-Scale Viscous Fingering in Porous Media", *Phys. Rev. Lett.,* 55 (1985) 1892–1895.

Chen, Z. Y., Meakin, P., Deutch, J. H., "Comment on Hydrodynamic Behaviour of Fractal Agglomerates", *Phys. Rev. Lett* 59 (1987) 2121.

Cleary, T., Samson, R., Gentry, J., "Methodology for Fractal Analysis of Combustion Aerosols in Particle Clusters", *Aerosol Sci. Technol.* 12 (1990) 518–525.

Colbeck, I., "Dynamic Shape Factors of Carbonaceous Smoke", *J. Aerosol. Sci.* 21 (1990) Suppl. 1, S43–S46.

Colbeck, I., Hardman, E. J., Harrison, R. M., "Optical and Dynamical Properties of Fractal Clusters of Carbonaceous Smoke", *J. Aerosol Sci.* 20 (1989) 765–774.

Crum, S. V., "Fractal Concepts Applied to Bench Blast Fragmentation", *Rock Mechanics: Contributions and Challenges,* Hustrulid and Johnson (Eds.), Balkema, Rotterdam (1990) pp. 913–919.

Daley, C., "Ice Edge Contact: A Brittle Failure, Process Model", *Acta Polytech. Scand. Mech. Eng. Ser.* No. 100, Scandinavian Council for Applied Research, The Finnish Academy of Technology, Helsinki, Finland (1993).

Davy, P., Sornette, A., Sornette, D., "Some Consequences of a Proposed Fractal Nature of Continental Faulting", *Nature* 348 (1990) 56–58.

Eastman, J., Siegel, R. W., "Nanophased Synthesis Assembles Material from Atomic Clusters", *Res. Dev.* Jan. (1989) 56–60.

Edwards, S. F., Oakeshott, R. B. S., "The Transmission of Stress in an Aggregate", *Physica D* 38 (1989) 88–92.

Einar, F., Kusters, K. A., Pratsinis, S. E., Scarlett, B., "A Simple Model for the Evolution of the Characteristics of Aggregate Particles Undergoing Coagulation and Sintering", *Aerosol Sci. Technol.* 19 (1993) 514–526.

Fan, L. T., Boateng, A. A., Walawender, W. P., "Surface Fractal Dimension of Rice Hull Derived Charcoal from a Fluidized Bed", *Can. J. Chem. Eng.* 70 (1992) 388–390.

Feder, J., *Fractals,* Butterworth, Oxford (1989).

Fischman, J., "Falling into the Gap", *Discover* Oct. (1992) 57–63.

Freisen, W., Mikula, R. J., Canmet Divisional Report, ERP-CRL 86–128. Available from Energy Mines and Resources Canada, Canmet Technology Information Division, Technical Enquiries, Ottawa, Ontario K1A 0G1, Canada.

Glenny, R. W., Robertson, T., Yamashiro, S., Bassingthwaighte, J. B., "Applications of Fractal Analysis to Physiology", *J. Am. Physiol. Soc.* (1991) 2351–2367.

Grumbacher, S. K., McEwen, K. M., Halverson, D. A., Jacobs, D. T., Lindner, J., "Self-Organized Criticality: An Experiment with Sandpiles", *Am. J. Phys.* 61 (1993) 329–335.

Gurav, A., Kodas, T., Pluym, T., Xiong, Y., "Aerosol Processing of Materials", *Aerosol Sci. Technol.* 19 (1993) 411–452.

Hedley, D. G. F., "A Five Year Review of the Canada-Ontario Industry Rockburst Project", Canmet Special Report, SP90-4E. Available from Canadian Energy Mines, 55 Booth Street, Ottawa, Ontario, Canada.

Held, G. A., Solina, D. H., Keane, D. T., Haig, W. J., Horn, P. M., Grinstein, G., "Experimental Study of Critical Mass Fluctuations in an Evolving Sandpile", *Phys. Rev. Lett.* 65 (1990) 1120–1123.

Hirata, T., Satoh, T., Ito, K., "Fractal Structure of Spatial Distribution of Microfracturing in Rock", *Geophys. J. R. Astron. Soc.* 90 (1987) 369–374.

Kadanoff, L. P., "Fractals and Multifractals in Avalanche Models", *Physica D* 38 (1989) 213–214.

Kasper, G., Chesters, S., Wen, H. Y., Lundin, M., "Fractal-Based Characterization of Surface Texture", *Appl. Surf. Sci.* 40 (1989) 185–192.

Kaye, B. H., "Applied Fractal Geometry and the Fineparticle Specialist. Part I: Rugged Boundaries and Rough Surfaces", *Part. Part. Syst. Charact.* 10 (1993) 99–110.

Kaye, B. H., *Chaos and Complexity: Discovering the Surprising Patterns of Science and Technology,* VCH, Weinheim (1993).

Kaye, B. H., "Characterizing the Structure of Fumed Pigments Using the Concepts of Fractal Geometry", *Part. Part. Syst. Charact.* 9 (1991) 63–71.

Kaye, B. H., "Describing Filtration Dynamics from the Perspective of Fractal Geometry", Kona 9, Hosokawa Foundation, New York (1991).

Kaye, B. H., "Fractal Dimensions in Data Space: New Descriptions for Fineparticles Systems", *Part. Part. Syst. Charact.* 10 (1993) 191–200.

Kaye, B. H., "The Impact of Fractal Geometry on Fineparticle Characterization", in Proc. Particle Size Analysis Conf., N. Stanley-Wood, R. Lines (Eds.), Royal Society of Chemistry, London (1991), pp. 300–313.

Kaye, B. H., Clark, G. G., "Experimental Characterization of Fineparticle Profiles Exhibiting Regions of Various Ruggedness", *Part. Part. Syst. Charact.* 6 (1989) 1–12.

Kaye, B. H., Clark, G. G., "Formation Dynamics Information: Can it be Determined from the Fractal Structure of Fumed Fineparticles?" in *Particle Size Distribution II, Assessment and Characterization,* T. Provder (Ed.), Am. Chem. Soc., Washington, DC (1991), Chap. 24.

Kobayashi, T., "Reconstruction of Crack History from Conjugate Fracture Surfaces", in *Fatigue Crack Measurement: Techniques and Applications,* K. J. Marsh, R. A. Smith, R. O. Rithcie (Eds.), Engineering Materials Advisory Services Ltd., West Midlands UK (1991).

Krohn, C. E., "Fractal Measurements of Sandstones, Shales and Carbonates", *J. Geophys. Res.* 93 (1988) 3297–3305.

Krohn, C. E., "Sandstone Fractal and Euclidean Pore Volume Distributions", *J. Geophys. Res.* 93 (1988) 3286–3296.

Kruhl, J., in Proc. Int. Symp. on Fractals and Dynamic Systems in Geoscience, J. Kruhl (Ed.), Johann Wolfgang Goethe Universtiy, Frankfurt am Main, FRG (1993).

Lam, L., Freimuth, R. D., Lakkaraju, H. S., "Fractal Patterns in Burned in Hele-Shaw Cells of Liquid Crystals and Oils", *Mol. Cryst. Liq. Cyst.* 199 (1991) 249–255.

Leuenberger, H., Leu, R., Bonny, J. D., "Applications of Percolation Theory and Fractal Geometry to Tablet Compaction", *Drug Dev. Ind. Pharm.* 18 (1992) 723–766.

Liu, S. H., "Fractal Model for the A. C. Response of a Rough Interface", *Phys. Rev. Lett.* 55 (1985) 529–532.

Longley, P. A., Battey, M., "Fractal Measurement and Line Generalization", *Comput. Geosci.* 15 (1989) 167–183.

Lung, C. W., Zhang, S. Z., "Fractal Dimension of the Fractured Surface of Material", *Physica D* 38 (1989) 242–245.

MacKenzie, D., "Alpine Countries Seek Controls on Skiers, Builders and Roads", *New Sci.* 14 October (1989) 22.

Magill, H., "Fractal Dimension and Aerosol Particle Dynamics", *J. Aerosol Sci.* 22 (1991) Suppl. 1, S165–S168.

McCloskey, J., Bean, C. J., "Time and Magnitude Predictions in Shocks due to Chaotic Fault Interactions", *Geophys. Res. Lett.* 19 (1992) 119–122.

McKinnon, R. D., Smith, C., Behar, T., Smith, T. G., Jr., Dugois-Dalcq, M., "Distinct Effects of bFGF and PDGF on Oligodendrocyte Progenitor Cells", *Glia* 7 (1993) 245–254.

Meakin, P., in *The Fractal Approach to Heterogeneous Chenistry,* D. Avnir (Ed.), Wiley, Chichester, UK (1989), Sec. 3.12.

Middleton, G. V., *Nonlinear Dynamics, Chaos and Fractals with Applications to Geological Systems,* Short Course Notes, Vol. 9, G. V. Middleton (Ed.), Available from the Geological Association of Canada McMaster University, Hamilton, Canada (1992).

Mon, K. K., "Self-Similarity and Fractal Dimension of a Roughening Interface by Monte Carlo Simulations", *Phys. Rev. Lett.* 57 (1986) 866–868.

Muller, J., Hansen, J. P., Skjeltorp, A. T., McCauley, J., "Multifractal Phenomena in Porous Rocks", *Mater. Res. Soc. Symp. Proc.* 176 (1990) 719–723.

Nadis, S., "Sandbox Scholars", *Technol. Rev.* Feb./Mar. (1990) 21-22.

Neale, E. A., Bowers, L. M., Smith, T. G., Jr., "Early Dendritic Development in Spinal Chord Cell Cultures: A Quantitative Study", *J. Neurosci. Res.* 34 (1993) 54-66.

Nelson, J., "Fractality of Sooty Smoke: Implications for the Severity of Nuclear Winter", *Nature* 339 (1989) 611.

Niemark, A., "A New Approach to the Determination of the Surface Fractal Dimension of Porous Solids", *Physica A* 191 (1992) 258-262.

Petford, N., Bryon, D., Atherton, M. P., Hunter, R. H., "Fractal Analysis in Granitoid Petrology: A Means of Quantifying Irregular Grain Morphologies", *Eur. J. Mineral.* 5 (1993) 593-598.

Porter, R., Ghosh, S., Lange, G. D., Smith, T. G. Jr., "A Fractal Analysis of Pyramidal Neurons in Mammalian Motor Cortex", *Neurosci. Lett.* 130 (1991) 112-116.

Pratsinis, S. C., Mastrangelo, S. V. R., "Materials Synthesis and Aerosol Reactors", *Chem. Eng. Prog.* May (1989) 62-66.

Reichenbach, A., Siegel, A., Senitz, D., Smith, T. G. Jr., "A Comparative Fractal Analysis of Various Mammalian Astroglial Cell Type", *Neuro Image* 1 (1992) 69-77.

Reist, P. C., Hsieh, M. T., Lawless, P. A., "Fractal Characterization of the Structure of Aerosol Agglomerates Grown at Reduced Pressures", *Aerosol Sci. Technol.* 11 (1989) 91-99.

Rogak, S. N., Flagan, R. C., Nguyen, H. V., "The Mobility and Structure of Aerosol Agglomerates", *Aerosol Sci. Technol.* 18 (1993) 25-47.

Rossi, G., "Fractal Dimension Time Variations in the Friuli (North Eastern Italy) Seismic Area", *Bull. Geol. Teor. Appl.* 32 (1990) 175-183.

Sahimi, M., "Fractal Dimensions in a Percolation Model of Fluid Displacement", *Phys. Rev. Lett.* 55 (1985) 1698.

Sakai, T., Ramula, M., Gosh, A., Bradt, R. C., "Cascading Fracture in a Laminated Tempered Safety Glass Panel", *Int. J. Fract.* 48 (1991) 49-69.

Sakai, T., Ramulu, M., Ghosh, A., Bradt, R. C., "A Fractal Approach to Crack Branching (Bifurcation) in Glass", *Fractogr. Glasses Ceram.* 17 (1990) 131-146.

Schaefer, H. J., Pfeifer, J., "Sizing of Submicron Aerosol Particles by the Whisker Particle Collector Method", *Part. Part. Syst. Charact.* 5 (1988) 174-178.

Schaeffer, D. W., "Fractal Models and the Structure of Materials", *MRS Bull.* 13, Feb. (1988) 22-27.

Schaeffer, D. W., "Polymers, Fractal and Ceramic Materials", *Science* 243 (1989) 1023-1027.

Scholz, C. H., *The Mechanics of Earthquakes and Faulting,* Cambridge University Press, Cambridge (1990).

Schroeder, M., *Fractals, Chaos and Power Laws, Minutes from an Infinite Paradise,* W. H. Freeman, New York (1991).

Siegel, A., Reichenbach, A., Hanke, S., Senitz, D., Brauer, ,K., Smith, T. G. Jr., "Comparative Morphology of Bergmann Glial (Golgi Epithelial) Cells", *Anat. Embryol.* 183 (1991) 605-612.

Simons, P., "Après Ski le Deluge", *New Sci.* 14 January (1988) 49-52.

Smith, D. M., Johnston, G. P., Hurd, A. J., "Structural Studies of Vapour-Phase Aggregates via Mercury Porosimetry", *J. Colloid Interface Sci.* 135 (1990) 227-237.

Smith, T. G. Jr., Behar, T. N., Lange, G. D., Marks, W. B., Sheriff, W. H. Jr., "A Fractal Analysis of Cultured Rat Optic Nerve Glial Growth and Differentiation", *Neuroscience* 41 (1991) 159-166.

Smith, T. G. Jr., Brauer, K., Reichenbach, A., "Quantitative Phylogenetic Constancy of Cerebellar Purkinje Cell Morphological Complexity", *J. Comp. Neurol.* 331 (1993) 402-406.

Smith, T. G. Jr., Marks, W. B., Lange, G. D., Sheriff, W. H. Jr., Neale, E. A., "A Fractal Analysis of Cell Images", *J. Neurosci. Methods* 27 (1989) 173-180.

Smith, T. G. Jr., Marks, W. B., Lange, G. D., Sheriff, W. H. Jr., Neale, E. A., "Edge Detection in Images Using Marr-Hildreth Filtering Techniques", *J. Neurosci. Methods* 26 (1988) 75-82.

Sorensen, C. M., Cai, J., Lu, N., "Light Scattering Measurements of Monomer Size, Monomers per Aggregate and Fractal Dimension for Soot Aggregates in Flames", *Appl. Opt.* 31 (1992) 6547–6557.

Stix, G., "Finding Fault", *Sci. Am.* Dec. (1992) 52.

Strahlar, A. N., "Dimensional Analysis Applied to Fluvially Eroded Landforms", *Geol. Soc. Am. Bull.* 60 (1985) 279–300.

Tarbotan, H., "Fractal Nature of River Networks", *Water Resour. Res.* 24 (1988) 1317–1322.

Tence, M., Chevalier, J. P., Julien, R., "On the Measurement of the Fractal Dimension of Aggregated Particles by Electron Microscopy: Experimental Method, Corrections and Comparison with Numerical Models", *J. Physique* 47 (1986) 1989–1998.

Thompson, A. H., "Fractals in Rock Physics", *Annu. Rev. Earth Planet Sci.* 19 (1991) 237–262.

Tohno, S., Takahashi, K., "Shape Analysis of Particles by an Image Scanner and a Microcomputer: Application to Agglomerated Aerosol Particles", *Kona* 6 (1988) 2–14.

Tsoganakis, C., Price, B. C., Hatzikiriakos, S., "Fractal Analysis of the Sharkskin Phenomenon – Error in Polymer Melt Extrusion", *J. Rheol.* 37 (1993) 335–365.

Turcotte, D. L., "Fractals in Geology: What are they and what are they good for?" *GSA Today* 1, Jan. (1991) 2–4.

Van Saarlos, W., "On the Hydrodynamic Radius of Fractal Aggregates", *Physica A* 147 (1987) 280–296.

Velde, B., Dubois, J., Moore, D., Touchard, G., "Fractal Patterns of Fractures in Granites", *Earth Planet. Sci. Lett.* 104 (1991) 25–35.

Venkataraman, K. S., "Predicting the Size Distributions of Fine Powders During Comminution", *Adv. Ceram. Mater.* 3 (1988) 498–502.

Watson, A., "How to Tailor an Explosion to the Task", *New Sci.* 21, Dec. (1991) 11.

West, B. J., Shlesinger, M., "The Noise in Natural Phenomena", *Am. Sci.* 78, Jan./Feb. (1990) 45–45.

Whalley, W. B., Orford, J. D., "The use of Fractals and Pseudo-Fractals in the Analysis of 2-D Outlines: Review and Further Exploration", *Comput. Geosci.* 15 (1989) 1985–1999.

Wilzius, P., "Hydrodynamic Behaviour of Fractal Aggregates". *Phys. Rev. Lett.* 58 (1987) 710–713.

Wong, P., Howard, J., Lin, J., "Surface Roughening and the Fractal Nature of Rocks", *Phys. Rev. Lett.* 57 (1986) 637–640.

Wong, Po-Zen, "The Statistical Physics of Sedimentary Rock", *Phys. Today* Dec. (1988) 24–32.

Xie, H., "Fractal Effect of Irregularity of Crack Branching on the Fracture Toughness of Brittle Materials", *Int. J. Fract.* 40 (1989) 267–274.

Xie, H., "Fractal Nature on Damage Evaluation of Rock Materials", *Chin. J. Rock Mech. Eng.* 10 (1991) 697–704.

Xie, H., "Fractals in Rock Mechanics", in *Geomechanics Research Series,* Balkema, Rotterdam (1992).

Xie, H., "Studies on Fractal Model of the Microfracture of Marble", *Chin. Sci. Bull.* 34 (1989) 1292–1296.

Xie, H., "The Fractal Effect of Irregularity of Crack Branching on the Fracture Toughness of Brittle Materials", *Int. J. Fract.* 41 (1989) 267–274.

Xie, H., Chen, Z. D., "Fractal Geometry and Fracture of Rock", *Acta Mech. Sinica* 4 (1988) 255–264.

Zaltash, A., Myler, C. A., Dhodapkar, S., Klinzing, G. E., "Application of Thermodynamic Approach to Pneumatic Transport at Various Pipe Orientations", *Powder Technol.* 59 (1989) 199–207.

Zhang, H. X., Sorensen, C. M., Ramer, E. R., Olivier, B. J., Merklin, J. F., "In Situ Optical Structure Factor Measurements of an Aggregating Soot Aerosol", *Langmuir* 4 (1988) 867–871.

Zimmer, C., "Landslide Victory", *Discover,* Feb. (1991) 66–69.

Zimmer, C., "Sandman", *Discover,* May (1991) 58–59.

Author Index

Subject Index